SERIES ON SEMICONDUCTOR
SCIENCE AND TECHNOLOGY

Series Editors

H. Kamimura   R. J. Nicholas   R. H. Williams

SERIES ON SEMICONDUCTOR SCIENCE AND TECHNOLOGY

1. M. Jaros: *Physics and applications of semiconductor microstructures*
2. V. N. Dobrovolsky and V. G. Litovchenko: *Surface electronic transport phenomena in semiconductors*
3. M. J. Kelly: *Low-dimensional semiconductors*
4. P. K. Basu: *Theory of optical processes in semiconductors*
5. N. Balkan: *Hot electrons in semiconductors*
6. B. Gil: *Group III nitride semiconductor compounds: physics and applications*

Group III Nitride Semiconductor Compounds

Physics and Applications

Edited by

Bernard Gil

*Centre National de la Recherche Scientifique
Université de Montpellier II*

CLARENDON PRESS · OXFORD

1998

Oxford University Press, Great Clarendon Street, Oxford OX2 6DP
Oxford   New York
Athens   Auckland   Bangkok   Bogota   Bombay
Buenos Aires   Calcutta   Cape Town   Dar es Salaam
Delhi   Florence   Hong Kong   Istanbul
Karachi   Kuala Lumpur   Madras   Madrid   Melbourne
Mexico City   Nairobi   Paris   Singapore
Taipei   Tokyo   Toronto   Warsaw
and associated companies in
Berlin   Ibadan

Oxford is a trade mark of Oxford University Press

Published in the United States
by Oxford University Press Inc., New York

© *Oxford University Press, 1998*
Chapter 10, 'III–V nitride-based short-wavelength LEDs and LDs':
Shuji Nakamura, is © *Springer-Verlag, 1997.*

All rights reserved. No part of this publication may be
reproduced, stored in a retrieval system, or transmitted, in any
form or by any means, without the prior permission in writing of Oxford
University Press. Within the UK, exceptions are allowed in respect of any
fair dealing for the purpose of research or private study, or criticism or
review, as permitted under the Copyright, Designs and Patents Act, 1988, or
in the case of reprographic reproduction in accordance with the terms of
licences issued by the Copyright Licensing Agency. Enquiries concerning
reproduction outside those terms and in other countries should be sent to
the Rights Department, Oxford University Press, at the address above.

This book is sold subject to the condition that it shall not,
by way of trade or otherwise, be lent, re-sold, hired out, or otherwise
circulated without the publisher's prior consent in any form of binding
or cover other than that in which it is published and without a similar
condition including this condition being imposed
on the subsequent purchaser.

A catalogue record for this book is available from the British Library

Library of Congress Cataloging in Publication Data
Group III nitride semiconductor compounds: physics and applications /
edited by Bernard Gil.
(Series on semiconductor science and technology; 6)
Includes bibliographical references and index.
1. Compound semiconductors–Materials.   2. Gallium nitride–
Electric properties.   I. Gil, B. (Bernard)   II. Series.
QC611.8.C64G74 1998   537.6'223–dc21   97-41445
ISBN 0 19 850159 5

Typeset by Technical Typesetting Ireland

Printed in Great Britain by Bookcraft (Bath) Ltd., Midsomer Norton, Avon

Preface

A thunderous cheer erupts from a room in the 'Palais des Congrès' in Strasbourg, where, on the sunny Friday afternoon of 7 June 1996, a symposium devoted to semiconductor visible light emitters is taking place under the auspices of the European Materials Research Society. Inside the crowded room we are loudly congratulating Shuji Nakamura from Nichia Laboratories, a smiling young Japanese who is proudly presenting the portable display he has just realized by combining bright red, blue, green, and white light-emitting diodes. From second to second Nakamura's display presents to the attendees a rolling, eclectic series of beautifully colored Kanji symbols. 'Magnifique!', breathes the French session chairman, who seems to be the only one—with Nakamura—who is capable of speech. A few minutes later, in the darkness of the conference room, where all lights have now been extinguished, Nakamura excites fluorescence from a paper screen by using the narrow beam of the world's first UV semiconductor laser. Again the audience howls. For most of the participants, this was their first sight of the laser that had been developed in Nichia's research laboratories a few months before. The recent commercialization of blue and green light emitting diodes by several companies, and the worldwide diffusion of these breakthroughs in the scientific literature, encouraged scientists to attend Nakamura's talk that day: something momentous was on the cards, and seeing was indeed believing.

There is currently tremendous research activity in the area of nitride semiconductor compounds, which is motivated by expectations of a plethora of novel applications for the electronics industry in the years to come. During the ensuing coffee break, I noticed that excited conversations were taking place between well-known (at least to me!) senior researchers and PhD students, all so thirsty to participate actively in the nitrides story. It struck me immediately that there was a need for a research book on nitride compound semiconductors and their applications. Here it is, and I would like to thank, via Sönke Adlung and Julia Tompson, the Oxford University Press for producing it.

We decided to conceive a book accessible to the largest community, including students who, at the beginning of their PhD studies, are not yet familiar with the whole world of semiconductor physics. Nitrides are subtle, from the viewpoints of growth, characterization and device processing; but not drastically difficult to understand. When, two years ago, I caught up with the wagon of nitride research, I was delighted by the welcome that the nitride fathers, the 'true originals' made to us, the 'wagon-hoppers'. For this reason, I asked the individual contributors to this book to write each chapter with a tutorial purpose, so that learning all that you have to know about nitrides will seem like a refreshing journey rather than a tedious expedition. Hopefully we have managed to do it.

Welcome to the nitrides!

The book is organized in the following way: we begin with a general chapter which briefly introduces the main semiconductor families, including the nitrides, and their major applications. The demands of high structural quality for advanced electronic devices require nitrides to be grown as thin films. Growth methods are technically sophisticated. We allocate two chapters to describe in detail the two most important growth techniques: the epitaxy of nitride compounds on hetero-substrates by thermal cracking of precursor molecules either in high vacuum or in a carrier gas. Doping processes and materials science challenges are also discussed. These chapters are followed by one devoted to the application of advanced structural characterization methods. You will discover how to identify the structural lattice defects that are produced during growth, and become familiar with the actual and potential performance of these materials. Optical spectroscopy offers very powerful complementary techniques for electronic characterization. They can furnish information about band structure properties, doping, contamination, strain fields, and so on. The current status of optically pumped stimulated emission and lasing is reviewed, as well as nonlinear optical studies in group III nitrides. Three complementary chapters are allocated to show you the potential of these characterization methods and to give you a flavour of some of the physics the nitrides offer you the opportunity to encounter. It is the aim of one chapter to provide the basic theoretical tools used to understand quantitatively the band structure of nitride epilayers and the related quantum wells. Here will be shown, for the sake of illustration, that utilization of such tools, combined with information obtained during growth, structural, and electronic investigations, is essential to calculate and to understand the correlation between a quantum well design and its corresponding laser threshold. This chapter is a useful starting point if you wish to model modern advanced devices. Straightforward applications of technology appear more or less explicitly in the two detailed chapters written by electronics industry researchers who have developed and produced nitride-based devices. These chapters concern a wide range of devices: bipolar transitors, field-effect transistors, ultraviolet photoconductors, ultraviolet photovoltaic detectors, light emitting diodes, and laser diodes. Nitrides, which generally condense in the wurtzite phase, may also be forced to grow in the zincblende. This fact presents an extremely promising line of investigation. The book concludes with a chapter which summarizes what is known at present about cubic GaN.

This book is the result of fruitful discussions and collaborations we have enjoyed—and hopefully will continue to enjoy—on the occasion of various scientific meetings, during which we try both to work hard and to have a good time. I wish to thank first the contributors, in the order of their appearances in the book: Kevin O'Donnell, Galina Popovici, Hadis Morkoç, S. Noor Mohammad, Olivier Briot, Fernando Ponce, Fred. H. Pollak, Jin Joo Song, Wei Shan, Bruno K. Meyer, Axel Hoffman, Patrick Thurian, Masakatsu Suzuki, Takeshi Uenoyama, Jean Yves Duboz, M. Asif Khan, Shuji Nakamura, and

Oliver Brandt. I would also like especially to thank my colleagues and friends, Roger-Louis Aulombard, Professor at the University of Montpellier, Mathieu Leroux, Chargé de Recherche at CNRS, and Alain Friederich, head of department in the 'Laboratoire Central de Recherche' at Thomson-CSF, who are responsible in part for my scientific contribution to the physics of group III nitrides.

Gook luck with nitrides, all!

Montpellier   B.G.
March 1997

Contents

The plates section may be found between pages 398 and 399.

List of contributors   xii
List of plates   xvii

1 Beyond silicon: the rise of compound semiconductors
Kevin P. O'Donnell
    1.1 Introduction   1
    1.2 Historical overview—fifty years of commercial semiconductor physics   4
    1.3 Electronic and optoelectronic materials   7
    1.4 Semiconductor families   9
    1.5 Old, new, borrowed, blue: are the nitrides different?   12
    1.6 The role of defects in widegap semiconductors   16
    1.7 Conclusion   18
    Acknowledgements   18
    References   18

2 Deposition and properties of group III nitrides by molecular beam epitaxy
Galina Popovici, Hadis Morkoç, and S. Noor Mohammad
    2.1 Introduction   19
    2.2 III-N structural and chemical properties   20
    2.3 Molecular beam epitaxy (MBE) growth   23
    2.4 Substrates and buffer layers   32
    2.5 AlN   43
    2.6 Challenges in InN growth   44
    2.7 Growth of ternary and quarternary alloys   44
    2.8 Defects   45
    2.9 Doping during growth   51
    2.10 Conclusions and future work   60
    Acknowledgements   61
    References   61

3 MOVPE growth of nitrides
Olivier Briot
    3.1 MOCVD precursors   70
    3.2 Direct MOCVD growth of GaN on sapphire substrate   77
    3.3 Growth of GaN on sapphire substrates using a buffer layer   82
    3.4 Alternative substrates   99
    3.5 Doping of GaN   103
    3.6 Growth of ternary nitride alloys   107
    3.7 MOVPE growth of nitride heterostructures   114
    References   116

4 Structural defects and materials performance of the III–V nitrides
Fernando A. Ponce

4.1	Introduction	123
4.2	The crystalline structure of the nitrides	126
4.3	Structure of the epilayer/substrate interface	128
4.4	Microstructure of MOCVD thin films	135
4.5	Characterization of dislocations in GaN	139
4.6	Polarity determination	141
4.7	Nanopipes and inversion domains	143
4.8	Lattice vibrations and microstructure	149
4.9	Spatial variation of luminescence	153
4.10	Summary	154
	References	155

5 Modulation spectroscopy of the Group III nitrides
Fred H. Pollak

5.1	Introduction	158
5.2	Modulation techniques	159
5.3	Lineshape considerations	163
5.4	Results	168
5.5	Summary	179
	Acknowledgements	180
	References	180

6 Optical properties and lasing in GaN
Jin Joo Song and Wei Shan

6.1	Introduction	182
6.2	Optical transitions associated with excitons	183
6.3	Radiative recombinations associated with impurities and defect states	205
6.4	Stimulated emission and lasing	212
6.5	Optical nonlinearities of GaN	226
	Acknowledgements	235
	References	236

7 Defect spectroscopy in the nitrides
Bruno K. Meyer, Axel Hoffmann, and Patrick Thurian

7.1	Introduction	242
7.2	Shallow impurities in GaN	243
7.3	Deep structural defects in GaN	255
7.4	Transition metals in GaN	260
7.5	Transition metals in AlN	292
7.6	Conclusions	301
	Acknowledgements	301
	References	302

8 Electronic and optical properties of GaN-based quantum wells
Masakatsu Suzuki and Takeshi Uenoyama

8.1	Introduction	307
8.2	Electronic band structures of bulk GaN and AlN	308
8.3	$k \cdot p$ theory with strains for wurtzite	313
8.4	Subband structures of GaN/AlGaN quantum wells	322
8.5	Optical gains of bulk GaN and GaN/AlGaN quantum wells	331
8.6	Conclusions	340
	Acknowledgements	341
	References	341

9 Transistors and detectors based on GaN-related materials
Jean-Yves Duboz and M. Asif Khan

9.1	Bipolar transistors	343
9.2	Field-effect transistors	347
9.3	Ultraviolet photoconductors based on GaN	366
9.4	Ultraviolet photovoltaic detectors based on GaN	378
9.5	Conclusion and perspectives	387
	References	387

10 III–V nitride-based short-wavelength LEDs and LDs
Shuji Nakamura

10.1	Introduction	391
10.2	Blue InGaN/AlGaN double heterostructure (DH) LEDs	393
10.3	Blue/green/yellow InGaN single quantum well (SQW) structure LEDs	394
10.4	White LEDs	398
10.5	Emission mechanism of SQW LEDs	402
10.6	Bluish-purple InGaN multi-quantum-well (MQW) structure LDs	405
10.7	Summary	414
	References	415

11 Cubic group III nitrides
Oliver Brandt

11.1	Introduction	417
11.2	Growth equipment and *in situ* characterization	420
11.3	Nucleation and growth	421
11.4	Morphology	436
11.5	Structural properties	438
11.6	Optical properties	442
11.7	Electrical properties	449
11.8	Conclusion	455
	Acknowledgements	456
	References	456

Index   461

Contributors

M. Asif Khan
APA Optics, APA Inc.
2950 NE 84th Lane,
Blaine, MN 55449
USA
Tel: +1 612 784 4995
Fax: +1 612 784 2038
Email: asif@interserv.com

Oliver Brandt
Paul-Drude-Institut für Festkörperelektronik
Hausvogteiplatz 5–7
D-10117 Berlin
Germany
Tel: +49 30 20377 332
Fax: +49 30 20377 201
Email: brandt@pdi.wias-berlin.de

Olivier Briot
Groupe d'Etude des Semiconducteurs
Université de Montpellier II
Case courrier 074
F-34095 Montpellier Cedex
France
Tel: +33 4 67 14 36 04
Fax: +33 4 67 14 42 40
Email: obriot@epi.univ-montp2.fr

Jean-Yves Duboz
Laboratoire Central de Recherches de Thomson CSF
Domaine de Corbeville
F-91404 Orsay Cedex
France
Tel: +33 1 69 33 07 10
Fax: +33 1 69 33 07 40
Email: duboz@lcr.thomson.fr

Bernard Gil

Centre National de la Recherche Scientifique
Groupe d'Etude des Semiconducteurs
Université de Montpellier II
Case courrier 074
F-34095 Montpellier Cedex 5
France
Tel: +33 67 14 39 24
Fax: +33 67 14 37 60
Email: gil@ges.univ-montp2.fr

Axel Hoffmann

Institut für Festkörperphysik
Technische Universität Berlin
Sekr. PN 5-1
Hardenbergstrasse 36
D-10623 Berlin
Germany
Tel: +49 30 31422001
Fax: +49 30 31422064
Email: axel0431@mailszrz.zrz.tu-berlin.de

Bruno K. Meyer

1. Physikalisches Institut
Justus-Liebig-Universität Giessen
Heinrich Buff ring 16
D-35392 Giessen
Germany
Tel: +49 641 99 33100
Fax: +49 641 99 33119
Email: bruno.k.meyer@expl.physik.uni-giessen.de

S. Noor Mohammad

Howard University
Materials Science Research Center of Excellence and Department of Electrical Engineering
2300 Six St NW
Washington, DC 20059
USA
Tel: +1 202 806 6618
Fax: +1 202 806 5367
Email: snm@msrce.howard.edu

Hadis Morkoç
University of Illinois at Urbana-Champaign
Materials Research Laboratory and Coordinated Science Laboratory
104 South Goodwin Avenue
Urbana, IL 61801
USA
Tel: +1 217 333 0722
Fax: +1 217 244 2278
Email: morkoc@uiuc.edu, hmorkoc@vcu.edu

Shuji Nakamura
R & D Department
Nichia Chemical Industries Ltd
491 Oka, Kaminaka, Anan
Tokushima 774
Japan
Tel: +81 884 22 2311
Fax: +81 884 23 1802
Email: shuji@nichia.co.jp

Kevin P. O'Donnell
Department of Physics
University of Strathclyde
Glasgow G4 0NG
Scotland
UK
Tel: +44 141 548 3365, 3458
Fax: +44 141 552 2891
Email: k.p.odonnell@strath.ac.uk

Fred H. Pollak
Physics Department and New York State Center for Advanced Technology in Ultrafast Photonic Materials and Applications
Brooklyn College of the City University of New York
Brooklyn, NY 11210
USA
Tel: +1 212 780 5356
Fax: +1 212 780 5418
Email: fhpbc@cunyvm.cuny.edu

Fernando A. Ponce
Xerox Palo Alto Research Center
3333 Coyote Hill Road
Palo Alto, CA 94304
USA
Tel: +1 415 812 4199

Fax: +1 415 812 41440
Email: ponce@parc.xerox.com

Galina Popovici

University of Illinois at Urbana-Champaign
Materials Research Laboratory and Coordinated Science Laboratory
104 South Goodwin Avenue
Urbana, IL 61801
USA
Tel: +1 217 333 0722
Fax: +1 217 244 2278
Email: popovici@uiuc.edu, popoviciG@aol.com

Wei Shan

Materials Sciences Division
Lawrence Berkeley National Laboratory
1 Cyclotron Road
Berkeley, California 94720
USA
Tel: +1 510 486 4555
Fax: +1 510 486 5530
Emaril: Weishan@ux8.lbl.gov

J. J. Song

Oklahoma State University
OSU Center for Laser and Photonics Research
413 Noble Research Center
Stillwater, Oklahoma 74078
USA
Tel: +1 405 744 6575
Fax: +1 405 744 6406
Email: jjsong@okway.okstate.edu

Masakatsu Suzuki

Central Research Laboratories, Matsushita Electric Industrial Co. Ltd
3-4 Hikaridai
Seikacho, Sourakugun
Kyoto 619-02
Japan
Tel: +81 774 98 2519
Fax: +81 774 98 2576
Email: suzuki@crl.mei.co.jp

P. Thurian
Institut für Festkörperphysik
Technische Universität Berlin
Sekr PN 5-1
Hardengergstrasse 36
D-10623 Berlin
Germany
Tel: +49 30 31422001
Fax: +49 30 31422064
Email: thurian@physik.tu-berlin.de

Takeshi Uenoyama
Central Research Laboratories, Matsushita Electric Industrial Co. Ltd
3-4 Hikaridai
Seikacho, Sourakugun
Kyoto 619-02
Japan
Tel: +81 774 98 2519
Fax: +81 774 98 2576
Email: takeshi@crl.mei.co.jp

Plates

Plate I Hexagonal feature in homoepitaxial GaN/GaN grown on the [00$\bar{1}$] N-terminated side. (a) Secondary electron image taken at 30 kV; (b) Raman scattering image at 735 cm^{-1}; (c) background reference image at 765 cm^{-1}; and (d) flat-field image obtained by dividing the images at 735 and 765 cm^{-1}.

Plate II The full-color LED display that was demonstrated in Tokyo, Japan, in 1994 for the first time. The blue InGaN/AlGaN LEDs, green GaP LEDs, and red GaAlAs LEDs are used as three primary color LEDs.

Plate III The full-color LED display that was demonstrated in Tokyo, Japan, in 1996. The blue InGaN SQW LEDs, green InGaN SQW LEDs, and red GaAlAs LEDs are used as three primary color LEDs.

Plate IV The LED traffic light that was demonstrated in Sweden in 1996, using green InGaN SQW, yellow AlInGaP, and red AlInGaP LEDs.

1 Beyond silicon: the rise of compound semiconductors

Kevin P. O'Donnell

1.1 Introduction

This is a book about gallium nitride and its alloys. To those physicists, chemists, materials scientists, and technologists who are fortunate enough to be working with these materials at the present time, there can be no doubt that the nitrides are the most fascinating of all semiconductors. At the same time, the nitrides are the materials most likely to be exploited commercially in the years leading up to the millennium. Perhaps our degree of fascination is closely related to the marketability? After all, researchers are as aware as anyone of the commercial realities embedded in scientific activity. The purpose of this introductory chapter is to provide an answer to some basic questions concerning the place of nitrides among the families of semiconductors that lie *beyond silicon*.

We should begin by noting that silicon is the prime semiconductor, accounting for an estimated 99% of commercial production, in terms of raw material (this is only an estimate: you could probably beat me down to 95% on a bad day) and some 90% of research and development costs. Silicon is an elemental semiconductor, from Group IV of the periodic table. Its pre-eminent position as an electronic material was won in the later 1960s from another Group IV semiconductor, germanium, for three basic reasons: (1) raw material is readily available: silicon is an extremely common element, present in most rocks and easily extracted by chemical reduction from common sand; (2) silicon can be doped easily to form electron-rich (n-type) or electron-poor (p-type) modifications that are necessary to make electronic transport devices such as diodes and transistors; (3) silicon forms a stable non-conducting oxide that can be used to define masks for microlithography, the process that exploits the integration of many components on a single 'chip'. The three legs of availability, dopability, and integrability support a multibillion-dollar industry.

The non-elemental semiconductor materials fall into several distinct families. If we concern ourselves mainly with inorganic materials, the families emerge quite naturally from an examination of the periodic table. It is convenient to start at element 32, germanium. Moving one place to the right on the table (Fig. 1.1) brings us to arsenic, a well known poison, of course, but for our purposes a member of Group V, which also includes nitrogen and phosphorus. A step to the left from germanium brings us to the trivalent (Group III) metal gallium. Since germanium forms stable crystals, it seems reasonable to assume that the

IB	IIB	III	IV	V	VI	VII
		$_5$B	$_6$C	$_7$N	$_8$O	$_9$F
		$_{13}$Al	$_{14}$Si	$_{15}$P	$_{16}$S	$_{17}$Cl
$_{29}$Cu	$_{30}$Zn	$_{31}$Ga	$_{32}$Ge	$_{33}$As	$_{34}$Se	$_{35}$Br
$_{47}$Ag	$_{48}$Cd	$_{49}$In	$_{50}$Sn	$_{51}$Sb	$_{52}$Te	$_{53}$I

Fig. 1.1 Part of the periodic table of the elements.

compound of gallium and arsenic will do likewise. This is indeed the case, although it perhaps never happens in the natural world. In the laboratory, the III–V semiconductor GaAs (gallium arsenide) can be readily produced. This material is the exemplar of the III–V family of semiconductors. A glance at Fig. 1.1 shows that the compounds indium phosphide (InP), gallium antimonide (GaSb) and even gallium nitride (GaN) should also form stable compounds of the III–V family. And indeed they do.

Two steps to the left and right of element 32, we find zinc and selenium. The compound zinc selenide (ZnSe) is an exemplar of the II–VI semiconductor family, which also includes zinc sulphide (ZnS) and cadmium telluride (CdTe). Three steps right and left bring us to copper bromide (CuBr). Is this material a semiconductor? Well, it's not chopped liver. Although CuBr is not at the moment a serious contender for commercialization, it is still a semiconductor in that it possess a bandgap (it is transparent to light with a certain range of frequencies). We can therefore also talk about the I–VII family of semiconductors. Indeed, if we want to make the classification general, there are many more possible families, like the IIA–VI family (exemplified by MgO) and the IA–VII family (like NaCl). Such an extended classification may lead, however, to protracted arguments with theorists, which we will try to avoid.

Also important as semiconductor materials are the *mixed crystals* formed by making a solid solution of one compound in another from the same family. These alloys possess properties intermediate and in proportion to those of their components. $Al_xGa_{1-x}As$ is a useful III–V alloy with optoelectronic applications. Finally, the compound of the group IV elements Si and C produces a large number of different crystal forms: silicon carbide (SiC) is the only IV–IV compound semiconductor of note.

Although semiconductors were identified as a *class* of materials in the days of Faraday, and were named as such by Dvigubsky in an 1826 textbook, it is only recently that the word has entered the common language [1]. Unfortunately it has done so in two distinct senses: the general public, and some engineers, think of a semiconductor as an all-solid electronic *device*; most physicists think that a semiconductor is a *material*, which can of course be used to makes devices by carrying out the appropriate processing steps.

We shall adhere to the original sense of the word, which derives from a classification of materials according to their ability to conduct electricity. If it were possible to make standard-shaped samples of various materials, it would be a simple matter to arrange them in order of conductivity by carrying out a simple electrical experiment. Provided that the temperature of the sample remains constant, we expect the current density to increase with the electric field (Ohm's Law): the constant of proportionality being the specific conductivity of the material. The best conductors, such as copper and gold, are recognizable as metallic elements. The very poorest conductors (insulators) tend to be transparent materials, crystals, like mica or quartz, and the amorphous mixtures of materials that we know as common glass. The range of conductivities found in nature confronts experimentalists with a dream or a nightmare: silver is some 10^{22} times more conductive than Mica at room temperature.

In the original, electrical, definition, semiconductors are simply those materials that possessed conductivities that fall in the range between those of metals and insulators. Of course this definition, used in the 19th and early 20th centuries, was not at all satisfactory. The first problem is fundamental: with 22 orders of magnitude to play with, the meaning of the word intermediate is stretched to snapping point. It is conceivable that materials with very different conductivities will fall within the class of semiconductors. Where does one draw the line? The second problem lies with experimental variability: early experimenters quoted wildly different results for different samples of the same material. Sometimes these differences extended over several orders of magnitude. Was the highest or the lowest available figure most reliable? With the benefit of hindsight, we recognize that non-uniformity of the available samples played a major role in some of the earliest experiments. (Such problems are still with us.) However, with purer samples, new observations led to a clearer definition of how to recognize a semiconductor. In particular, it was noticed that semiconductors (in the original definition) usually become more conducting as the temperature rises. (This behaviour is *anomalous* if metals are considered to be normal.) At the same time, semiconductors showed clearly a variety of new physical effects which are difficult to observe in other materials, including thermoelectricity, piezoelectricity, photoconductivity, and a very large Hall effect. Now the definition of a semiconductor, based upon the results of many different kinds of experiments, was in danger of becoming impossibly complicated. A solution to this problem emerged in the 1930s, when Wilson and others applied the new methodology of quantum mechanics to the electronic structure of solids and invented band theory.

Electrons in free atoms occupy more or less discrete energy states. Transitions between such states produce sharp line spectra. In solids, electrons still occupy discrete states, but there are many more of them, of order 10^{23} for each atomic level. These states fan out into bands. The relative placement of the bands ultimately determines whether the solid will be a metal, an insulator or a semiconductor. A simplified band diagram for a semiconductor (Fig. 1.2) shows that there is a characteristic energy gap between a low-lying band, which is full of electrons (representing bonding orbitals of the solid), and an upper-lying empty band (corresponding to anti-bonds). The electronic bands of energy levels give rise to continuous spectra. Somewhat confusingly, these spectra are also called bands.

4 Beyond silicon: the rise of compound semiconductors

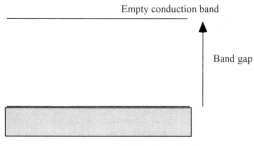

Fig. 1.2 A simplified band diagram for semiconductor.

With the ideas of band theory, we can readily explain why semiconductors become more conducting with increasing temperature, or with illumination by high-energy photons or radioactive particles. In fact it may be said that this great leap in our understanding makes possible the utilization of semiconductor materials in electronic devices. However, this naïve view ignores an important fact. Inventors do not need theories in order to bring new things into the world. Semiconductors were utilized long before they were understood. By way of a single example, it is clear that radio broadcasting came into being before anyone could reasonably explain the operating principle of many of radio's working components. Whatever the relationship between technology and theory may be, it is a fact that the last 50 years have seen the commercialization of semiconductor physics proceed at an ever-increasing pace. Semiconductors impact on practically every human life in one way or another. The complexity of devices has increased exponentially: new factories for their manufacture are now so expensive that multinational companies must combine in order to finance them.

The following section attempts to describe how electronics technology grew so quickly. The reminder of the chapter is organized along the following lines: Section 1.3 provides a beginners' guide to the physical characteristics of elemental and compound semiconductors which may form the bases of new technologies. Section 1.4 summarizes our progress in utilizing materials from the various semiconductor families alluded to above. In Section 1.5 we try to answer the question: in what ways are the nitrides different from other compound semiconductors? Section 1.6 examines the role of defects in semiconductors with large bandgaps. Section 1.7 concludes the chapter.

1.2 Historical Overview—fifty years of commercial semiconductor physics

The invention of the transistor in December, 1947 changed the world forever. At the time of writing (early 1997) it is estimated that more than 100 transistors are produced, every day, *for every person on the planet*. But 50 years ago, there were no transistors at all. Devices like the MOS transistor (1965), the LED or

1.2 Historical overview—fifty years of commercial semiconductor physics

light-emitting diode (1960), and the microprocessor (1971) are the necessary building blocks for the electronic consumer items that are now familiar to everyone: the pocket radio (1957), the home video recorder (1969), the personal computer and personal stereo (1980), the mobile phone (1979), the smart card (1990), the internet assistant (1996), and so on. Although earlier electronic technology, based upon vacuum components (tubes or valves), was capable of producing such items (in much the same way that certain cavemen were probably *capable* of conducting symphony orchestras) the resulting products would have been too expensive, too bulky or even just too ugly to appeal to a mass market. Putting the cart slightly before the horse, we can assert that it was the miniaturization of products, made possible by the new 'all solid state' devices, that allowed the personalization of electronics and a revolution in living style that has been called the second industrial revolution.

Let us now examine the development of three simple devices, the transistor, the LED, and the photodiode, in terms of the use of new materials. The very first transistor was a point contact device based upon polycrystalline germanium, and not much good! Its characteristics were not really good enough, or stable enough, for it to compete with the commercial triodes that it was designed to replace. Improvements necessary for this paradigm shift to occur included the use of single, rather than poly, crystal germanium in order to reduce electronic 'noise', the invention of diffusion doping, to replace the flaky point contacts, and, eventually, the substitution of silicon for germanium. All of this took about 20 years. (In fact it is said that the accountants at Bell Laboratories were so unsure of the commercial future of the transistor in 1950 that they sold the rights to reproduce the technology in Japan *in perpetuo* for a one-off payment of $50 000. Last year the Japanese electronics industry sold many billion dollars worth of product.) In the mid-1960s, the idea of integrating several transistors on a single chip of material took hold. This process was better suited to a different kind of transistor altogether, the field-effect transistor (FET), which requires the growth of an insulating layer on top of the semiconductor in order to operate. Silicon oxide was found to be excellent for this purpose. Hence silicon became established in the later 1960s as the material of choice for Small Scale Integration in metal-oxide-semiconductor (MOS) technology. Silicon is still king in these days of 0.18 micron technology and Ultra Large Scale Integration, when more than a hundred million transistors may share a few square centimeters of silicon surface.

Of course, transistors can be made from materials other than silicon. The very fastest transistors use epitaxially strained layers of III–V semiconductors based on GaAs or InP. Why have these materials not replaced silicon? Well, the three main reasons are cost, cost, and cost: the cost of a single silicon transistor in a microprocessor works out at slightly less than $0.01; the raw material is very cheap: an unprocessed silicon wafer costs less than $1; GaAs wafer is ten times and InP nearly a thousand times more expensive. It now seems quite unlikely that GaAs will replace silicon in large-volume electronics. Ever. In certain fast-growing niche markets, such as microwave-based telecommunications, on the other hand, the use of III–V semiconductors is mandatory.

The light-emitting diode, invented in 1955, suffered a long period of dormancy before commercialization [2]. Ironically, it was the use of silicon-based microprocessors (1968) in pocket calculators, wrist watches and electronic games that led to a revival of interest in (mainly) III–V light emitting devices. Even more ironic is the fact that calculator LED displays completely disappeared from these products following the introduction of liquid crystal displays in the early 1980s. But since they had become cheap and available, many other uses were found for LEDs. Their development continues.

An LED is an optoelectronic device that converts electrical current into light when electrons and holes recombine across the bandgap of a semiconductor. Hence there is an intimate connection between the colour of the light emitted and the material of which the LED is formed. GaAs LEDs produce infrared light, which is invisible to the human eye but is very well matched to silicon photodetectors. GaAs LEDs are used widely in security systems, remote controls, and fire alarms but we are more aware of LEDs that produce visible light. Red LEDs based upon GaAsP alloy were first to be commercialized. Yellow and green LEDs utilize GaP doped with certain impurities. Blue diodes based on SiC first appeared in the early 1990s, but have recently been superseded by devices that use GaN. Furthermore, by combining several diode heads within one package, it has been possible to produce displays in a wide variety of perceived colours. In particular, the correct mix of red, green, and blue emission produces a 'white' LED.

All of the LEDs mentioned above use compound semiconductors. For reasons made clear in the next section, one does not expect to be able to produce a light-emitting diode from silicon. But this feat has actually been accomplished recently by using a porous form of this material. The rather inefficient diodes that result emit red or infrared light, depending on the processing conditions.

The invention of the laser in 1960 encouraged several groups to realize a semiconductor version of the device based upon a GaAs LED. This was first achieved in 1962. Commercialization followed (after the obligatory time gap) to supply a market based upon the compact digital audio disc or CD (1979). While, at sufficiently low current density, every laser diode is an LED, not every LED can lase. In fact, it is rather difficult to produce useful laser diodes which meet the following specification: continuous wave operation (i.e. not pulsed) at room temperature for, say, 100 000 hours, with 10 mW of light output in a single longitudinal mode. Diodes which have met or bettered these standards include those based upon GaAs, which are mainly used to read CDs, and those based on the quaternary alloy $In_xGa_{1-x}As_yP_{1-y}$, which are used in fibre optic communications systems.

Photodetectors are the oldest semiconductor devices. It took only two years for Von Siemens to utilize W. Smith's discovery of selenium's sensitivity to light by inventing the photometer (1875), which is essentially also the first solar cell. Light absorption in a semiconductor creates electron–hole pairs which can contribute to its conductivity. By the same token, semiconductor materials are blind to light with photon energy insufficient to bridge their bandgap. Silicon is a useful detector material in the extended visible spectral range from 300 nm to

1100 nm. Germanium is used in the near infrared, from 700 nm or so to 1600 nm. Narrow gap materials such as PbS, InSb, and HgCdTe detect far-infrared light.

In terms of sales, the highest volume photodetector device is the CCD (Charged Coupled Devices) array used in the home video camera, introduced in the 1970s. Each element of such an array is a metal-oxide-silicon (MOS) capacitor. Light intensity information is stored in the form of trapped charges. CCD arrays are also used in computer-compatible digital still cameras (1991).

1.3 Electronic and optoelectronic materials

From the short history above, it can be seen that silicon and gallium arsenide are at present the best-developed semiconductors for electronic and optoelectronic applications. In this section, we will briefly summarize some of the characteristics of semiconductor materials which are important in their application to devices.

We find that there exist natural limits that encourage us to study more than one material. For example, silicon is not useful as a laser material because its band structure takes the form shown in Fig. 1.3.

Light absorbed by a semiconductor produces excited electron–hole pairs which tend to relax (lose energy) and recombine across the bandgap. In an indirect semiconductor like silicon, the electron relaxes into a state which has a different momentum (k) from the relaxed hole. (The hole's momentum and energy (E) are both conventionally rescaled to zero on the band diagram.) In consequence an unassisted recombination between an electron and a hole is very unlikely in silicon. Phonons with high momentum and low energy can assist recombination is silicon, in a sort of three-particle interaction, but phonons are rare, especially at low temperature. So what? Eventually, one would think, the electron–hole pair will recombine with the assistance of a passing phonon and light will emerge: the indirect gap will only slow things down. This is not true. While an electron is mobile in a semiconductor, it may encounter a defect and become trapped or otherwise de-excited. This process, non-radiative recombination, kills the luminescence. Hence, indirect semiconductors tend to be both

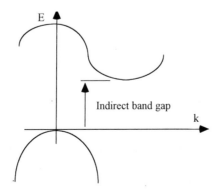

Fig. 1.3 The simplified band structure of an indirect semiconductor.

slow in response (we say that the carrier lifetime is long) and inefficient as light emitters. GaAs is in contrast a direct gap semiconductor. Its quantum efficiency, the ratio of radiative to total recombination, can approach 100% internally, although, due to internal reflection, only a fraction of the light escapes. The carrier lifetime is of order 1 ns in a typical device.

An optoelectronic parameter of some interest is the optical bandgap, which represents a threshold energy for light absorption if the semiconductor sample is perfect. Light absorption increases at photon energies above the bandgap. The rate of increase is relatively slow for indirect semiconductors such as silicon, but very abrupt for direct gap materials. At sufficiently low temperatures, electron−hole pairs bound together by the Coulomb interaction form quasi-particles called excitons. These add sharp lines to the absorption spectrum in the bandgap region. The size and nature of the optical bandgaps for representative semiconductors are as follows: silicon (indirect, 1.1 eV at room temperature) GaAs (direct, 1.4 eV), ZnSe (direct, 2.7 eV), GaN (direct, 3.4 eV). There are obvious chemical trends.

In terms of crystal structure, Si and GaAs are both cubic materials. Each comprises two interpenetrating face-centered cubic lattices that are displaced relative to one another along the cube diagonal (the [111] direction). While in Si the sublattices are of necessity chemically identical, compound semiconductors have metal atoms on one sublattice and non-metals on the other. The wider gap semiconductor materials have less stable structures than GaAs. In many cases, a hexagonal modification may occur. The hexagonal (or Wurtzite) structure differs from the cubic (zincblende) one only by the rotation of a bond. By controlling the growth conditions, in particular the substrate orientation and temperature, it is possible to engineer preferential growth of either phase for many compounds. For example, while most work has been done on growing wurtzite GaN, growth of the cubic modification is also practised. SiC has the largest number of stable crystal habits (polytypes) of any semiconductor. The modification from cubic (3C) to hexagonal (6H) silicon carbide changes the bandgap by 30%.

At one time, crystals for semiconductors applications tended to be sawn or cleaved, if that proved possible, from larger crystals that had been grown from the melt. But modern semiconductor devices tend to be much broader than they are deep. In the planar technology, a depth of 10 microns is probably large enough. For this reason, and in order to optimize crystal perfection, epitaxially grown material has taken over in the industry. At the forefront of technology, bulk crystals are thought to be really only good enough to be used as substrates. Epitaxy allows the growth of materials far from thermodynamic equilibrium. Temperatures as low as 150 °C, far below the melting point, have been used in the growth of ZnSe by molecular beam epitaxy. GaN epitaxy, on dramatically mismatched substrates, will be described in detail in later chapters.

In order to fashion devices, a semiconductor should be capable of being doped both n-type and p-type. An electron donor is an impurity which can release one or more electrons into the conduction band of a semiconductor at relatively low temperature. The energy required to ionize the donor should be comparable in magnitude with the thermal equipartition energy, $k_B T$, at the

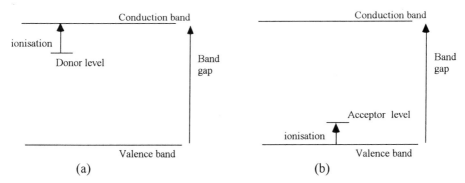

Fig. 1.4 In n-type and p-type semiconductors, electrons exchange between the allowed bands and levels in the bandgap. The arrows indicate possible electronic transitions.

operating temperature of the device, and a lot smaller than the bandgap. A semiconductor with more electrons than holes is said to be n-type. A p-type semiconductor is one in which conduction by holes is dominant. This is achieved by doping with an electron acceptor. Figure 1.4 compares band diagrams for n-type and p-type semiconductors.

Finally, we mention the temperature dependence of semiconductor devices. A distinction between n-type and p-type material is essential to the operation of transport devices. This distinction is lost at high temperatures because large numbers of electrons thermally excited across the bandgap swamp those that are introduced by dopants. (In addition, electronic noise increases with temperature for several different reasons.) Hence we expect silicon-based devices, with a rather small bandgap, to be rather susceptible to increases in temperature. In fact, commercial discrete silicon components do not operate well above 100 °C. For those situations that require devices to operate at higher temperatures, it is clear that we must turn our attention to other materials. Materials with larger bandgaps are preferable in this context. Similar strictures apply to light emitting devices. Lasers in particular tend to be extremely temperature-sensitive. The current required to push the device above laser threshold increases exponentially with absolute temperature.

1.4 Semiconductor families

In the following review of materials, we are particularly interested in those which provide new functionally, or improved performance compared with that offered by silicon and gallium arsenide. In particular, we turn our attention to semiconductors with larger bandgaps, brought about either by a decrease in the covalent radius of monotype atoms or an increase in the ionic component of the bonding between atoms of different type. These compounds may be grouped together under the name widegap semiconductors.

1.4.1 The group IV elements

Silicon is indisputably the most widely used semiconductor. As we have seen above, it dominates the electronics market and will continue to do so in the foreseeable future. What of the other elemental semiconductors? While germanium appears to have had its day, there is at present an increasing interest in the development of electronic devices based upon synthetic diamond [3].

Diamond occurs naturally, of course, in a limited number of world locations, in the form of smallish single crystals. These gemstones are mined, cut and polished for industrial applications as well as for use in jewellery. Diamond is the hardest material known (it will even cut other diamonds!), it is an excellent heat sink, and has an extremely high melting point. It is an indirect semiconductor with a bandgap of 5.5 eV. Some natural diamonds are p-type due to their contamination by boron. The acceptor ionization energy, 0.37 eV, is rather large. The single isolated nitrogen atom forms a shallow donor in diamond (it goes deep in silicon). Therefore, an electronics technology based upon diamond is certainly feasible. Epitaxial, rather than natural, diamond has offered the preferred route to commercialization since the 1970s. Bulk growth from the melt is impossible because of the high pressures and temperatures required, but small crystals can be produced commercially as a by-product of rather large explosions. Nowadays, the quality of material grown by chemical vapour deposition (CVD) readily matches that of natural diamond (which is not really very good). Methane is cracked in the presence of hydrogen and oxygen in order to grow polycrystalline carbon films at about 800 °C. Care must be taken to minimize the simultaneous deposit of graphite or amorphous carbon on hetcroepitaxial (non-diamond) substrates.

The first steps towards commercializing CVD diamond have been taken. Several research groups from industry and academia have produced prototype transistors and photodetectors. Device operation at up to 500 °C is possible. CVD diamond is also useful for X-ray and radioactive particle detection. At the moment, however, diamond's promise as an electronic material remains largely unfulfilled.

1.4.2 The IV-IV compound

SiC is a single compound with a formidable range of structural possibilities. The stacking sequence of the individual layers of Si-C pairs leads to a classification scheme of polytypes: the polytype 2H-SiC, with stacking sequence ABABAB..., is the wurtzite modification, 3C-SiC (ABCABC...) is cubic, and so on. More than 100 polytypes exist, although most work has been done on 2H, 3C, 4H, and 6H modifications.

SiC has in general many properties intermediate between those of silicon and diamond. For example, the room temperature bandgap is around 3 eV, depending on polytype. It is a rather hard material, familiar under the name carborundum as an industrial abrasive. Like silicon, and unlike diamond, SiC can be grown in the form of large bulk crystals, by the modified Lely method, invented in 1971. The first such samples were plagued by the presence of long hollow tubelike defects, about 1 micron wide. Micropipes are fatal to the operation of devices. Improvements in growth technique have led to a decrease in the density of these defects, and they may shortly be eliminated. However, commercially available 4H– and 6H–SiC substrates are at present several thousand times more expensive than their equivalent in silicon. SiC may be readily doped. Aluminium forms an effective-mass acceptor, while nitrogen acts as a donor, as in diamond. Doped material does not become intrinsic until 900 °C.

SiC blue LEDs produced by Cree Corporation were available commercially between 1991 and 1994, when they were superseded by brighter GaN diodes from Nichia Chemical Company. SiC is an indirect gap material. The efficiency of SiC diodes is therefore rather low, of order 0.1%. In 1995, Cree Corportion incorporated SiC as a conducting substrate in their own GaN LEDs. SiC has also been used successfully in prototype FETs that work at temperatures up to 500 °C, in power devices with breakdown voltages of order 10^4 V, and so on. Although, so far, the high price of substrates has discouraged commercial exploitation of the material, it certainly has great potential in several areas of electronics, in particular those which involve high temperatures, high voltages, and high powers.

1.4.3 The II–VI compounds

IIB–VI compounds are a rather diverse family of semiconductors, with bandgaps which range from less than zero (mercuric chalcogenides are semimetals) to nearly 4 eV. We shall focus here on those wide bandgap II–VI compounds, such as ZnSe (with a bandgap at 2.7 eV at room temperature), which have been developed since the later 1960s as materials for optoelectronics applications at shorter wavelengths. In particular, compounds of the Zn(Cd)S(Se) sub-family have bandgaps that span the entire visible spectral region. On account of the relatively long period during which II–VI semiconductors have been studied, observers of the field have become very aware of the fact that scientific advances tend to come in waves. Nor should we forget the fact that II–VI semiconductors have been commercialized for years in such diverse applications as phosphor screens (mainly polycrystalline ZnS with impurities) and gas sensors (mainly ZnO) [4].

Crystal growth is again the big issue. Difficulties encountered in producing large, single crystals of ZnSe by traditional techniques led growers towards expitaxy in the mid-1970s. The substrate of choice for II–VI compounds is... GaAs! Although this material has a slightly different lattice constant from ZnSe (0.3% mismatched at room temperature), the crystal structure is the same, the GaAs (001) surface is very well understood by crystal growers, and such substrates are relatively cheap. (Silicon substrates are non-polar; this causes problems in growing polar epilayers that have a stable phase.) Thin ZnSe films of very high quality can now be grown on GaAs by molecular beam or metalorganic vapour phase epitaxy (MBE or MOVPE). Doping with chlorine provides n-type conductivity, but the reproducibility and reliability of p-type doping of ZnSe continues to be a problem. In the late 1980s, lithium, nitrogen, and oxygen (?) all competed as acceptor species. Following the demonstration of reliable p-doping by use of a nitrogen plasma in 1990, rapid progress was made towards the development of the first II–IV laser diodes by a group from 3M Company in 1991. Unfortunately, reproducible p-doping has not yet been achieved, using this or any other method, by MOVPE growers.

The lattice mismatch problem becomes even more important when a device contains several layers of different semiconductor materials. An example of such a heterostructure is the laser diode announced by Sony in 1996, which operates continuously at room temperature for approximately 100 hours with 1 mW output. This device comprises eleven different layers of II–VI compounds and alloys on top of a GaAs substrate. While strain has been shown to aid the performance of certain devices, through its effects on the band structure, it may also have several deleterious mechanical effects, including the nucleation and propagation of ruinous extended defects in laser structures.

These and several other problems relating to crystal growth have been successfully tackled, at least to some extent, in the effort to produce working devices utilizing II–VI compounds. But companies are perhaps no closer to commercializing this technology than they were in 1991. In the meantime, substrates of ZnSe have been improved to the extent that they may soon compete in the market-place with GaAs. Will this development revive (once again) the fortunes of widegap II–VI compounds? We must wait and see.

1.5 Old, new, borrowed, blue: are the nitrides different?

The III-nitrides are partly ionic semiconductors with a very wide range of bandgaps. Boron nitride, with four stable polytypes, is usually considered as a special case. AlN in wurtzite form has a bandgap of 6.2 eV. This puts it beyond the reach of most laser excitation sources. Wurtzite GaN, the best studied semiconductor in the group, has a bandgap of 3.4 eV, in the near-ultraviolet spectral region. InN has a gap of 1.9 eV [5].

In 1994 the introduction of working LEDs based upon GaN technology took the semiconductor world almost by complete surprise. Most people were in fact looking in another direction, towards II–VI semiconductors, when the commercial breakthrough took place. Signs of increasing research activity were in fact plain to see from the beginning of the 1990s. Showing perfect hindsight, Figure 1.5 plots numbers of publications on GaN from the early 1980s to 1996. (If we include all contributions on nitride-related work, the numbers of papers per year more than double, but the trend is the same.) The nitrides are not new semiconductors in any sense. They are in fact fairly old. Serious study of GaN began in the mid-60s; LEDs with MIS structure were produced and briefly commercialized in the early 1970s. There were even reports of stimulated emisson and measurements of laser gain at that time. The full story of the development of the optical devices is presented in later chapters. Here we simply list the enabling steps.

Growth of bulk GaN again requires high temperatures and pressures. The size and quality of bulk crystals, produced by dissolving nitrogen in liquid gallium at around 10 kbar and 1600 °C, continues to improve. But, as is the case

1.5 Old, new, borrowed, blue: are the nitrides different? 13

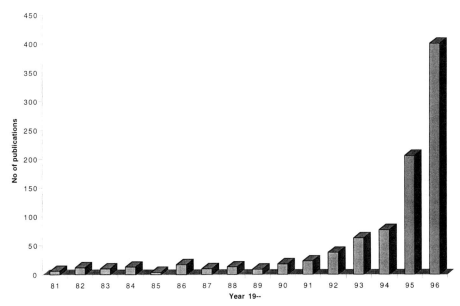

Fig. 1.5 The increase in nitride publication rate in the early 1990s (based on data from the SciSearch database, copyright the Institute for Scientific Information, 1997).

with other widegap materials, it is epitaxy that offers the preferred route to the growth of high-quality semiconductor. In contrast to the situation with II–VI material, the major advances in the growth of GaN and its alloys have been accomplished using MOVPE. In the conventional technique, trimethylgallium reacts with ammonia at temperatures of order 1000 °C to produce GaN. The substrate is usually Al_2O_3! Sapphire is a trigonal crystal which is lattice mismatched to wurtzite GaN by 13.8%. The surprising choice of substrate results from the unavailability or high cost of suitable alternatives, although many other crystalline materials continue to be tested.

The first enabling step [6] partly alleviates the lattice and structure mismatch problem: a very thin (few nm) polymorphic layer of AlN or GaN is deposited first on the bare substrate at fairly low temperatures (around 500 °C). The temperature is then ramped to around 1000 °C before growth proper begins. The buffer layer or accommodation layer is supposed to present a surface like that of bulk semiconductor to reactant species. Layers of GaN grown by this method are generally found to be n-type. The concentration of the residual donor, which may be a nitrogen vacancy or a silicon impurity, must be kept low, less than 10^{15} cm^{-3}, if controlled doping is the aim of the game. As with ZnSe, p-type doping turns out to be more of a problem than n-type doping.

The successful p-type doping of MOVPE-grown GaN [7] is the second enabling step towards a commercial LED. A suitable acceptor must first be identified. Of all the group-II elements that can substitute for native atoms on gallium sites, magnesium is found to have the lowest hole ionization energy. Doping with magnesium produces compensated material that can be made conducting either by low energy irradiation or annealing in nitrogen gas. (The compensating species appears to be an acceptor bound to a hydrogen atom.) The ability to dope material both ways allows a p–n junction diode to be produced. Now we have to consider how to introduce current into such a device; in other words, how to make contacts with the outside world.

It is always more difficult to make ohmic contacts to widegap semiconductors than to silicon or GaAs. Partly, this problem is associated with the relatively low doping levels that are possible in the widegap materials. A thin tunnelling contact forms between a metal and a semiconductor when the doping level is extremely high (greater than that required for device operation by several orders of magnitude.) If high levels of doping are not possible, the relative magnitude of the metal work function and the semiconductor electron affinity (the depth of its conduction or valence band) becomes important. Metals with small work functions, such as aluminium, should do the job for n-type semiconductors. p-type semiconductors should be able to contact to metals with large work functions, like platinum. Annealing the contact region after deposition often improves performance. However, these are only rules of thumb. In order to contact to p-type ZnSe, Sony uses an ingenious p-ZnTe/p-ZnSe multiple layer which is designed to allow holes to tunnel through a ladder of connected states. The strategy, which has proved to be successful with the nitrides, appears to be rather simple. The n-type contact is aluminium; the contact to p-type material uses an annealed multilayer of two metals (possibly an alloy) which may be Cr/Al or Au/Mg, borrowed from GaAs technology. This is the final enabling step that allowed Nakamura and co-workers at Nichia Chemical Company to commercialize nitride LED technology.

Technologies survive by lowering prices or by offering enhancements. The quality of commercial GaN-based LEDs continues to improve as the price falls. In Fig. 1.6, we compare the electroluminescence (EL) spectra of two different devices. Spectrum A belongs to a first-generation blue LED (1994), which comprises an InGaN/AlGaN double heterojunction in which the active medium is Zn-doped InGaN. The EL spectrum is rather broad. The diode emits light not only in the blue, near the peak at 430 nm, but also in the violet, green, and orange parts of the visible spectrum. Consequently, the perceived colour of the diode emission is bluish-white. In spectrum B, a much narrower spectral profile presents itself: the second generation diode (1995) has as its active region a very thin (2 nm) layer of undoped InGaN sandwiched between n-GaN and p-AlGaN buffer layers. The emission is very blue. In addition, the tuning of the indium content in the active region, between 5 and 80%, allows the spectral range to be tuned from violet to orange, although only blue, blue–green, and green diodes are available commercially at the time of writing (early 1997). The Nichia story continues with the extension of LED performance towards the eventual goal of a commercial working laser. Progress during 1996 was rapid: a first announcement of pulsed operation at 77 K in January was followed by room temperature (RT), pulsed operation; near-RT, continuous wave operation (September) and RT, CW, 35 hours lifetime operation in December.

So, are the nitride semiconductors different, as has often been claimed? Or perhaps a better question is: from what are the nitrides different?

1.5 Old, new, borrowed, blue: are the nitrides different?

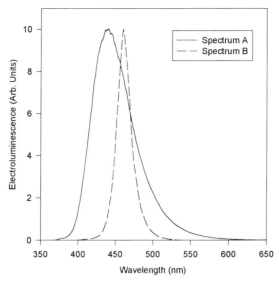

Fig. 1.6 Electroluminescence spectra of two successive generations of diodes.

At the moment, the research community who work on nitrides can be divided into three classes of worker. First, there are the True Originals who have been working in the field since its beginnings. The first beginning (bulk materials) came in the late 1960s; the epitaxial adventure began in the later 1980s. GaN has *several* fathers (and, to the best of my knowledge, no mother.) Then come the Wagon-Hoppers, including the present author, who were doing something different before 1994: this group has written most of the papers published in the last two years. Finally, there is a small number of graduate students who have never done anything else but work on nitrides, but who actually carried out most of the experiments that were discussed in the papers written by the second lot!

The nitrides have attracted workers in more traditional semiconductors as well as complete outsiders. People who know a lot about GaAs are likely to find GaN rather strange. Here are a few differences between GaN and GaAs:

1. The bandgap of GaN is a lot bigger. GaN is transparent, whereas GaAs is opaque, with a metallic lustre. It may seem strange at first that a transparent material can be doped and carry a current.
2. The crystal structures are different. Although cubic GaN exists and has potential for commercialization (see Chapter 11), the usual form of GaN is wurtzite (hexagonal). Both the electronic band structure and the crystal microstructure are strongly influenced by the crystal phase.
3. The donor and acceptor ionization energies, the exciton binding energies and the phonon energies are much larger in GaN than in GaAs. This tends to confound naïve expectations about temperature dependences of various physical properties.

Those who come to the nitride field from other widegap semiconductors, on the other hand, may be likely to notice more similarities than differences. The greatest single difference between III–N compounds and the other widegap compounds appears to be the fact that the nitrides have been successfully commercialized. We may think that this commercialization was very rapid, really an overnight success, if we carelessly forget the first 25 years of study.

1.6 The role of defects in widegap semiconductors

While the nitrides may not be too different, they offer scientists a fundamentally different kind of problem from those posed by other widegap materials. The question is no longer: how can we get these materials to work? With the nitrides, it is: how do these materials work so well? The technology is already with us, but the physics of the devices is not so well understood. Progress in devices is so rapid, there is so little time to think, that several potential solutions to the same problem may have to be debated at once. By the same token, the same physical solution to a given problem may occur simultaneously to different workers. In this section we examine a general problem, the influence of defects on device operation, in relation to GaN. Some comparisons will be drawn with ZnSe. These materials are rivals, it is true, but we should remember (1) that competition is a good thing and (2) that comparisons are odious.

Defects exist in all crystalline materials in a bewildering variety of types. We must first distinguish atomic-sized defects (point defects) from extended ones: line, area, and volume defects. Useful point defects include the (mainly substitutional) impurities that act as donor and acceptor species in semiconductors. Such shallow defects appear in band theory as states in the forbidden energy gap which lie close to their respective bands, with donors close to the conduction band. The shallowest possible levels, the effective mass states, have orbitals which are made up from a combination of band states. The extra electron or hole is localized within a Bohr radius of the impurity at very low temperatures. While donor and acceptor Bohr radii are large in silicon and GaAs, they tend to be much smaller in widegap semiconductors, such as ZnSe and GaN. Hence local effects on the donor and acceptor wavefunctions may be expected in the latter materials. In an extreme case, the electron or hole may localize on a single bond. The associated energy level is made deeper (closer to the middle of the gap) by such local interactions.

Successful strategies for p-type doping of ZnSe and GaN have emerged relatively recently, and are still in the process of being refined. Nitrogen doping of ZnSe is inhibited by a quite low level of activation: the saturated hole population falls well below the total number of incorporated nitrogen atoms. This is probably due to a compensation process whereby deep donors (formed by nitrogen complexes) are created in competition with the desired acceptors. A quite different problem occurs with GaN. Although p-type doping may be effected by incorporating both Zn and Mg as acceptors, the associated levels are rather deep. It is possible that a shallower dopant results from substitution of carbon for nitrogen, but it remains to be seen whether this dopant can be controlled, since carbon on a gallium site will readily form a deep compensating donor.

1.6 The role of defects in widegap semiconductors

Non-radiative recombination centres act to kill luminescence due to electron–hole recombination across the bandgap of a semiconductor. In the model originally proposed by Shockley, Read, and Hall, these defects form deep levels in the middle of the gap that sequentially capture an electron and a hole. Many intrinsic and impurity-related defects are known to form deep centres. The capture process does not usually result in light with a photon energy equal to half the bandgap as we might expect: instead, the electron–hole recombination energy dissipates into phonons. Non-radiative recombination lowers both the efficiency and the decay time of luminescence. In order to minimize its effects, spectroscopic measurements on light emitting materials are usually carried out at low temperature. However, devices are required to work at room temperature. It is interesting therefore to compare the photoluminescence efficiency of materials at low and high temperatures. We can expect some quenching of luminescence as the temperature rises, but how much? In the author's experience, the degradation measured in GaN samples is much less than that for any other semiconductor, with the sole exception of porous silicon. If we consider materials that are pure enough to show mainly free exciton luminescence at low temperature, the quenching ratios between 10 K and 300 K are a factor of 1000 for ZnSe, 30 for GaN. This result suggests that deep levels in GaN are either relatively rare or are ineffective as traps.

Point defects are zero-dimensional, like quantum dots. Higher dimensioned structural defects also occur in crystals. A dislocation is a line defect formed by the edge of an incomplete plane of atoms. The crystal surface is a rather important planar defect, without which light could not emerge from the semiconductor! A small-angle grain boundary may act as an internal surface. Neighbouring inversion domains also meet along a surface. Finally, a volume defect may be formed by an inclusion of foreign material or a circumscribed phase domain. The microscopy of such defects is discussed in later chapters. Here we briefly consider their effect on the operation of devices.

ZnSe optoelectronic devices, both LEDs and lasers, suffer fatally from defects that propagate and grow during device operation. A runaway degradation takes place. At first, small defects, perhaps even point defects, absorb light. The resultant local heating leads to the formation of new defects in some as-yet unexplained manner; in addition the defects may become mobile; in consequence, they grow in size. Eventually, the entire area of the emitting surface darkens and the device fails. It is perhaps encouraging to note that similar problems attended the development of GaAs optoelectronic devices. These problems were defeated by improvements in crystal growth techniques. Attempts are currently being made to reduce the susceptibility of the more ionic II–VI materials to defect interactions by alloying the active layer with beryllium selenide.

The state of affairs in GaN could not appear more different. Electron micrographs of epitaxial layers feature huge densities of dislocations, threading between the substrate and the surface, that appear to have no effect whatever on the efficiency of light emission. The solution to this conundrum is probably something like this: Mid-gap levels promote non-radiative recombination, as we saw earlier. Dislocations that do not possess mid-gap level do not inhibit radiative radiation. This pushes a final solution of the problem in the direction of the theorists.

1.7 Conclusion

Is it time already? This chapter has attempted, however briefly, to establish a context for the work currently being untaken in many laboratories, all over the world, on nitride semiconductors. Much work remains to be done in this field, in areas of basic science, crystal growth, device design and so on. It is remarkable that so much has been accomplished in the last two or three years, but it is also clear that commercialization would not have been possible in the absence of a substantial pre-history of scientific study.

Acknowledgements

While the mistakes here are all mine, I particularly benefited while writing this chapter from discussions with Bernard Gil, whose idea it was to write it, Bruno Meyer, Al A. Efros, Alan Collins, Ted Thrush, and Rob Martin. Paul Middleton performed a sanity check on the first draft manuscript.

References

1. See *Getting to Know Semiconductors* by M. E. Levinshtein and G. S. Simin (World Scientific, Singapore, 1992) for an entertaining introduction to semiconductor physics. See also W. Shockley's *Electrons and Holes in Semiconductors* (Van Nostrand, New York, 1951) for an historical account of transistor electronics.
2. An early summary of the development of LEDs is found in *Luminescence and the Light Emitting Diode* by E. W. Williams and R. Hall (Pergamon Press, Oxford, 1978).
3. For a recent review of the prospects of synthetic diamond as an electronic material see the article by A. T. Collins in *Materials Science and Engineering* **B11** (1992), 257–263.
4. General features of the II–VI compounds are reviewed in *Properties of Wide Bandgap II–VI Semiconductors*, edited by R. Bhargava *(EMIS Datareview* **12**, INSPEC, London, 1996).
5. See *Properties of Group III Nitrides*, edited by James H. Edgar *(EMIS Datareview* **11**, INSPEC, London, 1995).
6. I. Akasaki, H. Amano, Y. Koide, K. Hiramatsu and N. Sawaki (1989). *Journal of Crystal Growth*, **98**, 209–219.
7. I. Akasaki, H. Amano, M. Kito and K. Hiramatsu (1991). *Journal of Luminescence*, **48–9**, 666–670.

2 Deposition and properties of group III nitrides by molecular beam epitaxy

Galina Popovici, Hadis Morkoç, and S. Noor Mohammad

2.1 Introduction

Wide bandgap semiconductor materials extend the field of semiconductor applications to the limits where classical semiconductors such as Si and GaAs fail [1–7]. They can emit light at shorter wavelengths (blue and ultraviolet), and can operate at higher temperatures due to the large bandgap, high thermal conductivity, and chemical inertness.

In the past several years, the most studied of wide bandgap semiconductors have been the III-nitrides. Among them, GaN and its alloys with InN and AlN have attracted a great deal of attention since the successful commercialization of bright blue light-emitting diodes followed later by the demonstration of injection lasers [8–10]. GaN-based nitride semiconductors have several advantages over other wide bandgap semiconductors such as SiC and diamond. They can be doped both p- and n-type, have direct bandgaps, and form heterostructures conducive for device applications. They can be grown epitaxially over a number of substrates, and thus monocrystalline layers can be attained over large surfaces. Wurtzitic polytypes of GaN, AlN, and InN, and their ternary and quaternary alloys, have direct bandgaps which pave the way for lasers and efficient light-emitting diodes. As mentioned already, GaN can form continuous alloys with InN and AlN covering the spectral range from red to vacuum UV (1.9 to 6.2 eV). The ternary system InGaAlN has potential for attaining lattice-matched double heterojunctions. AlN and InN are still much less studied as compared to GaN, and their technology is much less advanced.

Remarkable progress in the growth of high-quality epitaxial III-nitride films by a variety of methods have recently been achieved. The most successful among all growth methods are metalorganic chemical vapor deposition (MOCVD) and molecular beam epitaxy (MBE). In spite of the rapid development, many problems remain to be overcome. Understanding of the fundamental mechanisms of impurity incorporation, particularly p-type, and native defect formation is in the incipient stage at best. The role of Ga and N vacancies in doping, the purported hydrogen passivation of acceptors, and the defects responsible for unintentional n-type doping represent some controversial issues that should be addressed in research for further progress in the device arena. The main technological issue has been and remains the lack of native substrates and resultant lack of high crystalline quality of films, the kind endemic to matched system.

20 *Deposition and properties of group III nitrides*

In this chapter, following the essential fundamentals, the progress and challenges concerning the MBE method in the growth of III (Al, Ga, In)-nitrides are reviewed. Since the bulk of the technological work was performed to data on GaN and its diluted alloys with In or Al (atomic fraction of In or Al in alloys generally does not exceed 0.2) this paper discusses especially GaN technological problems.

2.2 III-N structural and chemical properties

2.2.1 Common crystal structures

There are three common crystal structures shared by the group III–V nitrides: the wurtzitic, zincblende, and rock salt structures. At ambient conditions, the thermodynamically stable structures are wurtzitic for bulk AlN, GaN, and InN [11–13]. The zincblende structure for GaN and InN has been stabilized by epitaxial growth of thin films on the (001) crystal planes of cubic substrates such as Si, MgO, and GaAs. In these cases, the innate tendency to form the wurtzitic polytypes is overcome by topological compatibility. The rocksalt or NaCl structure can be induced in AlN, GaN, and InN at very high pressures. The wurtzitic polytype has a hexagonal unit cell and thus two lattice constants, c and a. It contains six atoms of each type. The space grouping for the wurtzitic structure is P6_3mc (C_{6v}^4). The wurtzitic structure consists of two interpenetrating hexagonal close-packed (HCP) sublattices, each of which is with one type of atoms, offset along the c-axis by 5/8 of the cell height ($5c/8$).

The zincblende structure has a cubic unit cell, containing four Group III and four nitrogen elements. The space grouping for the zincblende structure is F$\bar{4}$3m (T_d^2). The position of the atoms within the unit cell is identical to the diamond crystal structure in that both structures consist of two interpenetrating face-centered cubic sublattices, offset by one quarter of the distance along a body diagonal. Each atom in the structure may be viewed as situated at the center of a tetrahedron, with its four nearest neighbors defining the four corners of the tetrahedron.

The zincblende and wurtzitic structures are similar. In both cases, each Group III atom is coordinated by four nitrogen atoms. Conversely, each nitrogen atom is coordinated by four Group III atoms. The main difference between these two structures is in the stacking sequence of the closest packed diatomic planes. For the wurtzitic structure, the stacking sequence of (0001) planes is ABABAB in the ⟨0001⟩ direction. For the zinzblende structure, the stacking sequence of (111) planes is ABCABC in the ⟨111⟩ direction. Unless otherwise specified, the III-N wurtzitic growth is described.

The cohesive energy per bond is 2.88 eV (63.5 kcal/mol), 2.2 eV (48.5 kcal/mol) and 1.93 eV (42.5 kcal/mol) for AlN, GaN, and InN, respectively [14]. The fundamental properties of the GaN, AlN, and InN polytypes are listed in Tables 2.1–2.3, respectively.

2.2 III-N structural and chemical properties

Table 2.1 Properties of GaN.

Wurtzitic polytype		
Bandgap energy	E_g (300 K) = 3.39 eV	E_g (1.6 K) = 3.50 eV
Temperature coefficient	$\dfrac{dE_g}{dT} = -6.0 \times 10^{-4}$ eV/K	
Pressure coefficient	$\dfrac{dE_g}{dP} = 4.2 \times 10^{-3}$ eV/kbar	
Lattice constants	$a = 3.189$ Å	$c = 5.185$ Å
Thermal expansion	$\dfrac{\Delta a}{a} = 5.59 \times 10^{-6}$/K	$\dfrac{\Delta c}{c} = 3.17 \times 10^{-6}$/K
Thermal conductivity	$k = 1.3$ W/cm K	
Index of refraction	n (1 eV) = 2.33	n (3.38 eV) = 2.67
Dielectric constants	$\varepsilon_0 = 10$	$\varepsilon_\infty = 5.5$
Zincblende polytype		
Bandgap energy	E_g (300 K) = 3.2 – 3.3 eV	
Lattice constant	$a = 4.52$ Å	
Index of refraction	n (3 eV) = 2.9	

The lattice constants value of III-N is influenced by the growth conditions, impurity concentrations, and film stoichiometry. Probably as a result of increased interstitial defects, the lattice constants of GaN grown with higher growth rates was found to be larger. There occurred an expansion of the lattice when GaN was doped heavily with Zn and Mg. Due to the absence of good quality single crystal films, the crystal structure of InN has been measured mainly on non-ideal thin films, particularly the ordered polycrystalline films with crystallities in the range of 50–500 nm. The end results of the measurements [13] indicate that, although InN normally crystallizes in the wurtzitic (hexagonal) structure, occasionally it crystallizes also in the zincblende (cubic) polytype.

Table 2.2 Properties of AlN.

Wurtzitic polytype		
Bandgap energy	E_g (300 K) = 6.2 eV	E_g (5 K) = 6.28 eV
Lattice constants	$a = 3.112$ Å, $c = 4.982$ Å	
Thermal expansion	$\dfrac{\Delta a}{a} = 4.2 \times 10^{-6}$/K	$\dfrac{\Delta c}{c} = 5.3 \times 10^{-6}$/K
Thermal conductivity	$k = 2$ W/cm K	
Index of refraction	n (3 eV) = 2.15 ± 0.05	
Dielectric constants	$\varepsilon_0 = 8.5 \pm 0.2$	$\varepsilon_\infty = 4.68 - 4.84$
Zincblende polytype		
Bandgap energy	E_g (300 K) = 5.11 eV, theory	
Lattice constant	$a = 4.38$ Å	

Table 2.3 Properties of InN.

Wurtzitic polytype		
Bandgap energy	$E_g (300 \text{ K}) = 1.89 \text{ eV}$	
Temperature coefficient	$\dfrac{dE_g}{dT} = -1.8 \times 10^{-4} \text{ eV/K}$	
Lattice constants	$a = 3.548 \text{ Å}$	$c = 760 \text{ Å}$
Index of refraction	$n = 2.80 - 3.05$	
Dielectric constants	$\varepsilon_0 = 15.3$	$\varepsilon_\infty = 8.4$
Zincblende polytype		
Bandgap energy	$E_g (300 \text{ K}) = 2.2 \text{ eV}$, theory	
Lattice constant	$a = 4.98 \text{ Å}$	

2.2.2 Phase diagrams

The data on phase diagrams of GaN, AlN, and InN are limited and contradictory by reason of high melting temperatures (T_M) and high nitrogen dissociation pressures ($P_{N_2}^{dis}$). Dissociation pressure of MN, where M stands for Al, Ga, and

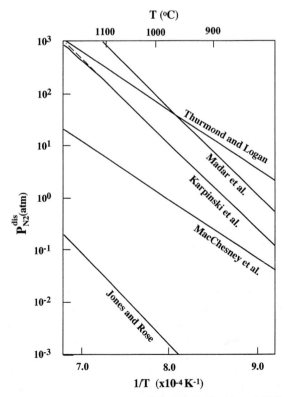

Fig. 2.1 Nitrogen pressure of GaN compiled in ref. [15].

In is defined as nitrogen pressure at the thermal equilibrium of the reaction [15]: MN(s) = M(l) + (1/2)N$_2$(g), where s, l, and g stay for solid, liquid, and gas. Reported values for P$_{N_2}^{dis}$ for GaN (Fig. 2.1) show large discrepancies. Sasaki and Matsuoka [15] concluded that the data of Madar *et al.* [16] and Karpinski *et al.* [17] are the most reliable. According to Fig. 2.1, the nitrogen dissociation pressure equals 1 atm at approximately 850 °C, and 10 atm at 930 °C. At 1250 °C the GaN decomposed even under pressure of 10.000 bar of N$_2$. It should therefore come as no surprise that the incorporation of nitrogen is not a trivial problem at high temperatures. For the pressures below equilibrium at given temperatures, the thermal dissociation occurs at a slow and apparently constant rate, suggesting a diffusion-controlled process of dissociation.

Due to the above-mentioned N dissociation, the data on the melting points are contradictory. Landolt and Börnstein [18] give T_M of AlN equal to 2400 °C at a pressure of 30 bar, T_M of GaN > 1700 °C at a pressure of 2000 bar, and T_M of InN equal to 1100–1200 °C at a pressure >10^5 bar. Thus, the nitrogen dissociation pressure of AlN is orders of magnitude smaller than that of GaN and InN. Massalski [19] cites T_M of AlN equal to 2800 °C. There are reports estimating T_M of AlN at 2200 °C [20] and 2450 °C [21]. Van Vechten [22] determined theoretically the T_M of AlN to be 3487 K. Porowski and Grzegory [23] estimate T_M of AlN to be larger than 3000 K at a dissociation pressure of a few hundred bar, and T_M of GaN larger than 2500 K at tens of kbar. Taking into account the high melting temperatures and high dissociation pressures of III-N compounds, no classical method can be expected to be successful for bulk crystal growth.

2.3 Molecular beam epitaxy (MBE) growth

2.3.1 MBE growth system

An MBE system [24] is a refined form of ultrahigh vacuum (UHV) evaporation, as illustrated in Fig. 2.2. Elements are heated in furnaces or Knudsen cells and directed beams of atoms or molecules are condensed onto a heated single-crystal substrate where they react. Because it is a UHV-based technique, it has the advantage of being compatible with a wide range of surface analysis techniques (reflection high-energy electron diffraction (RHEED), quadrupole mass spectrometry for gas analysis, and Auger electron spectrometry). Typically, the growth rate employed is one to three monolayers per second, approximately 0.3–1 μm/h, although much higher growth rates can be attained.

The MBE growth is usually performed at relatively low temperatures of 650–800 °C. The molecular nitrogen is inert and does not chemisorb on GaN surface below 950 °C, due to the strong N–N bond of the nitrogen molecule. Atomic nitrogen or nitrogen containing molecule with weaker bonds should, therefore, be provided. Several modifications to conventional MBE methods have been implemented for III-N growth. Among them, RF or ECR plasma sources are most commonly employed to activate the nitrogen species. A solid

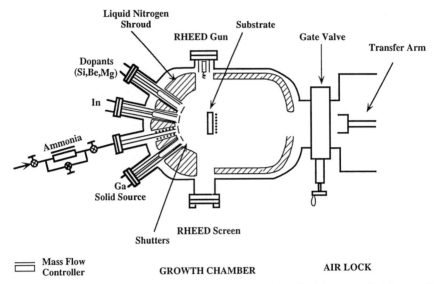

Fig. 2.2 A typical molecular beam epitaxy system configured with ammonia injector. For ECR and RF MBE methods, the ammonia injector is simply replaced with the applicable source.

source is usually used for Ga. Mg or Si are usually evaporated during the growth for p- and n-doping, respectively [25, 26]. Electron concentrations in the range of 10^{17}–10^{19} cm^{-3} and dependable hole concentration in the range 10^{17}–10^{18} cm^{-3} with good Hall mobilities have been succeeded without any post-growth treatment [27].

Before a nitride layer is grown by MBE, the substrates are degreased and etched. The degreasing procedure employed for various substrates is usually the same. The substrate is first dipped into a solution of TCE, kept at 200 °C, for 5 minutes. It is then rinsed for 1 minute each in acetone and methanol. This is followed by a 3 minute rinse in deionized water. The above process is repeated three times to complete the degreasing process. The substrates are then etched, the procedure for which is substrate-dependent.

For example, for SiC substrates, roughly 3 microns of the epilayer surface is removed using hot KOH solution (300–350 °C) as the etchant. This is followed by a rinse in deionized water for 20 minutes and the wafer is blown dried by N_2. The SiC substrate then undergoes oxidation and passivation procedure. The substrate is immersed for 10 minutes in a 5:3:3 solution of $HCl:H_2O_2:H_2O$ kept at 60 °C followed by a rinse by deionized water. The resulting oxide layer is then removed by dipping the substrate, for 20 seconds, in a 10:1 solution of $H_2O:HF$. These procedures are repeated after which the substrate can be exposed to the atmosphere for no longer than 30 minutes before another oxidation–passivation procedure would be required. For Al_2O_3 substrates, a 3:1 solution of hot $H_2SO_4:H_3PO_4$ is used as the etchant. The substrate is dipped in this solution at 300 °C for 20 minutes. This is followed by a rinse in D.I. water for 3 minutes. Prior to growth, the substrates undergo H_2 plasma treatment for about one hour.

2.3.1.1 ECR *and* RF MBE

Plasma excitation of nitrogen or nitrogen-containing molecules for III–V nitride growth has been studied for over twenty years [28–31]. Early attempts using radio-frequency (RF) plasmas suffered from a number of disadvantages, including high ion energies, which cause damage, and unfavorable plasma source geometry. Only recently there has been a revitalization of RF plasma-based nitride research which has been stimulated by the development and commercialization of MBE compatible compact electron cyclotron resonance (ECR) microwave and radio-frequency plasma sources. The emergence of the ECR source for excitation of nitrogen has, in fact, opened up new avenues for nitride growth by MBE. In an ECR-microwave plasma-assisted MBE system the GaN films are formed on the substrate through the reaction of atomic Ga provided by Knudsen cell, and radicals of atomic and to a lesser extent molecular and also again to some extent ionic nitrogen species formed by passing molecular nitrogen at a pressure of approximately 10^{-4} Torr through the ECR source. Approximately 10% of the molecular nitrogen gas is converted into atomic nitrogen.

Two commercial MBE compatible ECR designs are shown in Figs 2.3 and 2.4. The sources operate at 2.45 GHz instead of 13.56 MHz, thereby enabling the

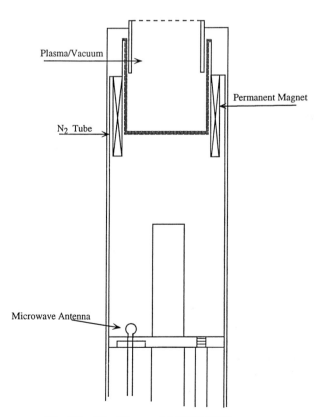

Fig. 2.3 The Wavemat MPDE ECR source.

26 *Deposition and properties of group III nitrides*

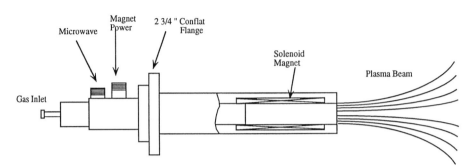

Fig. 2.4 The ASTEX CECR source.

plasma to be confined to a smaller volume. Each source is compact, allowing it to be inserted into the source flange and through the cryoshroud of an MBE chamber. The plasma enjoys a direct line of sight to the substrate so that recombination through collisions are minimized.

The Wavemat MPDR ECR source (Fig. 2.3) uses a coaxial cavity geometry. Wavemat uses a configuration of permanent magnets to provide a multicusp magnetic field which creates several ECR surfaces within the plasma. Because permanent magnets are used, the MPDR source requires only air cooling.

The ASTEX CECR source (Fig. 2.4) uses a geometry in which microwave energy propagates along a waveguide into the plasma discharge region. The magnetic field is provided by a water-cooled electromagnet. Due to the cylindrical symmetry, the electromagnets provide a uniform magnetic field so that the ECR condition exists throughout the entire plasma volume. Tuning is accomplished both by an external matching circuit and by adjusting the magnetic field and gas flow.

ECR sources have been applied towards the MBE growth of all of the III–V nitrides [25, 32–35]. Several problems have arisen which must be addressed in future designs. Unfortunately, commercially available compact ECR sources present a trade-off between the growth rate and ion damage which is a by-product of the source. Because of the stringent space restrictions imposed by the MBE system on plasma source, optimal plasma geometry is not always possible. Consequently, the energetic ions bombard the growing films along field lines. More specifically, to operate at low pressures in non-optimal geometry, high field conditions are introduced, leading to the creation of energetic ions which damage the bulk film. This forces ECR sources to be operated at low plasma power. Thus, the ion damage is significantly reduced. It results, however, in low growth rates, typically a few tenths of micrometer per hour. Figure 2.5 presents the growth rates obtained in an MBE ECR system as a function of Ga cell temperature [36]. Low growth rates are a serious limitation of plasma-enhanced MBE compared to MOCVD. Due to the lack of a substrate material which is thermally and lattice matched to the nitrides, thick layers are necessary to isolate the epilayer from the deleterious effects of the substrate. Subsequently, the rather long growth times required for the growth of good GaN have hindered the development of the MBE approach with respect to MOVPE-based techniques.

2.3 Molecular beam epitaxy (MBE) growth

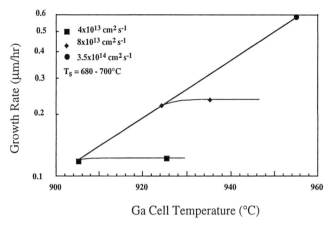

Fig. 2.5 Plot of the GaN growth rate as a function of Ga effusion cell temperature for three different nitrogen fluxes. Conditions where the growth saturates are defined as Ga-rich.

An alternative to the ECR, which has been used for MBE growth of GaN and InN, is an RF source manufactured by SVT Associates, EPI, and Oxford. The Oxford MPD 21 plasma source (Fig. 2.6). uses a 13.56 MHz RF coil to couple energy into the plasma discharge region. The plasma sheath effect confines ions and electrons within the plasma discharge regions allowing only low-energy ($<10\,eV$) neutral species to escape. RF MBE permits higher growth rates, up to a few micrometers per hour [37]. For high-quality crystals, however, the relatively low growth rate (below $0.5\,\mu m/h$) are required [38].

MBE growth employing a supersonic jet of nitrogen atoms generated by exciting a 1% nitrogen in He mixture with an RF discharge has been performed by Selidj et al. [39, 40]. GaN films have been grown on sapphire (0001) using supersonic jets of nitrogen atoms and a gallium effusion source. A growth rate of $0.65\,\mu m/h$ was obtained, independent of substrate temperature over the range 600–750 °C. The films were single crystalline wurtzitic GaN as determined

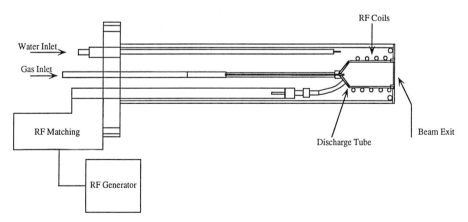

Fig. 2.6 Oxford MPD 21 plasma source.

by X-ray diffraction and *in situ* reflection high-energy electron diffraction (RHEED). The minimum peak width of the (0002) GaN reflection, as measured by high-resolution X-ray diffraction, was 38 arcmin. These results are in general agreement with other efforts exploiting seeded supersonic jet sources.

Several outstanding issues regarding all types of plasma sources should be resolved before the validity of this approach can be determined. These are exacerbated by the lack of availability of sources with larger volume to surface ratio which require among others, 6 inch ports on MBE systems. There is also the inadequate knowledge base regarding the molecular dynamics at the nitride growth surface. It it not known what type of species are most conducive to nitride growth, namely atomic nitrogen, ions, or another form of excited nitrogen. Fluences are measured by Langmuir probes, which measure only ions, making it difficult to evaluate neutral flux rates. MBE source flanges should be modified to allow the use of larger plasma sources, operating under more optimized geometry, for increased growth rates and lower ion energies. Finally, contamination must be minimized by eliminating the exposure of metal, quartz, and alumina components to the plasma discharge.

The influence of the nitrogen ion bombardment on the layer quality is controversial and not fully understood. It is widely observed that optical and electrical properties of the films grown at high power can degrade. The damage threshold of GaN was estimated to be 24 eV [41]. The ion removal magnets were used to reduce ion damage [42, 43]. By this method, high-quality epitaxial films were grown. X-ray rocking curve FWHM of 75 arcseconds of the film grown on sapphire was reported [44].

It was shown that the main role in the ECR MBE growth is played by the neutral species. The growth rate was reduced only by 8% when 90% of ions were cut off by biasing the grid positively at 20 V [45]. The removal of the energetic nitrogen ions from plasma may, however, be detrimental to the crystal

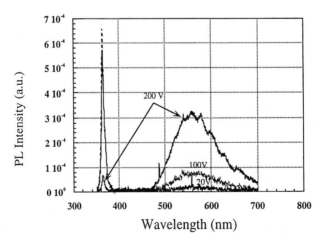

Fig. 2.7 Room temperature photoluminescence response of GaN film grown with grid potentials of 20, 100, and 200 V [45].

quality. Figure 2.7 shows the PL spectra of GaN films grown with grid potentials of 20, 100, and 200 V, which deflected ions. As the grid voltage was increased, the broad peak in the 520–280 nm region of PL spectra intensified. The PL results indicate that the energetic ions may be involved in minimizing, if not eliminating, defects that lead to the midgap states. High-energy ions were supposed to break up clusters, knock adatoms off the surface, and remove other growth-related defects. On the other hand, the deflection of ions of 100–200 V increases the growth rates, reducing the ion induced desorption of Ga atoms from the surface.

The effect of the nitrogen pressure and plasma power on the quality was studied by Maruyama *et al*. [46]. It was shown that nitrogen pressure affects the formation of dislocations, thus influencing the photoluminescence and XRD characteristics of the grown films. The plasma power, instead, influences strongly only photoluminesence, the XRD characteristics remaining unchanged. The latter suggests that the power is mainly responsible for the production of point defects.

2.3.1.2 *Reactive-ion* MBE *approach*

Several laboratories have attempted MBE growth of the III–V nitrides using non-plasma-based growth techniques. Low-energy nitrogen ions from a Kaufman ion source were adopted to grow GaN of quality purported to be comparable to that grown with ECR source [47]. Several laboratories in the past have attempted reactive molecular beam epitaxy (RMBE) growth in which N_2 or NH_3 was decomposed on the substrate surface [48–52].

Observing that ionization of source materials allows lowering of the growth temperature, Gotoh *et al*. [53] used ammonia in addition to ionized N_2. This approach reduced the density of nitrogen vacancies and other defects, and improved the electrical characteristics of GaN. Experiments performed in our laboratory unequivocally indicate that reducing/cracking ammonia before striking the substrate surface provides less reactive N for growth as well as accelerating undesirable reaction of N with the sources.

In RMBE ammonia is decomposed only on the surface of substrate by pyrolysis, which allows growth of GaN at temperatures as low as 600 °C. This growth temperature is much lower than the growth temperature in MOCVD. Kim *et al*. [54–56] showed that RMBE using ammonia as a nitrogen source can produce films of a very good quality. The study was performed on sapphire substrates, nitridated and with AlN buffer layer. In this research, the substrate temperatures varied between 610 and 820 °C. Growth rates up to 2.9 μm/h were obtained for temperatures between 725 and 820 °C. These growth rates are quite comparable to those by conventional MOCVD. The rocking curve full width at half maximum (FWHM) as a function of the substrate temperature is shown in Fig. 2.8; growth kinetic study has been carried out by employing various V/III ratio and substrate temperatures. It was found that the layer quality increases and compressive stress decreases, as the N/Ga ratio increases. Incorporation of the residual impurities as well as the creation of high density of native defects become less efficient at higher N/Ga ratio.

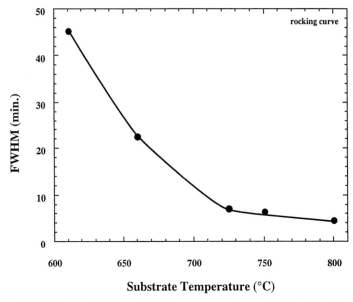

Fig. 2.8 Variation of X-ray rocking curve FWHM in GaN as a function of substrate temperature. The films were grown with reactive MBE using ammonia as the active nitrogen source [56].

The incorporation kinetics of gallium during gas source molecular beam epitaxy (GSMBE) of GaN using elemental Ga and NH_3 gas as source materials was studied by Jenny et al. [57]. Desorption mass spectrometry (DMS) was used to perform *in situ* quantitative measurements of GaN formation, Ga desorption, and Ga surface accumulation during growth. The rate of formation of GaN is reported as a function of incident Ga flux (0.1–0.75 ML/s) and incident NH_3 flux ($1-300 \times 10^{-7}$ Torr beam equivalent pressure) at a constant growth temperature of 725 °C. Three distinct growth regimes (Fig. 2.9) are observed: (a) GaN is formed where all of the incident Ga flux is consumed, (2) part of the incident Ga is consumed to form GaN while the excess is desorbed from the surface, and (3) GaN formation coexists with desorption and surface accumulation of Ga. It is generally observed that the Ga surface accumulation is inhibited by increasing the rate of incidence of NH_3 and/or by decreasing the rate of incidence of Ga at this temperature. In addition, the order of the reaction between Ga and NH_3 is determined to be unity, supporting the validity of the $Ga + NH_3 \rightarrow GaN + 3/2 H_2$ reaction.

2.3.1.3 MOMBE

There are possible extensions of the original RMBE which can reduce the required substrate temperature: hydrazine is believed to increase the reactivity of the nitrogen containing species. $Ga(C_2H_5)_3NH_3$, which already has a Ga–N bond, was also tried as a source material [58]. Bharatan et al. [59] tried to combine the advantages of MOCVD and ECR-MBE. Cubic and wurtzitic GaN

2.3 Molecular beam epitaxy (MBE) growth

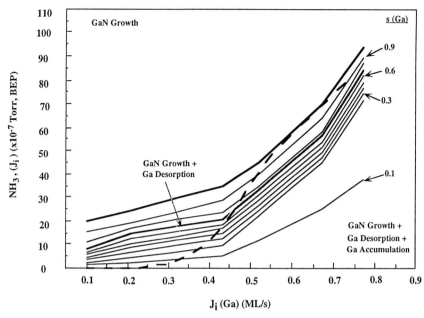

Fig. 2.9 Plot of Ga incorporation ratio, s(Ga), contours as a function of Ga and NH$_3$ fluxes. Labeled in the figure are the three primary regimes for Ga adatoms. (After Jenny et al. [57].)

was grown on GaAs, ZnO, and sapphire substrates adopting triethylgallium (TEG) and trimethylamine alane (TMAA) as sources. Further, monomethylhydrazine (MMHy) as nitrogen source along with triethylgallium (TEG) were also employed in MBE growth [60, 61]. It is difficult to assess the advantage of this source since only c-GaN was grown. And it is well known that good quality monocrystalline layers of c-GaN have not been obtained by any growth technique.

2.3.2 Modeling of the MBE-like growth

There is a little work published on modeling of III-N growth. The difficulty of the quantitative estimates for growth parameters residues in the lack of reliable data on surface migration rates, on nucleation parameters, and on sticking coefficients of III and N atoms to the different substrate surfaces, as well as other thermodynamic and kinetic parameters. Monte Carlo simulation of GaN crystal growth [62, 63] showed that the grown surfaces are likely to be Ga stabilized. The growth rate increases with the increasing of V/III ratio reaching a saturation value determined by Ga flux. At a given temperature the lower V/III ratio, the better the growth front, but the growth rates are correspondingly lower. The results indicate that there is a temperature window and an N/Ga ratio window in which a good growth front could be obtained while preserving a reasonable growth rate. Kinetic modeling of microscopic processes during electron cyclotron resonance microwave plasma-assisted molecular beam epitaxial growth of GaN/GaAs-based heterostructures was done by Bandic et al. [64]. Microscopic growth processes associated with GaN/GaAs molecular beam epitaxy (MBE) are examined through the introduction of a first-order kinetic model. The model is applied to the electron cyclotron resonance microwave plasma-assisted MBE (ECR-MBE) growth of a set of delta-GaN$_y$/As$_{1-y}$/GaAs strained-layer superlattices that consist of nitrided GaAs monolayers separated by GaAs spacers, and that exhibit a strong decrease of y with increasing T over the range 540–580 °C. This $y(T)$ dependence is quantitatively explained in terms of microscopic anion exchange, and thermally activated N surface-desorption and surface-segregation processes.

A theoretical model of MBE growth which accounts for a physisorption precursor of molecular nitrogen for the analysis of group III-nitride was proposed by Averyanova et al. [65]. The kinetics of nitrogen evaporation was found to be an essential factor influencing the MBE growth process. The high thermal stability of nitrides was explained to be related to the desorption kinetics resulting in a low value of the evaporation coefficient. The values of the evaporation coefficients as functions of temperature were extracted from the experimental Langmuir evaporation data of GaN and AlN. The process parameter-dependent growth rate and transition to extra liquid phase formation during the GaN MBE were calculated.

2.4 Substrates and buffer layers

The high melting temperatures and dissociation pressures are the main obstacles in obtaining large single crystals of III-N compounds, which can serve as substrates for the homoepitaxial growth. Therefore, single crystalline III–V films have been grown heteroepitaxially on a number of substrates which match more or less close lattice constants and thermal expansion coefficient of III-nitrides. There is no ideal substrate. For that reason, the results of III–V growth on different substrates will be discussed in this section. Table 2.4 presents the lattice constants and thermal expansion coefficient of the prospective substrates for III-N epitaxy.

The initial stage of growth is very important for the obtaining of heteroepitaxy and the quality of the resulting film [66]. The epitaxial growth can be two-dimensional (2D) layer-by-layer mode, three-dimensional (3D) island mode, or mixed (M) mode, layer-by-layer plus islands. The first mode results in the smooth surface, while the last two give a rough surface and lead to low quality of epitaxial layer. The mode of growth is determined by many parameters, such as interfacial energy of solid and vapor phase and interfacial energy of vapor phase and substrate, which in turn depend on growth temperature, the bond strength and bond lengths of the substrate and overgrowth atoms, the rate of impingement of species, surface migration rates of reactants, supersaturation of the gas phase, the size of critical nuclei, and others. Since the roughness of the nucleation layer is larger at higher growth temperatures, the epitaxial growth is

Table 2.4 Lattice parameters and thermal expansion coefficient of prospective substrates for III-N epitaxial growth. The thermal expansion coefficient values are taken at the temperature of approximately 800 °C, since this is the typical temperature for the MBE growth.

Crystal	Symmetry	Lattice constants (nm)	Thermal expansion coefficient $(a; c)$ $(\times 10^{-6}\ K^{-1})$
GaN	W	(0.1389; 0.5185)	(5.59; 3.17)
GaN	C	0.452	a
AlN	W	(0.3112; 0.4982)	(4.2, 5.3)
InN	W	(0.353, 0.569)	
Sapphire	H	(0.4758; 1.299)	(7.5; 8.5)
ZnO	W	(0.3250; 0.5213)	(8.25; 4.75)
6H SiC	W	(0.308; 1.512)	(4.2; 4.68)
3C SiC	C	0.436	
Si	C	0.54301	3.59
GaAs	C	0.56533	6
InP	C	0.5869	4.5
MgO	C	0.4216	10.5
MgAlO$_2$	C	0.8083	7.45
LiAlO$_2$	T	(0.5406, 0626)	
ScMgAlO$_4$	Th	(0.3240, 2.511)	(6.2, 12.2)

W: Wurtzitic, H: Hexagonal, C: cubic, T: tetragonal, Th: tetrahedral
[a] The linear expansion coefficient for cubic nitride is not available. However, since only the second nearest neighbor distance, and only in the c direction, differs between the wurtzitic and cubic phases, we can assume that the linear expansion coefficient for the cubic phase will be in the same bulk part as in the wurtzitic phase.

as a rule divided in two steps: smooth thin (~20 nm) buffer layer grown at low temperatures and main layer grown at higher temperatures. The current technology uses a very thin (tens of nanometers) GaN or AlN low-temperature buffer layer between the substrate and epitaxial layer in order to reduce the lattice and thermal mismatch and to promote nucleation sites which have the same orientation to the extent possible. The buffer layer decreases interfacial free energy, permitting two-dimensional growth of the III-N layer. The quality of the buffer layer is usually low.

2.4.1 Sapphire (Al$_2$O$_3$)

Sapphire is a ubiquitous substrate on which GaN was epitaxially grown for the first time. Sapphire remains the most frequently used substrate for GaN epitaxial growth so far, mostly due to its low price, the availability of large-area crystals of good quality, its transparent nature, stability at high temperatures, and a fairly mature growth technology. The orientation order of the GaN films grown on the main sapphire plans {basal, c-plane (0001), a-plane (11$\bar{2}$0), and r-plane (1$\bar{1}$02)} by ECR-MBE has been studied in great detail [67, 68].

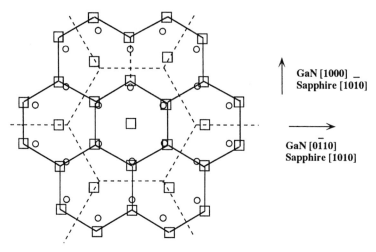

Fig. 2.10 Projection of bulk basal plane sapphire and GaN cation positions for the observed epitaxial growth orientation. The circles mark Al atom positions and the dashed lines show the sapphire basal-plan unit cells. The open squares mark gallium atom positions and solid lines show the GaN basal-plane unit cell. The Al atoms on the sapphire plan sit at positions approximately 0.5 Å above and below the plane position [67].

The calculated lattice mismatch between the basal GaN and the basal sapphire plane is larger than 30%. However, the actual mismatch is smaller ($\sim 15\%$), because the small cell of Al atoms on the basal sapphire plane is oriented 30° away from the larger sapphire unit cell, as shown in Fig. 2.10. It is on this plane that the best films were grown with relatively small in- and out-of-plane misorientation. In general, films on this plane show either none or nearly none of the cubic GaN phase.

Projection of a bulk a-plane sapphire and position of Ga atoms for the epitaxial growth are shown in Fig. 2.11. The mismatch between the substrate and film row spacing is only $\sim 0.7\%$, although many of the substrate and film cations have no good correspondence. The grain size and non-homogeneous stress of the film grown on the a-plane was the same as that grown on the c-plane. However, the film had approximately 1% of cubic phase. On the contrary, a comparative study of the properties of GaN grown by organometallic vapor phase epitaxy, using both a GaN and AlN buffer layer, as a function of sapphire orientation (c-plane vs. a-plane) gave different results. Growth on a-plane sapphire resulted in higher quality films (over a wider range of buffer thickness) than growth on c-plane sapphire [69]. Projection of a bulk r-plane sapphire and position of Ga atoms (GaN a-plane) for the epitaxial growth are shown in Fig. 2.12. The lattice mismatch is only 1.3% in the sapphire $[\bar{1}101]$ direction and $\sim 15\%$ in the $[11\bar{2}0]$ direction. The quality of the film grown on the r-plane was the lowest.

If GaN is grown directly on sapphire the quality of the crystals is poor due to large lattice and thermal expansion coefficient mismatch and non-ideal nucleation. The degradation of the quality was evidenced by a wide X-ray rocking

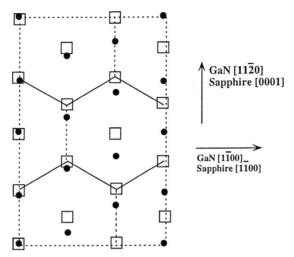

Fig. 2.11 Projection of bulk *a*-plane sapphire and basal plane GaN cation positions for the observed epitaxial growth orientation. The dots mark Al atom positions and the dashed lines show the sapphire *a*-plane unit cells. The open squares mark gallium atom positions and solid lines show the GaN basal plane unit cell [67].

curve (10 min typically), surface is not smooth (containing hillocks), high electron concentration (up to 10^{19} cm^{-3}), and considerable yellow emission in luminescence spectra. To improve the GaN layer quality, low-temperature AlN or GaN buffer layer is usually grown [70, 71]. Although the low-temperature AlN and GaN buffer layers were initially used solely for MOCVD [72], their use in the MBE process was shown to be highly beneficial also. The use of buffer layers became a standard technique for obtaining good quality GaN and AlGaN films.

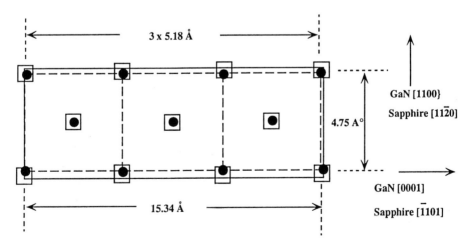

Fig. 2.12 Projection of bulk *r*-plane sapphire and *a*-plane GaN cation positions for the observed eptiaxial growth orientation. The dots mark Al atom positions and the dashed lines show the sapphire *r*-plane unit cells. The open squares mark gallium atom positions and solid lines show the GaN *a*-plane unit cell [67].

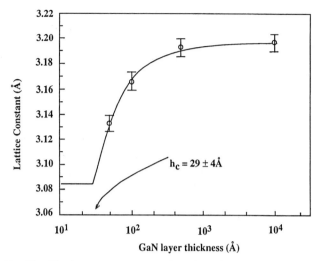

Fig. 2.13 Fit of lattice constant a as a function of GaN layer thickness [76].

A short period of nitridation of the sapphire substrate generally precedes the buffer layer growth [27]. During the nitridation process a thin $AlO_{1-x}N_x$ film grows on the substrate surface [73], transforming Al_2O_3, in a natural way, through $AlO_{1-x}N_x$ into AlN, and the epitaxial films grow on AlN, which has nearly the same lattice constant as GaN (mismatch 2% in c-plane). This is why the best films until now have been obtained on sapphire with an AlN buffer layer. It is obvious that nitridation parameters can greatly influence the properties of the epitaxial layer [74, 75].

The lattice constant a as a function of thickness of the GaN films grown on sapphire with AlN buffer layer is presented in Fig. 2.13 [76]. The values of a lie between those of the bulk GaN and the AlN buffer layer, down to the critical thickness h_c, which was estimated at 2.8 nm. Films with thickness smaller than h_c have a pseudomorphic epitaxial relationship with the AlN substrate. At a thickness larger than ~100 nm the lattice constant became equal to the bulk GaN value. A GaN defective layer grown at low temperature is also used as a buffer layer [77–81]. The properties of the GaN epitaxial layer are strongly influenced by the buffer layer thickness.

The structural film quality is commonly measured by a rocking curve full width at half maximum (FWHM) for the GaN(002) peak. The theoretical value of the FWHM for GaN in the ideal measurement conditions without thermal broadening is computed to be 33.2 arcsec [82]. The rocking curves become significantly broadened at high densities of dislocations (generally $> 10^5 \, cm^{-2}$). Effect of threading dislocation structure on the X-ray diffraction peak width in epitaxial GaN films was studied by Heying *et al.* [83]. Quantitative defect analysis showed that the threading dislocations are predominantly pure edge dislocations lying along the c-axis. The specific threading dislocations geometry will lead to distortions of only specific crystallographic planes. Edge dislocations will distort only (hkl) planes with either h or k non-zero. Rocking curves on off-axis (*hkl*) planes will be broadened, while symmetric (00*l*) rocking curves will be insensitive to pure edge threading dislocations content in the film. Screw threading dislocations with [001] directions have a pure shear strain field that distort all (*hkl*) planes with *l* non-zero. The dramatic broadening of the asymmetric (102) rocking curve evidences a high density of pure edge threading dislocations. Therefore the rocking curve widths for off-axis reflections are a more reliable indicator of the structural quality of GaN films. The structural properties of GaN film grown on sapphire differ for the in-plane and out-of-plane structural features [84]. The measured in-plane coherence lengths are smaller and the rocking curve widths are larger than in the plane-normal direction. It implies that optical and electrical properties are anisotropic along the film plane and plane normal directions. The in-plane structure measurements are more closely related to electronic mobility and optical properties than those in the plan-normal direction.

Kisielowski *et al.* [85] demonstrated the coexistence of biaxial and hydrostatic

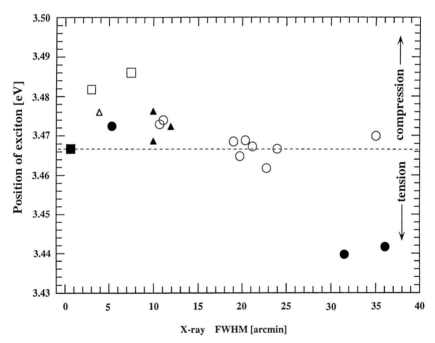

Fig. 2.14 Energetic position of the donor bound excitonic luminescence observed in films with different crystalline quality that degrades with increasing FWHM of the X-ray rocking curves. The film were grown on sapphire and typically on *c*-planes. It is seen that the position of the luminescence is determined by the choice of buffer layers, which greatly modifies the stress. Homoepitaxy provides a standard for stress free growth. A red shift is caused by tensile stress, such as on SiC, and blue shift by compressive stress, such as on sapphire substrates [85]. Symbols: ○ MBE films grown on a low-temperature GaN buffer layer; ● MBE films grown without buffer layer; △, ▲: MOCVD grown films on a low-temperature GaN buffer layer from two different sources; ■: homoepitaxially MBE grown film; □: MOVPE films grown on sapphire *a*-planes and on an AlN buffer layer.

strain present in GaN films grown on *a*- and *c*-plane sapphire and on SiC substrates. The external biaxial strain originates from growth on lattice mismatched substrates and the difference in the thermal expansion coefficient. The strain is altered by the presence of buffer layers and extended defects. Point defects introduce an internal strain that can be compressive or tensile depending on the size of the involved impurities. Figure 2.14 shows the PL data obtained from films grown on sapphire substrates with different buffer layers. The presence of the buffer layer greatly modifies the stress in the grown films. A low-temperature GaN buffer layer essentially has the effect of strain relaxation. In contrast, the films without buffer layers have large tensile stresses, while the films grown on AlN buffer layers exhibit large compressive stress.

High-quality layers grown on sapphire substrates by MBE permitted fabrication of a variety of devices, such as Schottky diodes [1] and modulation-doped FETs (MODFET) [1, 86–89], multiple InGaN/AlGaN quantum wells [90, 91], and observation of photopumped stimulated emission at 425 K [92]. Although the layer quality on sapphire is almost unmatched, sapphire (0001) substrates are not most suitable for GaN-based lasers, because the cleavage planes in sapphire substrate are not perpendicular to the surface. Thus, use of sapphire (0001) substrate requires the fabrication of laser mirror facets by reactive ion etching or other more complicated cleavage techniques [93]. The cleavage problem can be avoided by using an *a*-plane ($\bar{2}$110) sapphire substrate. However, the growth on this plane has been much less studied.

2.4.2 6H–SiC (1000)

The development of the growth technology on SiC is behind that on sapphire, because the large surface area SiC substrates became available commercially only a short time ago. As a substrate, 6H–SiC presents an advantage in that the lattice constant and thermal expansion coefficient are closer to that of GaN, compared with sapphire, and in the cubic form has the same stacking order as cubic GaN. The relatively small lattice mismatch for the basal plane is 3.5% for GaN and nearly perfect match for AlN. The problems with SiC are lack of stacking match and, more important at the moment, lack of the appropriate chemical etch for the surface. The later is making cumbersome the adequate preparation of the surface for the growth. Lin *et al.* [94] proposed a two stage cleaning of the SiC surface. In the first step, the surface is hydrogen passivated using an HF dip before introduction into vacuum. Second, the substrate is treated with a hydrogen plasma reducing the oxygen–carbon bonding to below the X-ray photoemission detection limit. Upon heating in the MBE chamber the SiC substrates were observed to have a sharp (1×1) surface reconstruction. GaN epilayers deposited on AlN buffer layers showed sharp X-ray diffraction and PL peaks.

GaN or AlN buffer layers are usually grown on SiC substrates [95]. The minimal defect density, the highest mobility and the best PL characteristics of GaN/SiC epitaxial layers were obtained on the optimized GaN buffer layers. The stress-related phenomena in GaN films grown on SiC are much smaller than those of films grown on sapphire [85]. The defect structure of GaN films grown with and without AlN buffer layer on 6H SiC by the ECR MBE was studied by Smith *et al.* [96]. Threading defects, identified as double position boundaries, originate at the substrate–film interfaces. The density of these defects was related to the smoothness of the surface.

A comparison of GaN epilayers grown on sapphire and SiC substrates with AlN buffer layers by ECR MBE revealed the superiority of SiC substrates [97]. Structural quality as measured by XRD rocking curve, photoluminescence, and mobility were superior for the films grown on SiC. The films grown on SiC had a higher mobility (580 cm^2 Vs) than on sapphire (230 cm^2/Vs), using AlN buffer layer for both cases. The electron mobility of 580 cm^2/Vs of film grown on SiC was the highest value ever reported by MBE.

2.4.3 ZnO

ZnO is considered as a promising substrate for III-N since it has a close match for GaN *c*- and *a*-planes and close stacking order [98, 99]. ZnO was used as a buffer layer for GaN growth, but the films were inferior to the quality of the films with AlN and GaN buffer layers [100, 101]. Matsuoka *et al.* have used ZnO substrates to grow lattice-matched In$_{0.22}$Ga$_{0.78}$N alloys [102]. The most comprehensive study to date of GaN growth on ZnO was accomplished by MBE [103]. The high quality of the films has been confirmed by photoluminescence and reflectance measurements. The width of excitonic transitions ($\sim$8 meV) was in the same range as the best results obtained on sapphire by MBE. The yellow luminescence was totally absent from PL spectra. X-ray measurements as well as polarized optical data confirm that the GaN epilayers are better oriented with respect to the substrate axis than those on sapphire and SiC. More work is needed to delineate intrinsic properties from extraneous issues, such as defective substrate surface morphology, which in time may be overcome.

2.4.4 Si

Si substrates are very attractive, not only because of the high quality, availability, and low price of Si wafers, but also for the possibility of integration of Si-based electronics with wide bandgap semiconductor devices. Cubic GaN is grown on Si(100) substrates [104], while both Si(100) and Si(111) are employed as substrates for wurtzitic GaN growth [105, 106]. Both phases, wurtzitic and zincblende, are usually detected, accompanied with a large number of extended defects such as dislocations, stacking faults, and twines [67].

GaN grown on Si(001) was predominantly cubic. A GaN 30 nm low-temperature buffer layer accommodated the 17% lattice mismatch between the film and substrate by a combination of misoriented domains and misfit dislocations [107]. Beyond the buffer layer, highly oriented domains separated by inversion domain boundaries were found by transmission electron microscopy (TEM). Stacking faults, microtwins, and localized regions of wurtzitic structure were major defects in the film. The film grown on Si(111) had a predominantly wurtzitic structure with a presence of twinned cubic phase.

A marked improvement of the wurtzitic GaN film quality grown by ECR MBE on Si(111) was observed when a surface preparation procedure that makes atomically flat terraces was used [108]. Wide atomically flat terraces were created by etching using solution of 7:1 $NH_4F:HF$, greatly increasing the distance between surface steps and thus decreasing their density. The latter reduced the number of stacking mismatch boundaries of the epitaxial GaN films, improving their crystalline quality, as was attested by XRD and PL measurements.

The buffer AlN layer was likewise grown on Si(111) substrate prior to the GaN epitaxy. The good quality of the epilayers was attested by presence of free exciton in low-temperature PL. The PL decay times of excitonic emission were in the range of those reported for high-quality GaN grown on sapphire [109].

2.4.5 GaAs and GaP

GaAs was extensively used as a substrate for both w-GaN and c-phases [110–117]. A comprehensive study of optical and structural properties of cubic GaN grown by MBE on (100) GaAs was performed by Strite et al. [33]. A lattice constant of 0.45 nm and bandgap of approximately 3.45 eV (low temperature, and may have been influenced by the presence of wurtzitic clusters) were determined. High-resolution transmission electron microscopy revealed a high density of planar defects propagating along {100} planes. The growth kinetic of cubic GaN on (001) GaAs was studied by Yang et al. [118]. They found that the GaN GaN surface exhibits three surface symmetries [(1×1), (2×2) and $c(2 \times 2)$] which correspond to Ga adatom coverage of 0, 0.5, and 1, respectively. An impinging Ga flux leads to reconstructing surface with $c(2 \times 2)$ symmetry, whereas with an impinging flux of active nitrogen a transision to a (1×1) reconstruction is observed. The above understanding was used to probe the surface stoichiometry during GaN growth. Smooth, highly resistive films with a strong bandgap emission in cathodoluminescence (CL) were grown. The XRD rocking curve showed 68 arcmin for (002) reflection.

GaAs(111) and GaP [119] substrates were utilized for the GaN cubic phase growth. Growth temperature plays a critical role in determining the phase of GaN deposited on (111) GaAs: cubic GaN growths below and wurtzitic GaN above 800 °C [118].

The wurtzitic phase was grown on these substrates as well, using AlN [120], ZnO, or GaN buffer layers [121]. ZnO buffer layers were deposited at 300 °C by RF magnetron sputtering with a sintered ZnO as a target and Ar as a sputtering gas. GaN buffer layer was deposited using MBE method. GaN thin films with thickness about 240 nm were grown on those buffer layers at 500–650 °C in MBE system. An interesting result was obtained on the layers grown on ZnO/GaAs(111). The samples grown at 500 and 600 °C had a very high charge carrier concentration, of order 10^{20} cm^{-3}. Hall effect and Zeebeck effect showed p-type conduction with a very high mobility for such concentrations, about 10 cm^2/Vs. Auger measurements indicated that p-type conductivity was not due to Zn thermal diffusion from the buffer layer. At 650 °C growth temperature the conductivity became of n-type. So the origin of p-type conductivity was not understood.

In view of the structural similarities of the zincblende and wurtzitic polytypes, a mixture of the two can occasionally be grown [122] by molecular beam epitaxy method. Cheng *et al.* noted that, when GaN is grown on (001) GaP or (001) GaAs substrate, an interesting growth kinetic mechanism plays a significant role: at a moderate substrate temperature of 620 °C and in the absence of an arsenic overpressure, GaN can be grown preferably in the stable wurtzite phase. On the other hand, in presence of arsenic overpressure ($\sim 2.4 \times 10^{-5}$ Torr) during the growth, GaN is forced to adopt the crystal structure of the underlying substrate which is the zincblende structure. This is accomplished by an interchange between the mobile active nitrogen species and the arsenic atoms occupying the Group V lattice of the zincblende structure.

Both GaAs and GaP present the problem of substrate decomposition at the temperatures of GaN growth.

2.4.6 Spinel ($MgAl_2O_4$)

Cubic $MgAl_2O_4$ has a spinel type structure (Fd3m) with the oxygen atoms forming a face-centered cubic sublattice and Mg and Al atoms occupying the tetrahedral and octahedral sites, respectively. Lattice mismatch is $\Delta d/d = 10\%$. The crystals are stable at the GaN growth temperature. Spinel substrate has an advantage over the sapphire substrate of obtaining mirror laser facets by cleaving [123, 124]. To our knowledge there are no reports of growth of III-N on spinel substrates by MBE. For completeness we are giving here the results obtained by MOCVD.

GaN crystals were grown on (100) and (111)-oriented $MgAl_2O_4$ substrates by low-pressure MOCVD [125, 126]. After a GaN buffer layer was deposited at 550 °C, films a few microns thick were grown at about 1000 °C. GaN films grown on (111) substrate were wurtzitic single crystals. The crystallinity of the film was comparable to that of films grown on sapphire substrates, while Hall mobility and PL were superior, a c-GaN growth was expected on (100) substrates due to fourfold substrate symmetry. A far more complicated structure was observed, instead, at the heterointerface. Besides c-GaN, several variants of wurtzitic structure were detected. With increasing film thickness the wurtzitic structure became dominant.

The crystalline characteristics of the best GaN layers grown on $MgAl_2O_4$(111) approached that of layers grown on the sapphire substrate. The rocking curve FWHM of 310 arcsec was reported. An optically pumped InGaN–GaN laser with cleaved cavity was realized on (111) oriented spinel substrate with a 50 nm GaN buffer layer [127]. The films were n-type with typical concentration of 10^{18} cm^{-3}.

2.4.7 MgO

MgO has NaCl structure and presents a lattice constant mismatch with c-GaN of 7.5%. Cubic GaN was grown on MgO (001) plane by reactive ion molecular beam epitaxy [128]. The XRD intensities of the wurtzitic (0002) peak was always less than 10^{-3} of the c-GaN (002) peak. The rocking curve FWHM was 28 min. The sub-band refractive index has been determined on the c-GaN samples grown on MgO over the energy range 0.8 to 3.1 eV [129]. The refractive index varied from 2.25 to 2.50, 3–4% smaller than previously reported values for w-GaN [130].

2.4.8 GaN

In the absence of suitable industrial applications for GaN substrates, basic research on homoepitaxial growth has been done by MOCVD [131, 132] and MBE [133] on GaN platelets grown from the liquid phase under high pressure. Such studies serve to establish benchmark values for the optoelectronic properties of thin GaN films.

Homoepitaxial films grown by MBE showed a superior quality of PL and structural characteristics with respect to heteroepitaxial layers grown on sapphire. The dislocation density was estimated to be 10^7–10^8 cm^{-2} for homoepitaxial film as compared to 10^9–10^{11} cm^{-2} for the films grown on sapphire. We should, however, mention that threading defects originating at the GaN substrate interface can in general propagate to the surface. When epitaxial GaN is attempted on this surface, these defects would propagate again through the layer [134, 135].

There are, in general, two possible approaches for suitable substrates for homoepitaxial growth. The first one is to grow a large bulk crystal, cut it, and polish slices from it, which is desirable. This method is widely used for conventional semiconductors. It cannot be applied easily to GaN bulk growth, since only small pieces of GaN can be grown only at high temperatures and very high pressures (tens of kbar), taking into account that GaN starts to decompose at 800 °C. The second possibility would be the following one. On the foreign substrate (sapphire, SiC, or Si) grow a thick (tens of microns) GaN film. It is well known that the quality of the heteroepitaxial film becomes better at larger thickness. The very popular methods, MOCVD and MBE, used now for epitaxial growth have very slow growth rates, a few microns per hour at best, and cannot be used for the growth of thick films. The inorganic CVD methods, however, have high growth rates (up to 100 μm/h). Thus the best way to grow homoepitaxial film is to use two-stage growth, a thick GaN substrate can be grown by inorganic CVD in 1–2 hours, and then the device can be grown on GaN substrate by MOCVD or MBE. A similar approach was employed by Johnson et al. [136]. MBE homoepitaxial growth of GaN and AlGaN films and GaN/AlGaN multiple quantum wells was performed on high-quality 3 μm thick GaN films grown by MOVPE on SiC. The film grown by MBE had a FWHM of the X-ray rocking curve equal to 156 arcsec. The room temperature PL spectrum was dominated by band-edge emission at 3.409 eV, with a FWHM of 29.7 meV.

2.4.9 Var der Waals substrates

The use of a new growth method called van der Waals epitaxy was proposed in order to solve the stress problem [137]. In this approach the substrate and the epitaxial film are separated by an intermediate epitaxial two-dimensional buffer material such as MoS_2, WS_2, or other materials, like III–V compounds (GaSe, InSe, etc.) having weak van der Waals bonding to the substrate and the film. Strain from lattice mismatch between the epitaxial film and substrate is completely relieved between layer and buffer region.

2.4.10 β-SiC coated Si

Pankove [138] suggested for the first time that realization of zincblende GaN would be possible if it is grown on a suitable cubic symmetric substrate, such as β-SiC. That the suggestion was worthwhile was proved by Liu *et al.* [139], who were successful in obtaining zincblende GaN on β-SiC coated (001) Si substrate by employing RF plasma discharge nitrogen free-radical source.

2.4.11 SiC-on-silicon-on-insulator compliant substrate

As indicated earlier, SiC can be an ideal substrate for minimizing the structural and thermal mismatch of the epitaxial GaN layers. Recently Yang *et al.* [140] showed that, if SiC can be overgrown on silicon-on-insulator (SOI) substrates, a high-quality GaN, SiC, and Si layer together will result in a new generation of heterojunction devices. For this, the SiC can be grown compliantly on SOI wafers up to 125 mm in diameter by MBE. Before the beginning of the carbonization process, SiO_2, Si, and C need to be deposited successively on Si substrate. By exposing the substrate to a flux of acetylene or carbon beam at 900 °C a thin layer (less than 50 nm) of Si (on SiO_2) will be partially or completely converted into SiC. GaN will then be grown on this SiC.

2.5 AlN

The most frequently method used for AlN growth in earlier research was inorganic [141–143] and MOCVD [144]. Morita *et al.* [145] grew for the first time epitaxial single crystalline AlN films on the basal plane of sapphire substrates employing MOCVD. This early work was reviewed by Duffy [146]. The more successful techniques used recently are MOCVD and MBE. These techniques have made strong progress in AlN and AlGaN epitaxial growth.

As sources, evaporated Al and NH_3 are frequently employed in MBE for AlN growth [147]. The substrate temperature is in the range 1000–1200 °C. Monocrystalline AlN (0001) films with few defects were grown by ECR MBE on (6H) SiC (0001), while films grown on Si in nearly the same conditions had a highly oriented polycrystalline structure [148].

The growth rate was of order of 0.24 μm/h. Sitar *et al.* [149] studied deposition of AlN/GaN superlattices by ECR plasma for N_2 activation. Polycrystalline and single crystalline AlN films have been grown on sapphire [150–153] and SiC [154]. The best quality AlN with rocking curve FWHM of 100 arcsec was realized on the sapphire [82].

2.6 Challenges in InN growth

To realize the full potential of the III–V nitride system, the growth of single crystalline InN, which is an important member of the nitride family, should be achieved. Because of the low dissociation temperature (550 °C), growth of high-quality InN films has proved to be difficult. The nitrogen equilibrium vapor pressure over InN is orders of magnitude greater than over AlN or GaN. To circumvent this difficulty a number of options, including low-temperature (less than 600 °C) deposition and various growth techniques [155–160], such as reactive evaporation, ion plating, reactive radiofrequency sputtering, reactive magnetron sputtering, vapor phase epitaxy, microwave-excited MOCVD, laser-assisted CVD, halogen transport, and MBE have been tried to obtain InN.

Most of the InN growths have been performed at about 500 °C. The best films obtained from these growths are those with high indium and nitrogen fluxes. Short thermal anneals (30 min, 450–500 °C) have been found to be useful for improving the crystallinity of the as-deposited InN films. InN films are generally polycrystalline with agglomerates of small columnar grains having different degrees of texture and epitaxy. The crystalline improvement is attributed to rearrangement of these crystallites in the films. The electrical properties of the as-deposited InN films are dominated by connectivity between the grains [161]. This connectivity is different for InN grown on different substrates. Buffer layers used for the growth also play a role. For example, use of an AlN buffer layer on sapphire substrate significantly alters the granular polycrystalline growth mode of InN, resulting in substantial improvements in growth morphology and electrical properties. However, the electrical transport properties are still dominated by intergranular interactions [162]. A zincblende InN polytype with an InAs interlayer (about 80 nm) was grown on GaAs (100) substrates by plasma-enhanced MBE. A TEM microstructural study revealed a high density of stacking fault defects from which wurtzitic domains of InN were nucleated [163].

2.7 Growth of ternary and quaternary alloys

By alloying InN together with GaN and AlN, the bandgap of the resulting alloy(s) can be increased from 2 eV to a value of 6 eV, which is critical for making high-efficiency visible light sources and detectors. The compositional dependence of the lattice constant, the direct energy gap, electrical, and CL properties of the AlGaN alloys were measured by Yoshida et al. [164]. The ternary compound semiconductor InGaN, which can have a direct energy bandgap ranging between 1.95 eV and 3.40 eV, and the quaternary InGaAlN semiconductor alloy, which can have a direct energy bandgap ranging between 2.0 and 6.2 eV, are promising materials for the active layer of double heterostructure light-emitting diodes (LEDs) and laser diodes. Unfortunately, incorporation of indium in these alloys is not easy. InGaN with the largest indium mole fraction has about 0.25 indium. To prevent InN dissociation, InGaN crystal was originally grown at low temperatures (about 500 °C) [165]. However, the use of a high nitrogen flux rate allowed the high temperature (800 °C) growth of high-quality InGaN and InGaAlN films on (0001) sapphire substrates. It was noted that the incorporation of indium into the growth of InGaN film is strongly dependent on the flow rate, N/III ratio, and growth temperature. The incorporation efficiency of indium was found to decrease with increasing growth temperatures to a value above 500 °C.

Crystalline quality of InGaN was observed to be superior when grown on the well-matched ZnO substrate than when grown on bare (0001) sapphire substrate [166]. Use of a buffer layer caused further improvement of the crystallinity quality of InGaN. For example, InGaN films grown on sapphire substrate using GaN as buffer layers exhibit optical properties much better than InGaN films grown directly on a sapphire substrate [167].

2.8 Defects

2.8.1 Improvement of epitaxial crystal structure

For device applications the structural quality of the epitaxial films is very important, since dislocations, stacking faults and other extended defects may significantly reduce carrier lifetimes and consequently device efficiencies.

As was discussed earlier, there exists no ideal substrate for epitaxial growth of the III-nitrides. In order to succeed in improving the quality of layers grown on sapphire, SiC, GaAs, Si, etc. substrates, much attention should, therefore, be devoted towards understanding the epitaxy of the nitrides grown on them. The epitaxial relationships between the various substrate orientations and the nitride films grown on them are now well established. It was verified by Madar et al. [168], who grew GaN on a hemispherical sapphire substrate. They observed that the epitaxial relationships with GaN quality hold in numerous orientations. The best epitaxy was obtained with the (0001) surface. Reasonable quality and better surface morphology was observed on the $(11\bar{2}0)$ planes tilted 10–20° towards the $[\bar{1}100]$ axis. The highest growth rate and reasonable crystal quality was observed on the $(01\bar{1}4)$ planes. This was confirmed by Sasaki and Zembutsu [169], who found improved surface morphology of layers grown on the $(11\bar{2}0)$ sapphire surface. It was also observed that background carrier concentrations and midgap emission are lower in layers with $(11\bar{2}0)$ sapphire surface orientation than in layers with (0001) sapphire surface orientation. However, the GaN grown on (0001) sapphire had superior crystallinity and higher incorporation of Zn dopant than GaN grown on $(11\bar{2}0)$ sapphire. To investigate the effect of misorientation on crystal morphology, Hiramatsu et al. [170] grew GaN on (0001) sapphire with a number of different misorientations towards both the [10$\bar{1}$0] and [1$\bar{2}$10] directions. It was found that the film quality was best when grown on nominally (0001) substrates, and the film morphology degraded rapidly with increasing misorientation. Madar et al. [171] and Manasevit et al. [172] detailed each of the epitaxial relationships observed between the various sapphire orientations and epitaxial GaN. They noted though that there are cases in which more than one relationship is possible. Similar characteristics were observed [173] when GaN was grown on $MgAl_2O_4$ or on (0001) sapphire [174]. It was apparent that the epitaxial GaN is (111)-oriented. This is consistent with the epitaxial relationship generally observed when the wurzitic phase is grown on (0001) sapphire, with the stacking order changed as described above.

TEM analysis of wurtzitic GaN grown on sapphire substrates was carried out by Humphreys et al. [174], and by Sitar et al. [175]. This analysis indicates that the major defects in the GaN were double positioning boundaries, inversion domain boundaries and dislocations. GaN grown on 6H SiC and ZnO suffered from similar types of defects, although the overall densities in these cases were substantially lower.

Sasaki and Matsuoka [176] investigated the epitaxial relationship between GaN and SiC by growing on both the (0001)Si and (0001)C faces. GaN grown on (0001)Si was found to have better crystallinity and a smoother surface morphology than GaN grown on (0001)C which exhibited hexagonally shaped islands. Photoelectron spectroscopy analysis was used to determine the polarity of the epitaxial GaN. This analysis showed that the polarity of the epitaxial GaN changes with the polarity of the SiC substrate face. Layers grown on the (0001)Si face were nitrogen-terminated, whereas layers grown on the (0001)C face were gallium-terminated and oxidized more heavily. Paisley et al. [25] studied the (001) zincblende GaN/β-SiC interface by transmission electron microscopy and noted an epitaxial relationship. The GaN films had many defects, mainly microtwins and stacking faults, whose density decreased away from the interface. The mismatch between the two crystals was mainly relieved by thin amorphous regions at the interface.

Epitaxial growth of zincblende (cubic) GaN has been performed on a number of substrates, such as (001) GaAs [177, 178], β-SiC on Si [179], and (0001) sapphire. It was observed that, if epitaxy is not achieved, the polycrystalline GaN film has the equilibrium wurtzitic structure [180]. For example, the crystal structure of epitaxial GaN grown on (001) GaAs by CVD was found to be critically dependent on the pretreatment of the GaAs surface before growth. In order to obtain the zincblende phase, the GaAs surface needed to be exposed to hydrazine for several minutes, which allowed a zincblende GaN template to form. If growth was begun after only 5 s of hydrazine exposure, wurtzitic GaN was obtained. This observation is consistent with those of Troost et al. [181, 182], who reported the formation of a similar 1.3 nm thick GaN layer resulting from the exposure of a cleaved (110) GaAs surface to atomic nitrogen.

Strite et al. [183] employed transmission electron microscopy (Fig. 2.15) and

Fig. 2.15 Electron micrograph of zincblende GaN on (001) GaAs interface. Disordered regions are evident at the interface and the majority of defects are stacking faults and microtwins propagating along the {111} directions. Some defects annihilate at their intersections, leading to improved quality in thicker films [33].

high-energy electron diffraction to study the zincblende (001) GaN/GaAs interface. Interestingly, the interface was found to be similar to the zincblende GaN/β-SiC and GaN/MgO interfaces observed by Paisley et al. [25] and Powell et al. [111] respectively. A large number of planar defects and disordered regions were located at the interface. However, the density of these defects decreases as the bulk was approached, away from the interface.

Lei et al. [26] found that a low-temperature initial growth step greatly improves the quality of GaN grown on (001)Si substrates. The low-temperature growth allowed full coverage of the substrate to be quickly achieved, and the remainder of the layer was then grown at the optimal homoepitaxial conditions. The growth kinetics was probably quite similar to that of GaAs on Si, in which the low-temperature step is necessary to reduce the surface mobility of the Ga adatoms, promoting two-dimensional growth in favor of island growth.

2.8.2 Effect of strain and lattice mismatch on crystal structure

Because of the large thermal mismatch between the nitrides and their substrate materials, most epitaxial nitride films are subject to some degree of strain, which is introduced during the post-growth cooling. A number of workers have observed residual effects from the thermal mismatch between epitaxial GaN and various substrate materials. GaN is a hard material, and a sufficiently thick film can actually crack the substrate material. That this is true at least for a number of substrates, as is evident from an observation of Chu [184] on SiC and of Grimeiss and Monemar [185] on sapphire. Transmission electron microscopy has been used to observe directly defects in GaAs substrates caused by thermal strain from a comparitively thin GaN film [186].

Amano et al. [187] studied the effect of the thermal mismatch between GaN and sapphire substrates. The GaN was strained under biaxial compression by an amount which depended on the substrate orientation. From the shift in the lattice constant and the photoluminescent peak energy, the deformation potential for GaN was found to be 12 eV. Naniwae et al. [188] studied the strain as a function of epitaxial layer thickness for GaN grown on sapphire. The perpendicular lattice constant was observed to be $c = 5.187$ Å up to GaN thicknesses of 50 μm. Thicker films showed a gradual relaxation to $c = 5.185$ Å as the thickness increased to 300 μm. The photoluminescence peak energy was observed to have a similar downward trend in energy as a function of thickness. The authors demonstrated reduced residual strain in a GaN film which was grown on a thick GaN layer previously grown on a sapphire substrate. Strite et al. [186] observed a 1% distortion of the lattice constant along the (001) growth axis in very thin zincblende GaN layers sandwiched between GaAs, which was interpreted as due to thermally induced compressive in-plane strain. Because of the poor thermal match between the nitrides and their substrate materials, and the rather high growth temperature used in CVD growth processes, nearly all of the nitrides which have been reported were probably under varying amounts of strain. This fact has been largely ignored in the literature.

Attention has also been paid to the optimization of the initial nitride overgrowth in order to minimize the defect density resulting from the substrate lattice mismatch. Yoshida et al. [189, 190] were the first to observe an improvement in GaN grown on sapphire when an AlN buffer layer was used. Akasaki et al. [191, 192] extensively studied the effect of an AlN buffer layer. The electron concentration in the GaN decreased by two orders of magnitude while the mobility increased by a factor of ten. The near bandgap photoluminescence was two orders of magnitude stronger and the X-ray diffraction peak width was four times smaller in layers having an AlN buffer layer. GaN grown directly on sapphire nucleated in tiny microcrystallites which led to hexagonal islands on the surface of the GaN films. When a buffer layer was used, the AlN was more highly oriented and the growth became two-dimensional more quickly, allowing improved GaN morphology to be obtained. An optimum AlN thickness of 500 Å was determined with the GaN tending to become polycrystalline when grown on thicker AlN layers. Kistenmacher and Bryden [193] have noted a similar improvement in the morphology and electrical characteristics of InN grown on sapphire when an AlN buffer layer is used.

2.8.3 Role of the stacking fault defect

Stacking faults are a common form of strain relief in face-centered cubic crystal structures since their formation energy is fairly low. Davis *et al.* [194] grew zincblende GaN on 3C–SiC. Employing TEM analysis they observed that many defects, mainly microtwins and stacking faults, propagate into the bulk of the crystal. Films grown on other substrates are not free from these defects either. Large quantities of stacking faults have been observed [195] in GaN films grown on sapphire, GaAs, 6H–SiC and ZnO, although the overall density of these stacking faults vary from substrate to substrate. For example, presumably due to the better lattice match the densities of stacking faults were significantly lower in layers grown on 6H–SiC and ZnO.

Recent works have attempted to elucidate the role of stacking faults in III–V nitrides. Based on observation of a small zincblende component in a bulk wurtzitic film, Seifert and Tempel [196] suggested the existence of these faults in GaN. Powell [197] has since reported the existence of small regions of wurtzitic GaN in bulk zincblende GaN crystals. Lei *et al.* [198] used X-ray diffraction to determine that zincblende domains exist in bulk wirtzitic GaN. They concluded that these domains have their (111) axis parallel to the (0001) wurtzitic axis, and that the zincblende domains were nucleated at a stacking fault, which allowed the wurtzitic stacking to shift to the zincblende stacking. This view was verified by TEM images showing the nucleation of a wurtzitic InN domain from a stacking fault in zincblende InN (Fig. 2.16(a)). The microstructure of a typical polytype domain boundary is shown in Fig. 2.16(b). From this figure it is apparent that, if the polytype domain boundary runs between the (111) zinzblende and (0001) wurtzitic crystal faces, a coherent interface is possible. This strongly suggests the idea of GaN polytype heterojunctions.

Both zincblende InN [199] and AlN [200] have been observed to suffer more from stacking faults than GaN. This indicates that wurzitic InN and AlN have probably a much lower enthalpy than their zincblende polytypes. Thus it becomes energetically possible to introduce a stacking fault defect in exchange for the lower bulk energy of the resulting wurtzitic domain. At present, there is conflicting evidence regarding the preference of GaN between the wurtzitic and zincblende structures. On the one hand, monocrystalline GaN grown on zincblende (111) substrates which provide no template for a preferential stacking sequence, as well as all polycrystalline GaN, has been observed to be wurtzitic. However, when zincblende GaN is grown on basal plane sapphire, which is a hexagonal substrate the nucleation of bulk zincblende GaN in wurtzitic films is observed. The tendency of the III–V nitrides to suffer from high densities of stacking faults is a major hindrance to the growth of high-quality material of either polytype. An in-depth understanding of the nature of this phenomenon is highly important.

2.8.4 Presence of dislocations

The use of non-ideal substrates for the growth of thin film nitrides accompanies

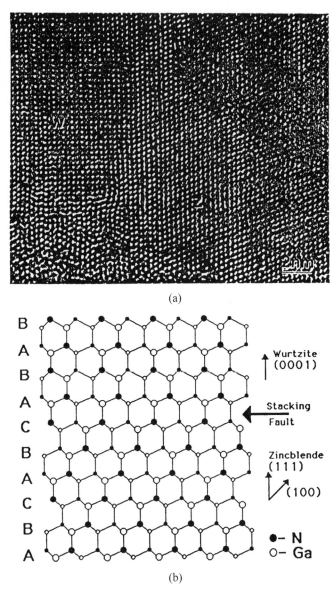

Fig. 2.16 (a) High-resolution TEM micrograph of the InN film showing the (111) stacking fault dividing the wurtzitic (upper left) and zincblende (lower right) domains [183]. (b) Coherent GaN polytype heterojunction formed by the presence of a stacking fault at the interface of (111) 3C and (0001) 2H GaN.

a large number of dislocations, which are formed in the epitaxial layer to alleviate the lattice mismatch and the strain of postgrowth cooling. These dislocations result also from thermal expansion mismatches propagated through all layers. Powell *et al.* [128] grew wurtzitic and zincblende GaN employing reactive-ion MBE method. The growth conditions involved the ratio of the N_2^+ incident flux J_N and the Ga incident flux J_G, and the N_2^+ kinetic energy E_N. It was noted that the dislocation image contrast was greatly reduced if the N_2^+ kinetic energy E_N exceeds a certain threshold value of about 60 eV. XTEM analysis showed that b-GaN contains high densities of (111) microtwins, and relatively few threading dislocations. On the other hand, both XTEM and XRD indicate that a-GaN (0001) films always have fewer dislocation defects than a-GaN ($2\bar{1}\bar{1}0$) layers, both grown on sapphire substrate. This may partially reflect the natural tendency for c-axis growth on low-surface-energy basal planes. The number density of threading dislocations with Burgers vector = $(a_0/3)\langle 11\bar{2}\rangle$ was about 2×10^{10} cm^{-2}.

2.9 Doping during growth

2.9.1 Native defects in the nitrides

The native defects are the structural defects such as cation and anion vacancies, antisites, and interstitials. Nitrogen vacancies were calculated to be shallow donors in InN and GaN [201, 202]. On the nitrogen vacancy in AlN different calculations yielded different results: it was predicted to be a relatively shallow donor, and also a deep donor for $x > 0.5$ in $Al_xGa_{1-x}N$ alloys [203]. Nitrogen vacancies are probably responsible for unintentional n-type doping of GaN and InN. The activation energy of the unintentionally GaN-doped samples was observed experimentally in the range 30–40 meV [204–206] Ga vacancy was predicted to be a shallow acceptor [198], while Neugebauer and Van de Walle [207, 208] calculated that a Ga vacancy can form a deep acceptor level, and was proposed to be responsible for the yellow luminescence in GaN. All other native defects are predicted to form deep levels.

Boguslawski et al. [202] demonstrated that the stoichiometry and the doping efficiency are strongly influenced by the large value of bandgap, since doping of wide bandgap semiconductors lead to a strong compensation of defects under conditions of thermal equilibrium. Thus, defect formation is strongly dependent on the growth conditions, temperature, Ga- or N-rich environment, and n- or p-type doping. As the growth conditions and growth temperature used for MOCVD and MBE are quite different, we can expect different results on doping efficiencies obtained by above growth methods.

2.9.2 Donor and acceptor binding energy in nitrides

The donor (E_d) and acceptor (E_a) binding energy calculated by hydrogen model is given in Table 2.5. The formula $E = 13.6 \, (m^*/m_0)\varepsilon^2$ (eV) was used, where ε is electric permitivity, m^* is electron (m_e) or hole (m_h) effective mass, and E is the donor or acceptor binding energy, respectively. In this formula, the static permitivity ε_0 is commonly accepted for covalent semiconductors. For polar semiconductors, choice of the permitivity ε, is determined by the relationship

Table 2.5 Donor and acceptor energies calculated by a hydrogen-like model.

	InN	GaN	AlN
ε_∞	8.4	5.5	4.7
ε_0	15.3	10	8.5
m_e^*	0.11	0.23	0.33
m_h^*	0.50	0.8	
E_d (mV) (ε_∞)	23	90	200
E_d (mV) (ε_s)	6.3	31	62
E_a (mV) (ε_∞)	96	363	
E_a (mV) (ε_s)	29	108	

Values of dielectric constant and effective mass of electron and holes for InN, AlN and GaN were taken from *Properties of Group III Nitrides*, ed. J. H. Edgar, INSPEC, IEE, London, UK (1994).

between E and transversal optical phonon energy $h\omega_{TO}$. If $E_a \ll h\omega_{TO}$, then static permitivity ε_0 is used. Alternatively, if $E_a \gg h\omega_{TO}$, then the high-frequency permitivity ε_∞ should be used. Since the transversal optical phonon energy $h\omega_{TO} = 69$ meV for GaN is near the ionization energy of donors and is smaller than ionization energy of acceptors, one may have to use ε_∞ [209]. However, using ε_∞ for InN and GaN gives too large ionization energies, larger than those measured. That should be reasonable, because III-N semiconductors have not a polar but a mixed type of covalent–polar bonding. Donor ionization energy calculated for AlN by hydrogen formula looks too small for both cases.

AlN is always an insulator and it is not clear at this time if it can be doped. Two causes may be responsible for the failure of AlN doping. The first is that shallow donor and acceptor levels are not formed in AlN: this might not be true, because the AlN neighbors, diamond and c-BN, can be doped by relatively shallow impurities. The second, more probably, cause is that the strong compensation due to a wide bandgap hinders doping, the lattice providing a compensating defect to any impurity added during growth. If the second is true, AlN doping will be achieved following the understanding of compensation, which is the main doping problem of all wide bandgap semiconductors. GaN and its dilute alloys with In and Al can be doped n- and p-type. As-grown InN, on the other hand, is highly n-type, possibly due to nitrogen vacancies.

2.9.3 Autodoping

Maruska and Tietjen [210] argued that autodoping is due to native defects, probably N vacancies, because impurity concentration was at least two orders of magnitude lower than electron concentration in their samples. Since then the attribution of the 'unintentional n-type doping in GaN to N vacancies' became a crutch statement, because there has not been an unequivocal experiment to support it or discount it. Neugebauer and Van de Walle [207] suggested that the formation of N vacancies in n-type material is highly improbable based on their first-principles calculations: too energy costly. Consequently, the N vacancy argument was found the less likely cause for the unintentional n-type doping. Instead, contaminants such as silicon or oxygen were suggested as possible sources for autodoping. On the other hand, the first-principles calculations of Perlin *et al.* [211] suggest that the dominant donors in GaN are N vacancies. Along the same lines, Jenkins *et al.* [201] predicted the N vacancy-related state which is shallow in GaN to get deeper in $Al_xGa_{1-x}N$ with increasing AlN mole fraction. They assigned the dramatic change from naturally n-type to semi-insulating behavior for $x = x_c \sim 0.5$. Another, yet extensive, theoretical study of native defects in hexagonal GaN carried out by Boguslawski *et al.* [212] suggests that the residual donor responsible for the n-type character of as-grown GaN is the nitrogen vacancy. However, the concentration of Ga interstitials under the equilibrium condition in Ga-rich material could become comparable to that of the N vacancies. Both n- and p-type doping efficiencies are substantially reduced by the formation of gallium vacancies, nitrogen vacancies, and gallium interstitials. In a recent study, Wetzel *et al.* [213] attempted to determine the position of the localized states in GaN with respect to the band edge employing infrared reflection and Raman spectroscopy analyses under large hydrostatic pressure. The observed reduction of the free carrier concentration to 3% at 27 GPa from its concentration at ambient pressure suggests that the defect concentration is as high as 10^{19} cm^{-3}. Further, this defect is strongly localized and has a state 126 ± 20 meV below the conduction band at 27 GPa. A resonant level of the neutral localized level was predicted to be 0.40 ± 0.10 eV above the conduction band edge at ambient pressure. Again, the N vacancy was suggested as a candidate for this defect.

In order to ascertain the likely cause of autodoping in GaN, undoped films with different ammonia flow rates were grown while keeping the other experimental conditions constant in an RMBE system [214]. The two causes of autodoping discussed above have been taken in consideration. On the one hand, the N-vacancy [210–213] is responsible for autodoping (referred to hereafter as the *N-vacancy argument*). On the other hand stands the impurity argument (Neugebauer and Van de Walle [207, 208]), such as Si and O—referred to hereafter as the *impurity argument*. The Hall effect and SIMS measurements were carried out to obtain the background doping levels and impurity levels, respectively. Figure 2.17 shows the change in the background doping level and the Hall mobility for the samples grown with different ammonia flow rates. In general, the background electron concentration decreases as the ammonia flow increases except for the lowest ammonia point. It is expected that the concentration of N-vacancies in the film should decrease as the ammonia flow rate increases. Thus the decreasing trend in the electron concentration of Fig. 2.17 supports the *N-vacancy argument*. The same trend can also be accounted for by the inclusion of compensating centers caused by the high ammonia flux during growth. Although declining mobility with increasing ammonia flow appears to support the latter argument, we must point out that the conduction mechanism in samples grown with high ammonia flow rates is through the hopping conduction with naturally lower electron mobility [56].

To verify the relationship between the autodoping level and the impurity level

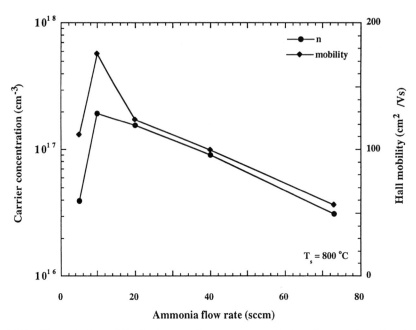

Fig. 2.17 The variation of the background carrier concentration for a series of undoped GaN films grown with different ammonia flow rates. The growth rate and film thickness were kept at 1.2 μm/h and 2.4 μm, respectively [214].

of Si and O, SIMS measurements were carried out for two samples grown with different ammonia flow rates. It was shown that Si impurity is not the dominant cause of autodoping in these GaN samples since Si concentration was lower than the electron concentration of the sample. Oxygen was also ruled out as a cause of autodoping because the sample having larger oxygen concentration by nearly one order of magnitude had a lower electron concentration, also by about one order of magnitude. Consequently, the *N-vacancy argument* is supported as the more plausible source of autodoping in GaN unless there are some yet to be determined compensation processes in play. In conclusion, the origin of the autodoping for undoped GaN films grown by RMBE technique proved to be neither Si nor O impurities by SIMS measurement. The SIMS result supports that the N-vacancy is responsible for autodoping of the undoped GaN films.

2.9.4 Silicon

Controlled n-type conductivity in InGaAlN alloys is generally achieved by Si doping [215]. Si substitutes Ga atom in the lattice, providing a loosely bounded electron. The ionization energy of the Si level in GaN was reported ∼27 meV [216, 217] and 12–17 meV [218] as measured by Hall effect temperature dependence. PL measurements yielded a value of 22 meV [218]. The solubility of Si in GaN is high, of order 10^{20} cm^{-3}. Therefore, Si is suitable for III-N doping and most frequently used.

2.9.5 Oxygen

As it was discussed above, oxygen is suspected by many authors to give n-type conductivity in unintentionally doped samples due to contamination during growth. Oxygen was found to form an energy level of 78 meV below the conduction band edge [219]. At high concentrations, it forms an impurity band which can merge with the conduction band. Therefore the oxygen unintentional doping can become important at high oxygen concentrations. At low donor concentrations the nitrogen vacancies are probably accountable for unintentional doping, as their level is much shallower (20–30 meV).

2.9.6 Hydrogen

It is predicted from the first principles calculation that H incorporation in n-type GaN is much lower than that in p-type films [220] due to the higher formation energy, while there are two contradicting experimental reports about deuterium incorporation after hydrogenation for both n- and p-GaN [221]. The H concentrations in MBE-grown n- and p-type GaN films were found to be almost the same [214]. Although a detailed study concerning the incorporation behavior of H into n- and p-type GaN films is necessary, the suggestion was made that the H concentration in GaN films grown by RMBE [222] is determined by the amount of ammonia during growth, regardless of the carrier type. This experimental result may not contradict the theoretical predictions if the additional impurity incorporation pathways, such as high density of defects, dominate the hydrogen incorporation.

2.9.7 p-type doping by Mg

Though a clear understanding is lacking, successful p-type doping in GaN, InGaN, and AlGaN has been achieved using Mg via MOCVD and MBE techniques. As-grown GaN:Mg samples by the MOCVD technique are highly resistive. Thermal annealing or electron beam irradiation in a hydrogen-free atmosphere is required to activate the Mg acceptors [223, 224]. Thermal annealing in an NH_3 atmosphere reverts the semiconductor to its highly resistive (10^6 Ω cm) state following the activation process [235]. Van Vechten et al. [226] suggested that hydrogen compensates for native defects, thus allowing p-type doping. Calculations by Neugebauer and Van de Walle [222] support the model of Van Vechten et al. [226]. Moreover, Neugebauer and Van de Walle pointed out that hydrogen is beneficial for p-type doping by Mg when compared to the hydrogen-free case, since the Mg concentration increases and nitrogen vacancy concentration decreases in the presence of hydrogen. Although it has been mentioned frequently that Mg and H form a complex in Mg-doped GaN films, the mechanisms of complex formation and release of H upon post-growth treatment have not been definitively cleared yet. Moreover, predictions about the position of H atom by the first principles calculations are not consistent. For example, in Neugebauer et al.'s calculation [227] H is stable at the nitrogen antibonding (AB_N) sites while it is stable at bond center (BC) sites according to Okamoto et al. [228]. Unlike the MOCVD-grown layers, Mg-doped GaN layers grown by reactive molecular beam epitaxy (RMBE) show as-grown p-type conductivity without any postgrowth treatments [229].

As-grown p-GaN films grown by RMBE have shown p-type conductivity even though those films were grown under H-containing environment [229]. The relatively small amount of ammonia (hydrogen) during this process compared with that of MOCVD was assumed responsible for the as-grown p-type conductivity. The amount of H atoms in RMBE-grown Mg-doped GaN films, of order 10^{18} to 10^{19} cm^{-3}, however, was comparable to the H amount in MOCVD-grown p-GaN films [230], which showed p-type conductivity after the thermal annealing. Actually, the amount of H in a p-GaN film in reference [230] as well as in RMBE samples [214] is much lower than the amount of Mg. One can then make the argument, in the context of Mg–H complex theory, that at least some of the Mg atoms would not form complexes with H.

The thermal annealing or rapid thermal annealing (RTA) processes did not change the electrical characteristics of Mg-doped RMBE films [229]. But the amount of H has decreased by about a factor of three after an RTA treatment for 2 min at 900 °C. It is apparent that the release of H atoms from as-grown GaN:Mg does not improve the conductivity of the already conducting RMBE-grown films. Although further investigations are needed to unravel the nature of as-grown p-type conductivity in RMBE-grown GaN:Mg films, the following explanation comes to mind. The ammonia flux is shut off upon the completion of the RMBE growth. Consequently, the film is vacuum-annealed, albeit for a short time, in the cooling process during which H may be released. The *in situ* annealing during the cooling procedure after the MBE film growth may achieve what annealing does for MOCVD grown samples.

It has been reported that annealing in an ammonia atmosphere may passivate the conductive (after post-growth treatment) p-GaN film and the conductivity can be restored by annealing the same sample in a nitrogen atmosphere again. It is, however, not clear whether the activation and deactivation process for already conductive (after post-growth treatment) p-GaN film is the same as that for highly resistive as-grown GaN:Mg film. The second, and perhaps more plausible, explanation is that the release of H might not be primarily responsible for the recovery of p-type conductivity. Rather, the Mg atoms in GaN film can be at some kind of metastable state due to the unique growth reaction or ionic crystal characteristics of GaN:Mg. They can be activated by energy from either electron irradiation or thermal annealing with H release being just a benign side effect of this process. This supposition is supported by the fact that the electrical conduction does not change after the RTA treatment, while H concentration decreases by a factor of about three. Even though there is some evidence supporting the above ideas, the necessity of further research on the behavior of Mg and H atoms at atomic scale in the film and their effect on the electrical properties of the Mg-doped GaN film still remains.

In another experiment [214], two Mg-doped GaN films were grown on *c*-plane sapphire with different Mg flux. The flow rate of ammonia was changed in

2.9 *Doping during growth* 57

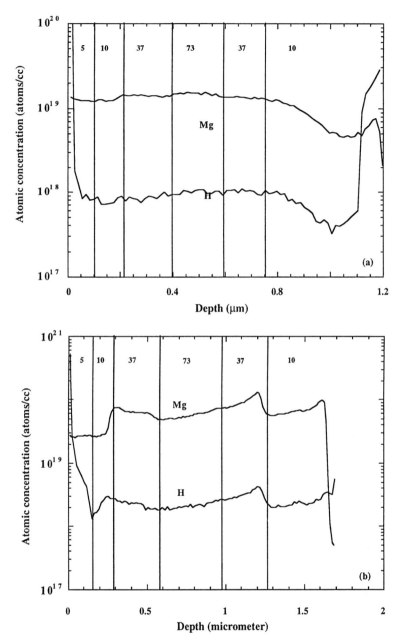

Fig. 2.18 The Mg and H profiles measured by SIMS for two Mg-doped samples grown with different Mg fluxes. The ammonia flow rate was changed in several steps throughout each growth. The numbers in each segment at the upper part of the figure represent the ammonia flow rates. (a) Low Mg flux; (b) high Mg flux [214].

several steps throughout each growth to observe the effect of the ammonia flow rate on the incorporation of Mg into the film. The depth profiles of H and Mg

58 Deposition and properties of group III nitrides

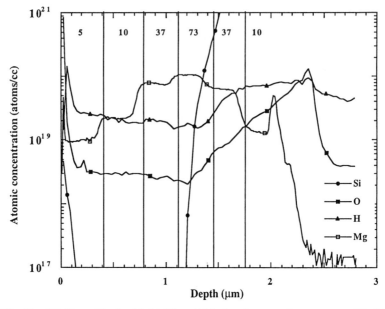

Fig. 2.19 The SIMS profile for Si, O, H, and Mg for the sample grown on 6H–SiC with the same scheme used in Fig. 2.18. The numbers in each segment at the upper part of the figure represent the ammonia flow rates. The steps in Mg concentrations are clearly seen for each segment [214].

atoms for these two samples were analyzed by SIMS. Figures 2.18(a) and (b) show the depth profile of H and Mg for these two samples. The numbers appearing on the top portion of each growth segment indicate the ammonia flow rate for each period. No clear steps are seen in Fig. 2.18(a), while two clear steps for both H and Mg are seen in Fig. 2.18(b) when the ammonia flow rate was changed between 10 and 37 sccm. When the depth profile of Mg and H is displayed using a linear scale for the sample grown with low Mg flux, the obvious steps also can be seen for the Mg profile while a broad peak is seen for H. It is noticeable that Mg and H concentrations change at the same position. Also, Mg concentration is more than one order of magnitude higher than H concentration, as was mentioned in the above discussion. It can therefore be concluded that only a portion of Mg atoms is activated and/or incorporates into the film with H atoms and forms complexes, while the majority of the Mg atoms incorporate without accompanying H atoms as long as the depth profile of Mg and H follows the same trend. In fact, the H profile does not match the Mg profile when the same experiment was carried out on a 6H–SiC substrate, as shown in Fig. 2.19. At this point it is not clear whether there is a difference in growth characteristics in terms of H and Mg incorporation when SiC substrates are used or H reaches the maximal solubility in GaN lattice. The H concentration for this sample, in general, was about one order of magnitude higher than the previous two samples. The growth characteristics related to SiC substrates such as the diffusion of impurities from the SiC surface into the growing film, might also be the cause of the different impurity distribution compared with the sapphire substrate. In any case, in Fig. 2.19 there is no such trend in the H profile as seen in Fig. 2.18(b) while steps in Mg concentration are clearly seen. The following argument, having to do with site selection, can explain the overall trend of Mg incorporation with changing ammonia flow rate: if films are grown with more active nitrogen species, the concentration of Ga vacancy in the film will be increased and, consequently, it will be more likely for Mg atoms to incorporate providing that there is sufficient Mg. One can suggest that the steps in Fig. 2.18(a) are not as obvious as those in Fig. 2.18(b) because the Mg flux was not sufficient to saturate the available Ga-vacancy sites provided by the growth condition. However, this does not exclude the presence of Mg atoms at other sites other than Ga sites, since a reasonable crystalline quality with very high Mg concentration cannot be maintained.

2.9.8 Beryllium

Earlier work on Be doping using CVD methods was unsuccessful, leading to incorporation of Be as a deep center [231, 232]. However, Be-doped GaN films grown by RMBE [233] indicate p-type conductivity as measured by hot probe. From the PL data a Be-related acceptor level was estimated to be a 250 meV above the valence band.

2.9.9 Carbon

Carbon can substitute N in the III-N lattice producing p-type conductivity. There has been one report of hole concentrations up to levels $\sim 3 \times 10^{17} \, \text{cm}^{-3}$ with mobility of $10^3 \, \text{cm}^2/\text{Vs}$ in CCl_4-doped GaN, grown by MOMBE [234]. Annealing at 800 °C did not increase the hole concentration, indicating that hydrogen passivation of acceptors was not significant.

2.9.10 Zinc

Attempts to achieve p-type conductivity by Zn doping of MOCVD grown films were unsuccessful. Highly resistive films with resistivities up to $10^9 \, \Omega \text{cm}$ were grown [235, 236]. Zn, however, causes a level that is about 0.5 eV deep both in GaN and InGaN with low InN mole fraction. In fact this deep level was exploited in the earlier versions of InGaN LEDs marketed by Nichia Chemical in an effort to keep the InN mole fraction in the emission layer small while achieving 450 nm emission required by the display community.

2.9.11 Rare earth

The strong emission sources in the range 1.3 and 1.54 μm are in demand for optical communications based on silica fibers. For these applications, wide bandgap semiconductors present an advantage in terms of room temperature stability compared with systems based on narrow bandgap semiconductors. AlN layers doped by Er present a strong PL at 1.54 μm. Doping levels in the range of 3×10^{17}–2×10^{21} were achieved in MOMBE systems using a solid Er source [237].

2.10 Conclusions and future work

Wide bandgap semiconductors have been coveted for decades for extending the wavelength of operation of optical devices to blue and beyond and for electronics operations at high temperatures and caustic environment. Recent advances in the science and art of GaN and its alloys with wide bandgaps have rendered them viable for the aforementioned device applications. In the past, the poor quality of materials and lack of p-type doping thwarted engineers and scientists from fabricating useful devices from these materials. Both of these obstacles have recently been sufficiently overcome to the point where high luminosity blue and blue–green light-emitting diodes are now available in the market-place. If properly exploited, the quaternaries from nitrides can particularly provide elaborate lattice-matched structures and buried ridges useful for lasers and LEDs.

None of the zincblende nitrides grown to date have rivaled the optical quality of the best wurtzitic material. In order to understand better the properties of the zincblende polytypes, the growth conditions must be optimized, and thicker films, which are less affected by defects resulting from the lattice mismatch with the substrate material, must be grown. One challenge that researchers in the zincblende phase of nitrides face is the metastability of the films, which has led to domain formation. However, with the discovery of the cubic AlN and InN phases, the zincblende nitrides have become a viable material system in their own right with a potential similar to the wurtzitic nitrides for device applications. The cubic nature of the crystal allow for the valence band to be engineered by appropriate strain to reduce the transparency current in lasers. As recent results may indicate, higher p-type doping levels with higher hole mobilities compared with the wurtzitic phase may be possible. Similar benefits may be possible for field-effect transistors as well. These new materials should provide an active and fruitful research topic for years to come as investigators attempt to quantify their potential.

For further development Group III nitride semiconductors, doping incorporation (especially p-type) must be improved. Improved substrate materials, ideally AlN or GaN itself, need to be aggressively pursued to further develop the GaN-based material system. Although p- and n-type doping of GaN have been achieved, many questions related to doping remain unanswered. Controlled Mg doping is not always possible, since increasing of Mg concentrations does not always result in increasing the hole concentration. The incorporation of Mg atoms during the growth of Mg-doped GaN films is dependent on the amount of active nitrogen species in such a way that the incorporation of Mg increases as the ammonia flow rate increases. And this was explained using a simple model associated with Ga vacancy concentration: if films are grown with a larger amount of active nitrogen species, the concentration of Ga vacancies in the films will be increased and, consequently, there will be more chance for Mg atoms to be incorporated into those Ga vacancy sites. The role of native defects and hydrogen in doping is not totally clear yet. Unintentionally doped InN has high electron concentrations up to 10^{18}–10^{19} cm^{-3}. Ways to successfully and reproducibly dope AlN have not been found. Under many growth and doping conditions AlN remains an insulator with resistivity of 10^9 to 10^{12} Ω cm. A branch of the recent MBE effort is being expended on exploring the ternary alloys of GaN with As and P, which have severe band bending, making it possible to obtain small bandgaps. Similarly, small amounts of N in InGaAs, which is coherent on GaAs, push its band edge to the same composition as that lattice matched to InP, which is remarkable. It will be interesting to follow the developments in this emerging field.

Acknowledgments

This work was supported by funds from ONR and AFOSR under the direction of Drs C. E. C. Wood, Y. S. Park, G. L. Witt, and M. Yoder. The authors would like to recognize many contributions by researchers in the field, including those by the MOCVD community and their collaborators. Discussions with Dr A. Botchkarev on growth problems were very helpful. The authors are also thankful to Dr Oliver Brandt for providing a copy of his paper on p-type cubic GaN. They would like to acknowledge the participation of Mr Wook Kim in doping problems and the earlier reviews written by our co-workers, Professor S. N. Mohammad and Dr S. Strite.

References

1. S. N. Mohammad, A. Salvador, and H. Morkoç, *Proc. IEEE*, **83**, 1306 (1995).
2. H. Morkoç, in *Proceedings of International Symposium on Blue Lasers and Light Emitting Diodes*, Chiba University, Japan, March 5–7, 1996, pp. 23–29.
3. H. Morkoç, S. Strite, G. B. Gao, M. E. Lin, B. Sverdlov, and M. Burns, *J. Appl. Phys*. **76**, 1363 (1994).
4. S. Strite and H. Morkoç, GaN, AlN, and InN: A Review, *J. Vac. Sci. Technol. B* **10**, 1237 (1992).
5. J. F. Schetzina, in *International Symposium on Blue Lasers and Light Emitting Diodes*, Chiba University, Japan, March 5–7, 1996, pp. 74–79.
6. I. Akasaki and H. Amano, *International Symposium on Blue Lasers and Light Emitting Diodes*, Chiba University, Japan, March 5–7, 1996, pp. 11–16.
7. M. E. Lin, S. Strite, and H. Morkoç, The physical properties of AlN, GaN and InN, in *The Encyclopedia of Advanced Materials*, eds. D. Bloor, M. C. Fleming, R. J. Brook, S. Mahajan, R. W. Cahn, pp. 79–96, Pergamon Press (1994).
8. S. Nakamura, T. Mukai, and M. Senoh, *Appl. Phys. Lett*. **64**, 1687 (1994).

9. S. Nakamura, T. Mukai, and M. Senoh, *Jpn. J. Appl. Phys.* **30**, 1998 (1991).
10. S. Nakamura, M. Senoh, S. Magahama, N. Iwasa, T. Yamada, T. Matsushita, H. Kiooku, and Y. Sugimoto, *Jpn. J. Appl. Phys.* **L74**, 1998 (1996).
11. W. J. Meng, in *Group III nitrides*, ed. J. H. Edgar, INSPEC, London, UK, pp. 22–29 (1994).
12. I. Akasaki and H. Amano, in Group III nitrides, ed. J. H. Edgar, INSPEC, London, UK, pp. 30–34 (1994).
13. T. L. Tansley, in Group III nitrides, ed. J. H. Edgar, INSPEC, London, UK, pp. 35–42 (1994).
14. W. A. Harris, *Electronic Structure and Properties of Solids*, Dover, NY, pp. 174–179 (1980).
15. T. Sasaki and T. Matsuoka, Analysis of two-step-growth conditions for GaN and AlN buffer layer, *J. Appl. Phys.* **77**, 192 (1995).
16. R. Madar, G. Jacob, J. Hallais, and F. Fruchart, High pressure solution growth of GaN, *J. Cryst. Growth* **31**, 197 (1975).
17. J. Karpinski, J. Jun, and S. Porowski, *J. Crystal Growth*, **66**, 1 (1984).
18. Landolt and Börnstein, *Numerical Data and Fundamental Relationships in Science and Technology, vol. 17, Semiconductors*, Springer, Berlin (1984).
19. T. B. Massalski, *Binary Alloys, Phase Diagrams*, 2nd edn, ASM International, USA, p. 176 (1990).
20. J. D. Latwa, *Metal. Progr.* **82**, 139 (1962).
21. B. E. Wayne, *Ceramics*, **15**, 48 (1964).
22. J. A. Van Vechten, Phys. Rev. **B7**, 1479 (1973).
23. S. Porowski and I. Grzegory, in Group III nitrides, ed. J. H. Edgar, INSPEC, London, UK, pp. 71–88 (1994).
24. M. Henini, *III-Vs Review*, **9**(3), 32 (1996).
25. M. J. Paisley, Z. Sitar, J. B. Posthill, and R. F. Davis, *J. Vac. Sci. Technol.* **A7**, 701 (1989).
26. T. Lei, M. Fanciulli, R. J. Molnar, T. D. Moustakas, R. J. Graham, and J. Scanlon, *Appl. Phys. Lett.* **59**, 944 (1991).
27. T. D. Moustakas, R. J. Molnar, and T. G. Menon, *Mat. Res. Soc. Symp. Proc.* **242**, 247 (1992).
28. M. T. Wauk and D. K. Winslow, *Appl. Phys. Lett.* **13**, 286 (1968).
29. B. B. Kosicki and D. Kahng, *J. Vac. Sci. Technol.* **6**, 593 (1969).
30. J. W. Trainor and K. Rose, *J. Electronic Mater.* **3**, 821 (1974).
31. K. Osamura, S. Naka, and Y. Murakami, *J. Appl. Phys.* **46**, 3432 (1975).
32. S. Zembutsu and T. Sasaki, *J. Cryst. Growth* **77**, 250 (1986).
33. S. Strite, J. Ruan, Z. Li, N. Manning, A. Salvador, H. Chen, D. J. Smith, W. J. Choyke, and H. Morkoç, *J. Vac. Sci. Technol. B* **9**, 1924 (1991).
34. T. Lei, M. Fanciulli, R. J. Molnar, T. D. Moustakas, R. J. Graham, and J. Scanlon, *Appl. Phys. Lett.* **59**, 944 (1991).
35. S. Strite, B. Sariel, D. J. Smith, H. Chen, and H. Morkoç, Seventh International MBE Conference, Schwäbisch Gmünd, Germany (1992). *J. Cryst. Growth* **127**, 204 (1993).
36. H. Morkoç, A. Botchkarev, and B. Sverdlov, *J. Cryst. Growth* **150**, 887 (1995).
37. A. Kikuchi, *Jpn. J. Appl. Phys.* **33**, 688 (1994). This type of source has been successfully used by groups at North Carolina State University (Prof. J. Schetzina), Siemens, Nottingham (Prof. C. T. Foxon), Meijo University (Prof. I. Akasaki), SVT Associates and most likely many others.
38. M. Mori, A. Kikuchi, and K. Kishino, *Inst. Phys. Conf. Ser.* **142**, 839 (1996).
39. A. Sellidj, B. A. Ferguson, T. J. Mattord, B. G. Streetman, and C. B. Mullins, *Appl. Phys. Lett.* **68**, 3314 (1996).

40. B. A. Ferguson, A. Sellidj, B. B. Doris, and C. B. Mullins, *J. Vac. Sci. Technol. A (Vacuum, Surfaces, and Films)* **14**, 825 (1996).
41. K. W. Boer, *Survey of Semiconductor Physics*, vol. 1, New York, Van Nostrand Reinhold (1990).
42. T. D. Moustakas and R. J. Molnar, *Mat. Res. Soc. Symp. Proc.* **281**, 753 (1993).
43. R. J. Molnar, R. Singh, and T. D. Moustakas, *J. Electron Mater.* **24**, 275 (1995).
44. T. D. Moustakas, R. P. Vaudo, R. Singh, D. Korakakis, M. Mistyra, A. Sampath, and I. D. Goepfert, *Inst. Phys. Conf. Ser.* **142**, 833 (1996).
45. A. Botchkarev, A. Salvador, B. Sverdlov, J. Myoung, and H. Morkoç, *J. Appl. Phys.* **77**, 4455 (1995).
46. T. Maruyama, S. H. Chao, and K. Akimoto, *Inst. Phys. Conf. Ser.* **142**, 851 (1996).
47. R. C. Powell, G. A. Tomasch, Y. W. Kim, J. A. Thornton, and J. E. Greene, *Mater. Res. Soc. Symp. Proc.* **162**, 525 (1990); M. Rubin, N. Newman, J. S. Chan, T. C. Fu, and J. T. Ross, *Appl. Phys. Lett.* **64**, 32 (1994).
48. S. Yoshida, S. Misawa, and A. Itoh, *Appl. Phys. Lett.*, **26**, 461 (1975).
49. S. Winsztal, B. Wauk, H. Majewska-Minor, and T. Niemyski, *Thin Solid Films* **32**, 251 (1976).
50. K. R. Elliott and R. W. Grant, *Rockwell Project Final Report MRDC41116.2FR* (1984).
51. H. U. Baier and W. Mönch, *J. Appl. Phys.* **68**, 586 (1990).
52. S. Yoshida, S. Misawa, and S. Gonda, *J. Vac. Sci. Technol. B* **1**, 250 (1983).
53. H. Gotoh, T. Suga, H. Suzuki, and M. Kimata, *Jpn. J. Appl. Phys.* **20**, L545 (1981).
54. W. Kim, O. Aktas, A. E. Botchkarev, A. Salvador, S. N. Mohammad, and H. Morkoç, *J. Appl. Phys.* **79**, 7657 (1994).
55. D. C. Look, D. C. Reynolds, R. L. Jones, W. Kim, Ö. Aktas, A. E. Botchkarev, A. Salvador, and H. Morkoç, *J. Appl. Phys.* **80**, 2960 (1996).
56. W. Kim, Ö. Aktas, A. E. Botchkarev, A. Salvador, S. N. Mohammad, and H. Morkoç, *J. Appl. Phys.* **79**, 7657 (1996).
57. J. R. Jenny, R. Kaspi, C. R. Jones, and K. R. Evans, to be published.
58. J. E. Andrews and M. A. Littlejohn, *J. Electrochem. Soc.* **122**, 1273 (1975).
59. S. Bharatan, K. S. Jones, S. J. Pearson, C. R. Abernathy, and F. Ren, *MRS Symp. Proc.* **339**, MRS, Pittsburgh PA, 1994, pp. 491–496.
60. H. Tsuchia, A. Takeuchi, M. Kurihara, and F. Hasegawa, *J. Cryst. Growth* **152**, 217 (1995).
61. A. Takeuchi, Y. Kumagai, H. Tsuchiya, M. Kurihara, M. Kawabe, and F. Hasegawa, *Inst. Phys. Conf. Ser.* **142**, 843 (1996).
62. K. Wang, J. Singh, and D. Pavlidis, in *Compound Semiconductors 1994. Proceedings of the Twenty-First International Symposium*, H. Goronkin and U. Mishra, eds, Bristol, UK, IOP Publishing, 1995, pp. 137–42.
63. K. Wang, J. Singh, and D. Pavidis, *J. Appl. Phys.* **76**, 3502 (1994).
64. Z. Z. Bandic, R. J. Hauenstein, M. L. O'Steen, and T. C. McGill, *Appl. Phys. Lett.* **68**, 1510 (1996).
65. M. V. Averyanova, S. Yu. Karpov, Yu. N. Makarov, I. N. Przhevalskii, M. S. Ramm, and R. A. Talalaev, *MRS Internet J. Nitride Semicond. Research* **1**, N31 (1996).
66. K. Wang, D. Pavidis, and J. Singh, *J. Appl. Phys.* **80**, 1823 (1996).
67. T. Lei, K. F. Ludwig, Jr., and T. Moustakas, *J. Appl. Phys.* **74**, 4430 (1993).
68. R. C. Powell, N. E. Lee, Y.-W. Kim, and J. Green, *J. Appl. Phys.* **73**, 189 (1993).
69. K. Doverspike, L. B. Rowland, D. K. Gaskill, and J. A. Freitas, Jr., *J. Electron. Mater.* **24**, 269–73 (1995).
70. H. Amano, I. Akasaki, K. Hiramatsu, and N. Sawaki, *Thin Solid Films* **163**, 415 (1988).

71. I. Akasaki, H. Amano, Y. Koide, K. Hiramatsu, and N. Sawaki, *J. Crystal Growth* **98**, 209 (1989).
72. J. N. Kuznia, M. A. Khan, D. T. Olson, R. Kaplan, and J. Freitas, *J. Appl. Phys.* **73**, 4700 (1993).
73. K. Uchida, W. Watanabe, F. Yano, M. Kougucci, T. Tanaka, and S. Minegava, *Proceedings of International Symposium on Blue Lasers and Light Emitting Diodes*, Chiba University, Japan, March 5–7, 1996, p. 48.
74. S. Keller, B. P. Keller, Y.-F. Wu, B. Heying, D. Kaponek, J. S. Speck, U. K. Mishra, and S. P. DenBaars, *Appl. Phys. Lett.* **68**, 1525 (1996).
75. N. Grandjean, J. Massies, and M. Leroux, *Appl. Phys. Lett.* **69**, 2071 (1996).
76. C. Kim, I. K. Robinson, J. Myoung, K. Shim, M.-C. Yoo, and K. Kim, *Appl. Phys. Lett.* **69**, 2358 (1996).
77. S. Nakamura, *Jpn. J. Appl. Phys.* **30**, L1705 (1991).
78. H. Amano, N. Sawaki, I. Akasaki, and Y. Toyoda, *Appl. Phys. Lett.* **48**, 353 (1986).
79. K. J. Fertitta, A. L. Holmws, J. G. Neff, F. J. Ciuba, and R. D. Dupius, *Appl. Phys. Lett.* **65**, 1823 (1994).
80. X. Z. Dang, G. Y. Zhang, B. Zhang, Z. Y. Yang, Y. Z. Tong, Z. I. Xu, S. X. Jin, W. Liu, M. Xu, and S. M. Wang, *Proceedings of International Symposium on Blue Lasers and Light Emitting Diodes*, Chiba University, Japan, March 5–7, 1996, p. 383.
81. X. H. Wu, D. Kapolnek, E. J. Tarsa, B. Heying, S. Keller, B. P. Keller, U. K. Mishra, S. P. DenBaars, and J. S. Speck, *Appl. Phys. Lett.* **68**, 1371 (1996).
82. P. Kung, A. Saxler, X. Zhang, T. C. Wang, I. Ferguson, and M. Razeghi, *Appl. Phys. Lett.* **66**, 2958 (1995).
83. B. Heyding, X. H. Wu, S. Keller, Y. Li, D. Kapolnek, B. P. Keller, S. P. DenBaars, and J. S. Speck, *Appl. Phys. Lett.* **68**, 643 (1996).
84. Q. Zhu, A. Botchkarev, W. Kim, Ö. Aktas, A. Salvador, B. Sverdlov, H. Morkoç, S. C. Y. Tsen, and D. J. Smith, *Appl. Phys. Lett.* **68**, 1141 (1996).
85. C. Kisielowski, J. Kruger, S. Ruvimov, T. Suski, J. W. Ager III, E. Jones, Z. Liliental-Weber, M. Rubin, E. R. Weber, M. D. Bremser, and R. F. Davis, *Phys Rev. B*, **54**, 17745 (1996).
86. M. A. Khan, M. S. Shur, Q. C. Chen, and J. N. Kuznia, *Electron. Lett.* **30**, 2175 (1994).
87. Ö. Aktas, W. Kim, Z. Fan, A. Bochkarev, A. Salvador, B. Sverdlov, and H. Morkoç, *Electron. Lett.* **31**, 1389 (1995).
88. Ö. Aktas, W. Kim, Z. Fan, F. Strengel, A. Bochkarev, A. Salvador, B. Sverdlov, and S. N. Mohammad, *IEEE Int. Electron Devices Meeting Techn. Digest*, pp. 205–208, Washington DC, 10–13 Dec (1995).
89. Z. Fan, S. N. Mohammad, Ö, Aktas, A. E. Botchkarev, A. Salvador, and H. Morkoç, *Appl. Phys. Lett.* **69**, 1229 (1994).
90. M. Smith, J. Y. Lin, H. X. Jiang, A. Salvador, A. Botchkarev, W. Kim, and H. Morkoç, *Appl. Phys. Lett.* **69**, 2453 (1996).
91. R. Singh, D. Doppalaudi, and T. D. Moustakas, *Appl. Phys. Lett.* **69**, 2388 (1996).
92. H. H. Yang, T. J. Schmidt, W. Shan, J. J. Song, and B. Goldenberg, *Appl. Phys. Lett.* **66**, 1 (1995).
93. R. K. Sink, S. Keller, B. P. Keller, D. I. Babic, A. L. Holmes, D. Kapolnek, S. P. deBaars, J. E. Bowers, X. H. Wu, and J. S. Speck, *Appl. Phys. Lett.* **68**, 2147 (1996).
94. M. E. Lin, S. Strite, A. Agarwal, A. Salvador, G. L. Zhou, N. Teraguchi, A. Rocket, and H. Morkoç, *Appl. Phys. Lett.* **62**, 702 (1993).
95. D. Byun, G. Kim, D. Kim, I.-H. Choi, D. Park, and D.-W. Kum, *Proceedings of International Symposium on Blue Lasers and Light Emitting Diodes*, Chiba University, Japan, March 5–7, 1996, p. 380.

96. D. J. Smith, D. Chandrasekhar, B. Sverdlov, A. Bochkarev, A. Salvador, and H. Morkoç, *Appl. Phys. Lett.* **67**, 1830 (1995).
97. M. E. Lin, B. Sverdlov, G. L. Zhou, and H. Morkoç, *Appl. Phys. Lett.* **62**, 3479 (1993).
98. T. Matsuoka, N. Yoshimoto, T. Sasaki, and A. Katsui, *J. Electronic Materials* **21**, 157 (1992).
99. F. Hamdani, A. Botchkarev, W. Kim, A. Salvador, and H. Morkoç, M. Yeadon, J. M. Gibson, S. C. T. Tsen, D. J. Smith, D. C. Reynolds, D. C. Look, K. Evans, C. W. Litton, W. C. Mitchel, and P. Hemenger, *Appl. Phys. Lett.* **70**, 467 (1997).
100. T. Detchprohm, H. Amano, K. Hiramatsu, and I. Akasaki, *J. Cryst. Growth* **128**, 384 (1993).
101. M. A. L. Johnson, S. Fujita, W. H. Rowland, W. C. Hughes, J. W. Cook, and J. F. Schetina, *J. Electron. Mater.* **25**, 855 (1996).
102. T. Matsuoka, N. Yoshimoto, T. Sasaki, and A. Katsui, *J. Electron Mater.* **21**, 157 (1992).
103. F. Hamdani, M. Yeadon, W. Kim, A. Botchkarev, J. M. Gibson, H. Morkoç, S. C. Y. Tsen, and D. J. Smith, *J. Appl. Phys.* **73**, January (1998).
104. T. Lei, T. D. Moustakas, R. J. Graham, Y. He, and S. J. Berkowitz, *J. Appl. Phys.* **71**, 4933 (1992).
105. K. Yokouchi, T. Araki, T. Nagatomo, and O. Omoto, *Inst. Phys. Conf. Ser.* **142**, 867 (1996).
106. A. Ohtani, K. S. Stevens, and R. Beresford, *MRS Symp. Proc.* **339**, MRS, Pittsburgh PA, 1994, pp. 471–476.
107. S. N. Basu, T. Lei, and T. D. Moustakas, *J. Mater. Res.* **9**, 2370 (1994).
108. G. A. Martin, B. N. Sverdlov, A. Botchkarev, H. Morkoç, D. J. Smith, S.-C. Y. Tsen, W. H. Thompson, and M. H. Nayfeh, *Proc. of Workshop on III–V Nitrides*, Nagoya, Japan, ed. I. Akasaki, p. 381 (1995)..
109. M. Godlewski, J. P. Bergman, B. Monemar, U. Rossner, and A. Barski, *Appl. Phys. Lett.* **69**, 2089 (1996).
110. J. Menninger, U. Jahn, O. Brandt, H. Yang, and K. Ploog, *Phys. Rev. B* **53**, 1881 (1996).
111. R. C. Powell, G. A. Tomasch, T. W. Kim, J. A. Thorton, and J. E. Greene, *Mater. Res. Soc. Symp. Proc.* **162**, 525 (1991).
112. J. N. Kuznia, J. W. Wang, Q. C. Chen, S. Krishankutty, A. M. Khan, T. Gerge, and J. Freitas, Jr., *Appl. Phys. Lett.* **65**, 2407 (1994).
113. S. Fujieda and Y. Matsumoto, *Jpn. J. Appl. Phys.* **30**, L1665 (1991).
114. S. Nakamura, T. Mukai, and M. Senoh, *J. Appl. Phys.* **60**, 5543 (1992).
115. K. Yasui, H. Yoshida, T. Harada, and T. Akahane, *Inst. Phys. Conf. Ser.* **142**, 871 (1996).
116. M. Sato, *Inst. Phys. Conf. Ser.* **142**, 875 (1996).
117. H. Yang, O. Brandt, M. Wassermeier, J. Behrend, H. P. Schonherr, and K. H. Ploog, *Appl. Phys. Lett.* **68**, 244 (1996).
118. J. W. Yang, J. N. Kuznia, Q. C. Chen, M. A. Khan, T. George, M. De Graef, and M. Mahajan, *Appl. Phys. Lett.* **67**, 3759 (1995).
119. C. T. Foxon, T. S. Cheng, S. V. Novikov, D. E. Lacklison, L. C. Jenkins, D. Johnston, J. W. Orton, S. E. Hooper, N. Baba-Ali, T. L. Tansley, and V. V. Tretiakov, *J. Crystal Growth* **150**, 892 (1995).
120. J. Ross, M. Rubin, and T. K. Gustafson, *J. Mater. Res.* **8**, 2613 (1993).
121. T. Suzuki, T. Matsui, H. Matsuyama, S. Kitamura, and H. Kaqmijo, *Proc. of International Symposium on Blue Lasers and Light Emitting Diodes*, Chiba University, Japan, March 5–7, 1996, p. 522.

122. T. S. Cheng, L. C. Jenkins, S. E. Hooper, C. T. Foxon, J. W. Orton, and D. E. Lacklison, *Appl. Phys. Lett.* **66**, 1509 (1995).
123. C. I. Sun, J. W. Yang, Q. Chen, M. A. Khan, T. George, P. Chang-Chien, and S. Mahajan, *Appl. Phys. Lett.* **68**, 1129 (1996).
124. J. W. Yang, Q. Chen, C. J. Sun, B. Lim, M. Z. Anwar, M. A. Khan, and H. Temkin, *Appl. Phys. Lett.* **69**, 369 (1996).
125. T. George, E. Jacobsohn, W. T. Pike, P. Chang-Chien, M. A. Khan, K. W. Yang, and S. Mahajan, *Appl. Phys. Lett.* **68**, 337 (1996).
126. A. Karamata, K. Horino, K. Domen, K. Shinohara, *Appl. Phys. Lett.* **67**, 2521 (1996).
127. M. A. Khan, C. J. Sun, J. W. Yang, Q. Chen, B. W. Lim, M. Z. Anwar, A. Osinski, and H. Temkin, *Appl. Phys. Lett.* **69**, 2418 (1996).
128. R. C. Powell, N.-E. Lee, Y.-W. Kim, and J. E. Green, *J. Appl. Phys.* **73**, 189 (1993).
129. M. A. Vidal, G. Ramirez-Flores, H. Navarro-Contreras, A. Lastras-Martinez, R. C. Powell, and J. E. Greene, *Appl. Phys. Lett.* **68**, 441 (1996).
130. P. Perlin, Y. Gorczyca, N. E. Christensen, Y. Grzegory, H. Teiserye, and T. Suski, *Phys. Rev. B* **45**, 13307 (1992).
131. F. A. Ponce, D. P. Bour, W. Gotz, N. M. Johnson, H. I. Helava, I. Grzegory, J. Jun, and S. Porowski, *Appl. Phys. Let.* **68**, 917 (1996).
132. F. A. Ponce, D. P. Bour, W. T. Young, M. Sounders, and J. W. Steeds, *Appl. Phys. Lett.* **69**, 337 (1996).
133. A. Gassmann, T. Susski, N. Newmann, C. Kiselowski, E. Jones, E. R. Weber, Z. Liliental-Weber, M. D. Rubin, H. I. Helava, I. Grezegory, M. Bockovski, J. Jun, and S. Porowski, *J. Appl. Phys.* **80**, 2195 (1996).
134. D. J. Smith, D. Chandrasekhar, B. Sverdlov, A. Botchkarev, A. Salvador, and H. Morkoç, *Appl. Phys. Lett.* **67**, 1830 (1995).
135. B. N. Sverdlov, G. A. Martin, H. Morkoç, and D. J. Smith, *Appl. Phys. Lett.* **67**, 2063 (1995).
136. M. A. L. Johnson, S. Fujita, W. H. Rowland, Jr., W. C. Hughes, Y. W. He, N. A. El-Masry, J. W. Cook, Jr., J. F. Schetzina, J. Ren, and J. A. Edmond, *J. Electron. Mater*, **25**, 793 (1996).
137. T. P. Pearsall, *III-Vs Review*, **9**(3), 38 (1996).
138. J. I. Pankove, *Mater. Res. Soc. Symp. Proc.* **162**, 515 (1990).
139. H. Liu, A. C. Frenkel, J. G. Kim, and R. M. Park, *J. Appl. Phys.* **74**, 6124 (1993).
140. Z. Yang, F. Guarin, I. W. Tao, and W. I. Wang, *J. Vac. Sci. Technol. B* **13**, 789 (1995).
141. A. J. Noreika and D. W. Ing, *J. Appl. Phys.* **39**, 5578 (1968).
142. T. L. Chu and R. M. Kelm, Jr., *J. Electrochem. Soc.* **122**, 995 (1975).
143. H. Komiyama and T. Osawa, *Jpn. J. Appl. Phys.* **24**, L795 (1985).
144. B. S. Sywe, Z. J. Yu, and J. H. Edgar, *Mat. Res. Soc. Proc. Symp.*, MRS, Pittsburgh PA, **242**, 1994, p. 463. G. Eichhorn and U. Rensch, *Phys. Stat. Sol. (a)* **69** K3 (1982).
145. M. Morita, N. Ueugi, S. Isogai, K. Tsuboushi, and N. Mikoshiba, *Jpn. J. Appl. Phys.* **20**, 17 (1981).
146. M. T. Duffy, *Heteroepitaxial Semiconductors for Electronic Devices*, eds G. W. Cullen and C. C. Wang, Springer-Verlag, New York, 1978, pp. 150–181.
147. S. Yoshida, S. Mizawa, Y. Fujii, S. Takada, H. Hayakawa, S. Gonda, and A. Itoh, *J. Vac. Sci. Technol.* **16**, 990 (1979).
148. B. Rowland, R. S. Kern, S. Tanaka, and R. Davis, *J. Mater. Res.* **9**, 2310 (1993).
149. Z. Sitar, M. J. Paisley, B. Yan, R. F. Davis, J. Ruan, and J. W. Choyke, *Thin Solid Films* **200**, 311 (1991).
150. W. M. Yim, E. F. Sofko, P. Zanzucchi, J. I. Pankove, M. Ettenberg, and S. L. Gilbert, *J. Appl. Phys.* **44**, 292 (1973).

151. H. M. Manasevic, F. M. Erdman, and W. I. Simpson, *J. Electrochem Soc.* **118**, 1864 (1971).
152. M. Morita, S. Isogai, N. Shimisu, K. Tsubuoshi, and N. Mikoshiba, *Jpn. J. Appl. Phys.* **20**, L173 (1981).
153. Z. J. Yu, J. H. Edgar, A. U. Ahmed, and A. Rys, *J. Electrochem. Soc.* **138**, 196 (1991).
154. T. L. Chu, D. W. Ing, and A. J. Noreika, *Solid-State Electron.* **10**, 1023 (1967).
155. For the method of ion plating for InN growth, see M. Katajima, M. Fukutomi, and R. Watanabe, *J. Electrochem. Soc.* **128**, 1588 (1981).
156. For the method of reactive radiofrequency sputtering for InN growth, see (a) T. J. Kistenmacher, W. A. Bryden, J. A. Morgan, D. Dayan, R. Fainchtein, and T. O. Pehler, *J. Mater. Res.* **6**, 1300 (1991); (b) K. Kubota, K. Kobayashi, and K. Fujimoto, *J. Appl. Phys.* **66**, 2984 (1989); (c) B. R. Natarajan, A. H. Eltoukhy, J. E. Greem, and T. L. Barr, *Thin Solid Films* **69**, 201 (1980).
157. For the technique of reactive magnetron sputtering for InN growth, see (a) K. L. Westra, R. P. W. Lawson, and J. Brett, *J. Vac. Sci. Technol. A* **6**, 1730 (1988); (b) B. T. Sullivan, R. R. Parsons, K. L. Westra, and M. J. Brett, *J. Appl. Phys.* **64**, 4144 (1988); (c) T. J. Kistenmacher, W. A. Bryden, J. S. Morgen, and T. O. Poehler, *J. Appl. Phys.* **68**, 1541 (1990); (d) T. J. Kistenmacher, W. A. Bryden, J. S. Morgen, D. Dayan, R. Faintchen, and T. O. Poehler, *J. Mater. Res.* **6**, 783 (1970); (e) J. S. Morgan, W. A. Bryden, T. J. Kistenmacher, S. A. Ecelberger, and T. O. Poehler, *J. Mater. Res.* **5**, 2627 (1990).
158. For microwave-assisted CVD technique for InN growth, see Q.-X. Guo, T. Yamamura, A. Yoshida, and I. Itoh, *J. Appl. Phys.* **75**, 4927 (1994); Q.-X. Guo, O. Kato, and A. Yoshida, *J. Appl. Phys.* **73**, 7969 (1993); Q.-X. Guo, O. Kato, M. Fujisawa, and A. Yoshida, *Solid-State Commun.* **83**, 721 (1992); Q.-X. Guo, O. Kato, and A. Yoshida, *J. Electrochem. Soc.* **139**, 2008 (1992).
159. For laser-assisted CVD technique for InN growth, see Y. Bu, L. Ma, and M. C. Lin, *J. Vac. Sci. Technol. A* **11**, 2931 (1993).
160. For the technique for halogen transport for InN growth, see O. Igarashi, *Jpn. J. Appl. Phys.* **31**, 2665 (1992).
161. T. J. Kistenmacher, S. A. Ecelberger, and W. A. Bryden, *J. Appl. Phys.* **74**, 1684 (1993); Q.-X. Guo, T. Yamammura, A. Yoshida, and N. Itoh, *J. Appl. Phys.* **75**, 4927 (1994).
162. T. J. Kistenmacher, S. A. Ecelberger, and W. A. Bryden, *J. Appl. Phys.* **74**, 1684 (1993). See also W. A. Bryden, J. S. Morgan, R. Fainchtein, and T. J. Kistenmacher, *Thin Solid Films* **213**, 86 (1992).
163. S. Strite, D. Chandrasekhar, D. J. Smith, J. Sariel, H. Chen. N. Teraguchi, and H. Morkoç, *J. Cryst. Growth* **127**, 204 (1993).
164. S. Yoshida, S. Misawa, and S. Gonda, *J. Appl. Phys.* **53**, 6844 (1982).
165. T. Nagamoto, T. Kuboyama, H. Minamino, and O. Omoto, *Jpn. J. Appl. Phys.* **28**, L1334 (1989).
166. N. Yoshimoto, T. Matsuoka, T. Sasaki, and A. Katsui, *Appl. Phys. Lett.* **59**, 2251 (1991); T. Matsuoka, N. Yoshimoto, T. Sasaki, A. Katsui, *J. Electron. Mater.* **21**, 157 (1992).
167. S. Nakamura and T. Mukai, *Jpn. J. Appl. Phys.* **31**, L1457 (1992).
168. R. Madar, D. Michel, G. Jacob, and M. Boulou, *J. Cryst. Growth*, **40**, 239 (1977).
169. T. Sasaki and S. Zembutsu, *J. Appl. Phys.* **61**, 2533 (1987).
170. K. Hiramatsu, H. Amano, I. Akasaki, H. Kato, N. Koide, and K. Manabe, *J. Cryst. Growth*, **107**, 509 (1991).
171. R. Madar, D. Michel, G. Jacob, and M. Boulou, *J. Cryst. Growth* **40**, 239 (1977).

172. H. M. Manasevit, F. M. Erdmann, and W. I. Simpson, *J. Electrochem. Soc.* **118**, 1864 (1971).
173. A. Tempel, W. Seifert, J. Hammer, and E. Butter, *Krist. Tech.*, **10**, 747 (1975).
174. T. P. Humphreys, C. A. Sukow, R. J. Nemanich, J. B. Posthill, R. A. Rudder, S. V. Hattangady, and R. J. Markunas, *Mater. Res. Soc. Symp. Proc.*, **162**, 531 (1990).
175. Z. Sitar, M. J. Paisley, B. Yan, and R. F. Davis, *Mater. Res. Soc. Symp. Proc.*, **162**, 537 (1990).
176. T. Sasaki and T. Matsuoka, *J. Appl. Phys.* **64**, 4531 (1988).
177. M. Mizuta, S. Fujieda, Y. Matusumoto, and T. Kawamura, *Jpn. J. Appl. Phys.* **25**, L945 (1986).
178. G. Martin, S. Strite, J. Thornton, and H. Morkoç, *Appl. Phys. Lett.* **58**, 2375 (1991).
179. P. M. Dryburgh, *J. Cryst. Growth* **94**, 23 (1989).
180. D. K. Gaskill, N. Bottka, and M. C. Lin, *Appl. Phys. Lett.* **48**, 1449 (1986).
181. D. Troost, H. U. Baier, A. Berger, and W. Mönch, *Surf. Sci.* **242**, 324 (1991).
182. A. Berger, D. Troost, and W. Mönch, *Vacuum*, **41**, 669 (1990).
183. S. Strite, D. S. L. Mui, G. Martin, Z. Li, D. J. Smith, and H. Morkoç, *Proc. of GaAs and Related Compounds Conf.*, Seattle (1991), Inst. Phys. Conf. Ser., IPO Ser., 1992, pp. 89–94.
184. T. L. Chu, *J. Electrochem. Soc.* **118**, 1200 (1971).
185. H. G. Grimmeiss and B. Monemar, *J. Appl. Phys.* **41**, 4054 (1970).
186. S. Strite, D. S. Miu, G. Martin, Z. Li, D. J. Smith, and H. Morkoç, *Proc. of GaN and Related Compounds Conf.*, Seattle (1991), Inst. Phys. Conf. Ser., IPO Ser., pp. 89–94.
187. H. Amano, K. Hiramatsu, and I. Akasaki, Jpn. J. Appl. Phys., **27**, L1384 (1988).
188. K. Naniwae, S. Itoh, H. Amano, K. Itoh, K. Hiramatsu, and I. Akasaki, *J. Cryst. Growth* **99**, 381 (1990).
189. S. Yoshida, S. Misawa, and S. Gonda, *J. Vac. Sci. Technol. B* **1**, 250 (1983).
190. S. Yoshida, S. Misawa, and S. Gonda, *Appl. Phys. Lett.* **42**, 427 (1983).
191. H. Amano, N. Sawaki, I. Akasaki, and Y. Toyoda, *Appl. Phys. Lett.* **48**, 353 (1986); Y. Koide, N. Itoh, K. Itoh, N. Sawaki, and I. Akasaki, *Jpn. J. Appl. Phys.* **27**, 1156 (1988).
192. I. Akasaki, H. Amano, Y. Koide, K. Hiramatsu, and N. Sawaki, *J. Cryst. Growth*, **98**, 209 (1989); H. Amano, I. Akasaki, K. Hiramatsu, N. Koide, and N. Sawaki, *Thin Solid Films* **163**, 415 (1988).
193. T. J. Kistenmacher and W. A. Bryden, *Appl. Phys. Lett.* **59**, 1844 (1991).
194. R. F. Davis, Z. Sitar, B. E. Williams, H. S. Kong, H. J. Kim, J. W. Palmour, J. A. Edmond, J. Ryu, J. T. Class, and C. H. Carter, Jr., *Mater. Sci. Eng.* **B1**, 77 (1988).
195. For a description of observed stacking faults, see, for example, R. F. Davis, *Proc. IEEE* **79**, 702 (1991).
196. W. Seifert and A. Tempel, *Phys. Status Solidi A* **23**, K39 (1974).
197. R. C. Powell, *PhD Thesis*, University of Illinois at Urbana-Champaign, 1993.
198. T. Lei and T. D. Moustakas, *Mater. Res. Soc. Symp. Proc.* **242**, 433 (1992).
199. K. Kawabe, R. H. Tredgold, and Y. Inushi, *Elect. Eng. Japan* **87**, 62 (1967).
200. D. Elwell, R. S. Feigelsdon, M. M. Simkins, and W. A. Tiller, *J. Cryst. Growth* **66**, 45 (1984).
201. D. W. Jenkins, J. D. Dow, and M.-H. Tsai, *J. Appl. Phys.* **72**, 4130 (1992).
202. P. Boguslawski, E. L. Briggs, and J. Bernholc, *Phys. Rev. B* **51**, 17255 (1995).
203. T. L. Tansley and R. L. Egan, *Phys. Rev. B* **45**, 10942 (1992).
204. I. Akasaki, H. Amano, Y. Koide, K. Hiramatsu, and N. Sawaki, *J. Crystal Growth* **98**, 209 (1989).
205. J. I. Pankove, S. Bloom, and G. Harbeke, *RCA Review* **36**, 163 (1995).
206. R. J. Molnar, T. Lei, and T. D. Moustakas, *Appl. Phys. Lett.* **62**, 72 (1993).
207. J. Neugebauer and C. G. Van de Walle, *MRS Proc.* **339**, 693 (1994).

208. J. Neugebauer and C. G. Van de Walle, *Appl. Phys. Lett.* **69**, 503 (1996).
209. B. K. Ridley, *Quantum Processes in Semiconductors*, Oxford Science, Oxford, UK, 1988, pp. 62–67.
210. H. P. Maruska and J. J. Tietjen, *Appl. Phys. Lett.* **15**, 327 (1969).
211. P. Perlin, T. Suzuki, H. Teisseyre, M. Leszczynski, I. Grezgory, J. Jun, S. Porowski, P. Boguslavski, J. Bernholc, J. C. Chervin, A. Polian, and T. D. Moustakas, *Phys. Rev. Lett.* **75**, 296 (1995).
212. P. Boguslawski, E. L. Briggs, and J. Bernholc, *Phys. Rev. B* **51**, 17255 (1995).
213. C. Wetzel, W. Walukiewich, E. E. Haller, J. Ager III, I. Grezgory, S. Porowski, and T. Suski, *Phys. Rev. B* **53**, 1322 (1996).
214. W. Kim, A. E. Botchkarev, A. Salvador, G. Popovici, H. Tang, and H. Morkoç, *J. Appl. Phys.* **82**, 219 (1997).
215. N. Koide, H. Kato, M. Sassa, S. Yamasaki, K. Manabe, M. Hashimoto, H. Amano, K. Hiramatsu, and I. Akasaki, *J. Cryst. Growth*, **115**, 639 (1991).
216. D. K. Gaskill, A. E. Wickenden, K. Doverspike, B. Tadayon, and L. B. Rowland, *J. Electron. Mater.* **24**, 1525 (1995).
217. P. Hacke, A. Maekava, N. Koide, K. Hiramatsu, and N. Sawaki, *Jpn. J. Appl. Phys.* **33**, 6443 (1994).
218. W. Götz, N. M. Johnson, C. Chen, H. Liu, C. Kuo, and W. Imler, *Appl. Phys. Lett.* **68**, 3144 (1996).
219. B.-C. Chung and M. Gershenzon, *J. Appl. Phys.* **72**, 651 (1992).
220. J. Neugebauer and C. G. Van de Walle, *Appl. Phys. Lett.* **68**, 1829 (1996).
221. W. Götz, N. M. Johnson, J. Walker, D. P. Bour, H. Amano, and I. Akasaki, *Appl. Phys. Lett.* **67**(18), 2666 (1995).
222. H. Amano, M. Kito, K. Hiramatsu, and I. Akasaki, *Jpn. J. Appl. Phys.* **28**, L2112 (1989).
223. S. Nakamura, T. Mukai, M. Senoh, and N. Iwasa, *Jpn. J. Appl. Phys.*, **31**, L139 (1992).
224. S. Nakamura, N. Iwasa, M. Senoh, and T. Mukai, *Jpn. J. Appl. Phys.* **31**, 1258 (1992).
225. J. A. Van Vechten, J. D. Zook, R. D. Hornig, and B. Goldenberg, *Jpn. J. Appl. Phys.* **31**, 3662 (1992).
226. J. Neubauer and C. G. Van de Walle, *Appl. Phys. Lett.* **68**, 1829 (1996).
227. J. Neugebauer and C. G. Van de Walle, *Phys. Rev. Lett.* **75**(24), 4452 (1995).
228. Y. Okamoto, M. Saito, and A. Oshiyama, Jpn. J. Appl. Phys. **35**(7A), L807 (1996).
229. W. Kim, A. Salvador, A. E. Botchkarev, Ö. Aktas, S. N. Mohammad, and H. Morkoç, *Appl. Phys. Lett.* **69**, 559 (1996).
230. C. Yuan, T. Salagaj, A. Guray, P. Zawadzki, C. S. Chern, W. Kroll, R. A. Stall, Y. Li, M. Schurman, C.-Y. Hwang, W. E. Mayo, Y. Lu, S. J. Pearton, S. Krishnankutty, and R. M. Kolbas, *J. Electrochem. Soc.* **142**(9), L163 (1995).
231. M. Ilegems and R. Dingle, *J. Appl. Phys.* **44**, 2434 (1973).
232. J. I. Pankove and J. A. Hutchby, *J. Appl. Phys.* **47**, 5387 (1976).
233. A. Salvador, W. Kim, Ö. Aktas, A. Botchkarev, Z. Fan, and H. Morkoç, *Appl. Phys. Lett.* **69**, 2692 (1996).
234. C. R. Abernathy, J. D. MacKenzie, S. J. Pearton, and W. S. Hobson, *Appl. Phys. Lett.* **66**, 1969 (1995).
235. J. I. Pankove, E. A. Miller, and J. E. Berkeyheiser, *RCA Rev.* **32** 383 (1971).
236. H. Amano, S. Sawaki, I. Akasaki, and Y. Toyoda, *J. Cryst. Growth* **93**, 79 (1988).
237. J. D. MacKenzie, C. R. Abernathy, S. J. Pearton, U. Hommerich, X. Wu, R. N. Schwartz, R. G. Wilson, and J. M. Zavada, *Appl. Phys. Lett.* **69**, 2083 (1996).

3 MOVPE growth of nitrides

Olivier Briot

3.1 MOCVD precursors

3.1.1 Group III precursors

Finding a proper group III precursor for growth of nitrides is an easy thing, since well-suited molecules have been developed, purified, and tested for the growth of the other III–V materials. The 'simplest' trimethyl, triethyl molecules are widely available, in very high purities, but more complicated molecules were also developed for specific purposes, which can be tested to grow nitrides.

The advantage with the trimethyl and triethyl species is that their chemistry has been extensively studied, and we summarize here the main results which can be helpful for the growth of nitrides.

3.1.1.1 *Vapor pressure data*

The vapor pressure is usually given in the form $\log_{10}(P) = A - B/T$, with P in torr and T in kelvin. The A and B coefficients for the most common group III organometallic are given in Table 3.1.

Concerning gallium, it may be seen that TEGa will be easier to control than TMGa, due to its much lower vapor pressure. As an example, for a molar flow of 20 μmol/min, which is a typical value, it is necessary to use 4.8 sccm at a temperature of 0 °C for TMGa, which is a low value, tricky to control with classical mass flow controllers. With TEGa, a flow of 100 sccm, much easier to control, at a bubbler temperature of 20 °C leads to the same molar flow. TEAl is not often used, due to its extremely low vapor pressure. An alternative was proposed by Khan *et al.* [85]: the trimethylamine alane, which can help to reduce carbon contamination (this is often the case with higher order organic radicals,

Table 3.1 Vapor pressure data for common organometallics (from Epichem Ltd. product data sheets).

Organometallic	A	B	Remarks
TMAl	10.475	2780	9.7 torr @ 20 °C
TEAl	10.784	3625	0.026 torr @ 20 °C
TMGa	8.495	1825	64.5 torr @ 0 °C
TEGa	9.165	2530	3.4 torr @ 20 °C
TMIn	9.735	2830	1.2 torr @ 20 °C
TEIn	8.935	2815	0.44 torr @ 30 °C

3.1 MOCVD precursors 71

which are more stable), while having a vapor pressure of 1.23 Torr at 20 °C, a quite convenient value.

3.1.1.2 *Pyrolysis data*
There is a clear trend of increasing stability from In to Ga and to Al. The heavier organic radicals exhibit a lower bonding to the metal atom in the metalorganic molecule. This results in TEIn being the least stable compound and TMAl the most stable one.

The ethyl radicals have a much lower reactivity compared with the methyl radicals, and it has been demonstrated for the growth of GaAs and GaAlAs that a reduced carbon contamination resulted from the use of TEGa, since the highly reactive CH_3 radicals are believed to be the main source of carbon in the epilayers. Al compounds are extremely reactive with oxygen, moisture, and carbon, and this problem is particularly important with TMAl, which can even produce aluminum carbide upon decomposition.

The pyrolysis of the organometallic can be homogeneous, taking place in the gas phase, like for TMIn, with an activation energy close to 40 kcal/mole for decomposition in an H_2 ambient, leading to a complete pyrolysis around 380 °C [27]. Under hydrogen, TMGa seems to decompose homogeneously through an H radical attack. The activation energies for the removal of the first methyl group is around 60 kcal/mole while it is 35 kcal/mole for the second methyl group. From the results of Lee *et al.* [107], the pyrolysis of TMGa in H_2 is completed at 450 °C while TEGa is fully decomposed in H_2 at 350 °C [195]. Park and Pavlidis [150] studied TMGa decomposition in presence of ammonia and found an important decrease of the activation energy for the pyrolysis to 34.5 kcal/mode (1.5 eV). In this case, TMGa can be considered fully pyrolized at 575 °C (at a V/III ratio of 3300). These authors also demonstrated that increasing the V/III ratio, by increasing the ammonia molar flow, promoted the decomposition of TMGa.

The pyrolysis of TMAl is different since it occurs mostly heterogeneously, with an activation energy of 13 kcal/mode, much below the average bond strength of 66 kcal/mole [172].

The electrical properties of GaN grown using either TMGa or TEGa were compared, using DLTS measurements by Lee *et al.* [108] and Chen *et al.* [30]. They reported the identification of three different deep levels in the bandgap of GaN when using TMGa, while only one of these levels was detected in the material grown using TEGa.

3.1.2 Nitrogen precursors: ammonia versus other molecules
3.1.2.1 *What makes ammonia different?*
For the growth of nitrides, we need to supply nitrogen. In contrast with other III–V material systems, here the group V species is already gaseous under usual conditions. Unfortunately, the bond strength in the dimer molecule is extremely high, resulting in a very high thermal stability. The dissociation reaction:

$$N_2 \rightleftharpoons 2N^* \tag{3.1.1}$$

which provides very reactive nitrogen radicals, is characterized by an equilibrium constant K_1:

$$K_1 = \frac{P_{N^*}^2}{P_{N_2}} \qquad (3.1.2)$$

Using thermodynamical data taken from [15], we obtain for K_1 at 1300 K (a typical GaN growth temperature): $K_1 = 2 \times 10^{-32}$.

This indicates that the dissociation rate for N_2 will be of the order of 10^{-16}, which prevents the use of N_2 as a nitrogen precursor[1]. For the growth of other III–V materials, the usual (and successful) approach consists in using an hydride of the group V element. For nitrogen, the hydride is ammomia (NH_3), which is widely used in the chemical industry, so it is widely available, relatively low-cost, and is much less toxic compared with other group V hydrides.

The thermal decomposition of ammonia proceeds as:

$$NH_3 \rightleftharpoons \tfrac{1}{2}N_2 + \tfrac{3}{2}H_2 \qquad (3.1.3)$$

thus leading to gaseous nitrogen, which is not useful for growth, as outlined above. A simple thermodynamical analysis leads to results given in Table 3.2, where the equilibrium dissociation rate of ammonia has been calculated for a closed system containing ammonia at various pressures and temperatures, using data in [15].

From this table, we notice that the ammonia should be fully decomposed, even at low temperatures. Usually, thermodynamical equilibrium considerations apply only roughly to MOCVD, where the processes are more driven by kinetical aspects, but may be used as a first approach. Unfortunately, in the case of the decomposition of ammonia, the reaction kinetics is so slow that the equilibrium state is not reached, by far.

After Ban [13], no more than 4% of the ammonia is decomposed up to 950 °C. The reaction can be accelerated by various catalysts [59], including platinum (Pt) and tungsten (W), and to a much lesser extent, by graphite and silica [13].

In this situation, the undecomposed ammonia is in a metastable state and is highly reactive. It will contribute to the growth *through a surface decomposition mechanism* [73, 126, 199].

Table 3.2 Dissociation rate of ammonia at the thermodynamical equilibrium at atmospheric and reduced pressure.

T (°C)	$P = 1\,atm$	$P = 0.1\,atm$
400	0.98452	0.99842
1000	0.99968	0.99997

[1] There is a discrepancy in the data concerning the formation enthalpy of N radicals. Birge [21] and Gaydon [45] have determined a value of 112 kcal/mol (value used here) while Kaplan [74], Herzberg and Sponer [57] and Hagstrum [55] give a lower value of 85 kcal/mol. This leads to a K_1 value of 5×10^{-24}, which does not significantly modify the demonstration.

3.1 MOCVD precursors

On the other hand, *this departure from equilibrium will result in an important effect: GaN surface stabilization.*

As has been discussed by McChesney et al. [116], Thurmond and Logan [186], Madar et al. [117], and Karpinski et al. [76], the equilibrium pressure of N_2 over GaN is extremely high (hundreds of bars under typical MOCVD growth conditions).

This should result in the decomposition of GaN into gaseous nitrogen and gallium droplets, following the equation:

$$\text{GaN} \rightleftharpoons \text{Ga}_{\text{liq}} + \tfrac{1}{2}N_2 \tag{3.1.4}$$

However, combining eqns (3.1.3) and (3.1.4) to describe the reaction of liquid gallium in an ammonia ambient, we obtain:

$$NH_3 + \text{Ga}_{\text{liq}} \rightleftharpoons \text{GaN} + \tfrac{3}{2}H_2 \tag{3.1.5}$$

This equation describes the surface stabilization effect which occurs: the GaN surface should decompose into liquid gallium, according to eqn (3.1.4), but eqn (3.1.5) indicates that liquid gallium will react with ammonia to grow GaN. As a result, *ammonia is preventing the decomposition of the GaN surface.*

Quantitatively, we may calculate the free enthalpy change of reaction (3.1.5) by writing:

$$\Delta G = \Delta G_0 + RT \ln\left(\frac{P_{H_2}^{3/2}}{P_{NH_3}}\right) \tag{3.1.6}$$

$$\Delta G = \Delta G_0 + RT \ln(\sqrt{P}) + RT \ln\left(\frac{X_{H_2}^{3/2}}{X_{NH_3}}\right) \tag{3.1.7}$$

assuming a negligible solubility of liquid Ga in GaN and a negligible solubility of GaN in liquid Ga. In eqn (3.1.7), X_{H_2} and X_{NH_3} stand for the molar fractions of hydrogen and ammonia in the gas phase, respectively, and P is the total reactor pressure.

Using the thermodynamical data in [15], we observe that the standard reaction enthalpy ΔG_0 has a value close to -12 kcal/mole at the growth temperature (close to 1000 °C), so the reaction enthalpy change in eqn (3.1.5) is largely negative under typical MOCVD growth conditions, assuming that NH_3 is almost undecomposed. This indicates that there is a strong driving force for reaction (3.1.5), thus effectively preventing GaN decomposition. Equation (3.1.7) demonstrates a quite surprising effect: a decrease of the reactor pressure favors the surface stabilization under a nitrogen ambient.

Should ammonia decomposition occur according to the equilibrium, the partial pressure of ammonia would drop to a very low value and the driving force calculated in (3.1.6) would be negligible, leading to GaN decomposition in an NH_3 ambient.

As we have seen, ammonia will decompose on the surface. In addition to this, due to its extremely low decomposition kinetics, ammonia is mostly undecomposed, in a metastable state, at common growth temperatures. *This highly reactive state corresponds to a high chemical potential which results in an efficient stabilization of the GaN surface, preventing its decomposition.* We believe that this behavior is not so common, and will not be encountered in many other nitrogen precursors, resulting in the formation of gallium droplets if the GaN surface is not stabilized at the growth temperature. In our opinion, this distinguishes ammonia among the other precursors. It should also be noted that the surface of AlN is much more thermally stable, compared with GaN, so that we can deduce that it will be much easier to find alternative precursors for the growth of AlN.

3.1.2.2 *Alternative precursors*

One can wonder why new nitrogen precursors would be needed, since extremely good results are obtained with ammonia. Some of the motivations were:

- to increase the efficiency of the nitrogen precursor, in order to handle smaller quantities, and possibly to reduce the density of nitrogen vacancies, which were believed to be responsible for the residual donor in GaN;
- to have an efficient nitrogen source for the low-temperature growth of cubic phase GaN.

Here are the main results regarding alternative molecules for the growth of nitrides:

1. **Adducts** The major drawback in that case is that the use of a single source precursor does not allow stoichiometry control.

    Ammonia adducts and t-butylamine (t-BuNH$_2$) adducts with TEGa or TMAl have been studied by Roberts *et al.* [159] for deposition of GaN and AlN, at both atmospheric and reduced pressure. Gallium droplets were obtained when trying to grow GaN, but polcrystalline AlN could be grown. It must be noted that Andrews and Littlejohn [12] were successful in growing GaN at low temperature (600–800 °C) using triethylgallium-monoamine adduct, but the films had poor morphologies and low mobilities.

2. **Hydrazine and its derivatives (monomethylhydrazine, dimethylhydrazine)** The use of hydrazine (N$_2$H$_4$) was studied by Gaskill *et al.* [43] and Mizuta *et al.* [42, 127]. The main potential of hydrazine is that its pyrolysis should proceed by breaking the NH$_2$–NH$_2$ bond first (with an activation energy of 71 kcal/mode), thus producing highly reactive NH$_2$ radicals. Such high chemical potential species are expected to stabilize the GaN surface. Due to the low value of the bond energy, efficient pyrolysis should occur at moderate temperatures. Smooth films were obtained on GaAs substrates at temperatures close to 600–650 °C, under low V/III ratios (typically 2–50). These films had mosaic structures, and contained both cubic and hexagonal phases.

    Monomethylhydrazine [188] and dimethylhydrazine [103] were also successfully used to grow cubic GaN onto CaAs substrates at 600–650 °C.

Hydrazine (and derivatives) seems to be successful for the growth of c-GaN, but few characterization results are reported which would allow comparison with the results obtained with ammonia (see Section 3.3).

3. **Hydrazonic acid (HN_3)** Hydrazoic acid is highly reactive. Freedman and Robinson [40] have shown that it is prone to produce the highly active radical NH, with a bond energy of only 0.5 eV, compared with the bond dissociation energies of 9.8 and 4.5 eV in nitrogen and ammonia, respectively.

These authors also demonstrated that exposing GaAs to hydrazoic acid at $T = 500\,°C$ leads to the formation of a GaN surface layer of 20 Å in thickness.

Bu *et al.* [26] have succeeded a GaN film onto a sapphire substrate using HN_3 at an optimal growth temperature of 600 °C. The V/III molar ratios used were ranging from 1 to 5, much below the values reported for NH_3. Due to its potential to produce nitrogen-based radicals with a high chemical potential, a stabilization of the GaN surface must be obtained. Moreover, the growth temperature is sufficiently low as to limit the thermally induced damages. However, the layers were slightly polycrystalline, with broad photoluminescence features.

4. **Diethylaluminum and diethylgallium azides** These molecules have been tried by Ho *et al.* [62] to grow GaN in the 400–600 °C range, but the films obtained had extremely poor crystallinity (amorphous–polycrystalline).

3.1.3 Parasitic reactions between ammonia and Group III precursors

The formation of gas phase adducts between the Group III and V precursors is a classical phenomenon which has been troublesome for MOVPE growth of various III–V materials [16, 101, 114], affecting the growth uniformity and efficiency.

The formation of an adduct arises from the fact that Group III organometallics are electron acceptors (Lewis acids), while ammonia, with a spare electron pair on the nitrogen atom, is an electron donor (or Lewis base).

Some authors have suggested that this should happen in the $TMGa/NH_3$ system [126]. Thon and Kuech [185] have investigated by mass spectroscopy the interaction between TMGa and ammonia and proposed that an adduct $[(CH_3)_2Ga:NH_2]_x$ was formed, which was stable up to 500 °C. A consequence was that the pyrolysis of TMGa was shifted towards higher temperatures in an H_2/NH_3 ambient, compared to the H_2 ambient. However, such an adduct should be fully decomposed at the high growth temperatures used for the growth of GaN.

Chen *et al.* [29] made a phenomenological analysis of the formation of adducts in the growth of GaN and AlN using the concept of reactor efficiency. The reactor efficiency is defined as the ratio between the growth rate and the molar flow rate of the group III element, and is expressed in micrometers per mole of reactant (μm/mole). Although this parameter is not of universal meaning, since it is linked to the reactor design, it is useful to compare, in a given reactor, the growth efficiency for various precursors. Chen *et al.* mention as a yardstick that an efficiency of several thousands of μm/mole indicated minimal parasitic reactions, while a few hundreds of micrometers per mole indicate severe premature reactions.

In their experiments, they used a horizontal, rectangular reactor cell, and grew GaN and AlN epilayers, using TMGa and TEGa, with typical MOCVD growth conditions. They determined the growth efficiency for GaN and AlN, at both atmospheric and low pressure (85 Torr).

Their conclusions were as follows: the GaN efficiency is almost constant (about 1000 μm/mole) versus reactor pressure between 85 Torr and atmospheric pressure, and is constant in the temperature range where the growth rate is diffusion limited. This indicates that negligible parasitic reactions occur between ammonia and TMGa. On the other hand, the AlN growth efficiency drastically decreases with increasing reactor pressure, and decreases at high growth temperatures. This indicates that strong premature reactions occurs between TMAl and NH_3. This is not connected to a possible reaction between TMAl and the residual moisture content in the ammonia (see Section 3.3.6.1), since an H_2O content of 10 ppm in NH_3 (much above the real value) would only consume a small fraction of the TMAl, under the V/III molar ratios commonly used.

This has a strong effect on the growth of $Al_xGa_{1-x}N$ alloys, since Al will only be efficiently incorporated if the contact time between TMAl and ammonia is minimal (either by using low pressure growth, or a special inlet nozzle to increase the linear velocity of the reactants).

The same group [28] also used TEGa for GaN growth and TMIn for GaInN growth, and concluded, using the same approach, that parasitic reactions between TEGa or TMIn and ammonia are not significant.

3.1.4 Dopant precursors

3.1.4.1 *n-type doping*

We will see in Section 3.5.1 that silicon and germanium are the most convenient dopants for the growth of n-type GaN. The electronics industry makes extensive use of their hydrides, silane (SiH_4) and germane (GeH_4), so they are available with a very high purity (electronic grade).

Since the incorporation efficiency of silicon and germanium in GaN is very high, we will need to use them in a dilute mixture with hydrogen (typically 1000 ppm in H_2). However, such dilute mixtures are not highly stable, since silane or germane will adsorb in the bottle and/or react with it within a few months, resulting in a concentration change.

Since molar flow rates of the order of the nmole/min (10^{-9} mole/min) are required, this would correspond to a mixture flow of 2.2×10^{-2} sccm. Of course, this cannot be directly realized and a dilution line is needed for silane and germane. All this is commonly used for the n-type doping of classical III–V materials. Similar good results were obtained with disilane (Si_2H_6) [160]. On the other hand, an organometallic n-type dopant was recently employed [147]:

tetraethylsilane (TeESi), which proved to be also an efficient precursor. Unfortunately, this precursor seems to have a very high vapor pressure, since it had to be cooled down to $-77\,°C$ during the doping experiments.

3.1.4.2 p-type doping

The most commonly used p-type dopant for GaN is magnesium. Two organometallic compounds have been used in MOCVD for magnesium: bis-cyclopentadienyl magnesium (Cp_2Mg) and bis-methylcyclopentadienyl magnesium (MCp_2Mg). Up to now, Cp_2Mg has been the main precursor for the growth of p-type GaN: although it is solid (melting point of $176\,°C$), it has a higher vapor pressure than MCp_2Mg. Cp_2Mg vapor pressure is described by the equation [109]:

$$\log_{10}(P) = 25.14 - \frac{4198}{T} - 2.18\ln(T) \qquad (3.1.8)$$

with P in torr and T in kelvin. As an example, this corresponds to 0.027 torr at $20\,°C$. According to Ohuchi *et al.* [147], Bis-ethylcyclopentadienyl magnesium ECp_2Mg could be a convenient alternative to Cp_2Mg, since it is liquid at room temperature and seems to have a higher vapor pressure (unfortunately not reported by Ohuchi *et al.*). These authors used ECp_2Mg at $25\,°C$, with hydrogen flows around 100 sccm to achieve Mg concentrations of $10^{20}\,cm^{-3}$ in GaN.

3.2 Direct MOCVD growth of GaN on sapphire substrate

3.2.1 Growth mechanisms

The MOCVD technique was born at the end of the 1960s with the work of Manasevit [118, 120]. Soon after, Manasevit himself applied the technique to the growth of GaN onto sapphire substrates [119]. The choice of sapphire substrate results from the facts and that (i) no bulk GaN substrates are available and (ii) a very stable material is required to deal with the high temperatures required for GaN growth (around $1000\,°C$).

Although different orientations have been tried, the most common to date is (0001). The epitaxial relationship in that case is such that:

$$(0001)GaN//(0001)sapphire \text{ and } (01\bar{1}0)GaN//(\bar{2}110)sapphire$$

corresponding to a $30°$ rotation of the crystal axis of GaN relatively to those of sapphire in the basal plane. In this case, the apparent lattice mismatch in the (0001) plane will be 13.8%, which is still huge.

The growth rate in MOCVD can be controlled by two types of mechanism [173]:

1. By surface reactions, corresponding to the chemistry of the crystal growth. In that case, the growth rate will be exponentially dependent on the growth

temperature (thermal activation), this behavior being characterized by a parameter, the activation energy, which reflects the limiting step in the chain of reactions occurring during the crystal growth. Then the growth rate will be very sensitive to the surface; in particular, different substrater orientations should lead to different growth rates in similar growth conditions.
2. In the high temperature range, provided the crystal does not decompose, the chemical reactions will be fast enough and growth will be limited by the slow supply of reactants to the interface. Since reactant mass transport occurs in MOCVD by diffusion of the species through the carrier gas, the resulting behavior is termed the diffusion limited case. The growth rate is then (almost) independent of the growth temperature and of the substrate orientation.

Although both regimes can usually be achieved by changing the growth temperature, the optimum value of the growth temperature is dictated by the bond strength in the crystal, its thermal stability, the precursor chemistry, and so on.

As outlined in Section 3.1.1, the most convenient precursors for the growth of GaN are TMGa or TEGa and ammonia (NH_3). Chen *et al.* [29] have measured the growth efficiency (see Section 3.1.1) in their reactor for the (TMGa + NH_3) and (TMAl + NH_3) systems at atmospheric and reduced pressure, and observed the two regimes described above for GaN, while the behavior for AlN growth is complicated by the occurrence of premature reactions. In Fig. 3.1, we report the growth rate of GaN, using TEGa and ammonia as precursors, at a reactor

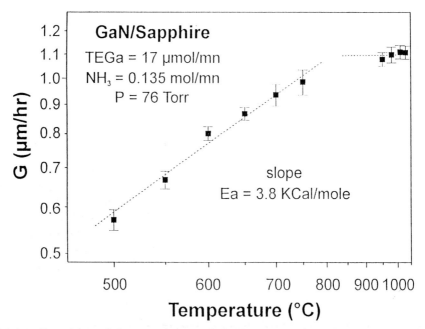

Fig. 3.1 Logarithm of the growth rate of GaN versus reciprocal temperature. Two regimes are evidenced: the kinetic mode at low growth temperature and the diffusion limited mode at high temperature.

pressure of 76 torr. The kinetically limited case is visible at low growth temperature, while the growth rate is diffusion limited above 850 °C.

Moreover, Sasaki and Zembutsu [163] reported a linear variation of the growth rate with the gallium molar flow, which was independent of the substrate orientation ((0001) and (01$\bar{1}$2)), at $T = 900$–950 °C. These results clearly demonstrate that the growth of GaN is controlled by the diffusion of gallium at usual (900–1050 °C) growth temperatures.

3.2.2 Discussion of characterization results and position of the problem

From the discussion above, the growth of GaN onto sapphire appears to proceed like more conventional III–V materials. However, the huge lattice mismatch between GaN and sapphire, as well as the chemical dissimilarities between both materials, result in an appalling layer quality:

- in X-ray diffraction, the full width at half maximum of the rocking curves is in the range 15–30 arcmin [163, 179].
- The residual electron concentrations are extremely high (10^{19}–10^{20} cm^{-3}), with poor mobilities (10–50 cm^2/Vs).
- The morphologies are rough, from cauliflower-like to large hexagonal pyramids [77], with the film not always being continuous on the whole substrate surface. Even the color of the film may change from transparent to whitish to yellow/brown, evidencing the presence of deep levels in the bandgap.

It appears that all the problems result from the fact that the growth is three-dimensional, the film being constituted of large islands, which may coalesce or not. The shape of the islands changes with the growth conditions, as well as their density.

This can be ascribed to a 'wetting' problem between GaN and sapphire: due to both the lattice mismatch and the surface chemical differences, the contact angle between the heteroepitaxial nuclei and the substrate is finite, leading to the formation of islands, in a mode termed the 'Volmer–Weber' island growth mode.

We will see, Section 3.3, that this problem can be drastically reduced by the use of a buffer layer which improves the wetting between the surface and the material to be grown. However, the problem is only reduced, not eradicated, by the use of a buffer layer, and it is interesting to study the direct growth first, since the discussion will still be qualitatively valid for the growth onto a buffer layer.

3.2.3 Morphology control: influence of the growth conditions

In the situation described above, the morphology will be mainly influenced by two factors:

- The density of initial nucleation, which will set the average distance between nuclei. The shorter this distance, the sooner the coalescence of islands will occur, leading to a continuous film.

- The ratio between lateral growth and vertical growth. The coalescence of islands will occur faster if the lateral growth is fast, compared with the on-axis growth. The amount of lateral growth will be directly reflected by the shape of the islands (more or lest isotropic or flat).

3.2.3.1 Control of the density of initial nucleation

The first parameter affecting it is the growth temperature. At higher growth temperatures, the desorption of adatoms is enhanced. Thus, the nucleation density decreases with increasing growth temperature.

The gallium precursor partial pressure controls the growth rate. Higher Ga partial pressures will result in higher nucleation densities.

The effect of the carrier gas is also extremely important and has been established experimentally by Kawabata et al. [77], who noticed that the nucleation was enhanced when nitrogen was used as carrier gas instead of hydrogen. Sasaki [161] demonstrated that the hydrogen carrier gas flow could precisely control the density of nucleation, and his findings are reported in Fig. 3.2: in this figure, the effect of an increasing flow of hydrogen on the morphology is reported for atmospheric and low pressure, and two different distances (labeled d in Fig. 3.2) between the inlet nozzle and the substrate.

In our opinion, the effect of an increasing distance between nozzle and susceptor results in a lower gas velocity at the susceptor level (due to the spreading of the gas flow), increasing the boundary layer thickness and decreasing the Ga molar flow rate at the surface. So this should have, qualitatively, the same effect as a variation of the Ga molar flow. The notations are NFD: No Film Deposition, DIG: Discontinuous Island Growth and CFG: Continuous Film Growth.

In summary, Sasaki observed that:

- At low pressure, increasing the H_2 flow decreases the initial nucleation (less coalescence and even no film growth).

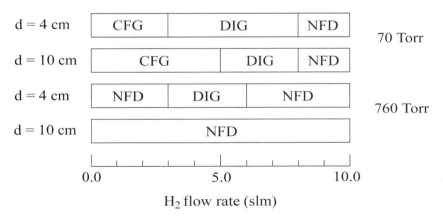

Fig. 3.2 Effect of the H_2 carrier gas flow rate on the GaN epilayers morphology, from [161].

- At atmospheric pressure, the problem is worse, no continuous film can be obtained, reflecting a lower nucleation density at higher pressure.

Our interpretation of this effect of the hydrogen carrier gas differ from Sasaki's one: we think that the nuclei are 'etched' by the hydrogen. A simple thermodynamical analysis does not favor the direct etch of GaN by H_2, following the reaction:

$$GaN + \tfrac{3}{2}H_2 \rightleftharpoons Ga + NH_3 \tag{3.2.1}$$

The formation of highly reactive monoatomic hydrogen radicals at the growth temperature is also apparently not quantitatively sufficient, so we turn to the hypothesis of the formation of volatile gallium hydrides like GaH or GaH_2 radicals, or GaH_3. Simple thermodynamic calculations demonstrate that such species may have substantial partial pressures at the high growth temperatures required for GaN.

3.2.3.2 Control of the lateral growth

Concerning the island shapes, two shapes are commoly observed: the Hexagonal Pyramid (HP) and the Flat-Top Pyramid (FTP). The flat-top pyramids correspond to a higher lateral growth rate, compared with the HP. Such a shape favors a rapid coalescence and a flatter morphology.

To understand the growth conditions which enhance the lateral growth, we will make use of a criteria, known as the Peclet number (Pe) [187], and defined as:

$$Pe = \frac{L^2 j}{D(T)} = \frac{L^2/D(T)}{1/j} = \frac{\tau_{\text{diffusion}}}{\tau_{\text{arrival}}} \tag{3.2.2}$$

L is the mean distance between steps on the surface, $D(T)$ is the surface diffusion coefficient, j is the arrival flux of species on the surface (atom/seconds). The Peclet number can be seen as the ratio between the diffusion time to the edge of a step on the surface and the arrival time between atoms.

If $Pe \ll 1$, the surface diffusion dominates and is fast compared to the arrival rate of the atoms, the growth occurs by propagation of the steps (classical 'step-flow epitaxy'), or in our case here, by lateral growth of the initial nuclei.

If $Pe > 1$, the surface mobility is low and the on-axis growth will be important (the atoms arriving on top of an island will stick there, while desorption is important between the islands, as a result of the low bonding between the sapphire and the arriving species).

This interpretation of the growth modes corresponding to the different values of the Peclet number is slightly different from the standard case, due to the particular 'wetting' problem which exists between the sapphire surface and the growing layer.

We notice that the condition $Pe \ll 1$ (enhancement of the lateral growth) corresponds to a low value of j (low Ga molar flow) or to a high value of $D(T)$ (high growth temperature).

Table 3.3 Effect of growth conditions on GaN epilayers morphologies.

Nucleation → ↓ Lateral growth	Low (high Tg, Low Ga)	High (low Tg, high Ga)
Low (low Tg, high Ga)	Discontinuous islands Hexagonal pyramids	Continuous films Hexagonal pyramids
High (high Tg, low Ga)	Discontinuous islands Flat-top pyramids	Continuous films Flat-top pyramids

If we now exclude the carrier gas problem, and consider only the influence of the growth temperature and the Ga molar flow rate, we may summarize our conclusions in a table (Table 3.3).

As one can see, the conditions for improving the initial nucleation and for improving the lateral growth are opposite. It will not be possible to find growth conditions that allow the two dimensional growth of perfectly smooth films in the case of direct growth onto sapphire.

3.3 Growth of GaN on sapphire substrates using a buffer layer

From the discussion above, it is obvious that the material quality is determined mostly by the beginning of the growth on the sapphire substrate. In 1983, in order to overcome the nucleation problems of GaN, Yoshida et al. [197] deposited an AlN buffer layer on the sapphire substrate prior to GaN growth. This technique appeared to be very effective in improving the overall material quality, so it was adapted to the MOCVD growth in 1986 by Amano et al. [10] and then extensively studied by the same group [3, 5, 94]. Later, it was demonstrated by Nakamura [130] that GaN could be employed with similar success for the realization of the buffer layer.

To summarize the effect of the introduction of a low-temperature buffer layer on the GaN epilayer quality, we will mention typical results:

- The crystalline quality, as probed by X-ray diffraction peaks width, improves from about 8 arcmin (MOCVD, no buffer) down to 1.9 arcmin (MOCVD AlN buffer) [10]. Similar results are obtained (1.6 arcmin) using GaN buffers [130]. These results should be compared with the results obtained by the VPE growth method [171]: typically 12 arcmin.
- The electrical properties of the GaN layers are greatly enhanced: the residual electron concentration, which was usually in the 10^{18}–10^{20} cm^{-3} range is decreased to the mid-10^{16} cm^{-3} [130]. This strongly suggests a correlation between the residual donor and the crystalline defects, either the donor being itself an intrinsic defect (nitrogen vacancies are often mentioned as a possible candidate) or because the growth conditions related to the creation of vacancies increase the probability of extrinsic impurities incorporation.

3.3 Growth of GaN on sapphire substrates using a buffer layer

The mobility in the layers is also consistently improved, a value of 600 cm^2/Vs at room temperature is mentioned by Nakamura [130].

3.3.1 Description of the process and role of the buffer layer

3.3.1.1 Description of the MOCVD growth process

Although Sasaki and Matsuoka [162] described the growth process as the double step process in order to emphasize the importance of the buffer layer, it is admitted now that other steps of the growth process play a key role in the obtention of high-quality material, like the substrate surface pre-treatment, prior to the buffer layer deposition and the buffer heat treatment, following its growth. In order to clarify the different steps, a typical sketch of the growth process for a GaN layer deposited onto a sapphire substrate is depicted in Fig. 3.3.

We will describe the growth process and identify the main growth parameters:

- Substrate pre-treatment: the substrate is heated at elevated temperatures: (above 1000 °C) to reorganize and improve its surface. Some groups perform this under a hydrogen ambient while many groups have reported improved results by using an NH$_3$ ambient to perform a nitridation of the Al$_2$O$_3$ surface. The parameters here are the substrate temperature, the nature of the ambient (gas flow, pressure) and the treatment time.

- Buffer layer deposition: after the pre-treatment, the substrate is cooled to a temperature ranging between 500 and 800 °C, at which the deposition of the GaN or AlN buffer occurs. In addition to the temperature, the important

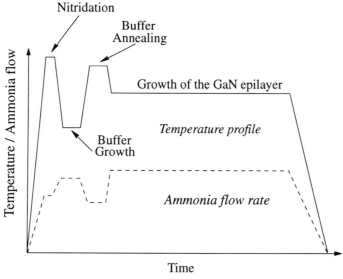

Fig. 3.3 Sketch of the double-step process: temperature and precursors flow versus time.

84 *MOVPE growth of nitrides*

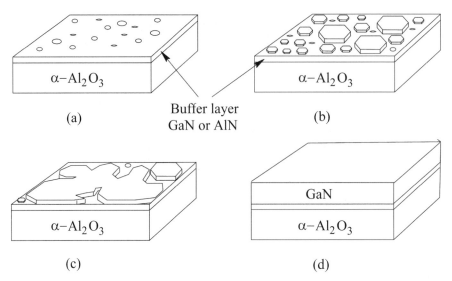

Fig. 3.4 Mechanism of hexagonal islands coalescence, leading to 2D growth for GaN grown on a buffer layer.

parameters are the buffer thickness, the precursors flow (controlling the growth rate), the V/III molar ratio.
- Buffer layer heat treatment: since the crystalline quality of the buffer layer is very poor, a 'recrystallization' heat treatment is performed. It consists in ramping up the temperature to an elevated value (at least equal to the GaN growth temperature and often above) for a given amount of time, under a given ammonia flow to stabilize the surface.
- Growth of the 'main' GaN layer: the GaN layer is grown. The parameters are the growth temperature, the growth rate, the V/III molar ratio, and the epilayer thickness.

3.3.1.2 *Role of the buffer layer*

The role of the buffer layer in the MOCVD growth of GaN has been thoroughly studied by Akasaki's group for AlN buffers [3, 60] and by Nakamura for GaN buffers [131], using *in situ* infrared transmission experiments. Their conclusions are summarized in Fig. 3.4.

When the main GaN layer is grown onto a buffer, due to the reduced interfacial energy between GaN and its buffer (compared to GaN/Al$_2$O$_3$), a high density of nucleation is obtained: Fig. 3.4(a). Then the nuclei grow to small hexagonal islands, as in the case of the direct growth of GaN on sapphire. But, at an early stage of the island growth, the lateral growth mode is enhanced (again as a result of the reduced interfacial energy between GaN and the buffer layer). The islands expand laterally and coalescence occurs rapidly (Fig. 3.4(b) and (c)). This can be seen in Fig 3.5, which displays an SEM picture, taken during this coalescence period. Up to this stage, RHEED experiments peformed

3.3 Growth of GaN on sapphire substrates using a buffer layer

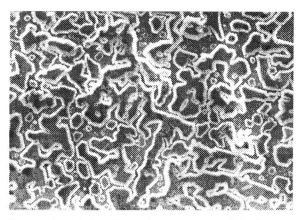

Fig. 3.5 SEM picture taken during the coalescence stage of the hexagonal GaN islands.

on the growing layers exhibit spotty patterns, indicating a three-dimensional growth. Finally, the whole substrate area is covered by the GaN film, a flat surface can be observed in SEM experiments, and the RHEED becomes streaky, indicating a flattened surface (Fig. 3.4(d)).

After Hiramatsu *et al.* [60], TEM experiments reveal that the AlN buffer layer is crystallized by solid phase epitaxy during the heat treatment into a highly oriented, columnar structure. GaN nuclei would grow on top of such columns and the high-density nucleation of GaN is directly linked to the high density of the AlN columns. These results were obtained by examining the defects structure versus distance from the AlN buffer in the layer. In the samples studied by Hiramatsu *et al.*, it took about 3000 Å before flat surfaces were obtained. We may expect that higher densities of GaN nucleation will cause the island coalescence to occur sooner, and that this density of nucleation will depend upon the structure of the buffer, which is greatly affected by the buffer growth conditions. It may thus be anticipated that the buffer growth conditions will have a drastic effect on the GaN epilayer quality.

3.3.2 Effect of the substrate pre-treatment (nitridation)

Most of the studies in the literature mention the use of a substrate heat treatment prior to the buffer growth. There are currently two approaches: the first consists in heating the substrate in the stream of hydrogen, while the second consists in heating the substrate under ammonia, in order to perform a nitridation of the surface. However, the influence of such treatments has been carefully studied only recently by different groups and the question is still open. Akasaki's group [3, 5, 10, 60, 94] and Nakamura [130, 134] reported excellent results without nitridation.

However, we have observed that substrate nitridation could improve the optical quality of the layers, as demonstrated in Fig. 3.6, which displays the 2 K photoluminescence spectra of GaN layers grown on GaN buffers [23, 25] for

86 *MOVPE growth of nitrides*

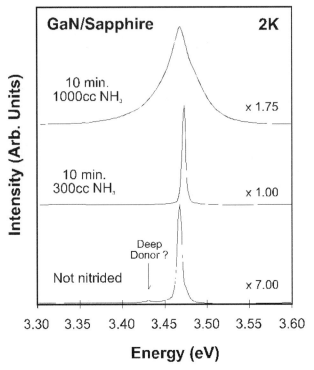

Fig. 3.6 2K photoluminescence of GaN layers grown in the same conditions, except for the nitridation. Top: 1000 sccm NH_3; center: 300 sccm NH_3; and bottom: no nitridation.

different nitridation treatments in MOCVD growth (different NH_3 flows at 1070 °C for 10 minutes). It may be seen that the layer grown onto a substrate nitrided for 10 minutes under an NH_3 flow of 300 sccm gives a much higher PL intensity ($\times 7$) compared to the un-nitrided layer.

This demonstrates that the un-nitrided layer contains a much higher density of non-radiative recombination centers, compared to the nitrided layer. The curve at the top of Fig. 3.6 also demonstrates that the nitridation has an optimum value: too much degrades the layer quality. Results published by Hwang et al. [66] also demonstrate that the electrical properties of GaN layers grown by MOCVD are enhanced by a nitridation of the substrate, compared to a non-nitrided sapphire surface.

Results obtained in MBE by Grandjean et al. [50, 51] help in understanding the mechanisms of the nitridation. These authors have observed by RHEED experiments that the surface lattice parameter changed during nitridation. Its mismatch with sapphire saturated within 10 minutes to a value of 13%, corresponding to the mismatch between AlN and sapphire. This would indicate that a very thin layer of crystalline AlN is formed on the surface during the nitridation. They also observed by AFM that the growth mode of a 600 Å GaN layer deposited on an un-nitrided substrate was three-dimensional, while a quite flat surface was obtained on nitrided sapphire.

3.3 Growth of GaN on sapphire substrates using a buffer layer

Table 3.4 Structural characterization of layers with different nitridation time (3 slm ammonia at 1050 °C). (After Keller *et al.* [82].)

Nitridation	(002) peak width (FWHM arcsec)	(102) peak width (FWHM arcsec)	Dislocation density (cm^{-2})
60 s	269	413	4×10^8
400 s	40	740	2×10^{10}

The situation is not so clear in MOCVD, since Uchida *et al.* [189] observed the formation of an amorphous Al–O–N layer by TEM due to the nitridation process. This intermediate layer was very smooth and allowed the two-dimensional growth of a GaN layer. They also observed that an overexposure of the substrate to NH_3 led to the appearance of protrusions, in densities increasing with the ammonia exposure time, which finally resulted in layers with poor morphologies. This last observation is in good agreement with out own results. Keller *et al.* [82] compared samples nitrided with 3 slm of ammonia at 1050 °C for 60 and 400 seconds. Then a GaN buffer (190 Å) was grown at 600 °C and a 1.2 µm thick GaN layer was grown on top at 1080 °C. Their results regarding the structural characterization of the layers are reported in table 3.4.

From these results, we observe that the nitridation apparently improves the X-ray full width at half maximum (FWHM) for the symmetric (002) reflections, but this corresponds to an artifact, discussed in [58]. This is evidenced by the dislocations densities and the FWHM of asymmetric reflections, like (102). Such results may explain the extremely narrow diffraction peaks which have been reported in the literature by some groups [38, 39, 63, 100].

In their paper, Keller *et al.* [82] report optical and electrical characterization results which clearly demonstrate that material with the lower dislocation densities have the better electronic properties.

It appears that more work is required to clarify the exact role of substrate nitridation and the mechanisms involved. The situation is complicated by the fact that a competition may exist in MOCVD at the beginning of the growth, between layer growth and substrate nitridation reactions, when both ammonia and Ga or Al precursors are flown into the reactor. However, there is strong evidence that a controlled nitridation step effectively improves the material properties.

3.3.3 Effect of the buffer layer growth parameters

3.3.3.1 *Growth temperature*

The buffer itself is deposited at low temperature, where uniform high nucleation densities are obtained as a result of higher supersaturation of the growth species above the substrate and because surface mobilities of the adatoms are much lower. Of course these low surface mobilities result in poor organization of the growing crystal, i.e. poor crystal quality. As a consequence, we understand that the growth temperature of the buffer will have a drastic influence and that it will be necessary to find a compromise between good surface coverage and sufficient crystalline quality.

Surprisingly, the influence of this parameter has not been investigated thoroughly and the results in the literature show some discrepancies, mainly concerning the use of AlN buffer layers. In [3] and [44], the X-ray diffraction peak widths of the GaN epilayer increase (from about 2–3 arcmin to 10 arcmin) when the growth temperature of the AlN buffer is increased (in the 450–800°C range). Scholtz et al. [167] reports that the crystalline quality the optical and electrical properties of GaN epilayers deposited on AlN buffers are optimum for an AlN deposition temperature of 800°C, and degrade abruptly below this temperature. These differences could possibly be due to the fact that the authors are comparing buffer layers of different thickness, since the growth rate of AlN is very dependent on growth temperature in this range. Our own results demonstrate that high-quality GaN can be grown on 500 Å AlN buffer layers deposited at 800°C.

Concerning GaN buffers, the situation is clearer: most of the studies [23, 25, 37, 44, 104, 130, 190, 191] mention an optimum buffer deposition temperature between 500 and 600°C.

In fact, the evolution of the buffer layer structure, as deposited, versus deposition temperature remains largely unknown and further studies are necessary to clarify this point.

3.3.3.2 Buffer layer thickness

This point has been more studied, and the results show a good agreement. The situation is different for GaN and AlN buffers.

Concerning GaN buffers, since the initial results of Nakamura [130] concerning the influence of the GaN buffer thickness on the electrical properties of GaN layer, , other teams (for example [20, 44, 104, 184]) reported that buffer thicknesses in the range 200–250 Å were necessary to achieve high mobilities and low carrier concentrations.

The same result is obtained regarding the crystalline quality of GaN deposited onto GaN buffers. Kapolnek et al. [75] reported excellent structural results (dislocation densities in the $10^8 \, \text{cm}^{-2}$ range) for an optimized buffer layer thickness of 190 Å.

Our own results concerning the evolution of the GaN epilayer morphology with the GaN buffer layer thickness are depicted in Fig. 3.7: when the buffer thickness is 150 Å, the morphology is dominated by large pyramids (about 40 μm in diameter). For a buffer thickness of 200 Å, the pyramids now have a flat top, indicating an enhancement of the lateral growth. When the buffer layer is 250 Å in thickness, the layer is mirror-like, and a rough, grainy morphology is observed for a buffer layer of 300 Å.

So, we can conlcude that for GaN buffers, all the GaN layer properties are optimized for a buffer thickness around 200–250 Å.

The situation is less evident concerning AlN buffers. It seems that the buffer thickness is not so critical in this case. Except in the paper of Kuznia et al. [104], where the best electrical properties are obtained for a precise buffer thickness

3.3 Growth of GaN on sapphire substrates using a buffer layer

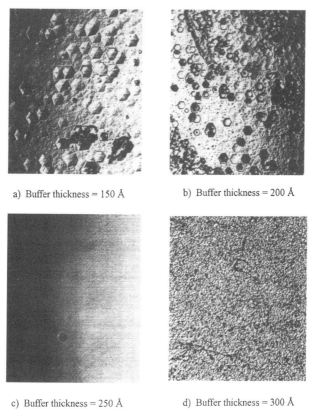

a) Buffer thickness = 150 Å
b) Buffer thickness = 200 Å
c) Buffer thickness = 250 Å
d) Buffer thickness = 300 Å

Fig. 3.7 GaN surface morphology versus GaN buffer layer thickness.

of 500 Å, excellent results are reported in the literature for AlN buffer layers in the range 300–1000 Å.

3.3.3.3 V/III molar ratio during buffer layer growth

We have observed [25] that the optical properties of the GaN layer could be influenced by the V/III molar ratio used during the buffer growth. As a matter of fact, it is not obvious that the same ammonia flow should be used during the GaN buffer growth and during the main GaN layer growth, since the incorporation efficiency of nitrogen is certainly different at the buffer and epilayer growth temperatures. So, we grew two different layers in the same growth conditions (reported in [25]), except for the V/III molar ratio used for the buffer growth, which was changed from 4000 to 6000 by changing the ammonia flow (Fig. 3.8). We observed that the 2 K photoluminescence intensity was improved by a factor of 30 when increasing the V/III ratio from 4000 to 6000. We attribute this to a reduction of the defect density in the buffer layer, due to the fact that the V/III molar ratio controls the stoichiometry of the growing layer. This, in turn, results in a lower defect density, and less non-radiative recombination centers in the GaN layer. From Fig. 3.8, it may seem that the lineshape of the photoluminescence is not affected, only the intensity.

90 *MOVPE growth of nitrides*

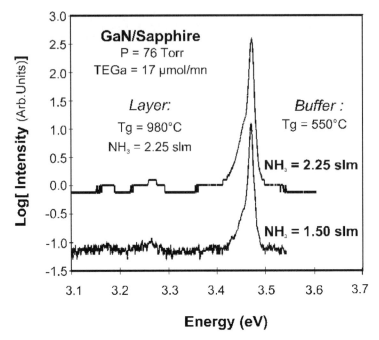

Fig. 3.8 2K photoluminescence versus buffer V/III molar ratio. The lineshape is the same but the intensities differ greatly.

3.3.4 Effect of the buffer layer heat treatment

Immediately following its growth, the buffer is heat treated. This heat treatment consists of an anneal, which, in its minimal version, occurs while ramping up to the GaN growth temperature (around 1000 °C). It is also possible to maintain the buffer layer at the growth temperature for some time, before effectively starting to grow the GaN layer, or even to heat the buffer layer at a higher temperature than the GaN growth temperature.

Before discussing the annealing effect of the buffer layer, we will try to summarize the elements concerning the structure of the buffer as-deposited.

It was first thought that the buffer layer was amorphous (for example [60]), because of the very low deposition temperature used. In some early papers, it was already suspected that the buffer could be partly crystalline (see [3]). In 1993, Wickenden *et al.* [191] demonstrated by X-ray measurements that the buffers were highly textured. In a more recent paper, Kapolnek *et al.* [75] observed by AFM that the GaN nucleation layers were constituted of densely packed uniform islands, and demonstrated by RHEED experiments that the islands were predominantly cubic phase GaN with the [111] direction normal to the substrate (0001) surface.

This work was further developed by Wu *et al.* [194], who found that the GaN buffer was predominantly cubic, with a 1–3° in-plane mosaic spread. Their TEM

3.3 Growth of GaN on sapphire substrates using a buffer layer

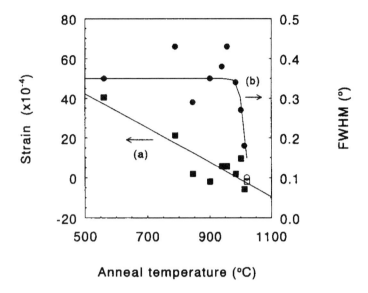

Fig. 3.9 X-ray diffraction for buffer layers annealed in various conditions [190].

observations revealed that it was highly faulted, with faceted grain morphology, the grain size changing from 25 to 33 nm when the buffer thickness was increased from 20 to 200 Å. In the same time, the buffer rms roughness evolved from 3.11 nm to 6.18 nm.

The effect of the annealing on the buffer layer structure was studied by Wickenden *et al.* [190, 193], who studied by transmittance experiments the appearance of the absorption edge corresponding to crystalline GaN, in the 360–365 nm region.

A first analysis versus annealing temperature (Figure 2 of their paper) reveals that the absorption edge appears for annealing temperature above 950 °C. A quantitative analysis of these data led them to calculate an activation energy of 0.9 eV for the crystallization of the GaN buffer layer, which compares to typical activation energies for steady state self-diffusion in alloys. They also measured by X-ray diffraction a lattice parameter which revealed a tensile strain situation (see Fig. 3.9). This strain decreased with increasing annealing temperature, leading to a fully relaxed lattice parameter at an annealing temperature of 900 °C.

The evolution versus annealing time reveals that most of the crystallization occurs during the initial temperature ramp. Long annealing times (above 30 minutes) even degrade the buffer quality.

Wu *et al.* [194] performed annealing experiments at 1080 °C on the GaN nucleation layers described above. They observe that the grains increase in average size upon annealing from 33 nm to 77 nm. PHEED indicates that the grains are now composed of equal proportions of c-GaN and h-GaN, with the

92 MOVPE growth of nitrides

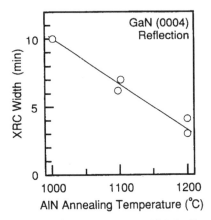

Fig. 3.10 X-ray linewidths for GaN grown on AlN buffers annealed at different temperatures [162].

hexagonal phase predominant near the free surface of the grains. The grains have lost their facets and appear rounded. Such evolution of the grain size corresponds to a mechanism similar to *Ostwald ripening*: the balance between surface and volume energy leads to stability above a given critical radius: the smaller nuclei will dissolve, feeding the bigger ones by diffusion. As a result, the mean grain size increases with time.

To our knowledge, there are no similar results concerning AlN buffers, and further work is required in order to compare with the results obtained for GaN buffers. We will reproduce here the results of Sasaki and Matsuoka [162] in Fig. 3.10, who studied the influence of the AlN buffer annealing temperature on the X-ray diffraction width of the GaN (0004) peak. We observe an improvement when the annealing temperature is increased to 1200 °C. This is a much higher value than those usually reported, especially for the annealing of GaN buffers, and is probably related to the stronger bond strength in AlN compared with GaN.

3.3.5 Influence of the 'main' GaN layer growth parameters

3.3.5.1 *Growth temperature*

GaN can be grown in a wide range of temperatures, the lowest values being reported by Dissanayake *et al.* [35], T_g = 400–650 °C, for epitaxial material grown by MOCVD using TEGa and ammonia. The material obtained at these temperatures displays very poor electronic properties, the photoluminescence at low temperature being a broad band of 1 eV in width for GaN grown at 500 °C.

But usually, device grade material is grown at temperatures close to 1000 °C. When comparing the available data in the literature, it appears that there is a slight difference depending upon the choice of TEGa or TMGa as the gallium precursor. In the case of TEGa, the optimum temperature is typically

3.3 Growth of GaN on sapphire substrates using a buffer layer 93

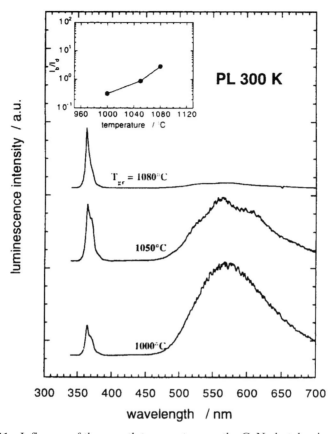

Fig. 3.11 Influence of the growth temperature on the GaN photoluminescence.

980–1000 °C, while it is a little higher (1030–1080 °C) when TMGa is used, independently of the reactor pressure. This difference cannot be explained in terms of different bond strength between the gallium atom and the organic radical (ethyl or methyl), since we explained in the beginning of this chapter that these bonds are both pyrolyzed below 400 °C. The typical effect of growth temperature on the low-temperature photoluminescence is illustrated in Fig. 3.11 (from reference [78]).

It can be seen that the optical quality of GaN is strongly affected by a temperature change of 30 °C over 1000 °C, i.e. a 3% change. The so-called 'yellow luminescence' at 550–600 nm is very sensitive to the growth temperature, and we observe that donor–acceptor pair lines also rise in the 22 K PL when the growth temperature is further lowered. This indicates that crystal defects are created when lowering the growth temperature, allowing an increased incorporation efficiency of impurities. The effect of growth temperature on the crystalline quality is depicted in Fig 3.12 (from reference [162]). These elements suggest that the influence of the growth temperature on the layer properties is more likely explained by changes of the surface mechanisms contributing to the

94 *MOVPE growth of nitrides*

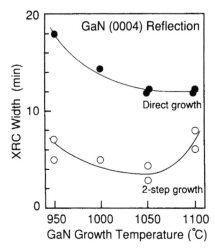

Fig. 3.12 Effect of the growth temperature on the X-ray rocking curve linewidths, after [162].

growth and organization of the crystal, rather than changes in the gas phase chemistry above the substrate. However, few is known about the growth mechanisms of GaN, which are complicated by the fact that the decomposition of ammonia occurs on the surface.

Concerning the electrical properties, a decrease of the carrier concentration versus growth temperature is most often observed for growth temperatures from 900 to 1080 °C. An increase of the electron mobility is also observed in that case, which rules out compensation to explain the observed decrease of the carrier concentration. Khan *et al*. [84] argued that this observation is not consistent with the fact that nitrogen vacancies are the dominant residual donor in GaN, since their density should increase with growth temperature.

3.3.5.3 *V/III molar ratio*

The V/III molar ratio imposed during GaN growth mainly controls the material stoichiometry, i.e. the densities of vacancies and/or interstitial atoms. When the density of vacancies of a given species increases, the probability for impurity incorporation on the given sublattice increases accordingly. The nature of the impurities which can enter the considered sublattice is usually determined by the fact that the difference of electronegativity and covalent radii should be the smallest possible. The elements prone to enter the nitrogen sublattice will be oxygen and carbon, mainly while germanium and silicon will easily substitute gallium. In addition to that, the intrinsic defects may be electrically active by themselves, the nitrogen vacancy should be a single donor, while the gallium vacancy should be a triple acceptor [141]. As a consequence, we understand that the V/III molar ratio should have a pronounced effect on the electronic properties of the material, but the interpretation of such effects in terms of the nature of the active species is particularly tricky. We will now comment on the

3.3 Growth of GaN on sapphire substrates using a buffer layer

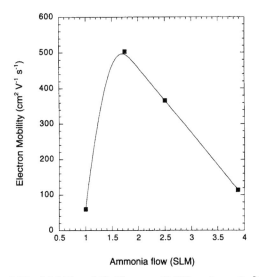

Fig. 3.13 Mobility of GaN versus V/III molar ratio [37].

influence of the V/III molar ratio on the properties of GaN epilayers, from typical results of the literature.

Concerning the effect of the V/III molar ratio on the electron concentration few data are available, but it seems that it tends to decrease monotonically when the V/III ratio is increased [78, 113]. On the other hand, the mobility exhibits a peak value when the V/III ratio increases [37, 44], as illustrated by Fig. 3.13, taken from [37]. These authors explain their results by a decrease of the gas phase reactions between TMGa and NH_3, leading to a decrease of the crystalline quality at high V/III ratios, explaining a reduced mobility although the density of donor decreases. Another possible interpretation could be that the amount of nitrogen vacancies, or donors incorporating on the nitrogen sublattice (like oxygen originating from the ammonia, see below), is decreased when increasing the V/III molar ratio, while high V/III ratios lead to the formation of Ga vacancies (acceptors) or favor the incorporation of extrinsic acceptors on the Ga site. That way, a minimum net impurity concentration is realized somewhere in between, resulting in a peak mobility.

The effect of the V/III molar ratio on the optical properties of the layers is depicted in Fig. 3.14, where 2 K photoluminescence for GaN grown at various V/III ratios (1000 to 10 000) is plotted. A drastic effect is observed for the lowest V/III ratio (1000), where the photoluminescence near the band edge is broadened, and extends to high energies, with strong donor–acceptor pair lines and their phonon replicas around 3.1–3.3 eV. At higher V/III ratios, narrow excitonic lines (donor-bound excitons) are observed. One may notice that the peak position is red shifted, a trend which is also found in the reflectivity transitions of these samples, indicating a decreasing residual strain in the layers when the V/III ratio increases. This effect is unclear at this stage, but clearly the residual strain in the epilayers have multiple origins [7, 24, 48, 183] (the

96 MOVPE growth of nitrides

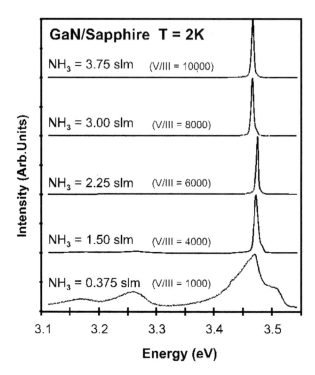

Fig. 3.14 2K photoluminescence of GaN versus V/III molar ratio.

most evident being the thermal expansion coefficient mismatch with the substrate, but not the only one, since nitridation buffer heat treatment also contribute to it).

As mentioned in the beginning of this paragraph, V/III molar ratio have a great effect on the incorporation of impurities. This has been demonstrated for the incorporation of carbon by Niebuhr et al. [143]. This is reported in Fig. 3.15, where the SIMS profile for carbon in GaN is plotted for different V/III molar ratios. The carbon level decreases by a factor 4 when increasing the V/III ratio from 150 to 1400, which is consistent with the fact that carbon should incorporate on the nitrogen sublattice.

3.3.6 Contamination effects in the MOCVD process

3.3.6.1 *Oxygen from ammonia*

It is known that the main residual contaminant in the ammonia is moisture and oxygen. Typical levels in electronic grade ammonia are in the ppm range. Since very high V/III molar ratios are used in the MOCVD growth of GaN (typically 1000 to 10 000), this means that the oxygen/gallium ratio in the gas phase can be as high as 10^{-3} to 10^{-2}. Depending upon the incorporation efficiency, this could easily lead to high levels of oxygen in the layers.

3.3 Growth of GaN on sapphire substrates using a buffer layer

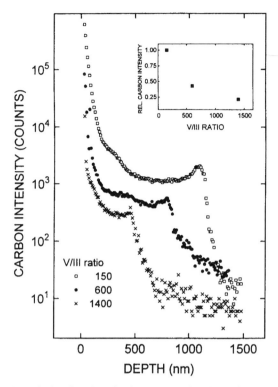

Fig. 3.15 SIMS analysis of carbon in GaN versus V/III molar ratio (from [143]).

In 1983, Seifert et al. [170] used Mg_3N_2 to remove moisture from the ammonia gas and observed a significant decrease of the residual electron concentration in GaN. Similar results on the effect H_2O/O_2 on the electrical properties of GaN were also previously reported in the halide VPE growth of GaN [22]. A detailed study of the incorporation of oxygen in GaN and on its effects on the electrical and optical properties of GaN was made by Chung and Gershenzon in 1992 [31]. These authors introduced moisture in the growth ambient by using a water bubble or, on the opposite, purified the ammonia using a liquid nitrogen trap and an Al–In–Ga eutectic bubbler. They observed that the incorporation of oxygen decreased with increasing growth temperature and demonstrated that both electrical and optical properties were adversely affected by an increasing amount of oxygen in the epilayers (increase of the residual electron concentration, broadening of the photoluminescence lines). They proposed that oxygen is acting as a relatively shallow level, its energy being 78 meV below the conduction band of GaN at 4.2 K. Oxygen and nitrogen have very similar covalent radii, so oxygen is prone to substitute nitrogen in the GaN lattice.

This study was conducted on samples grown without any buffer layer, with high-residual electron concentrations. The use of AlN or GaN buffer layers has led to material of improved crystalline quality, with lower residual electron concentrations, without the use of further ammonia purification. This could be explained by a reduction of the oxygen incorportayion efficiency in the material of improved crystalline quality, and further experiments are necessary to check whether ammonia purification will still bring improvements in the properties of GaN epilayers. It should be noted that efficient commerical purifiers based on organic resins are now widely available for the purification of ammonia, but they are still costly and have a relatively short lifetime for the usual flow conditions corresponding to the MOCVD growth of GaN (several litres per minute).

3.3.6.2 Transition metals

Important concentrations of transition metals have been systematically evidenced in MOVPE and HVPE GaN layers [17–18 19, 158]. Iron, chromium, and vanadium were detected through infrared luminescence lines corresponding to intra d-shell transitions.

The effect of such impurities on the properties of GaN is still unclear at the present stage. The possible sources for such contamination include the precursors and the stainless steel parts of the reactor. It is likely that ammonia, which is a highly corrosive chemical, can attack the stainless steel cylinder in which it is contained, thus contaminating itself with the transition metals included in the stainless steel. If this is the case, it is possible to use aluminum bottles instead of stainless steel cyclinders to reduce this contamination.

3.3.6.3 Carbon

Carbon is a classical impurity which is always present in MOCVD growth. It originates mainly from the organometallic molecules. For example, it is now well known that in the growth of GaAs by MOCVD, the use of TEGa instead of TMGa leads to epilayers with a much lower carbon content. In that case, carbon is known to be an acceptor. In the case of GaN, it has been reported [1] that p-type conductivity was observed in C-doped MBE grown material, using CCl_4 as a dopant. Due to its low difference of covalent radii with nitrogen (0.02 Å), carbon is effectively prone to incorporate on the nitrogen sublattice, where it should behave as an acceptor. However, Sato [166] made a systematic investigation of carbon incorporation, using TEGa in a low-pressure plasma enhanced MOCVD apparatus [164, 165]. In this case, due to their high chemical potential, the nitrogen radicals produced in the plasma should break the C–C bond within the organic radicals, yielding high concentrations of carbon in the layers. This is effectively observed and changing the growth conditions allows the carbon concentration in the layers to be changed by a factor of 60. Unfortunately, no p-type conduction was observed: the materials were highly resistive. A possible interpretation is that carbon is compensated by hydrogen, since the SIMS concentrations of hydrogen were correlated to the concentrations of carbon, suggesting the existence of a C–H complex within the GaN crystal. Even annealing at 900 °C under nitrogen could not turn the material p-type, like in the case of magnesium doping. The data in Sato's paper [166] suggest that carbon incorporation decreases with increasing growth temperature, and this could participate (along with other effects) in the improvement in electrical properties which is generally observed when increasing growth temperature.

3.4 Alternative substrates

Since the lattice mismatch and thermal expansion coefficient mismatch with sapphire is very large, alternate substrates have been studied by many groups. We should recall here that in addition to lattice and thermal coefficient mismatch considerations, the MOCVD growth of high-quality GaN requires the use of high temperatures, restricting the possible choice of substrates to those with a good thermal stability.

In addition to the lure of lower defect densities, alternative substrates may exhibit cleavage planes, which would make the realization of laser devices easier.

3.4.1 Homoepitaxy

The growth of GaN bulk crystals is extremely difficult, due to the huge partial pressure of nitrogen over GaN at elevated temperature [76]. A first attempt to realize the homoepitaxy was to use 'pseudo-bulk' substrate: this consisted of growing very thick epilayers (in excess of 100 μm) of GaN by VPE on ZnO coated sapphire [32]. The sapphire substrate was then removed to obtain the GaN pseudo substrate. Homoepitaxial Mg doped films were grown by MOVPE on these pseudo-bulk substrates [33]. Since the films were heavily doped with magnesium, it is difficult to assess if the method leads to improved, or even equal quality compared to the films grown onto sapphire.

Table 3.5 Lattice parameters and thermal coefficient of GaN and its substrates (after [176, 142, 121]).

Material	Lattice parameter a (Å)	Lattice parameter c (Å)	Thermal expansion coefficients (K^{-1})
GaN	3.189	5.185	$\alpha_a = 5.59 \times 10^{-6}$
			$\alpha_c = 3.17 \times 10^{-6}$
Al$_2$O$_3$	4.758	12.991	$\alpha_a = 7.5 \times 10^{-6}$
			$\alpha_c = 8.5 \times 10^{-6}$
6H–SiC	3.08	15.12	$\alpha_a = 4.2 \times 10^{-6}$
			$\alpha_c = 4.7 \times 10^{-6}$
3C–SiC	4.36	–	$\alpha = 2.7 \times 10^{-6}$
ZnO	3.252	5.213	$\alpha_a = 2.9 \times 10^{-6}$
			$\alpha_c = 4.75 \times 10^{-6}$
GaAs	5.653	–	$\alpha = 6.0 \times 10^{-6}$
Si	5.43	–	$\alpha = 3.59 \times 10^{-6}$
LiGaO$_2$	$a = 5.402, b = 6.372$	5.007	
MgAl$_2$O$_4$	Spinel structure, mismatch with GaN using (111) surface: 9–10%		

Real bulk GaN substrates are grown in the group of Porowski [157], at 1600 °C, under a nitrogen pressure of 15 kbar, and although they have a small size (few mm squared), they are sufficiently large for the deposition of homoepitaxal layers by MOCVD. Ponce et al. [155] reported such homoepitaxial growth. Their layers were of high crystalline quality (dislocation densities of 10^8cm^{-2}, while the substrates are below 10^6cm^{-2}), but the photoluminescence data was similar to that obtained on high-quality epilayers grown on sapphire. Unfortunately, electrical properties could not be measured, due to the fact that the substrate is highly conductive.

At this stage, although these results are encouraging, further experiments are required to assess whether the use of GaN substrates really makes a decisive difference over sapphire.

3.4.2 SiC substrates

SiC is potentially an ecellent substrate for GaN: it has a hexagonal structure (4H and 6H polytypes) with a low lattice mismatch (about 3%) and thermal expansion coefficient mismatch to GaN. It is extremely stable versus temperature and has an excellent thermal conductivity, which is a very important property in view of high-power applications.

The draw-backs are that SiC is a very expensive material compared with sapphire (at the time being), and obtaining clean surfaces prior to growth appears to be a major obstacle [112]. George et al. [47] have studied the initial deposition of a buffer layer on 6H–SiC. They used both AlN and GaN as buffer layers. They observed that the crystallinity of AlN buffers deposited on SiC was surprisingly poor, and was only slightly improved by annealing. Better results were obtained with as-grown GaN buffers, probably due to a higher surface mobility of Ga on SiC compared with Al. However, they observed by TEM and AFM that the GaN layer totally desorbed from the SiC surface upon annealing, which was attributed to the weak interfacial bonding. In contrast, Sun et al. [178] reported that GaN deposited on 6H–SiC at 900–1000 °C was stable under annealing at 1000 °C.

Tanaka et al. [181] have reported low dislocations density (10^8cm^{-2}) when using an ultra-thin buffer (15 Å) of AlN on 6H–SiC, while high densities of dislocations were obtained with thicker buffers. They calculated the critical thickness for AlN on 6H–Sic to be 50 Å, and attributed their results to the fact that a relaxed AlN buffer would degrade the structural quality of the GaN grown on top.

Dmitriev et al. [36] reported high-quality GaN grown by MOCVD onto 6H–SiC substrates, on the (0001) Si face. Now growth details are given in this paper, but the deposited GaN layers have low dislocation densities (10^8cm^{-2}) and good electrical properties ($= 380 \text{cm}^2/\text{Vs}$ for $n = 7 \times 10^{17} \text{cm}^{-3}$). Doped layers were also realized, as well as an LED device.

From these results, it seems that the properties of the GaN layers grown on SiC are equivalent to those of GaN grown on sapphire, in the best cases.

The results are disappointing in view of the low lattice and expansion coefficient mismatch between SiC and GaN. However, it can be argued that much less effort has been devoted to the growth of GaN on SiC, compared to the use of sapphire substrates, and the bonding properties between GaN and SiC during the initial stages of growth should be studied in order to check whether SiC will really be able to fulfill its promises.

3.4.3 ZnO substrates

The lattice mismatch between ZnO and GaN is only about 2%, both having the same wurtzite structure. Unfortunately, there are three problems that hamper the successful use of ZnO for the growth of GaN.

First, ZnO is not available commercially yet. Second, it is necessary to chemically etch ZnO, in order to obtain a smooth surface ready for the growth, but it is rather difficult. The etchants which have been reported are not fully satisfying [122, 98], the etched surface is not mirro-like [125]. And, last, but not least, the ZnO surface is not stable in a hydrogen/ammonia ambient. An important degradation of the surface occurs by reaction with the ammonia under typical MOCVD growth conditions.

However, a report of successful growth of high-quality GaN on ZnO has been made by Detchprohm *et al.* [32]. They used a sputtered ZnO buffer, deposited on sapphire. The GaN films were grown by HVPE, using ammonia and nitrogen as carrier gases.

Further investigations are required to clarify whether particular MOCVD growth conditions can be found to stabilize the ZnO surface prior to the growth of GaN.

3.4.4 GaAs substrates

GaAs substrate is not very stable at elevated temperatures, and usually requires an As overpressure to avoid degradition above 500–600 °C. It has a lattice mismatch of about 20% with cubic GaN. However, since the growth of cubic GaN has to be performed at low temperatures, below 650 °C [64], GaAs may be a good candidate for the growth of sphalerite GaN. Similarly to the 'GaN on sapphire' process, a low-temperature buffer layer was used [72, 105], and substrate surface nitridation [41] also improved the results. Although predominantly cubic material can be grown, it appears that a residue of the stable hexagonal phase is always present in the films [102]. Electronic properties are affected by the high densities of structural defects; we may cite here the results of Kuznia *et al.* [105], who measured a Hall mobility of 20 $\mu m^2/Vs$ for a carrier concentration of 4×10^{17} cm^{-3}, a value significantly lower than those to be expected at this carrier concentration.

3.4.5 Alternative oxide substrates

Oxides are usually materials of very high thermal stability. Some of these oxides have crystallographic planes in which the atomic arrangement is hexagonal, with an 'apparent' lattice parameter close to the *a* parameter of GaN.

3.4.5.1 *Lithium gallate $LiGaO_2$(LGO)*

LGO is an orthorhombic crystal [121], with an apparent hexagonal surface symmetry in the (001) plane, in which the distance between cations and anions lead to a 1–2% mismatch with GaN. Little is known about LGO substrate properties, but it may be less stable than sapphire, due to its lower melting point of 1600 °C, and may have a high sensitivity to moisture. Kung et al. [99] have grown GaN onto (001) LGO by low-pressure MOCVD. They used no buffer layer and investigated the growth temperature in the 600–1000 °C range. The structural quality of GaN deposited on LGO increases with growth temperature. A rocking curve width of 300 arcsec is obtained for $T_g = 900$ °C. In the same time, the electronic properties of the material decrease with increasing growth temperature; a decrease of the near band edge emission in photoluminescence is observed, with a corresponding increase of the yellow luminescence, which dominates the spectra above $T_g = 800$ °C. The near band edge luminescence is blue shifted, maybe due to the high residual carrier concentrations (in the 10^{20} cm^{-3} range).

These results appear to be due to a deterioration of the LGO substrate at high growth temperatures, resulting in a contamination of the GaN epilayer.

On the basis of these results, we may fear that LGO substrates have to be restricted to lower growth temperature techniques, such as those derived from the MBE.

3.4.5.2 *Magnesium aluminate $MgAl_2O_4$*

$MgAl_2O_4$ has a spinel structure, in which the oxygen atoms form a face centered cubic structure, with the magnesium and aluminum atoms occupying the tetrahedral and octahedral sites, respectively. With the (111) plane of the spinel structure parallel to the (0001) plane of GaN, one may expect a 9% lattice mismatch, which is almost 50% less than in the case of GaN and sapphire. Sun et al. [180] and George et al. [46] studied the low pressure MOCVD of GaN on (100) and (111) $MgAl_2O_4$. Due to the still high lattice mismatch, they used a GaN buffer layer (between 200 and 300 Å, deposited at 550 °C). In the case of the (111) substrate, they observed the following epitaxial relationship: [0001]GaN//[111] $MgAl_2O_4$ and [11$\bar{2}$0]GaN//[110] $MgAl_2O_4$. Since these last planes are cleavage planes in both materials, this opens promising perspectives for the realization of lasers. Moreover, the electronic properties of the layers grown on (111) $MgAl_2O_4$ were similar to those of high-quality GaN grown on sapphire. On the other hand, (100) spinel yielded mixed phase GaN, of poor quality. It seems that such spinel substrates are a promising alternative for the MOCVD growth of GaN.

3.5 Doping of GaN

3.5.1 n-type doping

Among the potential n-type dopants for GaN, there is silicon and germanium, which are well known to give excellent results for the n-type doping of other classical III–V materials. Although they should be able to act as both donors or acceptors depending upon which atom they substitute (Ga or N), they mainly replace the gallium, due to the low covalent radii difference between Si and Ga (0.15 Å) or Ge and Ga (0.04 Å), compared with the radii difference with nitrogen.

Murakami et al. [128] studied the incorporation of silicon into GaN using silane (SiH_4) as a precursor, but noticed the occurrence of cracks and pits at high Si concentrations. Nakamura et al. [135] also used silane to dope GaN layers grown on GaN buffers and observed neither cracks nor pits even at the higher concentrations (corrresponding to a carrier concentration of 2×10^{19} cm^{-3}). In Fig 3.16 (data from reference [135]), we display the carrier concentration in GaN layers grown on GaN buffers versus silane and germane flow rate. As can be seen, the incorporation efficiency of silicon from silane is extremely high. The solid vapor ratio, defined as: (dopant/Ga)$_{solid}$/(dopant/TMGa)$_{gas}$, is a convenient figure to discuss the relative efficiency of different dopants. Assuming an ionization efficiency of 100%

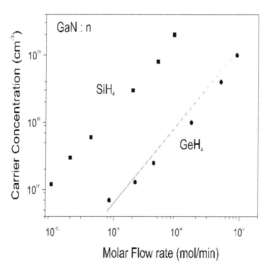

Fig. 3.16 Incorporation of Si and Ge in GaN, using silane and germane as precursors [135].

for silicon in GaN (sample with $n = 2 \times 10^{19}$ cm^{-3}), and assuming 3.79×10^{22} Ga atom cm^{-3} (calculated from the lattice parameters for the wurtzite structure), we obtain a Si/Ga ratio of 5.28×10^{-4} in the sample. The ratio of silane to TMGa in the gas phase is 1.67×10^{-4}. So, the solid–vapor ratio is 3.16, and there is, relatively, more silicon in the solid than in the gas phase, which indicates that silicon incorporates even more readily in GaN than Ga! Assuming a lower ionization efficiency for Si would just give a higher figure.

Although all the Si-doped layers were mirror0like, the FWHM of the (0002) X-ray diffraction peak was broadened by increasing Si content, from 4 to 6 arcmin for SiH$_4$ flow rates from 0.1 nmole/min up to 10 nmole/min.

Figure 3.16 also displays Nakamura et al. [135] results for germanium doping. It appears that germane is 10 times less efficient compared to silane. Good morphologies (mirror-like) were obtained up to electron concentrations of 1×10^{19} cm^{-3}. Above this value, many pits are observed on the surface. An increase in the widths of the X-ray rocking curves is also observed for increasing Ge doping. Moreover, in both cases (Si or Ge doping) the photoluminescence is dominated by the deep levels at 550 nm (the so-called 'yellow luminescence'). The overall luminescence intensity also increases with the doping level.

Silicon and germanium are thus suitable and convenient dopants for GaN, with germane being less efficient than silane.

3.5.2 p-type doping

The case of p-doping in GaN is much more complicated than that of n-type doping, as it frequently happens to be in wide gap materials. The most widely used dopant for p-doping of GaN is magnesium, although Ca has also been proposed, with a proposal that Ca could be a shallower acceptor than Mg [174]. However up to now. GaN layers implanted with Ca have failed to demonstrate high hole concentrations [106], as can be obtained with magesium.

Magnesium is convenient for MOCVD, since high purity organometallic sources of magnesium exist commercially, and have low vapor pressure, well suited for doping.

Magnesium was tried early as a possible p-dopant for GaN, but the layers doped with magnesium were resistive, with no reproducible measurable hole concentrations. Then, in 1988, Amano et al. [6] observed that the blue luminescence in magnesium-doped GaN samples was strongly affected by exposure to a low-energy electron beam. They called this treatment LEEBI, for 'Low-Energy Electron Beam Irradiation'. Then they performed Hall effect measurements and discovered that the LEEBI treatment had activated the magnesium dopant, turning the sample p-type [8]. The samples were GaN grown on AlN on sapphire, doped with 10^{20} Mg atoms/cm^3. The LEEBI treatment was typically performed with 10 kV electrons, under a beam current of 60 μA and a spot size of 60 μm. As a result, the film had a resistivity of 35 Ωcm, corresponding to a hole concentration of 2×10^{16} cm^{-3} and a mobility $\mu = 8$ cm^2/Vs. This corresponds to an activation ratio of the dopant of 2×10^{-4}. Taking into account the fact that Mg is not really a shadow acceptor, we should expect an activation

3.5 Doping of GaN

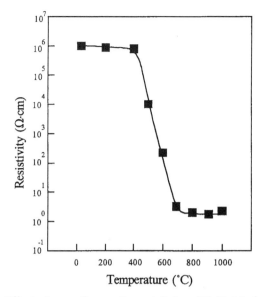

Fig. 3.17 Effect of annealing on the resistivity of GaN:Mg (after [136]).

ratio of 10^{-2}, if all the dopant was activated. From these arguments, we notice that the LEEBI treatment had a low efficiency in the layers grown on an AlN buffer. A much better result was obtained in 1991 from GaN layers deposited onto GaN buffers and LEEBI treated: Nakamura et al. [138] obtained a p doping level of 7×10^{18} cm^{-3}, for a mobility of 3 cm^2/Vs and a resistivity of 0.2 Ωcm. Although the Mg concentration was not reported in this case, this indicates an improved activation by the LEEBI treatment, at least by a factor of 10. An interesting point here is that Nakamura et al. noticed p-type conduction in their films even before LEEBI treatment ($p = 2 \times 10^{15}$ cm^{-3}) and the above-mentioned results were achieved using only 5 kV electrons for the LEEBI treatment. As a consequence, LED devices were realized [134] with improved performances over that realized on AlN buffers [8, 2]. In 1992, Nakamura et al. [136] suggested that the effect of the LEEBI treatment could be due to a heating of the sample in the high vacuum by the electron beam, and performed thermal annealing experiments which were successful to activate Mg-doped GaN. In a second, detailed paper, Nakamura et al. [132], clearly demonstrated that magnesium was hydrogen compensated, and that the dopant could be activated by removing the hydrogen upon annealing in a proper, hydrogen-free, ambient (either vacuum or nitrogen). The decrease of resistivity in Mg-doped GaN samples observed by Nakamura et al. [132] after annealing (20 minutes at each temperature, under nitrogen) is reported in Fig. 3.17.

Although the layer resistivity is constant and does not decrease further above 700 °C, high annealing temperatures have an adverse effect on the material quality: Fig. 3.18 displays the variation of the intensity of the room temperature photoluminescence of GaN:Mg samples annealed at various temperatures (from reference [132]). It can be seen that the blue emission (450 nm) intensity

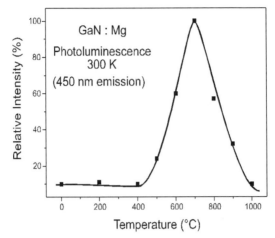

Fig. 3.18 Intensity of the room temperature near band edge photoluminescence of GaN:Mg versus annealing temperature (after [136]).

increases for annealing temperatures up to 700 °C, then decreases. Nakamura et al. [132] have demonstrated that this phenomenon is due to a thermal degradation of the layer surface (loss of nitrogen), as discussed in Section 3.1.2.1. However, it is not possible to prevent the surface decomposition by using a NH_3 ambient, since the removal of the compensating hydrogen would then not occur. It appears that an annealing temperature is the best compromise between efficient diffusion of hydrogen out of the sample and limited degradation of the sample.

This blue emission at 450 nm, although correlated with levels introduced by magnesium, are not directly linked to the p-type conductivity in the samples [33, 34, 49].

The activation of magnesium by heat treatment has several advantages over the LEEBI treatment. First, it is much simpler to realize, It can be done *in situ*, in the MOCVD equipment. Then, in LEEBI, only a thin layer is really activated, since low-energy electrons have a small penetration depth in GaN.

Concerning the incorporation of magnesium in GaN using Cp_2Mg as precursor, Amano et al. [9, 2] reported that it is almost independent of the growth temperature in the range 850–1040 °C, reflecting the fact that no re-evaporation of the dopant from the surface occurs (sticking coefficient close to unity). The incorporation of magnesium is linear with the Cp_2Mg molar flow. Although all the data are not available in the above-mentioned papers to calculate precisely the solid–vapor ratio for Mg in GaN, it can be evaluated to be close to a few 10^{-2} for an Mg concentration of about 10^{19} atoms/cm^3. This indicates that the magnesium incorporates a hundred times less than silicon in GaN.

Since Mg has a larger covalent radius ($r = 0.1$–0.14 Å) than gallium, which it substitutes, we may anticipate that high magnesium concentrations will degrade the crystal quality. As a matter of fact, the morphology of the layers is mirror-like up to a few 10^{19} Mg atoms/cm^3, but cracks occur at higher concentrations. Unfortunately, due to the large ionization energy of the acceptor, high magnesium concentrations will be necessary to achieve high p-type doping.

Ohba and Hatona [144] used Ar carrier gas instead of hydrogen (this is not supposed to have a positive incidence on the concentration of compensating hydrogen incorporated in the layers, since it mainly originates from ammonia and the metalorganics, where the bond between hydrogen and N or metal atoms is weaker than on H$_2$). They reported poor layer morphologoes (cracks and pits) even for low Mg concentrations (below 10^{18} atoms/cm^3). They also mentioned that magnesium could lead to memory effects in the MOCVD reactor [145].

3.6 Growth of ternary nitride alloys

3.6.1 AlGaN

3.6.1.1 *The GaN/AlN solid solution*

GaN and AlN both have a wurtzite structure. Their lattice parameters and thermal expansion coefficients, as well as those of the sapphire substrate, are given in Table 3.6.

This gives a lattice mismatch between GaN and Al of 2.5% along the *a*-axis and 4% along the *c*-axis. Moreover, the covalent cation radii are 1.26 Å for Ga and 1.18 Å for Al, leading to a covalent radii difference of 0.08 Å, which is low. Following the analysis of Pikhtin *et al.* [152], we may thus expect that GaAlX (X = N, P, As, Sb) solid solutions will be close to ideality.

In fact, Al$_x$Ga$_{1-x}$N solid solutions were realized for the first time in 1973, by Lyutaya *et al.* [115], the first epitaxy on sapphire being realized in 1978 by Hagen *et al.* [54] and Baranov *et al.* [14] by Vapor Phase Epitaxy (VPE), and then by Reactive Molecular Beam Epitaxy (RMBE) by Yoshida *et al.* [196] in 1982.

3.6.1.2 *MOVPE growth of Al$_x$Ga$_{1-x}$N*

Composition control. The control of the solid composition in MOCVD can be complicated by the occurrence of parasitic reactions, which can severely deplete the gas phase. In Section 3.1.1.3 we have seen that TMAl is prone to reacting homogeneously with ammonia.

Table 3.6 GaN, AlN, and α-Al$_2$O$_3$ lattice parameters and expansion coefficients (from [67]).

	a (Å)	c (Å)	α_a (10^{-6} K^{-1})	α_c (10^{-6} K^{-1})
GaN	3.189	5.185	5.59	7.75
AlN	3.111	4.980	5.3	4.2
Al$_2$O$_3$	4.758	12.991	7.3	8.1

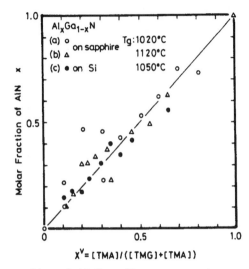

Fig. 3.19 Solid composition of $Al_xGa_{1-x}N$ versus gas phase composition for atmospheric pressure growth [92].

At low reactor pressures, the contact time between the precursor molecules is greatly reduced, resulting in a lower probability of reaction before reaching the surface. Khan et al. [86], using low-pressure MOVPE (5–100 torr), with TEGa, TMAl, and NH_3 precursors, were able to grow the ternary alloy on the whole range of composition. Khan et al. [85] have also employed trimethylamine alane (TMAAl) at low reactor pressure, but a low Al incorporation is obtained: solid compositions below $x = 0.2$ are obtained for gas phase compositions up to 0.7. However, for many purposes low x values are required, and TMAAl could be an interesting alternative.

At atmospheric pressure, Koide et al. [92] have encountered two situations: in a first experiment, due to their reactor design and gas flow rates, the gases had a low velocity (0.7 to 2 cm/sec). This resulted in a lack of control over the solid composition, and the authors observed white, Al-rich deposits on the layers. They attributed these deposits to involatile adduct formation, which depleted the gas phase in Al. Then they changed the reactor design to mix the precursors just before the reactor and inject them at a much higher velocity (110 cm/sec) to reduce the contact time between the precursor molecules. In that way, good control of the solid composition was obtained, as depicted in Fig. 3.19:

Matloubian and Gershenzon [123] mention that TMA decomposes preferentially on the reactor walls, when they are heated above 400°C. This is likely to be the case in a reactor with walls which are not water-cooled. The typical growth parameters are very similar to those of GaN: a high growth temperature (900–1050°C) and a large V/III molar ratio. The growth rate decreases with increasing reactor pressure, and increases with increasing Group III molar flow [86]. The growth rate was found to be almost independent of the substrate temperature, in the above-mentioned range. This clearly demonstrates that the growth rate of $Al_xGa_{1-x}N$ is limited by Group III element mass transport by diffusion through the gas phase.

Although a roughly linear relationship between the solid and gas composition is obtained, there is a noticeable dispersion in the experimental values which prevent further detailed analysis of this data.

Structural quality. Most reports in the literature indicate that the layers are monocrystalline up to $x = 0.4$. For higher Al concentrations, microscopic grains are observed on the whole surface.

In 1987, Koide *et al.* [91, 93] introduced the use of an AlN buffer layer for the growth of the AlGaN ternary compound. The buffer layer consisted of a 500 Å thick AlN layer, deposited at 800 °C and recrystallized at higher temperature prior to the main layer growth. As a consequence, the morphology is greatly improved: the surface is mirror-like up to $x = 0.4$, while hexagonal platelets are obtained when AlGaN is grown directly on sapphire.

However, even for low Al compositions the crystalline quality is poorer than in the case of GaN. To further analyze the crystalline quality through X-ray diffraction experiments, Koide *et al.* [93] and Akasaki *et al.* [4] followed the approach of Itoh and Okamoto [70], which states that the width of the diffraction peaks can be decomposed into three separate contributions:

- $\Delta\theta_1$, due to the orientation distribution (mosaicity) in the films;
- $\Delta\theta_2$, due to the distribution of lattice spacing;
- $\Delta\theta_3$, due to the finite size of the crystallites in the layer.

By performing ω-scan experiments as well as $\theta-2\theta$ scans, with slits or open window for the X-ray detector, it is possible to separate these contributions.

In the paper of Koide *et al.*, this analysis was applied to a GaN layer and to a $Ga_{0.9}Al_{0.1}N$ ternary layer, both grown directly on the substrate or on a buffer layer (bl). They obtained the values given in Table 3.7.

The values in Table 3.7 clearly demonstrate that both GaN and $Ga_{0.9}Al_{0.1}N$ have a mosaic structure. The quality of the $Ga_{0.9}Al_{0.1}N$ film is already lower than in the GaN film, although the Al content is low. Large values of $\Delta d/d$ indicate vertical inhomogeneities (along the c-axis). One can see that $\Delta d/d$ values are about 10 times greater in the alloy compared to GaN.

Another approach was developed by Wickenden *et al.* [192] which consists in

Table 3.7 FWHM of X-ray diffraction profiles, variations in lattice spacing $\delta d/d$, and average grain size D of AlGaN films.

		$\Delta\theta_\omega$	$\Delta\theta_1$	$\Delta\theta_2 + \Delta\theta_3$	$\Delta d/d$	D (Å)
$Ga_{0.9}Al_{0.1}N$	without bl	13.6'	10.2'	3.4'	3.2×10^{-3}	570
	with bl	6.4'	4.5'	1.9'	1.8×10^{-3}	1020
GaN	without bl	8.2'	7.6'	0.6'	5.6×10^{-4}	3200
	with bl	1.9'	1.5'	0.4'	3.7×10^{-4}	4800

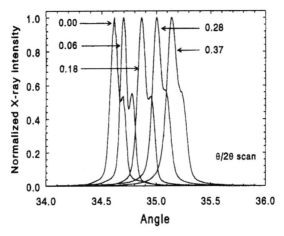

Fig. 3.20 X-ray diffraction profiles for $Al_xGa_{1-x}N$ up to $x = 0.4$ [192].

using a buffer layer of the same composition as the film to be grown ('self-nucleated' layers). This may reduce the mismatch between the buffer and the layer, but on the other hand, the crystalline quality of the AlGaN buffer will certainly be lower than that of an AlN buffer, due to the alloy disorder. The buffer layers were grown at 540 °C and the overlayers were grown at 1025 °C, with N_2 carrier gas, under V/III molar ratios ranging from 250 to 330, which are quite low values compared with others commonly reported in the literature.

The X-ray diffraction profiles for compositions up to $x = 0.37$ for the 'self-nucleated' layers, are shown in Fig. 3.20. The FWHM values, extracted from Fig. 3.20, are given in Table 3.8 below.

This illustrates the degradation of the material quality which occurs while increasing the Al content. Unfortunately, we cannot compare the value obtained here with the values mentioned above to compare the 'self-nucleated' layers with the layers deposited on an AlN buffer, since the recording techniques for the X-ray profiles were very different. We also note that little information exists in the literature, concerning the material quality of $Al_xGa_{1-x}N$ with x above 0.4. It seems that there is a quite sharp transition in the material quality around $x = 0.4$–0.5, which has not, to date, been thoroughly studied.

The electrical properties of undoped AlGaN are often reported up to $x = 0.4$.

Table 3.8 X-ray profile FWHM versus composition in $Al_xGa_{1-x}N$ (after [192]).

x	FWHM (arcmin)
0	3.6
0.06	3.6
0.18	4
0.28	5
0.37	5.8

The carrier concentration in $Al_xGa_{1-x}N$ usually decreases from $n = 10^{18}-10^{19}$ cm^{-3} at low Al content to below 10^{17} cm^{-3} for $x = 0.4$. Above this value, the material becomes extremely resistive, which is probably correlated with the poor quality of the material at high Al content.

The doping is performed in $Al_xGa_{1-x}N$ similarly to GaN. Silane is usually used and allows good control of n doping at low Al contents. Good results have been reported with germane ($n = 3.7 \times 10^{19}$ cm^{-3} for $x = 0.2$). A reduction of the deep level emission around 2.2 eV in photoluminescence is also noted [201]. There is very little information available concerning the efficiency of doping versus Al molar fraction. It seems that n-type doping can be performed up to $x = 0.4$ (using silicon), while p-type doping (using magnesium) is reported up to 10^{18} cm^{-3} in $Al_{0.08}Ga_{0.92}N$ by Tanaka et al. [182]. In their papers concerning nitride laser diodes, Nakamura et al. [140, 139] mention p-type $Al_{0.2}Ga_{0.8}N$, but no details are given concerning the doping level.

3.6.2 GaInN

The realization of the ternary solid solution $In_xGa_{1-x}N$ will be complicated by numerous problems:

- The covalent radii of the cation species, In and Ga are, respectively 1.44 Å and 1.26 Å, corresponding to a difference of 0.18 Å [175]. We may anticipate that this difference will create large internal strains, and that this ternary compound will differ noticeably from an ideal system.
- the lattice mismatch between GaN and InN is 10% along the *a*-axis and is about the same along the *c*-axis.
- The vapor pressure of nitrogen in equilibrium with the InN crystal is extremely large [124], a situation which is even worse than in the GaN case. This will lead to a difficulty in controlling the stoichiometry, requiring the use of very large V/III molar ratios, to limit the possible occurrence of In droplets on the surface.
- The elemental vapor pressure of In is more than 10 times larger than the vapor pressure of Ga at usual GaInN MOCVD growth temperatures (vapor pressures data from [149] are replotted in Fig. 3.21). This will lead to an important evaporation of In from the growing surface, and will make the composition control of the ternary compound difficult [81, 200].
- In a recent paper, I. H. Ho and G. B. Stringfellow [61], used a modified valence-force-field (VFF) model, allowing lattice relaxation up to the sixth nearest neighbor. From the values at the dilute limits, they calculated an average interaction parameter for the regular solution like the composition dependence of the enthalpy of mixing.

From these calculated values, they predicted that GaInN is prone to phase separation; a miscibility gap exists in the alloy solution. Figure 3.22 displays the calculated binodal and spinodal curves for $In_xGa_{1-x}N$. The solubility of In in GaN is limited to about 6% at 800 °C.

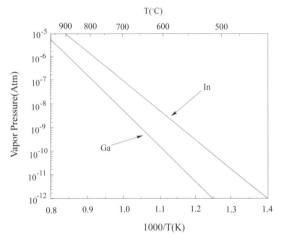

Fig. 3.21 Indium and gallium vapor pressure (from Panish and Temkin [149]).

As a matter of fact, phase separation was evidenced early (1975) by Osamura et al. [148], during heat treatments of GaInN. However, this situation is also encountered with other III–V materials and the modern growth techniques (MBE, MOCVD), which are not operating at the thermodynamical equilibrium, are able to produce metastable materials.

MOCVD growth. From the discussion above, it may be anticipated that the typical growth temperature for $In_xGa_{1-x}N$ will be lower than the typical GaN growth temperature. Most of the studies report growth temperatures between 500 °C and 850 °C, with a strong decrease of the In content as the growth temperature increases [129, 125, 198].

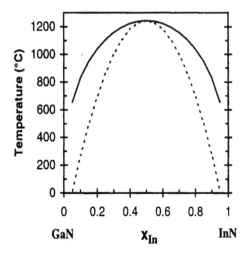

Fig. 3.22 Binodal and spinodal demixition curves for $In_xGa_{1-x}N$ alloy (after [61]).

3.6 Growth of ternary nitride alloys

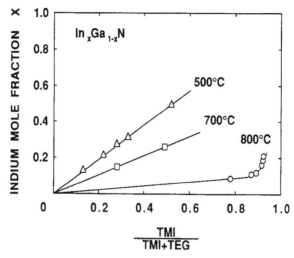

Fig. 3.23 Solid composition of $In_xGa_{1-x}N$ versus gas phase composition at different growth temperatures [125].

This situation is well depicted in Fig 3.23, where the solid–gas composition relationship is given at different growth temperatures.

At low temperatures (500 °C) a linear relationship is obtained [125, 198] between the gas phase composition and the solid phase composition, while at 800 °C, the solid composition (x_s) is almost constant until the gas phase composition (x_g) reaches $x = 0.8$, with an abrupt increase of solid composition for x_s above 0.9, making the composition control very difficult. A thermodynamic analysis which fits the experimental data, and accounts for this 'bowing', has been proposed by Koukitu *et al.* [96].

The main problem here consists in finding a good compromise between composition control and a satisfying crystal quality, which degrades very rapidly at lower growth temperatures. The layers grown by Matsuoka *et al.* [125] on sapphire substrates have X-ray diffraction (0004) linewidths from 30 arc minutes for $In_{0.2}Ga_{0.8}N$ layers grown at 800 °C, which increase to 100 arc minutes for $In_{0.2}Ga_{0.8}N$ layers grown at 500 °C. Drastic improvements of crystalline quality were obtained by Nakamura and Muka [133], who deposited InGaN onto a GaN layer. The GaN layers were about 2 μm in thickness and were deposited on a 300 Å GaN buffer layer on (0001) sapphire substrates. Then $In_xGa_{1-x}N$ layers (0.3 μm in thickness) were grown at 780 °C and 830 °C under a gas phase composition of $x_g = 0.92$. Narrow X-ray (0002) diffraction peaks were obtained (6 arcmin for $In_{0.14}Ga_{0.86}N$ grown at 780 °C and 8 arcmin for $In_{0.24}Ga_{0.76}N$ grown at 830 °C).

From the growth conditions reported in the literature, it appears that high V/III molar ratios are to be used to obtain high-quality materials. In the high growth temperature range (750–850 °C), V/III molar ratios similar to those used for the growth of GaN are reported (typically 1000 to 10 000). At the lowest growth temperatures (500–700 °C), Matsuoka *et al.* [125] have reported that

V/III molar ratios in excess of 15 000 were necessary to prevent the occurrence of In droplets.

There is also an important effect of the carrier gas nature on the In incorporation [169]. The carrier gases usually employed for the growth of GaInN are either H_2 or N_2, or a mixture of both. It has been observed that the incorporation of In is improved under N_2 [200, 168].

3.7 MOVPE growth of nitride heterostructures

Most device applications require the use of heterostructures. However, we may anticipate that the growth of nitride heterostructures can be complicated by different problems:

- The high growth temperatures required to obtain good crystalline materials will favor interdiffusion, possibly resulting in non-abrupt interfaces.
- The different optimum growth temperatures for the different materials involved can lead to degradation of the material with the lower growth temperature.
- Although considered as 'small' compared with the mismatch with substrates, the lattice mismatch between the different layers is actually important, introducing large defects densities for layers above the critical thickness.

3.7.1 GaN/Al_xGa_{1-x}N heterostructures

The first demonstration of quantum confinement in GaN/Al_xGa_{1-x}N structures was due to Khan et al. [88] in 1990. The photoluminescence lines from their structures were rather broad, but this is likely due to the low growth temperature used (850 °C). In GaAlN, the Al–N bond is stronger than the Ga–N bond, so the Al interdiffusion from the barriers into the wells should be limited. In a later paper, Khan et al. [83] realized GaN/AlN short-period superlattices and observed that even in a superlattice period containing two monolayers of GaN, no interdiffusion could be observed by TEM. The samples were realized by switched Atomic Layer Epitaxy (ALE).

The main problem for the realization of high-quality GaN/Al_xGa_{1-x}N heterostructures will thus be the material crystalline quality. On the one hand, the optimum growth temperature for AlN is very different from that of GaN; concerning GaN, as mentioned in Section 3.3.5.1, the optimum is around 1000 °C, while for AlN Ohba and Hatano [146] observed that AlN quality improves at high temperature. The width of the AlN epilayer X-ray rocking curves decrease from 45 arcmin at a growth temperature of 1000 °C down to 3 arcmin at a growth temperature of 1300 °C.

On the other hand, the lattice parameter difference between GaN and Al_xGa_{1-x}N causes a large dilation strain in the alloy layers grown on GaN. With increasing GaAlN thickness deposited onto GaN, relaxation occurs, followed by the appearance of macroscopic cracks [68]. For $Al_{0.1}Ga_{0.9}N$, this occurs between 1000 and 3000 Å.

The relaxation effects are evidenced in superlattices realized by Itoh *et al.* [69]: they realized $Al_{0.1}Ga_{0.9}N$/GaN superlattices with periods ranging from 45 Å to 600 Å. The super periodicity was evidenced by observation of satellite peaks, up to the second order, in X-ray diffraction. They observed that the 77 K photoluminescence peaks become sharper with decreasing period value. A possible interpretation for this is that defect density decreases for short periods, since coherent growth may be realized if the individual layers in the superlattice are below their critical thickness.

3.7.2 GaN/$In_xGa_{1-x}N$ heterostructures

From Section 3.6.2, we can imagine that the main problem here will be the difference of optimum growth temperature, since GaInN has to be grown around 800 °C, since In strongly re-evaporates at higher temperatures, which are required to obtain high-quality GaN.

To overcome this problem, Keller *et al.* [79, 80, 177] used a temperature ramping procedure between the barrier and well layers. This resulted in graded structures which exhibited carrier confinement for well thicknesses below 34 Å (for a well composition of 14%).

However, a marked increase of the photoluminescence linewidths was observed when the well widths were decreased. In these graded structures, the luminescence was more intense compared with simple heterostructures. This could be due to the pseudo-electric field, resulting from the compositional gradient, which improves carrier injection efficiency from the graded barriers into the well.

GaInN/GaN superlattices were studied by Li *et al.* [110] by X-ray diffraction experiments. Although these samples are of similar quality to the best reported to date, only the −2 satellite peak, with a very weak intensity, is observed although the sample has a period of 510 Å. The same observation can be made for similar published data [90, 137]. This could be due to a moderate crystalline quality of the GaN barriers, grown at low temperature. In the $Al_xGa_{1-x}N$/GaN system, satellite peaks are more easily observed.

A very interesting result was reported by these authors: a 2D mapping of the reciprocal space of such superlattice sample, for the asymmetric reflection $(10\bar{1}5)$ is displayed in Fig. 3.24.

The contour plots correspond to the reciprocal point of the GaN thick buffer layer and to the zero order satellite peak. The vertical dashed line corresponds to a constant in-plane lattice parameter (case of pseudomorphic growth) while the inclined dashed line corresponds to a fully relaxed superlattice. It can be clearly be seen that the superlattice has grown coherently. Since the individual thicknesses for both the $In_{0.06}Ga_{0.94}N$ and GaN layers are 255 Å, and the elastic

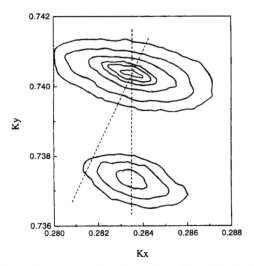

Fig. 3.24 ($10\bar{1}5$) reciprocal space mapping for an $In_{0.06}Ga_{0.94}N$/GaN superlattice (period 255 + 255 Å) grown on a 1.6 μm GaN layer. K_y and K_x are the reciprocal lattice vectors normal and parallel to the sample surface [110].

constant of GaN are large, it means that the critical thickness in this system is much higher compared to classical III–V semiconductors.

Such a result is consistent with the observation by Nakamura *et al.* [137] that the zeroth-order peak was stationary when the $In_{0.06}Ga_{0.94}N/In_{0.22}Ga_{0.78}N$ superlattice period was varied between 60 and 200 Å.

References

1. C. Abernathy, J. MacKenzie, S. Pearton, and W. Hobson. *Appl. Phys. Lett.* **66**:1969, 1995.
2. I. Akasaki, H. Amano, M. Kito, and K. Hiramatsu, *J. Lumin.*, **48–49**:666, 1991.
3. I. Akasaki, H. Amano, Y. Koide, K. Hiramatsu, and N. Sawaki. *J. Cryst. Growth*, **98**:209, 1989.
4. I. Akasaki, H. Amano, Y. Koide, K. Hiramatsu, and N. Sawaki. *J. Cryst. Growth*, **98**:209, 1989.
5. H. Amano, I. Akasaki, K. Hiramatsu, N. Koide, and N. Sawaki. *Thin Solid Films*, **163**:415, 1988.
6. H. Amano, I. Akaski, T. Kozawa, K. Hiramatsu, N. Sawaki, K. Ikeda, and Y. Ishii. *J. Lumin.*, **40–48**:121, 1988.
7. H. Amano, K. Hiramatsu, and I. Akasaki. *Jpn. J. Appl. Physics*, **27**(8):L1384, 1988.
8. H. Amano, M. Kito, K. Hiramatsu, and I. Akasaki. *Jpn. J. Appl. Physics*, **28**(12):L2112, 1989.
9. H. Amano, M. Kitoh, K. Hiramatsu, and I. Akasaki. *J. Electrochem. Soc.*, **137**(5):1639, 1990.
10. H. Amano, N. Sawaki, I. Akasaki, and Y. Toyoda. *Appl. Phys. Lett.*, **48**:353, 1986.
11. O. Ambacher, R. Dimitriv, D. Lentz, T. Metzger, W. Rieger, and M. Stutzmann. *J. Cryst. Growth*, **167**:1, 1996.

12. J. Andrews and M. Littlejohn. *J. Electrochem. Soc.*, **122**(9):1273, 1975.
13. V. Ban. *J. Electrochem. Soc.*, **119**(6):761, 1972.
14. B. Baranov, L. Daweritz, V. Gutan, G. Jungk, H. Neumann, and H. Raidt. *Phys. Statu. Solidii(a)*, **49**:629, 1978.
15. I. Barin. *Thermochemical data of pure substances.* VCH, 1994.
16. S. Bass, C. Pickering, and M. Young. *J. Cryst. Growth*, **64**:68, 1983.
17. J. Baur, U. Kaufmann, M. Kunzer, J. Schneider, H. Amano, I. Akasaki, T. Detchprohm, and K. Hiramatsu. *Appl. Phys. Lett.*, **67**(8):1140, 1995.
18. J. Baur, K. Maier, M. Kunzer, U. Kaufmann, and J. Schneider. *Appl. Phys. Lett.* **65**:2211, 1994.
19. J. Baur, K. Maier, M. Kunzer, U. Kaufmann, J. Schneider, H. Amano, I. Akasaki, T. Detchprohm, and K. Hiramatsu. *Appl. Phys. Lett.*, **65**:857, 1994.
20. B. Beaumont and P. Gilbart. In B. Gil and R. L. Aulombard, editors, *Semiconductor heteroepitaxy*, p. 258. World Scientific, 1995.
21. Birge. *Nature*, **122**:842, 1928.
22. P. Born and D. Robertson. *J. Mater. Sci.*, **15**:3003, 1980.
23. O. Briot, J. Alexis, B. Gil, and R. Aulombard. *Mat. Res. Soc. Symp. Proc.*, **395**:207, 1996.
24. O. Briot, J. Alexis, B. Gil, and R. Aulombard. *Mat. Res. Soc. Symp. Proc.*, **395**:411, 1996.
25. O. Briot, J. Alexis, M. Tchounkeu, and R. Aulombard. *Mat. Sic. Eng. B*, **43**:147, 1996.
26. Y. Bu, M. Lin, L. Fu, D. Chtchekine, G. Gilliland, Y. Chen, S. Ralph, and S. Stock. *Appl. Phys. Lett.*, **66**(18):2433, 1995.
27. N. Buchan, C. Larsen, and G. Stringfellow. *J. Cryst. Growth*, **92**:591, 1988.
28. C. Chen, H. Liu, D. Steigerwald, W. Imler, C. Kuo, and M. Craford. *Mat. Res. Soc. Symp. Proc.*, **395**:103, 1996.
29. C. Chen, H. Liu, D. Steigerwald, W. Imler, C. Kuo, M. Craford, M. Ludowise, S. Lester, and J. Amano. *J. Electron. Mater.*, **25**(6):1004, 1996.
30. J. Chen, N. Chen, W. Huang, W. Lee, and M. Feng. *Jpn. J. Appl. Physics*, **35**(7A):L810, 1996.
31. B. Chung and M. Gershenzon. *J. Appl. Phys.*, **72**(2):651, 1992.
32. T. Detchprohm, K. Hiramatsu, H. Amano, and I. Akasaki. *Appl. Phys. Lett.*, **61**(22):2688, 1992.
33. T. Detchprohm, K. Hiramatsu, N. Sawaki, and I. Akasaki. *J. Cryst. Growth*, **137**:170, 1994.
34. T. Detchprohm, K. Hiramatsu, N. Sawaki, and I. Akasaki. *J. Cryst. Growth*, **145**:192, 1994.
35. A. Dissanayake, J. Lin, H. Jiang, Z. Yu, and J. Edgar. *Appl. Phys. Lett.*, **65**(18):2317, 1994.
36. V. Dmitriev, K. Irvine, G. Bulman, J. Edmond, A. Zubrilov, V. Nikolaev, I. Nikitina, D. Tsvetkov, A. Babanin, A. Sitnikova, Y. Musikhin, and N. Bert. *J. Cryst. Growth*, **166**:601, 1996.
37. K. Doverspike, L. Rowland, D. Gaskill, and J. Freitas. *J. Electron. Mater.*, **24**(4):269, 1995.
38. K. Fertitta, A. Holmes, J. Neff, F. Ciuba, and R. Dupuis. *Appl. Phys. Lett.*, **65**(14):1823, 1994.
39. K. Fertitta, A. Holmes, J. Neff, F. Ciuba, and R. Dupuis. *J. Electron. Mater.*, **24**(4):257, 1995.
40. A. Freedman and G. Robinson. *Mat. Res. Soc. Symp. Proc.*, **395**:73, 1996.
41. S. Fujieda and Y. Matsumoto. *Jpn. J. Appl. Physics*, **30**(9B):L1665, 1991.
42. S. Fujieda, M. Mizuta, and Y. Matsumoto. *Jpn. J. Appl. Physics*, **26**(12):2067, 1987.

43. D. Gaskill, N. Bottka, and M. Lin. *Appl. Phys. Lett.*, **48**(21)1449, 1986.
44. D. Gaskill, A. Wickenden, K. Doverspike, B. Tadayon, and L. Rowland. *J. Electron. Mater.*, **24**(11):1525, 1995.
45. Gaydon. *Nature*, **153**:407, 1944.
46. T. George, E. Jacobsohn, W. Pike, P. Chang-Chien, M. Khan, J. Yang, and S. Mahajan. *Appl. Phys. Lett.*, **68**(3):337, 1996.
47. T. George, W. Pike, M. Khan, J. Kuznia, and P. Chang-Chien. *J. Electron. Mater.*, **24**(4):241, 1995.
48. B. Gil, O. Briot, and R. Aulombard. *Phys. Rev. B. Rapid Communications*, **52**(24):R17028, 1995.
49. W. Gotz, N. Johnson, J. Walker, D. Bour, and R. Street. *Appl. Phys. Lett.*, **68**(5):667, 1996.
50. N. Grandjean, J. Massies, and M. Leroux. *Appl. Phys. Lett.*, **69**(14):2071, 1996.
51. N. Grandjean, J. Massies, P. Vennegues, M. Laugt, and M. Leroux. *Mat. Res. Soc. Symp. Proc.*, **449**:67, 1997.
52. P. Hacke, A. Maekawa, N. Koide, K. Hiramatsu, and N. Sawaki. *Jpn. J. Appl. Physics*, **33**(12A):6443, 1994.
53. P. Hacke, H. Nakayama, T. Detchprohm, K. Hiramatsu, and N. Sawaki. *Appl. Phys. Lett.*, **68**(10):1362, 1996.
54. J. Hagan, R. Metcalfe, D. Wickenden, and W. Clark. *J. Phys. C*, **11**:L143, 1978.
55. Hagstrum, *J. Chem. Phys.*, **16**:848, 1948.
56. V. Harle, H. Bolay, F. Steuber, B. Kaufmann, G. Reyher, A. Dornen, F. Scholz, and K. Dombrowski. In *EW-MOVPE VI*, Ghent, Belgium, June 1995.
57. Herzberg and Sponer. *Z. Physik. Chem.*, **26**:B1, 1934.
58. B. Heying, X. Wu, S. Keller, Y. Li, D. Kapolnek, B. Keller, S. DenBaars, and J. Speck. *Appl. Phys. Lett.*, **68**(5):643, 1996.
59. C. Hinshelwood. *The Kinetics of Chemical Change*. Clarendon Press, Oxford, 1940.
60. K. Hiramatsu, S. Itoh, H. Amano, I. Akasaki, N. Kuwano, T. Shiraishi, and K. Oki. *J. Cryst. Growth*, **115**:628, 1991.
61. I. Ho and G. Stringfellow. *Appl. Phys. Lett.*, **69**(18):2701, 1996.
62. K. Ho, K. Jenson, J. Hwang, W. Gladfelter, and J. Evans. *J. Cryst. Growth*, **107**:376, 1991.
63. A. Holmes, K. Fertitta, F. Ciuba, and R. Dupuis. *Electron. Lett.*, **30**(15):1252, 1994.
64. C. Hong, D. Pavlidis, S. Brown, and S. Rand. *J. Appl. Phys.*, **77**(4):1705, 1995.
65. J. Huang, T. Kuech, H. Lu, and I. Bhat. *Appl. Phys. Lett.*, **68**(17):2392, 1996.
66. C. Hwang, M. Schurman, W. Mayo, Y. Li, Y. Lu, H. Liu, T. Salagaj, and R. Stall. *J. Vac. Sci. Technol. A*, **13**(3):672, 1995.
67. K. Ito, K. Hiramatsu, H. Amano, and I. Akasaki. *J. Cryst. Growth*, **104**:533, 1990.
68. K. Ito, K. Hiramatsu, H. Amano, and I. Akasaki. *J. Cryst. Growth*, **104**:533, 1990.
69. K. Itoh, T. Kawamoto, H. Amano, K. Hiramatsu, and I. Akasaki. *Jpn. J. Appl. Physics*, **30**(9A):1924, 1991.
70. N. Itoh and K. Okamoto, *J. Appl. Phys.*, **63**:1486, 1988.
71. C. Johnson, J. Lin, H. Jiang, M. Khan, and C. Sun. *Appl. Phys. Lett.*, **68**(13):1808, 1996.
72. D. P. K. Wang and J. Singh. *A. Appl. Phys.*, **80**(3):1823, 1996.
73. M. Kamp, M. Mayer, A. Pelzmann, S. Menzel, H. Chung, H. Sternschulte, and K. J. Ebeling. In *Topical Workshop on Nitrides TWN'95*, Nagoya, Japan, September 1995.
74. Kaplan. *Phys. Rev. B*, **45**:898, 1934.
75. D. Kapolnek, X. Wu, B. Heying, S. Keller, B. Keller, U. Mishra, S. DenBaars, and J. Speck. *Appl. Phys. Lett.*, **67**(11):1541, 1995.
76. J. Karpinski, J. Jun, and S. Porowski. *J. Cryst. Growth*, **66**:1, 1984.
77. T. Kawabata, T. Matsuda, and S. Koike, *J. Appl. Phys.*, **56**(8):2367, 1984.

78. B. Keller, S. Keller, D. Kapolnek, W. Jiang, Y. Wu, H. Masui, X. Wu, B. Heying, J. Speck, U. Mishra, and S. DenBaars. *J. Electron. Mater.*, **24**(11):1707, 1995.
79. S. Keller, B. Keller, D. Kapolnek, U. Mishra, S. DenBaars, I. Shmagin, R. Kolbas, and S. Krishnankutty. *J. Cryst. Growth*, **170**:349, 1997.
80. S. Keller, B. Keller, D. Kapolnek, A. Abare, H. Masui, L. Coldren, U. K. Mishra, and S. DenBaars. *Appl. Phys. Lett.* **68**(22):3147, 1996.
81. S. Keller, B. Keller, D. Kapolnek, A. Abare, H. Masui, L. Coldren, U. Mishra, and S. DenBaars. *Appl. Phys. Lett.*, **68**:3147, 1996.
82. S. Keller, B. Keller, Y. Wu, B. Heying, D. Kapolnek, J. Speck, U. Mishra, and S. DenBaars. *Appl. Phys. Lett.*, **68**(11):1996.
83. M. Khan, J. Kuznia, D. Olson, T. George, and W. Pike. *Appl. Phys. Lett.*, **63**(25):3470, 1993.
84. M. Khan, J. Kuznia, J. VanHove, D. Olson, S. Krishnankutty, and R. Kolbas. *Appl. Phys. Lett.*, **58**(5):526, 1991.
85. M. Khan, D. Olson, and J. Kuznia. *Appl. Phys. Lett.*, **65**(1):64, 1994.
86. M. Khan, R. Skogman, R. Schulze, and M. Gershenzon. *Apply. Phys. Lett.*, **43**(5):492, 1983.
87. M. Khan, R. Skogman, R. Schulze, and M. Gershenzon. *Appl. Phys. Lett.*, **42**(5):430, 1983.
88. M. Khan, R. Skogman, J. VanHove, S. Krishnankutty, and R. Kolbas. *Appl. Phys. Lett.*, **56**(13):1257, 1990.
89. M. Khan, R. Skogman, J. VanHove, D. Olson, and J. Kuznia. *Appl. Phys. Lett.*, **60**(11):1366, 1992.
90. M. Khan, C. Sun, J. Yang, Q. Chen, B. Lim, M. Zubair-Anwar, A. Osinsky, and H. Temkin. *Appl. Phys. Lett.*, **69**(16):2418, 1996.
91. Y. Koide, H. Itoh, M. Khan, K. Hiramatsu, N. Sawaki, and I. Akasaki. *J. Appl. Phys.*, **61**(9):4540, 1987.
92. Y. Koide, H. Itoh, N. Sawaki, I. Akasaki, and M. Hashimoto. *J. Electrochem. Soc.* **133**(9):1956, 1986.
93. Y. Koide, N. Itoh, K. Ito, N. Sawaki, and I. Akasaki. *Jpn. J. Appl. Physics*, **27**(2):1156, 1988.
94. Y. Koide, N. Itoh, K. Itoh, N. Sawaki, and I. Akasaki. *Jpn. J. Appl. Physics*, **27**:1156, 1988.
95. M. Koide, S. Yamasaki, S. Nagai, N. Koide, S. Asami, H. Amano, and I. Akasaki. *Appl. Phys. Lett.* **68**(10):1403, 1996.
96. A. Koukitu, N. Takahashi, T. Taki, and H. Seki. *J. Cryst. Growth*, **107**:306, 1997.
97. A. Koukitu, N. Takahashi, T. Taki, and H. Seki. *Jpn. J. Appl. Physics*, **35**:L673, 1996.
98. I. Kubo, M. Fujii, and M. Hirose. *J. Appl. Phys.*, **8**:627, 1969.
99. P. Kung, A. Saxler, X. Zhang, D. Walker, R. Lavado, and M. Razeghi. *Appl. Phys. Lett.*, **69**(14):2116, 1996.
100. P. Kung, A. Saxler, X. Zhang, D. Walker, T. Wang, I. Ferguson, and M. Razeghi. *Appl. Phys. Lett.*, **66**(22):2958, 1995.
101. C. Kuo, J. Yuan, R. Cohen, J. Dunn, and G. Stringfellow. *Appl. Phys. Lett.*, **44**:550, 1984.
102. N. Kuwano, K. Kobayashi, K. Oki, S. Miyoshi, H. Yaguchi, K, Onabe, and Y. Shiraki. *Jpn. J. Appl. Physics*, **33**(6A):3415, 1994.
103. N. Kuwano, Y. Nagamoto, K. Kobayashi, K. Oki, S. Miyoshi, H. Yaguchi, K. Onabe, and Y. Shiraki. *Jpn. J. Appl. Physics*, **33**(1A):18, 1994.
104. J. Kuznia, M. Khan, D. Olson, R. Kaplan, and J. Freitas. *J. Appl. Phys.*, **73**(9):4700, 1993.
105. J. Kuznia, J. Yang, Q. Chen. S. Krishnankutty, M. Khan, T. George, and J. Freitas. *Appl. Phys. Lett.*, **65**(19):2407, 1994.

106. J. Lee, S. Pearton, J. Zolper, and R. Stall. *Appl. Phys. Lett.*, **68**(15):2102, 1996.
107. P. Lee, T. Omstead, D. McKenna, and K. Jensen. *J. Cryst. Growth*, **85**:165, 1987.
108. W. Lee, T. Huang, J. Guo, and M. Feng. *Appl. Phys. Lett.*, **67**(12):1721, 1995.
109. C. Lewis, W. Dietze, and M. Ludowise. *J. Electron, Mater.*, **12**:507, 1983.
110. W. Li, P. Bergman, I. Ivanov, W. Ni, H. Amano, and I. Akasaki. *Appl. Phys. Lett.*, **69**(22):3390, 1996.
111. X. Li, D. Forbes, S. Gu, D. Turnbull, S. Bishop, and J. Coleman. *J. Electron. Mater.*, **24**(11):1711, 1995.
112. M. Lin, B. Sverdlov, G. Zhou, and H. Morkoc. *Appl. Phys. Lett.*, **62**:3479, 1993.
113. D. Lu, D. Wang, X. Wang, X, Liu, J. Dong, W. Gao, C. Li, and Y. Li. *Mater. Sci. Eng. B*, **29**:58, 1995.
114. M. Ludowise. *J. Appl. Phys.*, **58**:R31, 1985.
115. M. Lyutaya and T. Bartnitskaya. *Inorg. Mater.*, **9**:1052, 1973.
116. J. MacChesney, P. Bridenbaugh, and P. O'Connor. *Mater. Res. Bull.*, **5**:783, 1970.
117. R. Madar, G. Jacob, J. Hallais, and R. Fruchart. *J. Cryst. Growth*, **31**:197, 1975.
118. H. Manasevit. *Appl. Phys. Lett.*, **116**:1725, 1969.
119. H. Manasevit, F. Erdmann, and W. Simpson. *J. Electrochem. Soc.*, **118**:1864, 1971.
120. H. Manasevit and W. Simpson. *J. Electrochem. Soc.*, **12**:156, 1968.
121. M. Marezio. *Acta. Crystallogr.*, **18**:481, 1965.
122. A. Mariano and R. Hanneman. *J. Appl. Phys.*, **34**:384, 1963.
123. M. Matloubian and M. Gershenzon. *J. Electron. Mater.*, **14**(5):633, 1985.
124. T. Matsuoka, H. Tanaka, T. Sasaki, and A. Katsui. *Inst. Phys. Conf. Ser.*, **106**:141, 1989.
125. T. Matsuoka, N. Yashimoto, T. Sasaki, and A. Katsui. *J. Electron. Mater.*, **21**(2):157, 1992.
126. D. Mazzarese, A. Tripathi, W. Conner, K. Jones, L. Calderon, and D. Eckart. *J. Electron. Mater.*, **18**(3):369, 1989.
127. M. Mizuta, A. Fujieda, Y. Matsumoto, and T. Kawamura. *Jpn. J. Appl. Physics*, **25**(12):L945, 1986.
128. H. Murakami, T. Asahi, H. Amano, K. Hiramatsu, N. Sawaki, and L. Akasaki. *J. Cryst. Growth*, **115**:648, 1991.
129. T. Nagamoto, T. Kuboyama, H. Minamino, and O. Omoto. *Jpn. J. Appl. Physics*, **28**(2):L1334, 1989.
130. S. Nakamura. *Jpn. J. Appl. Physics*, **30**(10A):L1705, 1991.
131. S. Nakamura. *Jpn. J. Appl. Physics*, **30**(8):1620, 1991.
132. S. Nakamura, N. Iwasa, M. Senoh, and T. Mukai. *Jpn. J. Appl. Physics*, **31**(5A):1258, 1992.
133. S. Nakamura and T. Mukai. *Jpn. J. Appl. Physics*, **31**(10B):L1457, 1992.
134. S. Nakamura, T. Mukai, and M. Senoh. *Jpn. J. Appl. Physics*, **30**(12A):L1998, 1991.
135. S. Nakamura, T. Mukai, and M. Senoh. *Jpn. J. Appl. Physics*, **31**(9A):2883, 1992.
136. S. Nakamura, T. Mukai, M. Senoh, and N. Iwasa. *Jpn. J. Appl. Physics*, **31**(2B):L139, 1991.
137. S. Nakamura, T. Mukai, M. Senoh, S. Nagahama, and N. Iwasa. *J. Appl. Phys.*, **74**(6):3911, 1993.
138. S. Nakamura, M. Senoh, and T. Mukai. *Jpn. J. Appl. Physics*, **30**(10A):L1708, 1991.
139. S. Nakamura, M. Senoh, S. Nagahama, N. Iwasa, T. Yamada, T. Matsushita, H. Kiyoku, and Y. Sugimoto. *Appl. Phys. Lett.*, **68**(15):2105, 1996.
140. S. Nakamura, M. Senoh, S. Nagahama, N. Iwasa, T. Yamada, T. Matsushita, Y. Sugimoto, and H. Kiyoku. *Appl. Phys. Lett.*, **69**(10):1477, 1996.
141. J. Neugebauer and C. VandeWalle. *Mat. Res. Soc. Symp. Proc.*, **395**:645, 1996.
142. J. Nichools, H. Gallagher, B. Henderson, C. Trager-Cowan, P. Middleton, K. O'Donnell, T. Cheng, and C. Foxon. *Mat. Res. Soc. Symp. Proc.*, **395**:535, 1996.

143. R. Niebuhr, K. Bachem, K. Dombrowski, M. Maier, W. Pletschen, and U. Kaufmann. *J. Electron. Mater.*, **24**(11):1531, 1995.
144. Y. Ohba and A. Hatano. *Jpn. J. Appl. Physics*, **33**(10A):L1367, 1994.
145. Y. Ohba and A. Hatano. *J. Cryst. Growth*, **145**:214, 1994.
146. Y. Ohba and A. Hatano. *Jpn. J. Appl. Physics*, **46**:3432, 1975.
147. Y. Ohuchi, K. Tadamoto, H. Nakayama, N. Kaneda, T. Detchprohm. K. Hiramatsu, and N. Sawaki. *J. Cryst. Growth*, **170**:325, 1997.
148. K. Osamura, S. Naka, and Y. Murakami. *J. Appl. Phys.*, **46** 3432, 1975.
149. M. Panish and H. Temkin. *Gas Source Molecular Beam Epitaxy*, Springer series in Materials Science, vol. 26. Springer-Verlag, Berlin, 1993.
150. Y. Park and D. Pavlidis. *J. Electron. Mater.*, **25**(9):1554, 1996.
151. S. Pearton, S. Bendi, K. Jones, V. Krishnamoorthy, R. Wilson, F. Ren, R. Karlicek, and R. Stall. *Appl. Phys. Lett.*, **69**(13):1879, 1996.
152. A. Pikhtin. *Soviet. Phys. Semicond.*, **11**:245, 1977.
153. A. Polyakov, M. Shin, J. Freitas, M. Skowronski, D. Greve, and R. Wilson. *J. Appl. Phys.* **80**(11):6349, 1996.
154. A. Polyakov, M. Shin, D. Grev, M. Skowronski, and R. Wilson. *MRS Internet J.*, **1**, 1996.
155. F. Ponce, D. Bour, W. Gotz, N. Johnson, H. Helava, I. Grzegory, J. Jun, and S. Porowski. *Appl. Phys. Lett.*, **68**(7):917, 1996.
156. F. Ponce, J. Major-Jr, W. Plano, and D. Welch. *Appl. Phys. Lett.*, **65**(18):2302, 1994.
157. S. Porowski, J. Jun, M. Bockowski, M. Leszczynski, S. Krukowski, M. Wroblewski, B. Lucznik, and I. Grzegory. In M. Godlewski, Ed., *8th International Conference on Semi-Insulating III−V Materials*, p. 61, Warsaw, Poland, 1994. World Scientific.
158. K. Pressel, R. Heitz, L. Eckey, I. Loa, P. Thurian, A. Hoffman, B. Meyer, S. Fisher, C. Wetzel, and E. Haller. *Mat. Res. Soc. Symp. Proc.*, **395**:491, 1996.
159. V. Roberts, J. Roberts, A. Jones, and S. Rushworth. *Math. Res. Soc. Symp. Proc.*, **395**:337, 1996.
160. L. Rowland, K. Doverspike, and D. Gaskill. *Appl. Phys. Lett.*, **66**:1495, 1995.
161. T. Sasaki. *J. Cryst. Growth*, **129**:81, 1993.
162. T. Sasaki and T. Matsuoka. *J. Appl. Phys.*, **77**(1):192, 1995.
163. T. Sasaki and S. Zembutsu. *J. Appl. Phys.*, **61**(7):2533, 1987.
164. M. Sato. In *Topical Workshop on Nitrides—TWN'95*, Nagoya, Japan, September 1995.
165. M. Sato. *J. Appl. Phys.*, **78**:2123, 1995.
166. M. Sato. *Appl. Phys. Lett.* **68**(12):935, 1996.
167. F. Scholz, V. Harle, H. Bolay, F. Steuber, B. Kaufmann, G. Reyher, Λ. Dornen, O. Gfrorer, S. Im, and A. Hangleiter. In *Topical Workshop on Nitrides, TWN'95*, Nagoya, Japan, September 1995.
168. F. Scholz, V. Harle, F. Steuber, H. Bolay, A. Dornen, B. Kaufmann, V. Syganow, and A. Hangleiter. *J. Cryst. Growth*, **170**:321, 1997.
169. F. Scholz, V. Harle, F. Steuber, A. Sohmer, H. Bolay, V. Syganow, A. Dornen, J. Im, A. Hangleiter, J. Duboz, P. Galtier, E. Rosencher, O. Ambacher, D. Brunner, and H. Lakner. In *MRS Fall Meeting*, **449**:3, 1997.
170. W. Seifert, R. Franzheld, E. Butter, H. Sobotta, and V. Riede. *Cryst. Res. and Technol.*, **18**:383, 1983.
171. A. Shintani, Y. Takano, S. Minagawa, and M. Mari. *J. Electrochem. Soc.*, **125**:2076, 1978.
172. D. Squire, C. Dulcey, and M. Lin. *J. Vac. Sci. Technol. B*, **3**:1513, 1985.
173. G. Stringfellow. *Organometallic Vapor Phase Epitaxy: Theory and Practice*. Academic Press, San Diego, USA, 1989.
174. S. Strite. *Jpn. J. Appl. Physics*, **33**:L699, 1996.

175. S. Strite and H. Morkoc. *J. Vac. Sci. Technol. B*, **10**:1237, 1992.
176. J. Sumakeris, Z. Sitar, K. Ailey-Trent, K. More, and R. Davis. *Thin Solid Films*, **225**:244, 1993.
177. C. Sun, S. Keller, G. Wang, M. Minsky, J. Bowers, and S. DenBaars. *Appl. Phys. Lett.*, **69**(13):1936, 1996.
178. C. Sun, P. Kung, A. Saxler, H. Ohsato, E. Bigan, M. Razeghi, and D. K. Gaskill. *J. Appl. Phys.*, **76**:236, 1994.
179. C. Sun and M. Razeghi. *Appl. Phys. Lett.*, **63**(7):973, 1993.
180. C. Sun, J. Yang, Q. Chen, M. Khan, T. George, P. Chang-Chien, and S. Mahajan. *Appl. Phys. Lett.*, **68**(8):1129, 1996.
181. S. Tanaka, S. Iwai, and Y. Aoyagi. *J. Cryst. Growth*, **170**:329, 1997.
182. T. Tanaka, A. Watanabe, H. Amano, Y. Kobayashi, I. Akasaki, S. Yamazaki, and M. Koike. *Appl. Phys. Lett.*, **65**(5):593, 1994.
183. M. Tchounkeu, O. Briot, B. Gil, J. Alexis, and R. Aulombard. *J. Appl. Phys.*, **80**(9):5352, 1996.
184. A. Thompson, C. Yuan, A. Gurary, R. Stall, and N. Schumaker. In B. Gil and R. L. Aulombard, eds, *Semiconductor Heteroepitaxy*, p. 266. World Scientific, 1995.
185. A. Thon and T. Kuech. *Appl. Phys. Lett.*, **69**(1):55, 1996.
186. Thurmond and Logan. *J. Electrochem. Soc.*, **119**(5):623, 1972.
187. J. Tsao, *Materials Fundamentals of Molecular Beam Epitaxy*. Academic Press, San Diego, USA, 1993.
188. H. Tsuchiya, A. Takeuchi, M. Kurihara, and F. Hasegawa. *J. Cryst. Growth*, **152**:21, 1995.
189. K. Uchida, A. Watanabe, F. Yano, M. Kouguchi, T. Tanaka, and S. Minagawa. *J. Appl. Phys.*, **79**:3487, 1996.
190. A. Wickenden, D. Wickenden, and T. Kistenmacher. *J. Appl. Phys.*, **75**(10):5367, 1994.
191. A. Wickenden, D. Wickenden, T. Kistenmacher, S. Ecelberger, and T. Poehler. *Mat. Res. Soc. Symp. Proc.*, **280**:355, 1993.
192. D. Wickenden, C. Bargeron, W. Bryden, J. Miragliotta, and T. Kistenmacher. *Appl. Phys. Lett.*, **65**(16):2024, 1994.
193. D. Wickenden, J. Miragliotta, W. Bryden, and T. Kistenmacher. *J. Appl. Phys.*, **75**(11):7585, 1994.
194. X. Wu, D. Kapolnek, E. Tarsa, B. Heying, S. Keller, B. Keller, U. Mishra, S. DenBaars, and J. Speck. *Appl. Phys. Lett.*, **68**(10):1371, 1996.
195. M. Yoshida, H. Watanabe, and F. Uesugi. *J. Electrochem. Soc.*, **132**:677, 1985.
196. S. Yoshida, S. Misawa, and S. Gonda. *J. Appl. Phys.*, **53**(10):6844, 1982.
197. S. Yoshida, S. Misawa, and S. Gonda. *Appl. Phys. Lett.*, **42**:427, 1983.
198. N. Yoshimoto, T. Matsuoka, T. Sasaki, and A. Katsui. *Appl. Phys. Lett.*, **59**(18):2251, 1991.
199. Z. Yu, J. Edgar, A. Ahmed, and A. Rys. *J. Electrochem. Soc.*, **138**:196, 1991.
200. C. Yuan, T. Salagaj, W. Kroll, R. Stall, M. Schurman, C. Hwang, Y. Li, W. Mayo, Y. Lu, S. Krishnankutty, and R. Kolbas. *J. Electron. Mater.*, **25**(4):749, 1996.
201. X. Zhang, P. Kung, A. Saxler, D. Walker, T. Wang, and M. Razeghi. *Appl. Phys. Lett.*, **67**(12):1745, 1995.

4 Structural defects and materials performance of the III–V nitrides

Fernando A. Ponce

4.1 Introduction

The collection of extended structural discontinuities in the lattice, such as interfaces, dislocations and stacking faults, is referred to as the microstructure of the material. The microstructure is sensitive to steps involved in crystal growth, such as the substrate used for epitaxy, the growth method itself, and variations in chemical composition during growth. The microstructure is expected to have a substantial effect upon other properties, such as electric transport and optical performance. It is known that in almost all semiconductors dislocations play a detrimental role in the electronic properties for applications ranging from integrated circuits to diode lasers. Achieving superior semiconductor performance has typically meant reducing the density of extended defects to levels below $10^3 \, \text{cm}^{-2}$.

The microstructure of the Group III nitrides is interesting in many ways, and its effect on the materials performance appears to follow opposite trends when compared to other III–V compounds. Although the nitrides were initially difficult to grow with specular morphology, the evolution of the growth method has led to exciting high optoelectronic performance in this decade, much to the surprise of many. Thus, breaking away from traditional thinking, the nitrides appear to shine in spite of large defect densities [1]. Current existing data suggest that the peculiar microstructure is somewhat responsible for the high optoelectronic performance of these materials. This chapter reviews the main structural characteristics associated with the growth of nitride thin films, including analysis of the substrate/thin film interface; the polarity of growth; the morphology of crystalline defects; the structure of dislocations, nanopipes, and inversion domains; and finally the correlation between microstructure and the spatial variation of electronic and optical properties.

4.1.1 Growth and microstructure

In contrast with other semiconductors, such as silicon and GaAs, bulk single crystals of the Group III nitrides are not currently available commercially. The difficulties in the growth of bulk single crystals of the nitrides are due in part to the high melting temperatures and elevated equilibrium vapor pressures (Table 4.1). Millimeter-size crystals of GaN and AlN have been produced in the

Table 4.1 Melting characteristics of the nitrides.

Nitrides	Melting Temp. (K)	Melting entropy (cal mol^{-1} K^{-1})	Melting enthalpy (kcal mol^{-1})	Reference
GaN	2791	16.01	44.68	Karpinski *et al.*
AlN	3487	16.61	57.92	Van Vechten
InN	2146	14.1		Grzegory *et al.*

J. Karpinski, J. Jun, and S. Porowski, *J. Cryst. Growth* **66**, 1 (1984).
J. A. Van Vechten, *Phys. Rev. B* **7**, 1479 (1973).
I. Grzegory, J. Jun, S. Krukowski, M. Bockowski, W. Wroblewski, and S. Porowski, *Physica B* **185**, 99 (1993).

laboratory. Single crystals of GaN are currently produced in the form of platelets with lateral dimensions of several millimeters and thickness of the order of 0.1 mm [2]. They are grown from gallium melt saturated with molecular nitrogen at 1600 °C under nitrogen vapor pressures of about 15 kbar; the crystals are typically highly conductive due to high levels of impurities (oxygen being the most abundant) and have regions with very low dislocation densities (under 10^6 cm^{-2}) [3, 4]. AlN bulk single crystals with dimensions of a few millimeters were reported by Slack and McNelly in 1976 [5]. They were grown by sublimation in closed tungsten crucibles, and the technique shows promise for growth of cm-size crystals. It appears that much progress is still necessary to achieve larger sizes and higher yields before Group III nitride substrates become commercially available.

In the absence of bulk nitride crystals, the use of foreign substrates is necessary. ZnO and SiC have characteristics which are similar to the nitrides. Wurtzitic ZnO crystals are available, but they are unstable in the highly corrosive environment and high temperatures needed for growth of the nitrides. SiC has proved to be a suitable substrate and further discussion of growth on silicon-terminated crystals is presented later. By far the most useful substrate to date is sapphire. This is surprising given the large differences in thermal expansion and in lattice parameter between sapphire and the nitrides. However, epitaxy of nitrides on sapphire has produced the best device performance to date. For this reason, most of the discussion in this chapter relates to the GaN/sapphire epitaxial system.

Epitaxial thin films can be produced in a variety of ways, the most common methods being hydride vapor phase epitaxy (HVPE), molecular beam epitaxy (MBE), and metalorganic chemical vapor deposition (MOCVD). The microstructure of these films vary markedly with technique and with the growth parameters. HVPE uses ammonia (NH$_3$) and GaCl (formed by HCl flowing over Ga melt) as sources with purified N$_2$ as carrier gas, and growth is typically performed at 1100 °C [6]. This method produces high growth rates and is suitable for the production of thick films (tenths of millimeter in thickness). MBE uses liquid Ga and atomic nitrogen as the source. The production of atomic nitrogen requires the use of one of various technologies, such as electron–cyclotron resonance [7], in order to atomize the molecular nitrogen.

This process produces highly energetic particles that result in somewhat poor growth efficiencies and the formation of subsurface defects due to atomic impact during growth. Various alternative methods for the production of atomic nitrogen such as ammonia (gas-source MBE) have been used with relative success [8]. MOCVD is widely used in the industry; virtually all commercially available optoelectronic devices use materials grown by this technique. It is no surprise that such is the case with the nitrides. Ammonia is used as the nitrogen source, and the tri-methyl forms of Ga, Al, and In are used as sources for the Group III elements [9]. Best optoelectronic performance has been achieved using low-temperature buffer layers (to achieve smooth films) typically deposited at around 550 °C, followed by film grown at temperatures in the range 700–1100 °C. The buffer layer plays a key role in determining all properties of the film. Initially with an amorphous structure, it undergoes solid phase crystallization as the temperature is increased. The resulting crystallites are closely related to the substrate and act as seeds for the growth of the film, the latter acquiring a columnar structure to be discussed in more detail later.

4.1.2 The evolution of GaN thin film morphology

It is at first surprising that GaN thin films grown on sapphire are suited for use in high-efficiency light-emitting devices. The high temperatures required for growth by vapor phase epitaxy, and the absence of 'suitable' substrates, make for poor predictions for the performance of epitaxial layers, a reason for which the nitrides were largely ignored until the beginning of this decade. The large differences in lattice parameter and in thermal expansion rates between GaN and sapphire forecast serious difficulties in the growth of high-quality films for optoelectronic applications. Thus their excellent device performance was unexpected. Many of the properties of the nitrides are related to the rich microstructure of the epitaxy, which is much different from other semiconductor systems.

The development of the current state of technology took much work and significant breakthroughs over the last three decades. First reported in 1907, the nitrides were among the first semiconductors ever discovered [10], their crystalline structure having been known since 1937 [11]. The early deposition of GaN on (0001) sapphire by Maruska and Tietjen in 1969 used ammonia and HCl flow for transport for Ga and resulted in colorless, single epitaxial crystalline layers [12]. Pankove reported the first electroluminescent GaN diodes in 1971 using metal–insulator–semiconductor devices [13]. That same year, Manasevit used metalorganic chemical vapor deposition (MOCVD) to grow GaN thin films [14]. The introduction of AlN as an intermediate layer by Yoshida *et al.* [15] in 1983 improved the morphology of the layers. It was the introduction of low-temperature nucleation layers by Amano and Akasaki [16] in 1986 that produced the first smooth films and allowed control of impurity incorporation during growth, a necessary step in order to achieve adequate doping in device structures. The achievement of p-type layers using low-energy electron beam irradiation by Amano and Akasaki in 1989 led to the first p–n junction LED [17].

Further improvements were achieved with the introduction of low-temperature GaN buffer layers by Nakamura [18] in 1991. Shortly after, Nakamura demonstrated p-type activation by thermal annealing [19], opening the road for increasingly better light-emitting diodes which culminated with the commercial introduction of blue and green LEDs in 1994. Some of these milestones had direct impact on the microstructure of the nitride films, although no systematic studies were recorded.

The first reports on the structural properties focused on the surface morphology (smoothness). X-ray diffraction rocking curve widths were used as a measure of crystal quality. Nakamura reported that the optoelectronic properties did not necessarily improve with crystalline quality as assessed by X-ray rocking curve widths [18]. Both are sensitive to buffer layer thickness, but optimum electrical mobility and light emission efficiencies did not occur for optimized crystalline parameters, as would be expected. Nakamura's best growth conditions happened for XRRC widths of about 6 arcmin. No mention was made about dislocation densities, although Akasaki et al. [20] did report large defect densities associated with the early stages of growth. Because other semiconductors, such as the arsenides and the phosphides, have shown a significant correlation between extended crystalline defects, such as dislocations, and light emission efficiencies in device structures, significantly more work was believed necessary before commercially viable devices could be produced. With this background, the announcement of candela-class LED performance [21] produced much excitement and enthusiasm. Measurements performed on those LEDs showed a characteristic high dislocation density, several orders of magnitude above those observed in other high-performance semiconductor thin films [22].

There has been much interest in establishing the connection between the crystalline defect structure and the optoelectronic properties of the nitrides. Of particular interest is the microstructure of GaN thin films grown by MOCVD on sapphire substrates, following the growth techniques initially conceived by Akasaki and Amano, and further developed by Nakamura.

4.2 The crystalline structure of the nitrides

Like most other semiconductor materials, the nitrides have a tetrahedrally coordinated atomic arrangements that results in either cubic (zincblende) or hexagonal (wurtzite) lattice structures. The wurtzite form is the easiest to grow and has given best results to date for optoelectronic applications. The atomic arrangement of the nitrides can be viewed as consisting of hexagonal double layers, one layer is occupied by nitrogen, while the other contains the Group III elements. The zincblende structure occurs when the hexagonal layers are stacked in a periodic ...ABCABC... sequence, while the wurtzite structure follows an ...ABAB... arrangement, shown in Fig. 4.1. Transition from one structure to the other can be produced by slip generated by motion of a dislocation along the basal plane, with the consequent change of the stacking sequence.

4.2 The crystalline structure of the nitrides

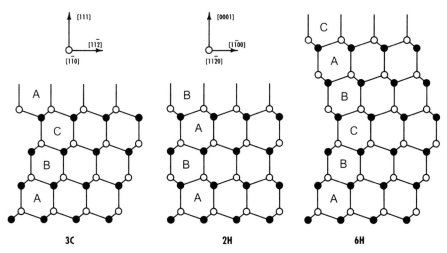

Fig. 4.1 Atomic arrangement in tetrahedrally coordinated structures. (a) cubic zincblende, (b) hexagonal wurtzite, and (c) 6H hexagonal lattice.

Figure 4.2 shows the bandgap plotted against the interatomic distance for the nitrides and other wide-bandgap semiconductors. In contrast with the arsenides, where the $Al_xGa_{1-x}As$ system has only small variations, the nitrides are characterized by large changes in lattice parameter associated with different chemical compositions (see also Table 4.2). Thus, large values of lattice mismatch are expected for heterojunctions involving the different Group III elements; e.g. 2.4% for GaN/AlN, 10.6% for GaN/InN, and 13% for AlN/InN.

Table 4.2 Lattice parameters and interplanar distances relevant to epitaxy on the basal planes. All hexagonal forms of SiC and the AlGaN system belong to the space group $P6_3mc$ (186). The coefficient of linear thermal expansion for the nitrides varies with temperature in a nonlinear fashion (for more complete data see e.g.: O. Madelung, ed. *Data Science and Technology, Semiconductors: Group IV Elements and III–V Compounds*, Springer-Verlag, Berlin, 1991). The lattice mismatches for basal plane interfaces are given.

Crystalline properties		GaN	AlN	InN	α-Al$_2$O$_3$	6H–SiC
Lattice Parameter (Å)	a	3.186	3.1114	3.5446	4.758	3.081
	c	5.178	4.9792	5.7034	12.991	15.092
	c/a	1.6252	1.6003	1.6090	2.730	1.633($\times$3)
Thermal expansion coefficient ($\times 10^{-6}$ C^{-1})	a	5.59	4.2	5.7	7.5	4.2
	c	3.17	5.3	3.7	8.5	4.68
Interplanar distances (Å)	Basal	2.59	2.49	2.85	2.165	2.516
	(1$\bar{1}$00)	2.760	2.695	3.070	1.374	2.669
	(11$\bar{2}$0)	1.593	1.556	1.772	2.379	1.541

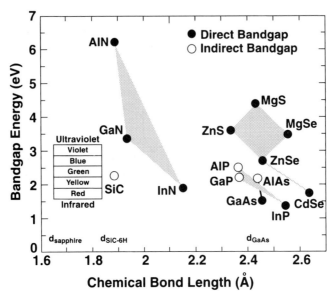

Fig. 4.2 Bandgap and chemical bond lengths of semiconductor compounds that emit in the visible range of the electromagnetic spectrum. The chemical bond length or interatomic separation is shown here, instead of the lattice parameter, to allow proper comparison of the hexagonal (wurtzite) nitrides with the cubic (zincblende) phosphides and arsenides.

Such large variations in lattice parameter should play a significant role in the properties of nitride heteroepitaxy. Within the quaternary $Al_xIn_yGa_{1-x-y}N$ system it is possible to obtain the desired bandgap over a range of lattice parameters, as shown in Fig. 4.2. It should also be noted that the coefficient of linear thermal expansion varies significantly within the nitrides and in all cases has a substantial difference with sapphire (e.g. 36% for GaN and 78% for AlN with respect to sapphire). The large difference in thermal expansion rates, together with the high temperature required to pyrolyze the MOCVD reactants, should result in large thermal strains at room temperature. Specifically, the change in size of the lattice parameter during cooling from 1050 °C augments the lattice mismatch by an equivalent of 0.2% for GaN and 0.33% for AlN on sapphire.

4.3 Structure of the epilayer / substrate interface

The quality of epitaxy on foreign substrates depends strongly on the nature of the interface between the substrate and the epilayer, since it is at this point where the film is nucleated, the relationship between the substrate and the layer is established, and any differences in atomic arrangement are resolved. The choice of substrate plays an important role in the resulting microstructure. In the case of the nitrides, one of the main restrictions is the chemical stability at the elevated temperatures and in the corrosive environments typical of growth with nitrogen sources. Sapphire (Al_2O_3) and SiC are the two most widely used substrates. As mentioned earlier, the initial difficulties in achieving smooth films were overcome when Amano et al. [16] demonstrated that low-temperature buffer layers produced epilayers with specular surfaces, thus overcoming the three-dimensional nucleation problems.

4.3.1 Interface between the nitrides and sapphire

The atomic arrangement between an AlN buffer layer and the sapphire substrate involves the transition from the aluminum oxide to aluminum nitride. A lattice image of such an interface is shown in Fig. 4.3. Sapphire (α-Al_2O_3) has a hexagonal lattice structure, and its (0001) basal plane is commonly used for epitaxy of the nitrides. The nitride layer has the hexagonal wurtzite structure with the basal planes lying parallel to the substrate surface. Lattice images indicate that the nitride/sapphire interface is planar and atomically sharp [23], meaning that the transition between the oxide to the nitride is abrupt and stable, without intermixing. A tendency towards local coherence at the AlN/AlO_3 interface is observed, with a slight distortion of the atomic planes towards a one-to-one correlation between lattice planes in the substrate and planes in the epilayer. Local coherence requires the presence of misfit dislocations at distances of 2.0 nm, corresponding to the 12.5% mismatch between the

Fig. 4.3 Lattice image of the AlGaN/AlN/sapphire interface region.

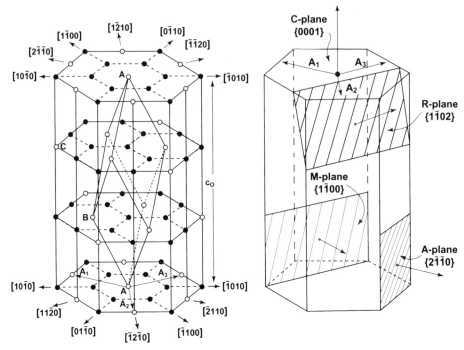

Fig. 4.4 Lattice structure of sapphire. The aluminum planes with array of vacancies are shown in the left. These planes are intercalated with hcp oxygen layers (not shown). The morphological rhombohedral unit cell is drawn connecting the aluminum vacancies. Sapphire planes used in epitaxy of nitrides are shown at the right.

bulk lattice parameter values. The critical thickness for generation of misfit dislocations at such large lattice mismatch is less than a monolayer, thus the stress associated with the lattice mismatch is expected to be immediately relieved during the solid phase crystallization process, and all misfit dislocations are observed to be located at the substrate interface.

In order to understand the interface between sapphire and the nitrides, it is necessary to discuss the atomic arrangement of sapphire. Sapphire can be expressed as a hexagonal unit cell in Fig. 4.4, which consists of hexagonal close packed planes of oxygen, alternated with a hexagonal array of aluminum planes [24]. The aluminum planes are also in a hexagonal close-packed arrangement, with one third of the sites vacant. The vacant sites produce the 2/3 stoichiometry ratio of Al/O in sapphire. The Al planes are arranged with the Al vacancies ordered with a three-fold symmetry axis along the [0001] direction.

An idealized transition from the oxide to the nitride is shown in Fig. 4.5. The aluminum atom is bonded to three oxygen atoms below and two nitrogen atoms on top. The bulk arrangements of nitrogen atoms in the nitride and oxygen atoms in the sapphire do not facilitate this 'ideal' configuration. The AlN/sapphire interface has to satisfy not only the period atomic array, but also the distribution of anions and cations in order to produce a neutral charge

4.3 *Structure of the epilayer / substrate interface* 131

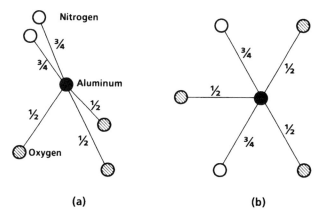

Fig. 4.5 Idealized transition from sapphire to AlN. Aluminum is bonded to oxygen (below) and nitrogen (above). (a) Lateral view; (b) [0001] projection (top view).

density at the interface [25]. Two possible arrangements are shown in Fig. 4.6. These give two different polar orientations to the nitride lattice. Simple arithmetic shows that model A requires an excess quantity of aluminum at the interface, whereas the reverse is true for model B.

Epitaxy with GaN buffer layers also shows an abrupt and planar interface with the sapphire substrate [26], as shown in Fig. 4.7. The coherence observed in the

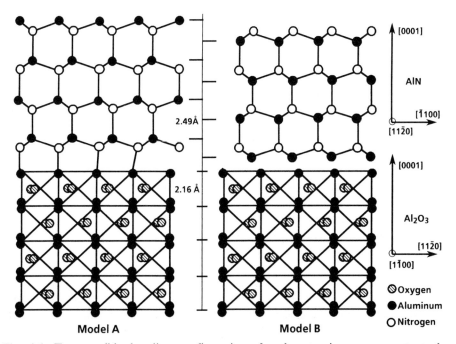

Fig. 4.6 Two possible bonding configurations for the atomic arrangement at the AlN/sapphire interface, yielding to two different polar orientations for AlN.

132 *Defects and performance of the III–V nitrides*

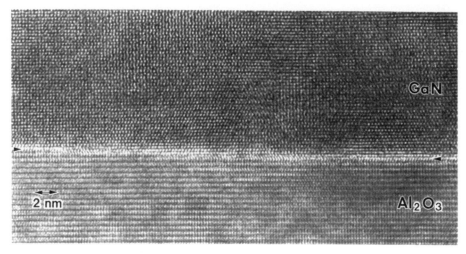

Fig. 4.7 Lattice image of the GaN/sapphire interface.

case of AlN is not easily observed in the case of GaN. This may be due to the presence of Ga and Al at the interface. The mixing of Al and Ga at the interface may introduce a means for relaxation of the local strain associated with lattice mismatch.

4.3.2 The AlN/SiC interface

Silicon carbide has a similar atomic arrangement to the nitrides. The tetrahedral coordination and the hexagonal symmetries are common in both materials, with a polar structure consisting of alternating silicon and carbon layers characteristic of the wurtzite and zincblende structures. The arrangement of the basal planes in SiC occurs in a variety of ways [25]. The polymorphism of SiC results in lattice structures being cubic (C), hexagonal (H), or rhombohedral (R). The lattice arrangements are typically described by the symmetry (C, H, or R) and the number of planes in a period in the direction normal to the basal planes. Thus, 2H is the wurtzite structure and 3C is the zincblende lattice. 6H–SiC is the most common commercially available form; its lattice arrangement is shown in Fig. 4.1(c). SiC and AlN have nearly identical linear thermal expansion characteristics, and their lattice mismatch is <1% (see Table 4.3). The better match at the basal planes between SiC and AlN is expected to facilitate the formation of nearly perfect junctions. Epitaxial layers grow best on the silicon-terminated surface, which happens to be morphologically more stable than the carbon-terminated surface. A high-resolution transmission electron micrograph of the interface region of AlN grown by MOCVD on SiC is shown in Fig. 4.8 [27]. This lattice image contains information regarding the atomic arrangement at the interface, which should be responsible for the lattice orientation and for the stable growth configuration of the nitride epilayer. The lattice orientation can be determined by careful analysis of the intensity contrast and interplanar

4.3 Structure of the epilayer / substrate interface

Table 4.3 The lattice mismatch and misfit dislocation separation corresponding to complete misfit relaxation at (0001) interfaces are given.

Crystalline properties		GaN	AlN	InN	α-Al$_2$O$_3$	6H–SiC
Lattice mismatch with	Sapphire	14.8%	12.5%	25.4%	–	11.5%
	SiC	3.3%	1.0%	14.0%	−11.5%	–
	GaN	–	−2.4%	10.6%	−14.8%	−3.3%
Dislocation distance on	Sapphire	17.2	20.3	10.6	–	21.9
	SiC	80.9	276.7	20.4	21.9	–
	GaN	–	114.4	27.3	17.2	80.9

separations observed in the lattice structure images. It has been shown that atomically abrupt planar interfaces correspond to two possible configurations:

C–Si—C–Si—C–Al—N–Al—N–Al   and   C–Si—C–Si—N–Al—N–Al,

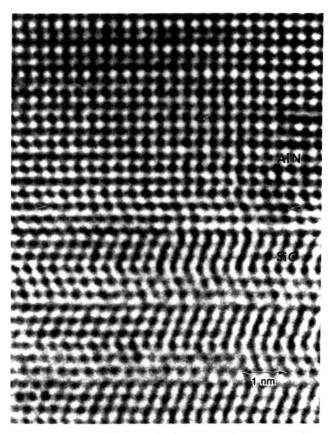

Fig. 4.8 Lattice image of the AlN/SiC interface. Note the transition between 6H–SiC to 2H–AlN, identified by the transition from zigzag to vertical columns of lattice spots.

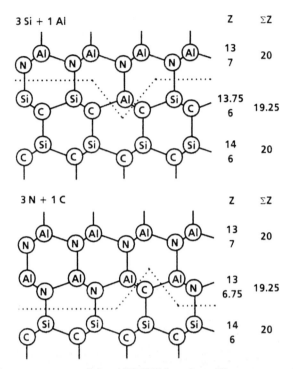

Fig. 4.9 Atomic arrangement of the AlN/SiC interface. These two models are consistent with lattice images and with a neutral charge interface.

which are indistinguishable by HRTEM [28]. Considerations of the anion–cation distribution and the requirement of local charge neutrality indicate the need for a model involving a combination of these two configurations, shown in Fig. 4.9, in a ratio of 3 to 1 [29]. Thus, it is possible in principle to grow high-quality nitride layers on (0001) SiC, with highly coherent interfaces and low dislocation densities. In practice, atomic steps in the basal plane of the 6H structure produce lateral offsets (see Fig. 4.1) which cannot be easily compensated by the wurtzite structure, causing the appearance of threading dislocations and stacking faults in nitride films grown by MOCVD. Such lateral shifts are not present in 2H–SiC or in sapphire.

4.3.3 Homoepitaxy

The ideal case for epitaxy is when the same material is used as the substrate. As discussed earlier, the technology for growth of bulk single-crystal GaN has recently been developed to a state-of-the-art production of crystalline platelets with lateral dimensions ranging from 3 to 5 mm, and thickness of the order of 100 μm. The faces of the platelets are closely parallel to the crystallographic (0001) GaN planes. The polar orientation of the faces were determined by convergent beam electron diffraction [30], as discussed in some detail later in the chapter. Polished platelets were used for epitaxy using MOCVD at the standard growth temperatures of 1050 °C.

4.4 Microstructure of MOCVD thin films

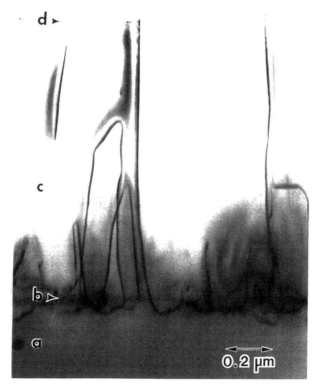

Fig. 4.10 TEM image of the substrate/epilayer region in homoepitaxial growth along the $(000\bar{1})$ direction. Dislocation loops originate from stacking faults that intersect the surface.

Growth on the A-face (gallium on the external position of the basal plane) of the bulk platelets produces specular surfaces with no dislocations visible by transmission electron microscopy, indicating a dislocation density of $<10^6$ cm^{-2}. Epitaxy on the B-face (N-terminated) of the platelets can also produce highly specular morphologies. The TEM image in Fig. 4.10 shows that some dislocation loops are generated at the substrate interface [31]; these dislocations originate at stacking faults which are common near the $(000\bar{1})$ surface (B-face) of bulk GaN crystals [32].

With the improved microstructure associated with homoepitaxial films, considerably superior optical properties have been reported [33, 34], demonstrating some of the benefits of homoepitaxy.

4.4 Microstructure of MOCVD thin films

As stated earlier, films grown by MOCVD have demonstrated high light emission efficiencies and have been used in commercial device manufacturing since late 1993. The microstructure of these thin films depends strongly on the nature of the buffer layer. Figure 4.11 is a cross-section transmission electron micrograph of a high performance Nichia blue LED [16]. The film was grown on

136 *Defects and performance of the III-V nitrides*

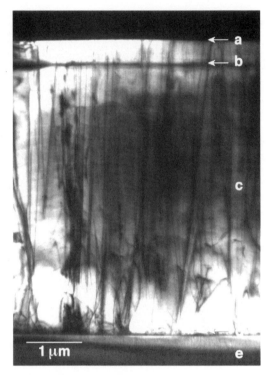

Fig. 4.11 Cross-section TEM image of high-efficiency blue LED. The dislocation density is 10^{10} cm^{-2}. Indicated in the figure are (a) the top surface of the film, (b) the active region, (c) the main GaN film, (d) the buffer layer, and (e) the sapphire substrate.

(0001) sapphire by MOCVD [35]. The device layers consist of ~4 μm of Si-doped n-type GaN, followed by an AlGaN:Si/InGaN:(Si + Zn)/AlGaN:Mg double heterostructure, and a p-type GaN:Mg cap layer (here ':Si' means 'doped with silicon'). The InGaN active layer was heavily doped with Si and Zn to provide recombination centers that emit at ~450 nm. The active areas of the LEDs were ~3×10^{-4} cm^2. The device had a maximum external quantum efficiency of 4% at 0.8 mA. The total output power increases steadily with drive current to about 3.3 mW at 100 mA.

The image of Fig. 4.11 was taken under two-beam bright-field conditions which highlights the dislocation distribution in the material. Dislocations are seen as dark lines propagating in a direction normal to the substrate (vertical direction in the figure). The distribution of dislocations stabilizes after the first fraction of a micron of growth. Beyond that point the density and distribution of dislocations remain constant. In addition to the dislocations that propagate in the [0001] direction, there are dislocations that lie on the basal planes, and can be observed with higher contrast under other imaging conditions. The vertical dislocations cross the active region of the device. They appear in clusters with a separation of about 0.3 μm, yielding a dislocation density of ~10^{10} cm^{-2}. A view from the top is shown in Fig. 4.12; the dislocations in this planar-view TEM image have a cellular arrangement corresponding to a columnar structure of

4.4 Microstructure of MOCVD thin films   137

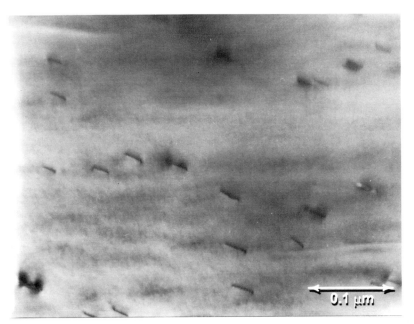

Fig. 4.12 View from the top of the film. Plan view of dislocation distribution in GaN/sapphire thin films. Dislocation segments are observed in a tilted configuration.

crystallites with a very small angular distribution in orientation, as shown in Fig. 4.13. There are two components to this distribution: one is the tilt of the c-axis with respect to the growth direction (full width at half maximum of ~6 arcmin),

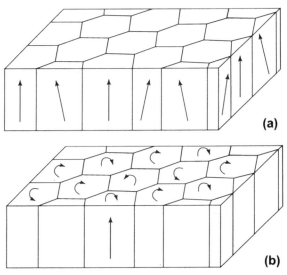

Fig. 4.13 Columnar structure model for the III–V nitrides. The films consist of columns with a slight distribution in the crystalline orientation with two components: (a) tilt (~5 arcmin) and (b) twist (~8 arcmin).

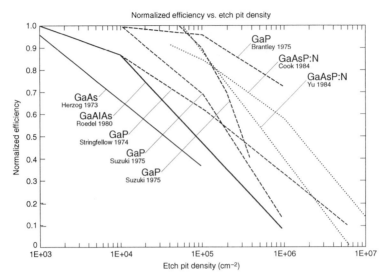

Fig. 4.14 Dependence of LED efficiency on dislocation densities for a variety of III–V materials used in commerical devices [22]. References shown on the figure are as follows. A. H. Herzog, D. L. Deune, and M. G. Craford, *J. Appl. Phys.* **43**, 600 (1972). R. J. Rodel, A. R. von Neida, R. Causo, and L. R. Dawson, *J. Electrochem. Soc.* **126**, 641 (1979). W. A. Brantley, O. G. Lorimor, P. D. Dapkus, S. E. Haszko, and R. H. Saul, *J. Appl. Phys.* **46**, 2629 (1975). G. B. Stringfellow, P. F. Lindquist, T. R. Cass, and R. A. Burmeister, *J. Electron. Mater.* **3**, 497 (1974). T. Susuki and Y. Masumoto, *Appl. Phys. Lett.* **26**, 431 (1975). L. W. Cook and J. G. Yu (unpublished).

and the other is the twist of the column's orientation about the c-axis. These low-angle tilt- and twist-boundaries require dislocations with separations of the order of 100 nm. TEM observations of such high densities of dislocations in highly efficient LEDs demonstrate that dislocations in the nitrides are not efficient nonradiative recombination centers as they are in other semiconductors.

Past experience with the AlGaAs and GaAsP systems indicate that LED efficiencies are very sensitive to the presence of dislocations in the thin films [22]. Figure 4.14 depicts the dependence of the LED efficiency on dislocation density of devices made with a wide range of III–V materials. Dislocation densities of the order of $10^3 \, \text{cm}^{-2}$ are known to adversely affect the light emission efficiencies of GaAs and AlGaAs. This is seven orders of magnitude below the observed densities in GaN/sapphire. The fact that efficient LEDs can be made from these III–V nitrides films with such high densities of structural defects indicates a critical difference between the nitrides and the phosphides and arsenides. This property can be attributed to the highly ionic character of bonding in these materials. Reports based on the I–V characteristics of metallized contacts indicate that the surface of GaN is unpinned [36], implying that minority carrier recombination is expected to be insensitive to the presence of surfaces in the III–V nitrides. It has been proposed that since dislocations are discontinuities which can be visualized as internal surfaces, it should follow that they are not efficient minority carrier recombination centers [36].

4.5 Characterization of dislocations in GaN

One of the main characteristics of a dislocation is its Burgers vector, which describes the net lattice displacement associated with the defect. To achieve unambiguous results in the determination of dislocation Burgers vectors in GaN, a combination of conventional diffraction contrast and large-angle convergent beam electron diffraction (LACBED) techniques has been used [37]. In the

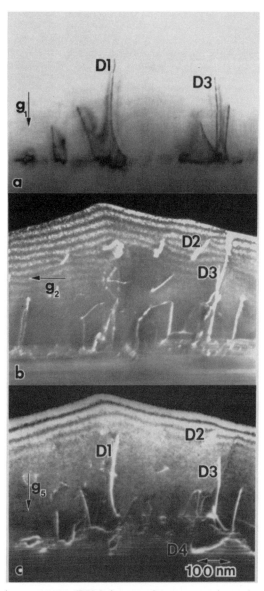

Fig. 4.15 Diffraction contrast TEM images in cross-section, showing a variety of dislocations in a GaN homoepitaxial film. (a) Bright field with **g** = [0002] close to the [1$\bar{1}$00] pole. (b) Dark-field image with **g** = [01$\bar{1}$0]. (c) Dark-field image with **g** = [$\bar{1}$$\bar{1}$20].

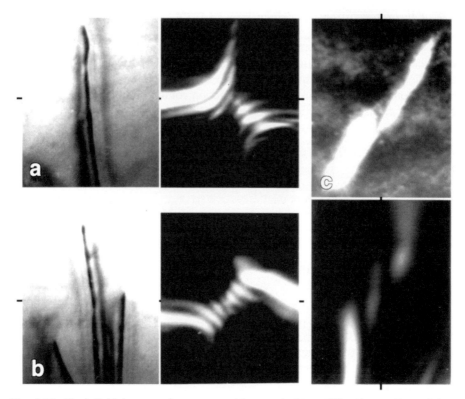

Fig. 4.16 Dark-field large-angle convergent-beam electron diffraction patterns taken from dislocations D1, D2, and D2 in Fig. 4.15. (a) The contour $\mathbf{g} = [0006]$ crossing D1 as indicated by the dotted line. (b) $\mathbf{g} = [0006]$ crossing D3. (c) $\mathbf{g} = [22\bar{4}0]$.

conventional diffraction contrast method, images taken under two-beam diffraction conditions are used to investigate the Burgers vector **b** using the invisibility criterion, i.e. $\mathbf{g} \cdot \mathbf{b} = 0$ when dislocations are out of contrast; thus, using values of **g** with different orthogonal components, the direction of **b** can be established. Cherns and Preston [38] demonstrated that an LACBED Bragg line may be split and displaced when intersecting a dislocation. The displacement and multiplicity of the splitting is directly related to the sign and the magnitude of the Burgers vector. Figure 4.15 is a TEM image taken in cross-section of a GaN film showing a variety of dislocations [37]. Three of the dislocations are followed in a series of images under different diffraction conditions; D1 and D2 disappear under some specific conditions, whereas D3 remains visible in all three conditions. From these images, the components of the Burgers vector are deduced. Figure 4.16 consists of LACBED images of the same dislocations showing the multiplicity and sense of the splitting of the Bragg lines, from where the magnitude and sign of the Burgers vector are determined. These observations performed on GaN epitaxy on GaN crystal and sapphire substrates show that three types of dislocations are present, with Burgers vectors **a**, **c**, and **a** + **c**, where **a** and **c** are the unit cell vectors of the hexagonal wurtzite lattice (Table 4.4). The magnitudes of **c** and **a** + **c** are relatively large for semiconductors, particularly if we

Table 4.4 Magnitude of the Burgers vectors for dislocations observed in the nitrides.

Burgers vectors	GaN	AlN	InN
$\mathbf{a} = \frac{1}{3}\langle 11\bar{2}0 \rangle$	3.186	3.111	3.545
$\mathbf{c} = [0001]$	5.178	4.979	5.703
$\mathbf{c} + \mathbf{a} = \frac{1}{3}\langle 11\bar{2}3 \rangle$	6.080	5.871	6.715

take into consideration the elastic moduli and bond energies typical of the nitrides (Table 4.5). The strain energy associated with screw dislocations is proportional to the factor μb^2, where μ is the shear modulus ($= C_{44}$), and b is the magnitude of the Burgers vector [39]. Table 4.6 shows μb^2 for dislocations in GaN, GaAs, and Si.

4.6 Polarity determination

The quality of epitaxy is expected to be sensitive to the growth direction. Surface absorption kinetics should play an important role in the growth velocity and in the rate of absorption of impurities. The wurtzite structure has a polar asymmetry along the (0001) direction, which is typically the growth direction. By convention, the [0001] direction is chosen in the direction of the bond from the cation (group III element) to the anion (nitrogen) as shown in Fig. 4.17. The polarity of growth for bulk single crystals and epitaxial layers has been determined by convergent beam electron diffraction techniques [30]. It has been reported that the smooth side of bulk single-crystal platelets correspond to Ga-on-top (0001) planes. It has also been shown that epitaxy on (0001) and

Table 4.5 Elastic moduli for GaN, GaAs, and silicon. The experimental values for the hexagonal wurtzite form of the nitrides are taken from Polian *et al.*[1] Experimental values for Si and GaAs are from Madelung.[2] C_1, C_{12}, and C_{44} are in 10^{11} dyn cm^{-2} at 300 K. The cohesive enegy is from Harrison.[3]

	GaN	GaAs	Si
C_{11}	39.0	11.9	16.577
C_{12}	14.5	5.38	6.393
$C_{44} = \mu$	10.5	5.95	7.962
E_{coh} (eV)	2.24	1.63	2.32

[1] A. Polian, M. Grimsditch, and I. Grzegory, *J. Appl. Phys.* **79**, 3343 (1996).
[2] O. Madelung, *Semiconductors*, Data in Science and Technology Series, Springer-Verlag, 1991.
[3] W. A. Harrison, *Electronic Structure and the Properties of Solids*, W. H. Freeman, San Francisco, 1980, p. 176.

Table 4.6 Gb^2 for various dislocations and materials. The relative energy of dislocations is proportional to Gb^2, calculated using values in Tables 4.3 and 4.4. Burgers vectors for GaAs and Si of 3.997 and 3.84 Å are used. Values are in 10^{-3} ergs/cm.

Burgers vectors	GaN	GaAs	Si
$\mathbf{a} = \frac{1}{3}\langle 11\bar{2}0 \rangle = \frac{1}{2}\langle 110 \rangle$	1.066	0.951	1.174
$\mathbf{c} = [0001]$	2.815	–	–
$\mathbf{c} + \mathbf{a} = \frac{1}{3}\langle 11\bar{2}3 \rangle$	3.881	–	–

($000\bar{1}$) bulk GaN surfaces retain the lattice direction of the substrate. It should not be taken for granted that GaN is as likely to grow in either direction, since for the range of growth parameters one orientation may be favored. That does not seem the case in MOCVD, however, where it seems possible to grow smooth films in both directions. It has also been observed that hillock formation tends to happen in the ($000\bar{1}$) direction, and are absent in (0001) epitaxial growth. Figure 4.18 and 4.19 are cross-section TEM images of [0001] films grown on bulk GaN and sapphire, respectively [30]. The convergent beam diffraction patterns are compared with calculations in order to identify the Miller indices of the reflections, and hence the polarity. These results show that [0001] homoepitaxy does not introduce new defects into the film; the dislocation density being below the detection limit by TEM ($<10^6$ cm^{-2}). Epitaxy in the [$000\bar{1}$] direction shows a number of dislocations which are related to the intersection of stacking faults with the surface; this is a typical characteristic of the ($000\bar{1}$) face of the bulk GaN substrates that we studied. Epitaxial layers on sapphire with high optoelectronic properties grow in the [0001] direction, with Ga at the top

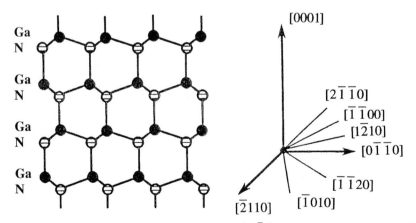

Fig. 4.17 Projection of the GaN lattice along the [$11\bar{2}0$] direction, with standard Miller indices notation. The [0001] is established in the direction of the cation-to-anion (Ga-to-N) bond between basal planes.

4.7 Nanopipes and inversion domains

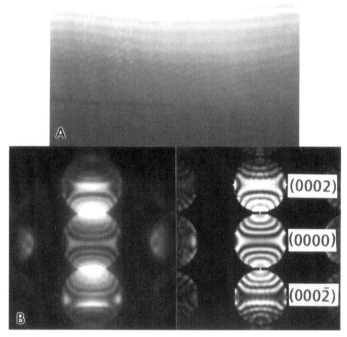

Fig. 4.18 Microstructure and polarity for homoepitaxy in the [0001] direction. (a) Cross-section TEM bright-field image showing absence of threading defects; a stacking fault is used to identify the substrate/film interface. (b) convergent beam electron diffraction pattern on left, calculated pattern on right.

position of the lattice and, as mentioned earlier, exhibit high dislocation densities ($\sim 10^{10}$ cm^{-2}).

4.7 Nanopipes and inversion domains

The line defects observed in Fig. 4.11 represent dislocations with the various Burgers vectors described in Section 4.5. In addition to these basic dislocations, others are also present with similar appearance of dislocations propagating in the [0001] direction of growth, and are not easily distinguished from the pure dislocations. Two types of such defects have been observed: coreless screw dislocations in the form of nanopipes and filamentary inversion domains [40].

4.7.1 Nanopipes

Figure 4.20 is a dark-field TEM image taken under two-beam diffraction conditions. In addition to pure dislocations (D), two other types of defect propagate in the [0001]GaN growth direction. These defects are labeled 'T' and 'I'. It is not easy to distinguish these defects in standard TEM cross-section

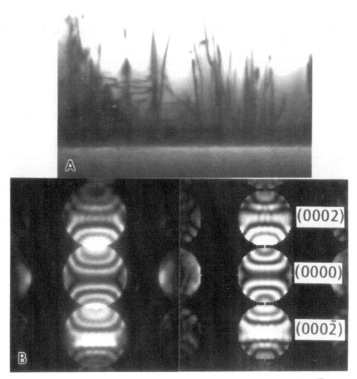

Fig. 4.19 Microstructure and polarity for homoepitaxy in the [000$\bar{1}$] direction. (a) Cross-section TEM bright-field image showing absence of threading defects; a stacking fault is used to identify the substrate/film interface. (b) convergent beam electron diffraction pattern on left, calculated pattern on right.

views. Closer inspection shows that the 'T' defects are actually empty tubes with dimensions of the order of 10 nm, as shown in Fig. 4.21. These hollow tubes have generally constant cross-section in the range 5 to 25 nm in diameter; they are faceted and exhibit an irregular hexagonal shape when viewed end-on.

Conventional TEM diffraction imaging analysis provides information about the direction of the Burgers vector. Large angle convergent beam electron diffraction (LACBED) imaging can accurately determine the magnitude, direction, and sense of the Burgers vectors. LACBED imaging has been performed on the nanopipes, demonstrating that these hollow tubes contain a Burgers vector in the [0001] direction with magnitude 0.52 nm (equal to the c constant). Thus, with a few notable exceptions, nanopipes can be described as coreless screw dislocations with elementary Burgers vectors of the **c** type [41]. Frank predicted that a dislocation whose Burgers vector exceeded a critical value should have a hollow tube at the core, and predicted an equilibrium core radius of $r_{eq} = \mu b^2 / 8\pi^2 \gamma$, where γ is the surface energy, μ is the shear modulus, and b is the Burgers vector [42]. Recent reports by Qian first noted that the contrast in two-beam bright-field TEM images from a tilted nanopipe was similar to dislocation contrast, but no proof was advanced [43, 44]. The difficulty lies in

4.7 *Nanopipes and inversion domains* 145

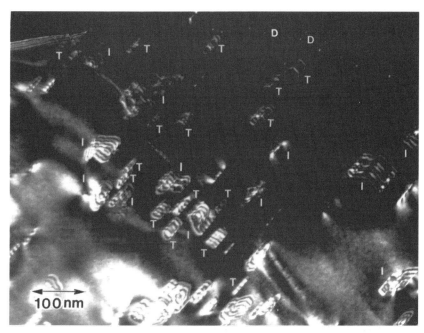

Fig. 4.20 Plan view transmission electron micrograph showing a region with a large density of nanopipes (T) and inversion domains (I). Dislocations (D) belonging to low-angle grain boundaries are also observed. This dark-field image was taken under two-beam diffracting conditions in $\mathbf{g} = 2\bar{2}02$.

being able to measure accurately the magnitude of the Burgers vector, since the actual value for the nanopipe could be an integral fraction of the Burgers vector

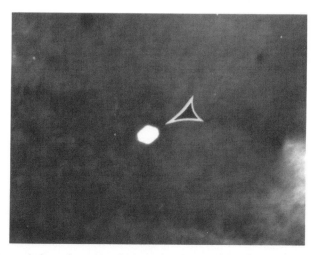

Fig. 4.21 Transmission electron micrograph of nanopipe observed along its axis. It presents an irregular hexagonal cross-section.

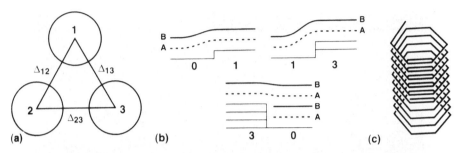

Fig. 4.22 Proposed mechanism for the generation of nanopipes in GaN/sapphire. (a) A dislocation is trapped if the island displacements $\Delta = \Delta_{12} + \Delta_{23} + \Delta_{31}$ do not sum to zero, the coalescence of the islands being associated with large elastic stress. (b) Steps on the sapphire surface in units of $c_{\text{sapphire}}/6$ can join to trap a screw dislocation, given that three atomic steps in sapphire are almost equal in height to two atomic layers (one lattice parameter) of GaN. (c) Addition of deposited material near a nanopipe containing a screw dislocation preserves the helicoidal nature of the defect about its central [0001] axis [40, 41].

of a perfect dislocation. The LACBED technique used in this study is a new technique which has found an important application in solving this problem.

The observed tube diameters are too large to agree with the simple Frank equation. An upper limit using existing theoretical data gives 0.5 nm for c-dislocations. Thus, coreless dislocations are not in an equilibrium state, but may arise by the trapping of dislocations at pinholes at the early stages of growth, as shown in Fig. 4.22(a). Trapped dislocations arise if the displacements of the islands do not sum to zero. A mechanism that favors the generation of c-dislocations is presented in Fig. 4.22(b), where it is shown that three-layer steps in the sapphire can couple to a two-layer step in the GaN lattice, thus giving rise to a c-type screw dislocation. Those dislocations with an a-component of the Burgers vector eventually close out, and only the pure c-type dislocations can survive to generate nanopipes of constant cross-sections [41]. The 3-to-2 ratio results from the 3-fold multiplicity in the stacking of lattice planes in sapphire and the two-layer period in wurzitic GaN, discussed in Sections 4.2 and 4.3, which happen to have very similar values.

4.7.2 Inversion domains

When the foil is tilted to excite a reflection with a c-component, a new set of defects appears which is not visible in reflections of type $hki0$ (i.e. in $\{10\bar{1}0\}$ or $\{11\bar{2}0\}$ reflections). Such a case is shown in Fig. 4.20, where in addition to dislocations and nanopipes, a large density of defects labeled 'T' is noted. The regions of alternating black–white fringe contrast vary in size up to ~50 nm across. The nature of the domains was clarified by comparing CBED patterns taken from the matrix and from regions within the domains which do not overlap the matrix.

Figure 4.23 compares CBED patterns taken along the $[1\bar{1}02]$ axis from the

4.7 Nanopipes and inversion domains 147

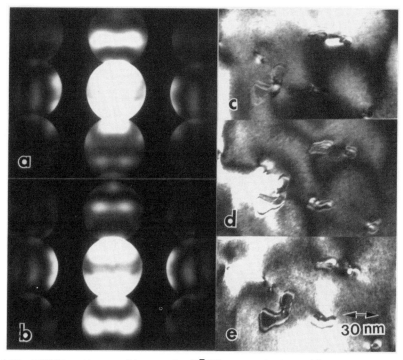

Fig. 4.23 CBED patterns taking at the [1̄102] zone axis from (a) inside an inversion domain, and (b) the surrounding film (matrix). Bright-field (BF) and dark-field (DF) TEM images taken at the [1̄102] zone axis of the same inversion domain: (c) BF image showing same contrast as the matrix, (d) DF image using the (1̄101) disk resulting in brighter contrast, and (e) DF image using the (1̄101̄) disk producing a darker contrast for the domain in comparison with the matrix.

matrix (a) and from an adjacent domain (b) in a region where the foil thickness was approximately constant. In each case, the patterns show an asymmetry between ±**g** which is due to the GaN polarity [30]. The reversal of the asymmetry between matrix and domain clearly confirms that the new defects are inversion domains. Using the same orientation as with the CBED patterns shown before, dark-field images can be obtained by using the asymmetric reflections. Figure 4.23(c) shows a bright-field image of the domain, taken at the [1̄102] axis. The contrast of the domain is the same as that of the matrix, as expected for images formed using the central (0000) beam. Figure 4.23(d) is a dark-field image, using the (1̄101) disk (top central disk in the CBED patterns), and gives a brighter contrast for the domain than for the matrix. Correspondingly, Fig. 4.23(e) is a dark-field image using the (1̄101̄) disk (lower central disk), giving a darker contrast for the domain than the matrix. These observations provide definitive evidence that the defects are inversion domains.

It is important to stress that the inversion domains are not easily seen under some commonly used conditions, namely with strong {101̄0} or {112̄0} reflections (see e.g. [45]), and that there is no evidence of significant strain associated with

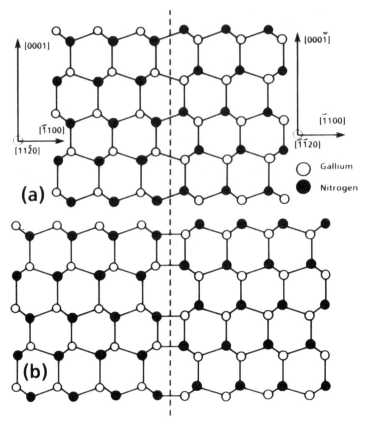

Fig. 4.24 Possible models for inversion domain boundaries along the prismatic plane of GaN. (a) Pure inversion boundary, (b) pure inversion plus $c/2$ shift boundary.

the presence of these defects. These observations indicate that any lattice displacement should be parallel to the c-axis. Further TEM studies show that the Ga sublattice is displaced by $3c/8$ or $7c/8$ along [0001] between the matrix and inversion domains [46]. Two possible models consistent with these observations are shown in Fig. 4.24. Figure 4.24(b) is an inversion boundary involving a $c/2$ shift of the lattice in addition to a pure inversion. The shift removes like-atom bonding but requires distortion of the bonds at the boundary itself. Theoretical and experimental evidence supports this model [46, 47].

In summary, in addition to pure dislocations the microstructure of high-quality GaN thin films exhibits the presence of nanotubes which are coreless dislocations with Burgers vector **c**, and inversion domains with orientation $\langle 000\bar{1}\rangle$ within the $\langle 0001\rangle$ matrix. The nature of these faults have been unequivocally determined by a combination of conventional transmission electron microscopy and convergent beam electron diffraction techniques. Models for the generation of coreless dislocations and for the interface of the inversion domain boundary have been proposed. The effects of these structural defects on the electrical and optical properties have not been studied at the present time. One possible speculation is that these defects could act as sources and sinks for point defects that could enhance the mobility of dislocations and thus provide a low-temperature mechanism for the relaxation of thermal stresses.

4.8 Lattice vibrations and microstructure

Raman spectroscopy probes the lattice vibrational modes. The crystal quality can be judged from the peak shape and the adherence to selection rules [48]. It has been recently shown that Raman scattering is a very sensitive and straightforward method for distinguishing the hexagonal and cubic phases in GaN [49]. Unexpected peaks in the Raman spectra may be the consequence of local modes, electronic transitions of dopants, or impurity atoms in the material [50, 51]. Crystal strain may be determined from shifts of the peaks of the Raman spectra [52, 53], and in the case of the ternary AlGaN such shifts may be used to determine the alloy composition [54]. Phonon–plasmon modes have been observed in n-type GaN [55], GaAs [56], and ZnSe [57], and a close correlation between these modes and the free carrier concentration has been observed. It has also been shown that it is possible to use Raman scattering in the determination of the surface space-charge layer [58]. In this section, the application of Raman imaging to establish correlation between microstructure and lattice vibrations is presented [59].

Longitudinal optical (LO) phonons interact strongly with a free carrier plasma when the plasma frequency is close to the LO phonon frequency. The coupled LO phonon plasma modes thereby created diminish the intensity of the LO phonon peak in a systematic fashion that can be used to determine the donor concentration in Si-doped GaN [60]. Recent improvements in Raman spectroscopic equipment have produced faster and more efficient means to acquire spectra and have made feasible Raman spectroscopic imaging [61, 62]. The preferred method is to tune by tilting a multilayer dielectric bandpass filter, transmitting a narrow spectral range ($10–20\,\text{cm}^{-1}$), and to image the sample directly to a two-dimensional array detector.

Raman spectra corresponding to undoped and silicon-doped GaN epitaxial layers on (0001) sapphire are shown in Fig. 4.25. The free carrier concentrations were $<5\times 10^{16}\,\text{cm}^{-3}$ and $\sim 2\times 10^{17}\,\text{cm}^{-3}$ at room temperature for the undoped and Si-doped cases, respectively. The measurements were performed in backscattering $z(xx)z'$ or $z(xy)z'$ geometry, with the z-direction parallel to the c-axis of the GaN wurtzite structure. In this notation, xx means that the polarization of the incident light and of the polarizer are parallel to each other; xy means that they are perpendicular. Note that the main difference in the Raman spectra is that the strong A1(LO) peak present in the undoped film is absent in the Si-doped case. As already mentioned, quenching of the A1(LO) phonon has been observed previously [60], and it has been explained in terms of phonon–plasmon coupled modes associated with the free electrons introduced by the silicon dopant.

A hexagonal crystallite, typical of GaN homoepitaxy on the [000$\bar{1}$] nitrogen-terminated surface of GaN bulk single crystals is shown in Plate I(a). These hexagonal hillocks have typical dimensions of the order of 20 to 50 μm, and are

150 Defects and performance of the III–V nitrides

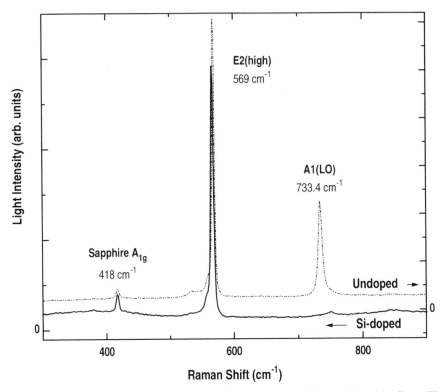

Fig. 4.25 Raman spectra of undoped, and silicon-doped GaN/sapphire thin films. The spectra were taken in the $z(xx)z'$ configuration, with incident and analyzer polarizations parallel. Note the absence of the A1(LO) phonon line in the silicon-doped film.

characterized by a faceted dome with six triangular facets and a pronounced tip in the center of the hexagon. The feature at the center of the hillock has been characterized by TEM and identified as an inversion domain column with opposite polarity to the matrix [63]. Raman spectra taken at the center of the hexagon and in the middle of the facets are shown in Fig. 4.26. The center has a strong Al(LO) peak, similar to the undoped case in Fig. 4.25, whereas the facets show a weaker intensity for this peak. The spectra have been normalized using the E2(high) peak at 567.5 cm^{-1} (which does not exhibit dependence on doping). Spectra taken at the vertex of the facets show no Al(LO) peak, in a very similar fashion to the silicon-doped case in Fig. 4.25. Using the E2(high) peak as a reference, the ratios Al(LO)/E2(high) for the center/facet/vertex of the hexagon are (5.7%)/(1.4%)/(0%), respectively.

The region about the Al(LO) was used to form an image, shown in Plate I(b). (For experimental details see [59]). A background reference was generated at 765 cm^{-1} and it is shown in Plate I(c). The normalized image in Plate I(d) shows a symmetric distribution of the 735 cm^{-1} Raman signal. The center of the hexagon has a strong signal, indicative of low free-electron densities (or absence of silicon and other n-type dopants). The edge between the triangular facets appears dark over a linear thickness of about 500 nm, indicating either high

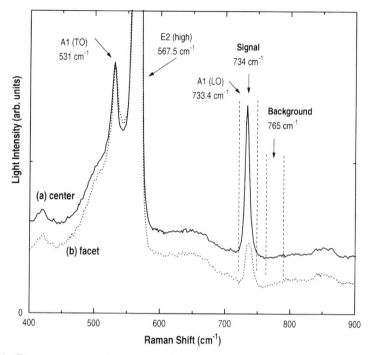

Fig. 4.26 Raman spectra taken from a hexagonal crystallite in homoepitaxial GaN. It shows a significant decrease in the Al(LO) line away from the center of the hexagon, and in particular at the edges between the triangular facets.

donor concentration or that the normalization procedure does not completely remove the effects of the surface on scattering collection efficiency. Full Raman spectra taken at various sample positions indicate that only the Al(LO) peak has a spatial variation. This fact makes unlikely the possibility of the geometric factors, such as reflections or waveguiding, being responsible for the image contrast, leading us to conclude that the variation of the image is due to local variations in the free electron concentration, likely produced by preferential incorporation of donor impurities during growth. The spatial variation in the plasmon–phonon coupling is responsible for the image characteristics of the Al(LO) line in GaN shown in Plate I.

The reduction in the LO phonon line, associated with an increase in carrier concentration in the range from 10^{16} to 10^{18} cm^{-2}, is accompanied by broadening and a shift towards the high-frequency side. This has already been observed previously by Kozawa *et al.* [60], who concluded that the dominant scattering mechanisms in GaN involve deformation potential and electro-optic mechanisms. Absence of the 735 cm^{-1} Al(LO) line may not be limited solely to silicon in GaN. It may be a property of donor impurities in general. Incorporation and segregation of oxygen are suspected in GaN and could be in part responsible for the observations reported here. Following Kozawa *et al.* [60] and our calibration shown in Fig. 4.25, we estimate that the free carrier concentration varies from $<10^{16}$ cm^{-3} (in the middle) to $\geq$ mid 10^{17} cm^{-3} in the edges.

It is interesting to note that the growth sense (polarity) seems to be associated with different levels of impurity distribution. The inversion domain in the center of the hillock has the same polarity as high-quality GaN films on sapphire substrates, such as shown in Fig. 4.11. The implication is that the incorporation of (donor) impurities such as silicon and oxygen is sensitive to the growth direction. MOCVD epitaxy is usually carried out under nitrogen-rich conditions, with the growth rate limited by the availability of Ga. In the [0001] case, gallium atoms are located at the top position of the basal planes, whereas in the [000$\bar{1}$] case nitrogen atoms are located at the top position of the basal plane. From our observations, it is clear that the Ga-on-top case is less reactive to the incorporation of this type of impurities than the N-on-top case. This is the first case that we know where variations caused by plasmon coupling are used to map out dopant distribution in a specimen at high spatial resolution. The technique is non-destructive and yields images with resolutions better than 1 micron. No vacuum methods or cryogenic techniques have been used, while the Raman data is gathered very rapidly—typically 1–10 sec per spectrum and between 10 and 20 minutes per image.

These results demonstrate that Raman imaging can be used for directly mapping the local variations of donor impurities in semiconductors, a technique that could eventually prove useful for device technology advancement. It should be noted that this is one of the first direct correlations between Raman imaging and microstructure.

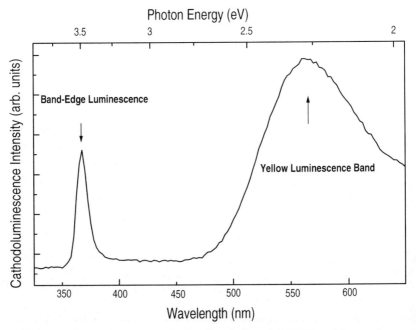

Fig. 4.27 Cathodoluminescence spectrum of an undoped GaN film with specular surface morphology.

4.9 Spatial variation of luminescence

It is also important to assess the effect of the microstructure on the electrical and optical properties of the III–V nitrides. The luminescent properties of MOCVD GaN thin films are important for the operation of optoelectronic devices. There are three basic components to the luminescence spectrum in high-quality films at room temperature: near band edge luminescence, yellow luminescence observed in undoped and silicon-doped epilayers, and dopant-related luminescence (donor–acceptor in p-type material, and donor–exciton in n-type material). Light emission can be estimulated by irradiation with photon energies higher than the bandgap in what is called photoluminescence (PL). Excitation with an electron beam can also produce light emission in a process called cathodoluminescence (CL). Both CL and PL spectra reflect optical transitions in the semiconductor. The CL spectrum of an undoped (unintentionally n-type) GaN film is shown in Fig. 4.27 [64]. Near band edge emission is observed at ~364 nm, and a yellow luminescence band is seen centered at ~560 nm. It is of interest to know the spatial variation of the luminescence and its relationship, if any, with the observed microstructure. An electron beam can be focused to a spot with diameter of the order of 5 nm, exciting a small region of the material, the diameter of which depends on the energy of the electron beam. Light emitted from small spots can be then gathered in a sequential fashion in a scanning electron microscope to produce a CL image. CL studies of undoped GaN thin films show that the luminescence intensity is quite inhomogeneous. In an attempt to observe possible correlation between luminescence and microstructure, we first looked at sufficiently large features such as hexagonal hillocks. Their morphology is seen in the secondary electron image in Fig. 4.28. TEM shows that these crystallites have low dislocation densities with some dislocations present in the central region of the crystal. The two CL images in the figure were taken at room temperature at the band edge and yellow band

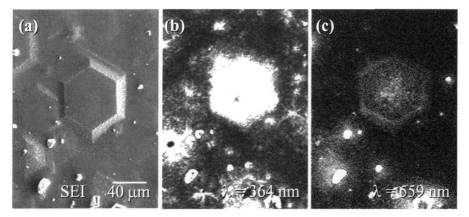

Fig. 4.28 Cathodoluminescence of undoped GaN with hexagonal hillocks. (a) Secondary electron image, (b) CL image at 364 nm, and (c) CL image at 559 nm.

154 *Defects and performance of the III–V nitrides*

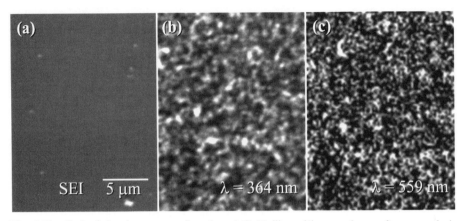

Fig. 4.29 Cathodoluminescence of undoped GaN film with specular surface morphology. (a) Secondary electron image, (b) CL image at 364 nm, and (c) CL image at 559 nm.

regions of the spectrum. From these images it is clear that near band edge emission originates in the bulk of the GaN hexagonal crystallites and that the yellow band emission is associated with the crystallite boundaries. Some yellow luminescence is also observed at the center of the crystallites. The presence of yellow emission appears to correlate with the presence of dislocations, observed in these hillocks to be present at their center and at the crystallite boundaries. Figure 4.29 contains equivalent observations for an undoped highly specular GaN film, grown under conditions similar to those used in devices like the one shown in Fig. 4.11. The surface planarity is observed in the secondary electron image. The CL image taken with the band edge luminescence has a spotty appearance, whose features are similar to the columnar structure observed by TEM. The image obtained using the yellow band emission has features with similar distribution although sharper in nature. These CL observations indicate that the spatial distribution of luminescence in undoped GaN thin films is highly inhomogeneous, and that it correlates with features observed by TEM, in particular the distribution of dislocations. This does not mean that yellow luminescence originates at the dislocation, but that the sources of yellow luminescence may be clustered at or around dislocations.

4.10 Summary

The nitrides have unique microstructural characteristics. This chapter surveys the structural defects that are typical of MOCVD GaN thin films used in the fabrication of high efficiency LEDs. We have observed that large densities of dislocations are common. These dislocations exhibit Burgers vectors equal to **a**, **c**, and **a** + **c**, where **a** and **c** are the unit cell vectors. Nanopipes and filamentary inversion domains are common defects as well. It is still surprising that GaN/sapphire works so well in spite of the large differences in lattice parameters and thermal expansion rates, and of the large density of structural defects.

Many important puzzles have not been completely solved, such as the relative absence of thermal strain. The nanopipes and inversion domains coupled with the large density of dislocations may be responsible for thermal stress relief. It is also surprising that the best optoelectronic results do not occur for a minimum value of X-ray rocking curves. Again this may indicate the need for the dislocation network in order to relieve thermal stresses. Some direct correlation between microstructure and electrical and optical properties has been observed using Raman imaging and cathodoluminescence.

The understanding of the microstructure of the nitrides is not complete by any means. More detailed studies are required, specially regarding the structure of nitride heterojunctions and of quantum wells. Recent studies of the microstructure of InGaN quantum wells indicate that roughening in these materials is quite common, in particular for high-indium content films, which may lead under some conditions to localized optical transitions [65–67].

References

1. F. A. Ponce and D. P. Bour, *Nature* **386**, 351 (1997).
2. J. Karpinski, S. Porowski, and S. Miotkowska, *J. Cryst. Growth* **56**, 77 (1982).
3. Z. Liliental-Weber, C. Kisielowski, S. Ruvimov, Y. Chen, J. Washburn, I. Grzegory, M. Bockowski, J. Jun, and S. Porowski, *J. Electron. Mat.* **25**, 1545 (1996).
4. F. A. Ponce, D. P. Bour, W. T. Young, M. Saunders, and J. W. Steeds, *Appl. Phys. Lett.* **69**, 337 (1996).
5. G. A. Slack and T. F. McNelly, *J. Crystal Growth* **34**, 263 (1976).
6. T. Detchprohm, K. Hiramatsu, H. Amano, and I. Akasaki, *Appl. Phys. Lett.* **61**, 2688 (1992).
7. T. D. Moustakas, *Mat. Res. Soc. Symp. Proc.* **395**, 111 (1996).
8. R. F. Davis, T. W. Weeks, Jr., H. D. Bremser, S. Tanaka, R. S. Kern, Z. Sitar, R. S. Alley, W. G. Perry, and C. Wang, *Mat. Res. Soc. Symp. Proc.* **395**, 3 (1996).
9. I. Akasaki and H. Amano, *J. Electrochem. Soc.* **141**, 2266 (1994).
10. F. Fichter, *Z. Anorg. Chem.* **54**, 322 (1907).
11. J. V. Lirman and H. S. Zhdanov, *Acta Physicochim. U.S.S.S.* **6**, 306 (1937).
12. H. P. Maruska and J. J. Tietjen, *Appl. Phys. Lett.* **15**, 327 (1969).
13. J. I. Pankove, E. A. Miller, and J. E. Berkeyheiser, *RCA Rev.* **32**, 383 (1971).
14. M. Manasevit, F. M. Erdmann, and W. I. Simpson, *J. Electrochem. Soc.* **118**, 1864 (1971).
15. S. Yoshida, S. Misawa, and S. Gonda, *Appl. Phys. Lett.* **42**, 427 (1983).
16. H. Amano, N. Sawaki, I. Akasaki, and Y. Toyoda, *Appl. Phys. Lett.* **48**, 353 (1986).
17. H. Amano, M. Kito, K. Hiramatsu, and I. Akasaki, *Jpn. J. Appl. Phys.* **28**, L2112 (1989).
18. S. Nakamura, *Jpn. J. Appl. Phys.* **30**, L1705 (1991).
19. S. Nakamura, T. Mukai, M. Senoh, and N. Iwasa, *Jpn. J. Appl. Phys.* **31**, L139 (1992).
20. I. Akasaki, H. Amano, Y. Koide, K. Hiramatsu, and N. Sawaki, *J. Crystal Growth* **98**, 209 (1989).
21. S. Nakamura, T. Mukai, and M. Senoh, *Appl. Phys. Lett.* **64**, 1687 (1994).
22. S. D. Lester, F. A. Ponce, M. G. Craford, and D. A. Steigerwald, *Appl. Phys. Lett.* **66**, 1249 (1995).
23. F. A. Ponce, J. S. Major, W. E. Plano, and D. F. Welch, *Appl. Phys. Lett.* **65**, 2302 (1994).

24. M. L. Kronberg, *Acta Met.* **5**, 507 (1957).
25. R. W. G. Wyckoff, *Crystal Structures*, Vols 1–4, Interscience Publishers, 1958.
26. K. G. Fertitta, A. L. Holmes, F. J. Ciuba, R. D. Dupuis, and F. A. Ponce, *J. Electron. Mat.* **24**, 257 (1995).
27. F. A. Ponce, B. S. Krusor, J. S. Major, W. E. Plano, and D. F. Welch, *Appl. Phys. Lett.* **67**, 410 (1995).
28. F. A. Ponce, M. A. O'Keefe, and E. C. Nelson, *Phil. Mag. A* **74**, 777 (1996).
29. F. A. Ponce, C. G. Van de Walle, and J. E. Northrup, *Phys. Rev. B*, **53**, 7473 (1996).
30. F. A. Ponce, D. P. Bour, W. T. Young, M. Saunders, and J. W. Steeds, *Appl. Phys. Lett.* **69**, 337 (1996).
31. F. A. Ponce, D. P. Bour, W. Götz, N. M. Johnson, H. I. Helava, I. Grzegory, J. Jun, and S. Porowski, *Appl. Phys. Lett.* **68**, 917 (1996).
32. F. A. Ponce, D. Cherns, W. T. Young, and J. W. Steeds, *Appl. Phys. Lett.* **69**, 770 (1996).
33. K. Pakula, A. Wysmolek, K. P. Korona, J. M. Baranowski, R. Stepniewski, J. Grzegory, M. Bockowski, J. Jun, S. Krokowski, M. Wrobleski, and S. Porowski, *Solid State Comm.* **97**, 919 (1996).
34. J. M. Baranowski, Z. Liliental-Weber, K. Korona, K. Pakula, R. Stepniewski, A. Wysmoiek, I. Grzegory, G. Nowak, S. Porowski, B. Monemar, and P. Bergman, *Mat. Res. Soc. Symp. Proc.* **449**, 393 (1997).
35. S. Nakamura, *Jpn. J. Appl. Phys.* **30**, L1705 (1991).
36. J. S. Foresi and T. D. Moustakas, *Appl. Phys. Lett.* **22**, 1433 (1969).
37. F. A. Ponce, D. Cherns, W. T. Young, and J. W. Steeds, *Appl. Phys. Lett.* **69**, 770 (1996).
38. D. Cherns and A. R. Preston, *Proceedings of the 11th International Congress on Electron Microscopy*, Japan Society of Electron Microscopy, Kyoto, 1986, p. 721.
39. J. P. Hirth and J. Lothe, *Theory of Dislocations*, 2nd edn, Wiley, 1982, p. 159.
40. F. A. Ponce, D. Cherns, W. T. Young, J. W. Steeds, and S. Nakamura, *Mat. Res. Soc. Symp. Proc.* **449**, 405 (1997).
41. D. Cherns, W. T. Young, J. W. Steeds, F. A. Ponce, and S. Nakamura, *J. Crystal Growth* **178**, 201 (1997).
42. F. C. Frank, *Acta Cryst.* **4**, 497 (1951).
43. W. Qian, M. Skowronski, K. Doverspike, L. B. Rowland, and D. K. Gaskill, *J. Cryst. Growth* **151**, 396 (1995).
44. W. Qian, G. S. Rohrer, M. Skowronski, K. Doverspike, L. B. Rowland, and D. K. Gaskill, *Appl. Phys. Lett.* **67**, 2284 (1995).
45. L. T. Romano and J. E. Northrup, *Mat. Res. Soc. Symp. Proc.* **449**, 423 (1996).
46. D. Cherns, W. T. Young, J. W. Steeds, F. A. Ponce, and S. Nakamura, submitted.
47. J. E. Northrup, J. Neugebauer, and L. T. Ramano, *Phys. Rev. Lett.* **77**, 103 (1996).
48. C. A. Argüello, D. L. Rousseau, and S. P. S. Porto, *Phys. Rev.* **181**, 1351 (1969).
49. H. Siegle, L. Eckey, A. Hoffmann, C. Thomsen, B. K. Meyer, B. Schikora, M. Hankein, K. Lischka, *Solid State Comm.* **96**, 943 (1995).
50. A. S. Barker, Jr. and M. Ilegems, *Phys. Rev. B* **7**, 743 (1973).
51. M. V. Klein, in *Light Scattering in Solids I*, ed. M. Cardona, Springer-Verlag, Berlin, 1975, p. 147.
52. W. Rieger, T. Metzger, H. Angerer, R. Dimitrov, O. Ambacher, and M. Stutzmann, *Appl. Phys. Lett.* **68**, 970 (1996).
53. T. Kozawa, T. Kachi, H. Kano, H. Nagase, N. Koide, and K. Manabe, *J. Appl. Phys.* **75**, 1098 (1994).
54. K. Hayashi, K. Itoh, N. Sawaki, and I. Akasaki, *Solid State Comm.* **77**, 115 (1991).
55. R. Ruppin and J. Nahum, *J. Phys. Chem. Solids* **35**, 1311 (1974).
56. D. Olego and M. Cardona, *Phys. Rev. B* **24**, 7217 (1981).

57. H. Shen, F. H. Pollak, and R. N. Sacks, *Appl. Phys. Lett.* **47**, 891 (1985).
58. J. Kraus and D. Hommel, *Semicond. Sci. Technol.* **10**, 785 (1995).
59. F. A. Ponce, J. W. Steeds, C. D. Dyer, and G. D. Pitt, *Appl. Phys. Lett.* **69**, 2650 (1996).
60. T. Kozawa, T. Kachi, H. Kano, H. Nagase, N. Koide, and K. Manabe, *J. Appl. Phys.* **77**, 4389 (1995).
61. I. P. Hayward, K. J. Baldwin, D. M. Hunter, D. N. Batchelder, and G. D. Pitt, *Diamond Rel. Mat.* **4**, 617 (1995).
62. N. C. Burton, J. W. Steeds, G. M. Meader, Y. G. Shreter, and J. E. Butler, *Diamond Rel. Mat.* **4**, 1222 (1995).
63. J.-L. Rouviere, M. Arlery, A. Bourret, R. Niebuhr, and K.-H. Bachem, *Mat. Res. Soc. Proc.* **395**, 393 (1996).
64. F. A. Ponce, D. P. Bour, W. Götz, and P. J. Wright, *Appl. Phys. Lett.* **68**, 57 (1996).
65. S. Chichibu, T. Azuhata, T. Sota, and S. Nakamura, *Appl. Phys. Lett.* **69**, 4188 (1996).
66. F. A. Ponce, S. A. Galloway, W. Götz, and S. Kern, *Mat. Res. Soc. Symp. Proc.* **482**, in press (1998).
67. F. A. Ponce, D. Cherns, W. Götz, and S. Kern, *Mat. Res. Soc. Symp. Proc.* **482**, in press (1998).

5 Modulation spectroscopy of the Group III nitrides

Fred H. Pollak

5.1 Introduction

Modulation spectroscopy is a powerful and versatile optical technique for obtaining valuable information about a large variety of semiconductor systems [1–9], including bulk/thin films, micro- and nanostructures (quantum wells, multiple quantum wells, superlattices, quantum wires, quantum dots), surfaces/interfaces (semiconductor/air (vacuum), semiconductor/semiconductor (hetero- and homojunctions), semiconductor/metal, semiconductor/electrolyte), and the effects of growth/processing, as well as the characterization of actual device structures (heterojunction bipolar transistors, pseudomorphic high electron mobility transistor, quantum well lasers).

Modulation spectroscopy is an analog method for taking the derivative of the optical spectrum (reflectance or transmittance) of a material by modifying in some manner the measurement conditions [1–9]. The basic idea of this optical technique is a very general principle of experimental physics. Instead of measuring the optical reflectance (or transmittance) of the material, the derivative with respect to some parameter is evaluated. This can easily be accomplished by varying some property of the sample or measuring system in a periodic fashion and measuring the corresponding normalized change in the reflectance (transmittance). The spectral response of the sample can be modified directly by applying a repetitive perturbation such as an electric field (electromodulation), a heat pulse (thermomodulation), or stress (piezomodulation). This procedure is termed 'external' modulation [1–7]. The change can also occur in the measuring system itself; for example, the wavelength [1, 3] or polarization conditions [8, 9] can be modulated or the sample reflectance (transmittance) can be compared to a reference sample [1, 3]. This mode has been labeled 'internal' modulation.

The periodic variation of the measurement conditions gives rise to sharp, differential-like spectra in the region of interband (intersubband) transitions. Therefore, modulation spectroscopy emphasizes relevant spectral features and suppresses uninteresting background effects. The ability to perform a lineshape fit is one of the great advantages of modulation spectroscopy. Since, for the modulated signal, the features are localized in photon energy, it is possible to account for the lineshapes to yield accurate values of important parameters such as the energies and broadening functions of interband (intersubband) transitions. For example, even at 300 K it is possible to obtain the energy of a particular feature to within a few meV.

Modulation spectroscopy also can be employed to investigate the effects of static perturbations, such as electric and magnetic fields, hydrostatic pressure, stress/strain (external or internal), temperature, and composition.

Because the changes in the optical spectra are typically small, in some cases 1 part in 10^6, phase-sensitive detection or some other signal processing procedure is required. Since modulation spectroscopy is a normalized technique, not every photon need be collected as long as there are enough to produce a good signal-to-noise ratio. Thus, this experimental method does not have stringent conditions on the morphology of the sample surface.

A particularly useful form of modulation spectroscopy is electromodulation (electric field modulated reflectivity (transmission)) since it is sensitive to surface/interface electric fields and can be performed in contactless modes that require no special mounting of the sample. Under appropriate conditions the electromodulation (EM) spectrum can exhibit above-band features, called Franz–Keldysh oscillations (FKOs) [1–6], which are a direct measure of the built-in electric field. The most widely used form of EM is photoreflectance (PR).

In spite of its considerable utility very little work using modulation spectroscopy has been performed on the Group III nitrides [10–19]. Giordana et al. reported the first application of modulation spectroscopy (electrolyte electroreflectance and PR) on these materials (wurtzite (WZ)-GaN) [10]. The temperature dependence of the A, B, and C excitons in WZ-GaN [11, 13, 18], zincblende (ZB)-GaN [12] and WZ-Ga$_{0.95}$Al$_{0.05}$N [17] have been measured using EM. Shikanai et al. [14] reported the effects of biaxial strain on the exciton resonances in WZ-GaN using PR. Several groups [14–16] have observed fine structure in the wavelength modulated and/or PR spectra associated with the excited states of the A and B excitons in W-GaN, thus making it possible to determine the exciton binding energy. The in-plane optical anisotropies of GaN, GaAlN, and AlN films using reflection anisotropy spectroscopy (RAS) have been reported by Rossow et al. [19].

5.2 Modulation techniques

5.2.1 Photoreflectance

Figure 5.1 is a schematic drawing of the Brooklyn College photoreflectance system [1–3]. In PR, modulation of the built-in electric field of the sample is caused by photo-excited electron–hole pairs created by the pump source, which is chopped/modulated at frequency Ω_m, typically ~100–200 Hz. The photon energy of the pump beam must be larger than the lowest energy gap of the material. Light from an appropriate light source (xenon arc, halogen, or tungsten lamp) passes through a scanning (probe) monochromator. The exit intensity at wavelength λ, $I_0(\lambda)$, is focused on the sample by means of a lens. The reflected light is collected by a second lens and is focused on an appropriate detector, generally a photodiode. For simplicity the lenses are not shown.

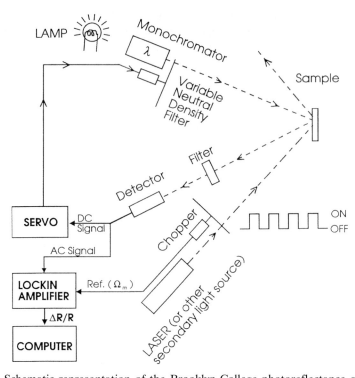

Fig. 5.1 Schematic representation of the Brooklyn College photoreflectance apparatus.

The light striking the detector contains two signals: the d.c. (or average value) is given by $I_0(\lambda)R(\lambda)$, where $R(\lambda)$ is the d.c. reflectance of the material, while the modulated value (at frequency Ω_m) is $I_0(\lambda)\Delta R(\lambda)$, where $\Delta R(\lambda)$ is the change in reflectance produced by the modulation source. The a.c. signal from the detector, proportional to $I_0 \Delta R$, is measured by the lock-in amplifier. Typically, $I_0 \Delta R$ is 10^{-4}–10^{-6} of $I_0 R$.

In order to evaluate the quantity of interest, i.e. the relative change in reflectance $\Delta R/R$, a normalization procedure must be used to eliminate the uninteresting common feature $I_0(\lambda)$. In Fig. 5.1 the normalization is performed by a variable neutral density filter (VNDF) connected to a servo mechanism. The d.c. signal from the detector, which is proportional to $I_0(\lambda)R(\lambda)$, is introduced into the servo, which moves the VNDF in such a manner as to keep $I_0(\lambda)R(\lambda)$ as a constant, i.e. $I_0(\lambda)R(\lambda) = C$. Under these conditions the a.c. signal $I_0(\lambda)\Delta R(\lambda) = C \Delta R(\lambda)/R(\lambda)$. Thus, the signal to the lock-in-amplifier is proportional to the quantity of interest, i.e. $\Delta R(\lambda)/R(\lambda)$.

A drawback of PR is the spurious modulated background signal reaching the detector because of (a) photoluminescence from the sample and/or (b) scattered light from the pump source. Photoluminescence can sometimes be a problem for measurements near the fundamental gap, particularly at low temperatures. Scattered pump light can be reduced by means of an appropriate long pass filter in front of the detector. If the overall spurious background signal is not too large in relation to $\Delta R/R$, it can be subtracted by the normalization method of Fig. 5.1 (the spurious light is not optically dispersed and shows up as a large, wavelength-independent background signal).

The spurious background signal can also be reduced or eliminated [1–3, 5, 6] by approaches such as the use of a double monochromator, a tuneable dye laser as the probe beam, sweeping PR, differential PR, or a properly phase-shifted signal from the reference source that is applied to the lock-in amplifier's input.

Another alternative is to eliminate the pump beam and perform the experiment using the technique of contactless electroreflectance (CER) [1, 3, 5, 6].

5.2.2 Contactless electroreflectance

The CER method utilizes a condenser-like system with the top electrode being either (a) a thin, transparent, conductive coating (indium-tin-oxide or 50–60 Å of a metal such as Au or Ni) on a transparent substrate (glass, quartz, etc.) or (b) a wire grid. A second electrode consisting of a metal strip is separated from the first electrode by an insulating spacer. The sample (~ 0.5 mm thick) is placed between these two capacitor plates. The dimensions of the spacer are such that there is a very thin layer (~ 0.1 mm) of air (or vacuum) between the front surface of the sample and the conducting part of the first electrode. Thus, there is nothing in direct contact with the front surface of the sample.

Figure 5.2 shows the sample holder and electrode arrangement. The a.c. modulating (~ 1 kV peak-to-peak) and d.c. bias voltages are applied between the metal strip and the transparent conductor. The probe beam is incident through

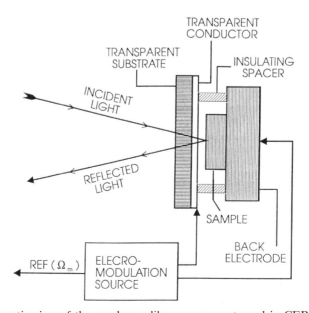

Fig. 5.2 Schematic view of the condenser-like arrangement used in CER experiments.

the first transparent electrode. This approach can also be employed in the transmission mode by replacing the metal electrode with a second transparent electrode/substrate.

5.2.3 Reflection anisotropy spectroscopy

Reflection anisotropy spectrosopy is a technique that relies on the reduced symmetry of surfaces/interfaces and bulk/thin film material [8, 9]. In RAS, the difference between near-normal-incidence reflectances of light polarized along two principal axes of the sample in the plane is determined. For example, in cubic materials RAS measures the difference in light linearly polarized along the [110] and [$\bar{1}$10] principal axes in the plane of the (001) surface. Such investigations can yield important information about symmetry-breaking effects due to surfaces (including crystal growth) [8, 9], surface electric fields [20] and the anisotropy produced by misfit dislocations [21].

Figure 5.3 is a schematic diagram of an RAS apparatus. It consists of an Xe short-arc lamp, front surface spherical and plane mirrors, MgF_2 and quartz Rochon polarizers, a 50 kHz photoelastic modulator, a monochromator, and a photomultiplier (although other detectors can be used).

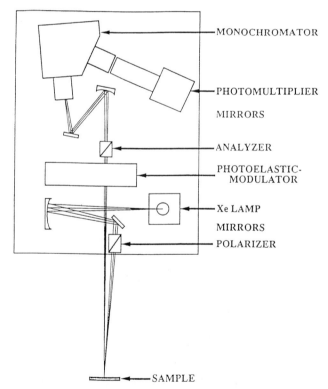

Fig. 5.3 Reflection anisotropy spectroscopy apparatus.

5.3 Lineshape considerations

Differential changes in the reflectivity can be related to the perturbation of the complex dielectric function [$\varepsilon(-\varepsilon_1 + i\varepsilon_2)$] expressed as [1–7]:

$$\Delta R/R = a(\varepsilon_1, \varepsilon_2)\Delta\varepsilon_1 + b(\varepsilon_1, \varepsilon_2)\Delta\varepsilon_2 \tag{5.1}$$

where a and b are the Seraphin coefficient, related to the unperturbed dielectric function, and $\Delta\varepsilon_1$ and $\Delta\varepsilon_2$ are the changes in the complex dielectric function due to the perturbation. The Seraphin coefficients a and b can be written as:

$$a = \frac{1}{R}\frac{\partial R}{\partial \varepsilon_1}; \qquad b = \frac{1}{R}\frac{\partial R}{\partial \varepsilon_2} \tag{5.2}$$

Near the fundamental gap of bulk materials, $b \approx 0$, so that $\Delta R/R \approx a\Delta\varepsilon_1$ is the only significant term. However, in multilayer structures interference effects are important, so that the Seraphin coefficients are modified, and both $\Delta\varepsilon_1$ and $\Delta\varepsilon_2$ may have to be considered. The quantities $\Delta\varepsilon_1$ and $\Delta\varepsilon_2$ are related by a Kramers–Kronig inversion. The functional form of $\Delta\varepsilon_1$ and $\Delta\varepsilon_2$ can be calculated for a given perturbation provided that the dielectric function and critical point are known.

5.3.1 Electromodulation

EM is a particularly useful form of modulation spectroscopy since it often yields the sharpest structure and is sensitive to surface or interface electric fields. It is also the most complex form, since, in certain cases (bulk materials, superlattices), the perturbation can destroy the translational symmetry of the material.

EM can be classified into two categories, i.e. low- and high-field regimes, depending on the relative strengths of certain characteristic energies [1–3]. In the first range $|\hbar\theta| \leq \Gamma$, where Γ is the broadening parameter and $\hbar\theta$ is the electro-optic energy given by:

$$(\hbar\theta)^3 = q^2\hbar^2 F^2/2\mu_\parallel \tag{5.3}$$

In eqn (5.3), F is the electric field and $\mu_\parallel$ is the reduced interband effective mass in the direction of the field.

In the high-field case $|\hbar\theta| \geq \Gamma$, but $qFa_d \ll E_g$, where a_d is the lattice constant (or an appropriate periodic length in microstructures) and E_g is the bandgap. In this situation, the band structure is unchanged. For the very high field the electro-optic energy is much greater than the broadening, but $eFa_d \sim E_g$ and the Stark shifts are produced. We are not interested in such a very high field regime and it will be ignored in the following discussions.

5.3.1.1 Low field regime: band-to-band transitions

In the case of band-to-band transitions in bulk materials (and superlattices), EM can destroy the translational symmetry of the material, and hence can accelerate unbound electrons and/or holes. The time-dependent Schrödinger equation for an electron in the presence of a uniform electric field F can be written as [22]:

$$(H_0 + e\mathbf{F} \cdot \mathbf{r})\psi = i\hbar \frac{\partial \psi}{\partial t} \quad (5.4)$$

where the crystal Hamiltonian H_0, with crystal potential $V(r)$, is:

$$H_0 = \frac{p^2}{2m_0} + V(r) \quad (5.5)$$

The time evolution of the wavefunction, which at the time $t = 0$ is the Bloch function $\psi_n(k_0, r)$, is given by:

$$\psi_n(k, r, t) = \exp\left(-i\frac{(H_0 + e\mathbf{F} \cdot \mathbf{r})t}{\hbar}\right)\psi_n(k_0, r) \quad (5.6)$$

The main effect of the electric field on a periodic system is to change the wave vector from $\mathbf{k}$ to $\mathbf{k} - e\mathbf{F}t/\hbar$. As a result, states with $\mathbf{k}$ vector along the field direction are mixed. The solution of the time-dependent Schrödinger equation can be approximately written as:

$$\psi_n(k, r) = \exp\left(-\frac{i}{\hbar} \int E_n(k) \, dt\right)\psi_n(k, r, t) \quad (5.7)$$

where $\mathbf{k}(t)$ is given by:

$$\mathbf{k}(t) = \mathbf{k}_0 - \frac{e\mathbf{F}t}{\hbar} \quad (5.8)$$

Now we can determine the optical constants in the presence of an electric field. By taking into account the time dependence of $\mathbf{k}$ and the broadening parameter Γ, for the imaginary part of the dielectric function we obtain:

$$\varepsilon_2(\omega) = \frac{4\pi^2 e^2}{m_0^2 \omega^2} \sum_{v,c} \int_{BZ} \frac{2 d^l k}{(2\pi)^l} |\hat{\boldsymbol{\varepsilon}} \cdot \mathbf{M}_{cv}(k)|^2 \, \delta\left[E_{cv}\left(k - \frac{e\mathbf{F}t}{\hbar}\right) + i\Gamma - \hbar\omega\right] \quad (5.9)$$

where l is the dimensionality of the system ($l = 1, 2,$ or 3). This equation can be rewritten as [22]:

$$\varepsilon_2(\omega) = \frac{4\pi e^2}{m_0^2 \omega^2} \sum_{v,c} \int_{BZ} \frac{2 d^l k}{(2\pi)^l} |\hat{\boldsymbol{\varepsilon}} \cdot \mathbf{M}_{cv}(k)|^2$$
$$\times \int dt \exp\left\{-i\left[E_{cv}\left(k - \frac{e\mathbf{F}t}{\hbar}\right) - i\Gamma + \hbar\omega\right]t\right\} \quad (5.10)$$

where

$$\hat{\boldsymbol{\varepsilon}} \cdot \mathbf{M}_{cv}(k) = \hat{\boldsymbol{\varepsilon}} \cdot \int_{\text{crystal volume}} \Psi_c(k, r)(-i\hbar\nabla)\Psi_v(k, r) \, d^l r \quad (5.11)$$

5.3 Lineshape considerations

In eqn (5.11), $\hat{\varepsilon}$ is the unit polarization vector of the incident radiation and $\Psi_c(k,r)$ and $\Psi_v(k,r)$ are the conduction and valence band wavefunctions, respectively.

In the low-field regime, we can expand $E_{cv}(\mathbf{k})$ in terms of the field and retain terms only to second-order:

$$\varepsilon_2(\omega) = \frac{4\pi e^2}{m_0^2 \omega^2} \sum_{v,c} \int_{BZ} \frac{2d^l k}{(2\pi)^l} |\hat{\varepsilon} \cdot \mathbf{M}_{cv}(k)|^2$$

$$\times \int dt \exp\left(-i[E_{cv}(k) - i\Gamma + \hbar\omega]t - i\frac{(\theta t)^3}{12}\right) \quad (5.12)$$

If the field is small or the broadening Γ large, such that $\Gamma \gg \hbar\theta$, we can expand the exponent as follows: $\exp[-i(\theta t)^3/12] \approx 1 - i(\theta t)^3/12$ and obtain an expression for ε_2 in the low-field limit:

$$\varepsilon_2(E,F,\Gamma) = \varepsilon_2(E,0,\Gamma) + \frac{(\hbar\theta)^3}{12} E^2 \frac{\partial^3}{\partial E^3}\left[E^2 \varepsilon_2(E,0,\Gamma)\right] + \cdots \quad (5.13)$$

where t is operationally equivalent to $[i\hbar(\partial/\partial E)]$ in quantum mechanics.

Using the Kramers–Kronig relation, it is also possible to obtain ε_1 in the presence of this weak electric field F:

$$\varepsilon_1(E,F,T) = \varepsilon_1(E,0,\Gamma) + \frac{(\hbar\theta)^3}{12} E^2 \frac{\partial^3}{\partial E^3}\left[E^2 \varepsilon_1(E,0,\Gamma)\right] + \cdots \quad (5.14)$$

Therefore the change in the complex dielectric function in the presence of a weak electric field is given by:

$$\Delta\varepsilon = \frac{(\hbar\theta)^3}{12 E^2} \frac{\partial^3}{\partial E^3}\left[E^2 \varepsilon(E,0,\Gamma)\right] \quad (5.15)$$

As can be seen, for low-field modulation ($|\hbar\theta| \leq \Gamma$) EM gives a third derivative spectroscopy for band-to-band transitions.

If the broadening in eqn (5.15) is Lorentzian, eqns (5.1) and (5.15) can be written in the simplified form:

$$\Delta R/R = A\,\mathrm{Re}\left[e^{i\phi}(E - E_g + i\Gamma)^{-m}\right] \quad (5.16)$$

where A is the amplitude, ϕ is a phase angle (which accounts for the mixture of real and imaginary components of ε in eqn (5.1) as well as the influence of non-uniform electric fields and interference and electron–hole interaction effects), and E_g is the bandgap energy. The parameter m in eqn (5.15) depends on critical point type. For a three-dimensional M_0 critical point, such as the direct gap of diamond-, zincblende-, and wurtzite-type semiconductors, $m = 2.5$.

5.3.1.2 Low field regime: excitons

For bound states such as excitons, either in the bulk or in quantum wells, the perturbing electric field does not accelerate electrons and/or holes. These types of particle do not have translational symmetry and are confined in space. Their energy spectrum is discrete and not continuous as in the case of the unbound particles. In this situation the modulating field can alter the binding energy of the particle [2].

Since, for confined systems, the energies (in the confinement direction for microstructures) are discrete and dispersionless, this results in an infinite effective mass (in the confinement direction). An applied electric field (along the confinement direction) adds a linear potential, which tilts the confining potential, changing its shape. The electrons and holes become spatially polarized, but still remain confined. This alters both the electronic energies and the wavefunction overlap (intensity (I)). Also the tilting of the potential can result in a change in lifetime Γ as a result of tunneling.

In terms of EM, the infinite mass means that eqn (5.15) is no longer applicable, since $\hbar\theta = 0$. Under these conditions the change in the dielectric function induced by the modulating field, $F_{a.c.}$, is first-derivative and not the third derivative as for 3D systems [1–3]. It can be shown that $\Delta\varepsilon$ can be expressed in a compact manner as a first derivative functional form for the modulated dielectric function [1–3]:

$$\Delta\varepsilon = \left[(\partial\varepsilon/\partial E_g)(\partial E_g/\partial F_{a.c.}) + (\partial\varepsilon/\partial \Gamma)(\partial \Gamma/\partial F_{a.c.})\right.$$
$$\left. + (\partial\varepsilon/\partial I)(\partial I/\partial F_{a.c.})\right] F_{a.c.} \tag{5.17}$$

where $\partial E_g/\partial F_{a.c.}$ is the change in energy due to the Stark effect, $\partial E_g/\partial F_{a.c.}$ is the change in broadening parameter due to a variation of the lifetime, and $\partial E_g/\partial F_{a.c.}$ is the change in intensity due to redistribution of charges.

If the broadening is Lorentzian and the third term in eqn (5.17) is neglected, eqns (5.1) and (5.17) can be rewritten as [1–3]:

$$\Delta R/R = ARe\left[e^{i\phi}(E - E_g + i\Gamma)^{-2}\right] \tag{5.18}$$

where A and ϕ have the same meaning as in eqn (5.16).

For the case in which the broadening is Gaussian (and we again neglect the third term in eqn (5.17)) the expression for the lineshape is more complicated in relation to eqn (5.18) and is given in Refs. 1 and 3.

5.3.1.3 High field regime (Franz–Keldysh oscillations)

For the unbound situation, in the event that the low-field criterion is not satisfied, then a full quantum mechanical treatment must be used [21]. For

5.3 Lineshape considerations

example, for a 3D critical point the dielectric function in the presence of a uniform field F (neglecting Γ) can be written as [22]:

$$\varepsilon(E,F,0) = 1 + \left(\frac{C\sqrt{\hbar\theta}}{E^2}\right)\{G(\eta) + iF(\eta)\} + D\left(2\sqrt{E_g} - \sqrt{E_g + E}\right)\Big/E^2 \quad (5.19)$$

where:

$$F(\eta) = \pi[Ai'^2(\eta) - \eta Ai^2(\eta)] - \left(\sqrt{-\eta}\, H(-\eta)\right) \quad (5.20)$$

$$G(\eta) = \pi[Ai'(\eta)Bi'(\eta) - \eta Ai(\eta)Bi(\eta)] + \sqrt{\eta}\, H(\eta) \quad (5.21)$$

with

$$\eta = \frac{E_g - E}{\hbar\theta} \quad (5.22)$$

The quantities $G(\eta)$ and $F(\eta)$ are referred to as the electro-optic functions, while $Ai(\eta)$, $Ai'(\eta)$, $Bi(\eta)$ and $Bi'(\eta)$ are Airy functions and their derivatives and $H(-\eta)$ is the unit step function [22]. The parameters C and D are related to matrix element effects and can be considered constants over the vicinity of the critical point.

In the presence of Lorentzian broadening (Γ), the field-induced change in the dielectric function $\Delta\varepsilon(E,F,\Gamma)$ can be obtained from the unbroadened change $\Delta\varepsilon(E,F,0)$, using [23]:

$$\Delta\varepsilon(E,F,\Gamma) = 1/\pi \int_{-\infty}^{\infty} \frac{\Delta\varepsilon(E',F,0)\Gamma}{(E-E')^2 + \Gamma^2}\, dE' \quad (5.23)$$

A contour integral of eqn (5.23) yields:

$$\Delta\varepsilon(E,F,\Gamma) = \Delta\varepsilon(E + i\Gamma, F) \quad (5.24)$$

Although the exact form of $\Delta R/R$ for the intermediate field case with broadening is quite complicated, Aspnes and Studna [24] have written down a relatively simple expression:

$$\Delta R/R \propto \exp\left[-2(E - E_g)^{1/2}\Gamma/(\hbar\theta)^{3/2}\right]$$
$$\times \cos\left[(4/3)(E - E_g)^{3/2}/(\hbar\theta)^{3/2} + \chi\right]\left[E^2(E - E_g)\right]^{-1} \quad (5.25)$$

where χ is a phase angle.

From eqn (5.25) the position of the nth extremum in the FKOs is given by:

$$n\pi = (4/3)\left[(E_n - E_g)/\hbar\theta\right]^{3/2} + \chi \quad (5.26)$$

where E_n is the photon energy of the nth extremum. A plot of $(4\pi/3)(E_n - E_g)^{3/2}$ vs. the index number n will yield a straight line with slope

$(\hbar\theta)^{-3/2}$. Therefore, the electric field (F) can be obtained directly from the period of the FKOs if $\mu_\parallel$ is known.

5.3.2 Thermo- and piezoreflectance

These two modulation mechanisms do not destroy the translational symmetry of the system and hence yield first-derivative spectroscopies. For Lorentzian broadening the lineshape can be written as eqn (5.16) with $m = 0.5$ or 2 for band-to-band or excitonic features, respectively. If the broadening is Gaussian the lineshapes are given in Refs. 1 and 3.

5.3.3 Reflection anisotropy spectroscopy

For cubic materials the RAS lineshape can be written as [8]:

$$\frac{\Delta\tilde{R}}{\tilde{R}} = \frac{\Delta R}{R} + i\Delta\theta = \frac{2\left(\tilde{R}_{[hkl]} - \tilde{R}_{[h'k'l']}\right)}{\left(\tilde{R}_{[hkl]} + \tilde{R}_{[h'k'l']}\right)} \quad (5.27)$$

where $\tilde{R}_{[hkl]}$ and $\tilde{R}_{[h'k'l']}$ are the complex near-normal-incidence reflectances of light polarized along the two orthogonal principal axes $[hkl]$ and $[h'k'l']$. For wurtzite materials the expression would be similar to eqn (5.27) with the appropriate notation for the relevant axes.

5.4 Results

5.4.1 Temperature dependence of the energies and linewidths in WZ-GaN, ZB-GaN and WZ-Ga$_{0.95}$Al$_{0.05}$N

Using CER the temperature dependence of the energies and linewidths of the A, B, and C excitons associated with the direct gap of wurtzite (WZ) GaN/sapphire (0001) [18] and Ga$_{0.95}$Al$_{0.05}$N/sapphire (0001) [17] have been measured in the range $15\,\text{K} < T < 450\,\text{K}$. The temperature variation ($15\,\text{K} < T < 300\,\text{K}$) of the bandgaps of the excitons in (a) WZ-GaN/sapphire (A, B, and C excitons) [11, 13] and (b) zincblende (ZB) GaN/MgO [12] have also been evaluated using PR.

Plotted by the open circles in Fig. 5.4 are the experimental CER spectra of the WZ-GaAlN sample [17] at 16 K, 181 K, and 434 K. The solid lines in Fig. 5.5 are least squares fits to eqn (5.18), which are appropriate for excitonic transitions with Lorentzian broadening. The obtained values of the energies are designated by arrows in the figure.

The CER spectrum at 16 K has three distinct resonances, as indicated by the vertical arrows. These correspond to the A, B, and C excitons of the direct gap and are related to the Γ_9^V–Γ_8^C, Γ_8^V (upper band)–Γ_7^C, and Γ_7^V (lower band)–Γ_7^C interband transitions in the wurtzite material. As the temperature increases, the

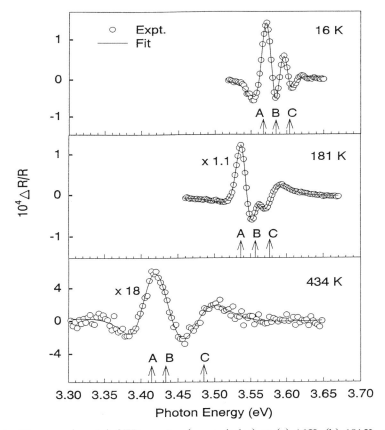

Fig. 5.4 The experimental CER spectra (open circles) at (a) 16 K, (b) 181 K, and (c) 434 K from WZ-Ga$_{0.95}$Al$_{0.05}$N/sapphire. The solid lines are fits to eqn (5.18) which yield the energies indicated by the arrows.

lineshapes of the CER spectral features shift towards lower energies and broaden. The sharp features on the lower energy side (corresponding to the A and B excitons) are not clearly resolved at 434 K, although the lineshape fit reveals that there are two structures. Also, as the temperature increases the splitting between the A, B, and C excitonic energies becomes larger. Similar results have been observed in a WZ-GaN/sapphire film [18]. The possible origin of this effect is the difference in the thermal expansion coefficients between the GaAlN and the sapphire substrate, which generates some compressive stress.

Plotted by the open circles, open squares and open triangles in Fig. 5.5 are the experimental values of the energies $E_A(T)$, $E_B(T)$, and $E_C(T)$ corresponding to the A, B, and C transitions, respectively. Representative error bars are shown. The solid lines in Fig. 5.5 are least squares fits to the semi-empirical Varshni relationship [25]:

$$E(T) = E(0) - \alpha T^2/(\beta + T) \qquad (5.28)$$

where $E(0)$ is the bandgap at $T = 0$. The obtained values of $E(0)$, α, and β for

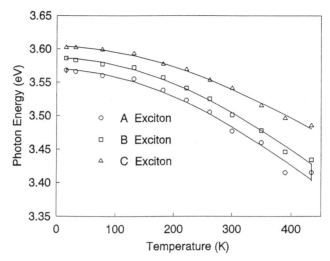

Fig. 5.5 The experimental temperature dependence of the A (open circles), B (open squares), and C (open triangles) exciton features from WZ-$Ga_{0.95}Al_{0.05}N$/sapphire. The solid lines are fits to the Varshni equation (5.28).

the three interband transitions in WZ-$Ga_{0.95}Al_{0.05}N$ are listed in Table 5.1. For comparison purposes we have also listed numbers for GaN (WZ and ZB) and WZ-CdSe.

The temperature dependence of the direct bandgap also can be described by a Bose–Einstein-type expression [26]:

$$E(T) = E(0) - 2a_B/[\exp(\theta_B/T) - 1] \tag{5.29}$$

where $E(0)$ is the bandgap at $T = 0$, a_B represents the strength of the exciton–average phonon (optical and acoustical) interaction, and θ_B corresponds to the average phonon temperature.

The temperature dependence of the A, B, and C features has also been fitted to eqn (5.29). The obtained values of the various parameters also are given in Table 5.1. For comparison purposes, also presented in Table 5.1 are the numbers for WZ-GaN and WZ-CdSe.

It should be mentioned that the value of $E(0)$ of the A exciton was also used to evaluate the Al composition of the sample.

The experimental values of $\Gamma(T)$ (half width at half maximum (HWHM)) of the A and B excitons as obtained from the lineshape fit are displayed by the open circles and squares, respectively, in Fig. 5.6. Representative error bars are shown. The temperature dependence of the broadening parameter of semiconductors can be expressed as [26]:

$$\Gamma(T) = \Gamma(0) + \Gamma_{LO}/[\exp(\theta_{LO}/T) - 1] \tag{5.30}$$

where the first term of eqn (5.30) is due to intrinsic effects (electron–electron interaction, impurity, dislocation, and alloy scattering) at $T = 0$. The second

Table 5.1 Values of the Varshni- and Bose–Einstein-type fitting parameters which describe the temperature dependencies of the energy band gaps of WZ-Ga$_{0.95}$Al$_{0.05}$N, GaN (WZ and ZB), WZ-In$_{0.14}$Ga$_{0.86}$N, and WZ-CdSe.

Material	$E(0)$ (eV)	α (10^{-4} eV/K)	β (K)	$E(0)$ (eV)	a_B (meV)	θ_B (K)
WZ-Ga$_{0.95}$Al$_{0.05}$N[a]						
A exciton	3.571 ± 0.004	13 ± 2	1145 ± 200	3.565 ± 0.004	138 ± 20	436 ± 100
B exciton	3.586 ± 0.004	14 ± 2	1338 ± 200	3.582 ± 0.004	137 ± 20	449 ± 100
C exciton	3.604 ± 0.008	9 ± 4	1041 ± 400	3.601 ± 0.008	97 ± 40	418 ± 200
WZ-GaN[b]						
A exciton	3.484 ± 0.002	12.8 ± 2.0	1190 ± 150	3.484 ± 0.002	110 ± 20	405 ± 100
B exciton	3.490 ± 0.002	12.9 ± 2.0	1280 ± 150	3.490 ± 0.002	112 ± 20	420 ± 100
C exciton	3.512 ± 0.004	6.6 ± 3.0	840 ± 300	3.512 ± 0.004	57 ± 30	340 ± 100
WZ-GaN[c]						
A exciton	3.486	8.32	835.6			
B exciton	3.494	10.9	1194.7			
C exciton	3.520	29.2	3698.9			
WZ-GaN[d,e]	3.510 (3)	8.58 (17)	700 (fixed)	3.489 (300)	236 (332)	692 (61)
WZ-GaN[e,f]	3.512	5.66	737.9	3.510	38	289
WZ-GaN[e,g]	3.476	9.39	722			
WZ-GaN[e,h]	3.470 ± 0.003	5.9 ± 0.5	600			
ZB-GaN[d]	3.200 (45)	6.06 (38)	800 (fixed)	3.225	126 (43)	607 (132)
ZB-GaN[i]	3.302	6.697	600			

Table 5.1—*Continued*

Material	$E(0)$ (eV)	α (10^{-4} eV/K)	β (K)	$E(0)$ (eV)	a_B (meV)	θ_B (K)
WZ-In$_{0.14}$Ga$_{0.86}$N[e,j]	3.1438	10	1196			
WZ-CdSe[k]						
A exciton	1.834 (3)	4.24 (20)	118 (40)	1.834 (3)	36 (50)	179 (4)
B exciton	1.860 (2)	4.17 (10)	93 (20)	1.860 (2)	31 (6)	152 (25)
C exciton	2.263 (4)	3.96 (20)	81 (35)	2.263 (4)	27 (8)	142 (40)

[a] Ref. 17
[b] Ref. 18
[c] Ref. 11
[d] J. Petalas, S. Logothetidis, S. Boultadakis, M. Alouani, and J. M. Wills, *Phys. Rev. B* **52**, 8082 (1995). The numbers in parentheses are 95% confidence limits
[e] A, B, and C excitons not resolved
[f] M. O. Manasreh, *Phys. Rev. B* **53**, 16425 (1996)
[g] H. Teisseyre, P. Perlin, T. Suski, I. Grzegory, S. Porowski, J. Jun, A. Pietraszko, and T. D. Moustakas, *J. Appl. Phys.* **76**, 2429 (1994)
[h] Ref. 27
[i] Ref. 12
[j] Ref. 28
[k] S. Logothetidis, M. Cardona, P. Lautenschlager and M. Garriga, *Phys. Rev. B* **34**, 2458 (1986). The numbers in parentheses are the error margins in units of the last significant figure

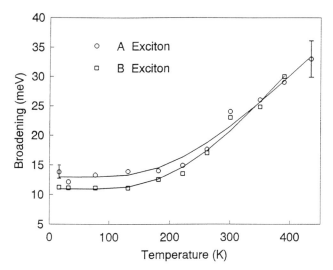

Fig. 5.6 The experimental temperature-dependent linewidths of the A (open circles) and B (open squares) exciton features from WZ-Ga$_{0.95}$Al$_{0.05}$N/sapphire. The solid lines are fits to eqn (5.30).

term is caused by the exciton–LO ($q \approx 0$) phonon (Fröhlich) interaction. The quantity Γ_{LO} represents the strength of the exciton–LO phonon coupling while θ_{LO} is the LO phonon temperature.

The solid lines in Fig. 5.6 are least squares fits to eqn (5.30), which made it possible to evaluate $\Gamma(0)$, Γ_{LO}, and θ_{LO} for the A and B transitions of this material. The obtained values of these quantities are listed in Table 5.2, together with numbers of WZ-GaN, WZ-CdSe, ZB-ZnSe, and ZB-ZnCdSe.

It is interesting to note that the exciton–phonon coupling parameters a_B and Γ_{LO} for WZ-GaN and WZ-GaAlN are considerably larger than those reported for most other III–V and II–VI semiconductors.

Even though Chichibu *et al.* [13] measured the temperature dependence (10 K < T < 300 K) of the energies and broadening parameters of the A, B, and C excitons of WZ-GaN/sapphire (0001) using PR, they did not fit their data to the appropriate terms, i.e. either eqn (5.28) or eqns (5.29) and (5.30). In the high temperature regime $T > 200$ K, where $E(T)$ is approximately linear, they obtained a temperature coefficient of 3.70×10^{-4} eV/K.

5.4.2 Effects of biaxial strain on the excitons on WZ-GaN

Shikanai *et al.* have conducted a detailed PR investigation at 10 K of the effects of biaxial strain on the A, B, and C excitons of WZ-GaN/sapphire (0001) grown by metalorganic vapor phase epitaxy [14]. The results were analyzed theoretically using the Luttinger–Kohn-type Hamiltonian for the valence bands under an in-plane biaxial stress. The authors obtained the shear deformation potential constants and energy gap in the unstrained material. To control the strain,

Table 5.2 Values of the parameters that describe the temperature dependence of Γ (in terms of HWHM) for the excitonic transitions of WZ-Ga$_{0.95}$Al$_{0.05}$N, GaN (WZ and ZB), WZ-CdSe, ZB-ZnSe, and ZB-Zn$_{0.56}$Cd$_{0.44}$Se.

Material	$\Gamma(0)$ (meV)	Γ_{LO} (meV)	θ_{LO} (K)
WZ-Ga$_{0.95}$Al$_{0.05}$N[a]			
A exciton	12 ± 2	91 ± 20	736 ± 100
B exciton	11 ± 2	122 ± 20	779 ± 100
WZ-GaN[b]			
A exciton	15 ± 2	60 ± 20	731 ± 100
B exciton	13 ± 2	74 ± 20	773 ± 100
WZ-GaN[c,d]	34.5	104	800 (fixed)
WZ-GaN[d,e]	8.6 ± 0.4	179	1108
ZB-GaN[c]	39.5	124	700 (fixed)
WZ-CdSe[f]			
A exciton	2.7 (6)	81 (20)	530 (85)
B exciton	4 (1)	139 (25)	775 (86)
C exciton	2.3 (3)	64 (25)	613 (160)
ZB-ZnSe[g]	6.5 ± 2.5	24 ± 8	360
ZB-Zn$_{0.56}$Cd$_{0.44}$Se[h]	6.0 ± 2.0	17 ± 6	334

[a] Ref. 17
[b] Ref. 18
[c] J. Petalas, S. Logothetidis, S. Boultadakis, M. Alouani, and J. M. Wills, *Phys. Rev. B* **52**, 8082 (1995)
[d] A, B excitons not resolved
[e] Ref. 27
[f] S. Logothetidis, M. Cardona, P. Lautenschlager and M. Garriga, *Phys. Rev. B* **34**, 2458 (1986). The numbers in parentheses are the error margins in units of the last significant figure
[g] W. Krystek, L. Malikova, F. H. Pollak, M. C. Tamargo, N. Dei, and A. Cavus, *Acta Physica*, Polonia, **A88**, 1013 (1995).
[h] L. Malikova, W. Krystek, F. H. Pollak, N. Dai, A. Cavus, and M. C. Tamargo, *Phys. Rev. B* **54**, 1819 (1996)

various types of samples including homo- and heterostructures were employed.

The lattice parameters a and c of GaN were measured using X-ray diffraction and the residual strain was calculated using the intrinsic values. Photoreflectance spectra were taken at 10 K using a mechanically chopped 325.0 nm line of a c.w. He–Cd laser as the pump source. To determine the exciton energies from the PR spectra accurately, the authors used eqn (5.18).

Typical PR spectra of WZ-GaN at 10 K are shown in Figs. 5.7(a)–(c). Three peaks, labeled A, B, and C are observed, which correspond to transitions from respective valence bonds to conduction band. The exciton resonance energies (E_A, E_B, and E_C) and broadening parameters were obtained from a lineshape

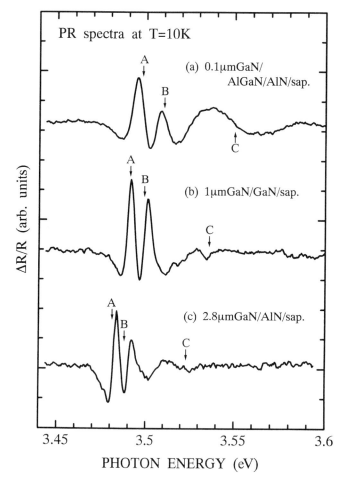

Fig. 5.7 PR spectra of WZ-GaN taken at 10 K. The resonance energy positions obtained from the lineshape fit are indicated by arrows.

fit. According to the polarization selection rules for the space group C_{6v}^4, the A and B transitions are allowed for the light $\vec{E}$ polarization perpendicular ($\perp$) to the c-axis ($\vec{E} \perp \vec{c}$), while the C transition is essentially allowed for $\vec{E}$ parallel ($\parallel$) to the c-axis ($\vec{E} \parallel \vec{c}$). Because the epilayers have their c-axis normal to the substrate plane, $\vec{E} \perp \vec{c}$ is predominantly obtained, leading to dominant A and B transitions in the spectra. The C transition is also recognized in the spectra. One of the reasons for this depolarization may be due to the polarization mixing originating from the strain.

To distinguish the C transition from others, the authors tried quasi-polarization-dependent PR measurements by setting the incidence angle to about 45–60°. Typical results are shown in Figs 5.8(a) and 5.8(b) for $\vec{E} \perp \vec{c}$ and $\vec{E} \parallel \vec{c}$, respectively. The structure at 3.522 eV in the $\vec{E} \parallel \vec{c}$ spectrum is substantially stronger than that in the $\vec{E} \perp \vec{c}$ one. Therefore the authors have assigned those structures as C exciton transitions. A weak structure often observed at the

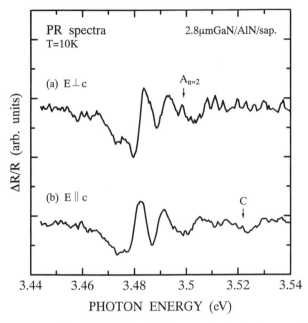

Fig. 5.8 Typical PR spectra at 10 K taken in the quasi-polarization configuration.

higher energy side of the B exciton transition (see Fig. 5.8(a)) has been assigned to a resonance of the first excited states of the A exciton ($A_{n=2}$) because its energy ($E_{n=2}$) agrees well with the photoluminescence (PL) peak energy of the first excited states of A exciton for various samples. The fact that the PL peak energy of $A_{n=2}$ has varied along with A-free-exciton emission with an increase of temperature also has supported this assignment.

The exciton energies are plotted in Fig. 5.9 as a function of the strain along the c-axis (ε_{zz}). The energies of E_A, E_B, E_C, and $E_{n=2}$ all increase with increasing biaxial compressive strain. E_A is a linear function of ε_{zz} and the ground state binding energy of the A exciton, $E_{ex,A}$, is not affected by the strain. From the difference between E_A and $E_{n=2}$, $E_{ex,A}$ is estimated to be 26 meV.

The dependence of $E_A(\varepsilon_{zz})$ on the strain ε_{zz} can be written as [14, 15]:

$$E_A(\varepsilon_{zz}) = E_{gA} - E_{ex,A} + \left[\Xi_D - \left(D_1 - \frac{C_{33}}{C_{13}}D_2\right) - \left(D_3 - \frac{C_{33}}{C_{13}}D_4\right)\right]\varepsilon_{zz} \quad (5.31)$$

where E_{gA} is the energy gap between the conduction and A-valence bands in the unstrained crystal, Ξ_D is a deformation potential related to the dilation, D_i are shear deformation potentials, and C_{jk} are elastic stiffness constants. Expressions for $E_B(\varepsilon_{zz})$ and $E_C(\varepsilon_{zz})$, which involve the crystal field (Δ_{cr}) and spin–orbit (Δ_{so}) splittings, also are given in Refs. 14 and 15.

From an analysis of the data, the authors have obtained values of $\Delta_{cr} = 22$ meV and $\Delta_{so} = 15$ meV, which are close to those reported previously. The value of Δ_{so} agrees well with that of cubic GaN (17 ± 1 meV). This analysis also yielded the following physical parameters: $[\Xi_D - (D_1 - (C_{33}/C_{13})D_2] = 38.9$ eV, $[D_3 - (C_{33}/C_{13})D_4] = 23.6$ eV, and the energy gap in unstrained crystal (E_{gA}) as

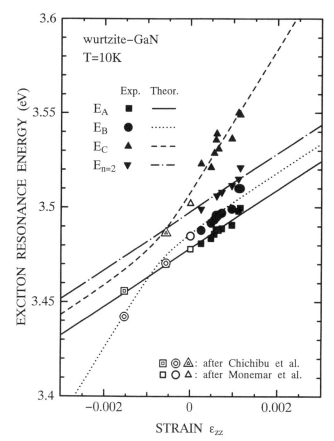

Fig. 5.9 Exciton resonance energies as a function of ε_{zz}. The closed plots are the experimental data of Ref. 14. Theoretical fit lines also are shown.

3.504 eV. Note that from this analysis it is not possible to evaluate Ξ_D and $(D_1 - (C_{33}/C_{13})D_2)$ separately.

Figure 5.9 also shows an interesting anti-crossing of the B and C exciton lines.

5.4.3 Excited states of the A, B and C excitons in WZ-GaN

Shan *et al.* have measured the PR spectra at 10 K from high-quality GaN/sapphire and GaN/SiC samples [16]. The observation of a series of spectral features has made it possible to evaluate the exciton binding energies of the A and B (21 ± 1 meV) as well as the C (23 ± 1 meV) excitons. Figure 5.10 shows the PR spectra (open circles) from these two samples. Using eqn (5.18), the data has been fitted (solid lines), yielding the indicated transition energies. The identifications of the various spectral features is given.

Excited states of the A and B excitons in WZ-GaN/sapphire also have been observed in wavelength modulated reflectivity and PR in Ref. 15.

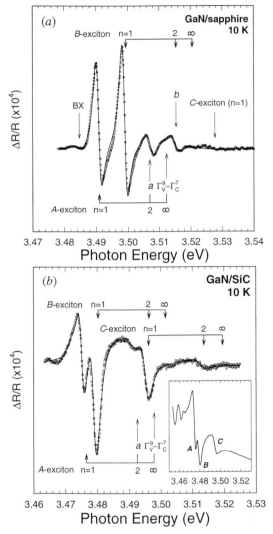

Fig. 5.10 PR spectra of WZ-GaN/sapphire and WZ-GaN/SiC at 10 K (open circles). The solid lines are the lineshape fits. The identities of the various spectral features are given by the annotations.

Also, of course, the first excited state of the A exciton in WZ-GaN/sapphire is shown in Fig. 5.8(a).

5.4.4 Reflection anisotropy spectroscopy investigation of GaN, GaAlN and AlN

Rossow *et al.* have used RAS to study the in-plane optical anisotropies of GaN, GaAlN, and AlN. The layers were grown by organometallic chemical vapor

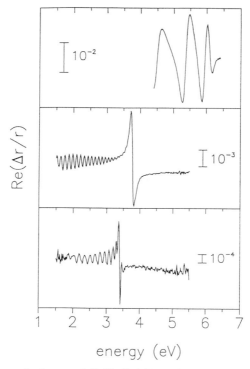

Fig. 5.11 RAS spectra for layers of GaN, GaAlN, and AlN, all grown on 6H–SiC.

deposition on both on-axis sapphire (0001) and offcut 6H–SiC (0001) substrates [19]. Representative RAS spectra are shown in Fig. 5.11. The observed interference fringes have been used to determine the thickness of the layers. In addition to the below bandgap interference fringes the data for the GaN and GaAlN also exhibits first-derivative type features at the bandgap. By extending the measurements above 6 eV the authors have also observed the bandgap of the AlN sample. The paper discusses several mechanisms by which the observed anisotropy can occur in the epilayer, including (a) anisotropic strain in the plane of the layer, (b) a tilt of the c-axis with respect to the surface normal, and (c) non-normal-incidence radiation.

5.5 Summary

This article has reviewed modulation spectroscopy of the Group III nitrides, including instrumentation and lineshape considerations. In relation to other III–V as well as II–VI semiconductors, very little work has been performed on this system. However, these initial investigations have already demonstrated the considerable value of this experimental technique in gaining important information about this semiconductor system. The sharp, derivative-like features of PR or CER and associated lineshape fits were employed (a) to observe the excited states of the A, B, and C excitons in WZ-GaN, making it possible to determine exciton binding energies, (b) to evaluate the temperature dependence of the energies and linewidths in WZ- and ZB-GaN and Wz-GaAlN, (c) to determine the Al composition of WZ-GaAlN, and (d) to measure the effects of strain in WZ-GaN. Reflection anisotropy spectroscopy was used to detect the in-plane anisotropies in WZ-GaN, WZ-GaAlN, and WZ-AlN.

To date, there have been no reports of the observation of FKOs in the Group III nitrides. FKOs in other semiconductors have been the prime experimental method to evaluate the built-in electric fields in a variety of systems, including bulk/thin films in addition to actual device structures.

Acknowledgements

This work was supported by National Science Foundation grant #DMR-9414209, PSC/BHE grant #666424 and the New York State Science and Technology Foundation through its Centers for Advanced Technology program.

Note added in proof

Y. Li *et al.* have measured the temperature dependence of the bandgap and broadening parameter in WZ-GaN using thermoreflectance [27]. Using PR W. Shan *et al.* have measured (a) the temperature dependence of the bandgap of $In_{0.14}Ga_{0.86}N$ and (b) the composition dependence of the bandgap of $In_xGa_{1-x}N$ at 10 K and 295 K [28]. At the former temperature they find $E(x) = 3.5 - 2.63x + 1.02x^2$ eV.

References

1. F. H. Pollak and H. Shen, *Mater. Sci. Eng.* **R10**, 275 (1993).
2. O. J. Glembocki and B. V. Shanabrook, in *Semiconductors and Semimetals*, vol. 36, ed. D. G. Seiler and C. L. Littler, Academic Press, New York, 1992, p. 221, and references therein.
3. F. H. Pollak, in *Handbook on Semiconductors*, vol. 2, ed. M. Balkanski, North-Holland, Amsterdam, 1994, p. 527, and references therein.
4. W. Krystek, M. Leibovitch, F. H. Pollak, M. L. Gray, and W. S. Hobson, *IEEE J. Selected Topics Quantum Electron.* **1**(4), 1002 (1995).
5. H. Shen and M. Dutta, *J. Appl. Phys.* **78**, 2151 (1995).
6. F. H. Pollak, in *Photonic Probes of Surfaces*, ed. P. Halevi, North-Holland, New York, 1995, p. 175.
7. A. K. Ramdas and S. Rodriguez, in *Semiconductors and Semimetals*, vol. 36, ed. D. G. Seiler and C. L. Littler, Academic Press, New York, 1992, p. 137, and references therein.
8. D. E. Aspnes, *Mat. Res. Soc. Symp. Proc.* **324**, 3 (1994); also *Proc. Society of Photo-optical Instrumentation Engineers* (SPIE, Bellingham, 1990) **1361**, 551.

9. O. Acher, S. M. Koch, F. Omnes, M. Defour, M. Razeghi, and B. Drévillon, *J. Appl. Phys.* **68**, 3564 (1990); W. Richter, *Phil. Trans. R. Soc. London A* **344**, 453 (1993).
10. A. Giordana, D. K. Gaskill, D. K. Wickenden, and A. E. Wickenden, *J. Electron. Mat.* **23**, 509 (1993).
11. W. Shan, T. J. Schmidt, X. H. Yang, S. J. Hwang, J. J. Song, and B. Goldberg, *Appl. Phys. Lett.* **66**, 985 (1995).
12. G. Ramírez-Flores, H. Navarro-Contrersa, A. Lastraz-Martinez, R. C. Powell, and J. E. Greene, *Phys. Rev. B* **50**, 8433 (1994).
13. S. Chichibu, T. Azuhata, T. Sota, and S. Nakamura, *J. Appl. Phys.* **79**, 2784 (1996).
14. A. Shikanai, T. Azuhata, T. Sota, S. Chichibu, A. Kuramata, K. Horino, and S. Nakamura, *J. Appl. Phys.* **81**, 417 (1997); S. Chichibu, A. Shikanai, T. Azuhata, T. Sota, A. Kuramata, K. Horino, and S. Nakamura, *Appl. Phys. Lett.* **68**, 3766 (1996).
15. M. Tchounkeu, O. Briot, B. Gil, J. P. Alexis, and R.-L. Aulombard, *J. Appl. Phys.* **80**, 5352 (1996).
16. W. Shan, B. D. Littles, A. J. Fischer, J. J. Song, B. Goldberg, W. G. Perry, M. D. Bremser, and R. F. Davis, *Phys. Rev. B* **54**, 16369 (1996).
17. L. Malikova, Y. S. Huang, F. H. Pollak, Z. C. Feng, M. Schurman, and R. A. Stall, *Solid State Comm.* **103**, 273 (1997).
18. C. F. Li, Y. S. Huang, L. Malikova, and F. H. Pollak, *Phys. Rev.* **B55**, 9251 (1997).
19. U. Rossow, N. V. Edwards, M. D. Bremser, R. S. Kern, H. Liu, R. F. Davis, and D. E. Aspnes, *Mat. Res. Soc. Symp. Proc.* **444**, 835 (1997).
20. S. E. Acosta-Ortiz and A. Lastraz-Martinez, *Phys. Rev.* **B40**, 1426 (1989); H. Tanaka, E. Colas, I. Kamiya, D. E. Aspnes, and R. Bhat, *Appl. Phys. Lett.* **59**, 3443 (1991).
21. B. R. Bennett, J. A. Del Alamo, M. T. Sinn, F. Peirò, A. Cornet, and D. E. Aspnes, *J. Electron. Mat.* **23**, 423 (1994).
22. D. E. Aspnes, in *Handbook on Semiconductors*, vol. 2, ed. M. Balkanski, North-Holland, New York, 1980, p. 109.
23. H. Shen and F. H. Pollak, *Phys. Rev. B* **42**, 7097 (1990).
24. D. E. Aspnes and A. A. Studna, *Phys. Rev. B* **7**, 4605 (1973).
25. Y. P. Varshni, *Physica (Utrecht)* **34**, 149 (1967).
26. P. Lautenschlager, M. Garriga, S. Logothetidis, and M. Cardona, *Phys. Rev. B* **35**, 9174 (1987); S. Logothetidis, M. Cardona, P. Lautenschlager, and M. Garriga, *Phys. Rev. B* **34**, 2458 (1986).
27. Y. Li, Y. Lu, H. Shen, M. Wraback, M. G. Brown, M. Schurman, L. Koszi, and R. A. Stall, *Appl. Phys. Lett.* **70**, 2458 (1997).
28. W. Shan, B. D. Little, J. J. Song, Z. C. Feng, M. Schurman, and R. A. Stall, *Appl. Phys. Lett.* **69**, 3315 (1996).

6 Optical properties and lasing in GaN

Jin Joo Song and Wei Shan

6.1 Introduction

It is well known that widegap Group III nitrides have enormous potential in device applications, particularly in UV–visible light emitters and high-temperature and high-power electronics. The study of optical properties in these materials has been hampered by the difficulties in obtaining high-quality nitride materials. The tremendous recent progress made in nitride growth technology, such as MBE and MOCVD as discussed in previous chapters, has renewed interest in optical investigations in GaN after a relatively long inactive period since the early work during the 1970s by Dingle (1971a–c), Pankove (1976), and Monemar (1974).

In this chapter, we present recent studies of the optical properties of GaN with emphasis on those aspects relevant to photonic applications. We discuss optical transitions associated with excitons and impurities, stimulated emission and lasing by optical pumping, and optical nonlinearities, with more focus on experimental aspects. The experimental data shown in the figures were selected to represent the most pertinent aspects of the chosen topics. Many are the first of their kind, especially in the optical pumping and nonlinear optical studies. The data presented were taken by a wide range of techniques, from simple CW absorption spectroscopy to femtosecond coherent transient spectroscopy. These data, taken mostly from GaN, will provide valuable reference bases for the study of alloys such as AlGaN and InGaN, as well as quantum structures such as quantum wells and superlattices containing AlN, GaN, InN, and their ternaries and quarternaries, in much the same role GaAs data played in studying its related structures in the past.

During the past few years, we have witnessed a significant increase in the number of publications in optical properties of GaN and related subjects. Even for the several topics we have selected in this chapter, it is a challenge to provide a comprehensive, as well as detailed, review in those areas. We have, therefore, chosen to discuss those we thought were quite relevant in practical applications and material characterizations, with the hope that what we present here will ultimately aid in the advancement of nitride growth techniques, photonic device development, and general theoretical understanding in nitrides, and help develop innovative approaches in experimental techniques.

The organization of this chapter is as follows. In Section 6.2, we concentrate on the optical transitions associated with excitons in epitaxially grown GaN. The effects of residual strain on the excitonic transitions and the binding energy for excitons, as well as recent work on exciton–phonon interactions, will be discussed. In Section 6.3, the discussion focuses on optical transitions associated with impurities and defect states observed in nominally undoped GaN and doped GaN with different doping species. In Section 6.4, the main topic is stimulated emission and lasing observed in GaN and related heterostructures. The optical pumping discussion is followed by a very brief summary of InGaN current injection laser diodes. The last section is devoted to a relatively unexplored but promising area: nonlinear optical studies of GaN, which not only provide additional understanding of optical properties of GaN but may open another avenue for new device applications.

6.2 Optical transitions associated with excitons

Many significant optical processes in semiconductors are associated with electronic transitions between the band edges of the valence bands and the conduction band: direct and indirect interband transitions; intrinsic free excitons and excitons bound to impurities (bound excitons); and the transitions related to impurity states associated with the bottom of the conduction-band valleys (donors or donor-like) and with the valence-band maxima (acceptors or acceptor-like). In this section, we focus on the optical transitions associated with free and bound excitons in GaN. The excitonic processes, including excitation, relaxation, formation, and annihilation, are important in semiconductor physics because they involve a number of basic interactions between excitons and other elementary excitations as well as impurities and defects. Exciton studies are very useful for the fundamental understanding of the physics involved in those basic interactions in the material systems and deriving some important parameters in terms of electronic structures and optical properties of the materials.

6.2.1 General aspects of excitons in GaN

Gallium nitride crystallizes in a hexagonal wurtzite structure. Its conduction band minimum locates at the center of the Brillouin zone (Γ-point, $k = 0$), and has a Γ_7-symmetry with a quantum number $J_z = 1/2$. The valence band of GaN has its maximum at the Γ-point as well, which leads to a direct fundamental bandgap. The top of the valence band is split into three sub-bands as a result of crystal–field and spin–orbit coupling due to the wurtzite structure of GaN, conveniently denoted by A, B, and C. The A-band has Γ_9-symmetry, while B and C have Γ_7-symmetry. Detailed reviews of the electronic band structure and the numerical values of some important parameters can be found in a number of recent review articles (Jenkins, 1989; Gorczyca, 1991, 1993; Min, 1992; Palummo, 1993; Rubio, 1993, 1995; Christensen, 1994; Kim, 1994; Capaz, 1995; Suzuki, 1995; T. S. Yang, 1995; Lambrecht, 1995; Sirenko, 1996; Chuang, 1996). We summarize only the structure and symmetries of the lowest conduction band and

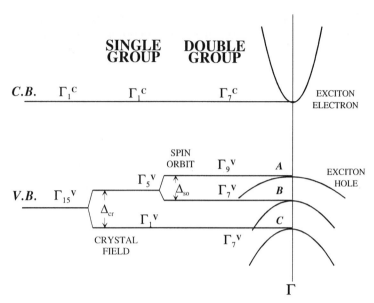

Fig. 6.1 Structure and symmetries of the lowest conduction band and the upmost valence bands in wurtzite GaN at the Γ point ($k \approx 0$).

upmost valence bands in GaN near $k = 0$ (the Γ point) in Fig. 6.1, which is directly related to the object of discussion in this section.

Intrinsic free excitons (FX) represent the lowest energy intrinsic excitation of electrons and holes in pure materials at low excitation density (Elliott, 1957; Knox, 1963). In GaN, the free-exciton state can be adequately described by the Wannier–Mott approximation, where the electrons and holes are treated as nearly independently interacting through their Coulomb fields. The interactions result in completely correlated motions of the paired electrons and holes and a reduction of the total energy of the bound state relative to that of the unrelated free carrier states by an amount corresponding to the exciton binding energy. In effective mass approximation, the total energy of the two-particle structure involving an electron and a hole bound together because of the Coulombic attraction is given by:

$$E_K = E_n + \hbar^2 K^2 / 2(m_e^* + m_h^*), \qquad (6.1)$$

where m_e^* and m_h^* are the effective masses of the electron and hole, respectively, and K is the wavevector of the exciton, and

$$E_n = e^4 \mu / 2(4\pi \hbar \epsilon)^2 n^2, \qquad (6.2)$$

where n is an integer, ϵ is the low-frequency dielectric constant, and μ is the reduced effective mass given by $1/\mu = 1/m_e^* + 1/m_h^*$. Equation (6.2) is a solution of the hydrogenic wave equation, suggesting that free excitons can exist in a series of excited states similar to the excited states of hydrogen-like atomic systems, with an ionization limit corresponding to the bottom of the conduction band. Due to the existence of excitons, optical transition such as absorption or emission from the discrete states occurs below the bandgap E_g at the exciton energy:

$$E = E_g - E_b/n^2. \tag{6.3}$$

The binding energy (E_b), which is the energy necessary to ionize an exciton, apparently is equal to $e^4\mu/2(4\pi\hbar\epsilon)^2$. In GaN, each valence band generates a hydrogenic series of exciton states (Dingle, 1971a, b). The free excitons associated with the Γ_9^V valence band (A-band), the upper Γ_7^V valence band (B-band), and the lower Γ_7^V valence band (C-band) in GaN are often referred to as A-, B-, and C-excitons, as indicated in Fig. 6.1.

Bound excitons are the formation of a complex involving an impurity and an exciton. The impurity might first capture an electron or a hole and then capture the carrier of the opposite charge, or the impurity might directly capture a free exciton (Lampert, 1958; Haynes, 1960). Excitons can be bound to neutral or ionized donors and acceptors. The annihilation of bound excitons is an important recombination process in the near-band-edge spectral region. At low temperatures, the bound exciton emission lines are prominent features of luminescence spectra of many semiconductors. The energy of the emitted photon is

$$\hbar\omega = E_g - E_b - E_{BX}, \tag{6.4}$$

where E_{BX} is the so-called exciton localization energy, that is, the energy required to dissociate the exciton from the impurity. Haynes (1960) first found the exciton localization energy in Si linearly depending on the impurity binding energy. The approximation of $E_{BX} \approx 0.1E_i$, where E_i is the impurity binding energy, is called the Haynes rule. A more general relation between the localization energy E_{BX} and the impurity binding energy E_i, in the case of neutral impurities, has been empirically found to be

$$E_{BX} = a + bE_i, \tag{6.5}$$

where a and b are constants, and in general, $a \neq 0$, b depends on whether the impurity is donor or acceptor (Dean, 1982). The variation of the exciton localization energy for different chemical species is clearly a central-cell effect (Merz, 1972): both the impurity binding energy and the exciton localization energy can be written

$$E_i = E_0 + PV, \tag{6.6}$$

and

$$E_{BX} = E_0' + P'V, \tag{6.7}$$

where E_0 and E_0' are the effective-mass binding energy, respectively, V is a square-well potential in the central cell of the impurity, which distinguishes the difference of the impurities, P is the probability that the impurity electron (or hole) is in the central cell, and P' is the probability of the bound-exciton electron in the cell. Combining these two equations gives

$$E_{BX} = [E_0' - (P'/P)E_0] + (P'/P)E_i = a + bE_i. \qquad (6.8)$$

In direct gap semiconductors, since the impurity binding energy for donors is generally smaller than that for acceptors, the transitions of the donor-bound excitons are expected to lie closer to that of the free excitons than that of the acceptor-bound excitons. It must be pointed out that caution should be taken when the Haynes rule is used to estimate the impurity binding energy based on the measured energy difference between the free-exciton and the bound-exciton recombinations, since it does not apply to shallow acceptors in direct gap semiconductors (Dean, 1982).

6.2.2 Excitonic recombinations in GaN

The near-band-edge luminescence spectra observed from most GaN samples are dominated by a strong, sharp emission line resulting from the radiative recombination of bound excitons. Fig. 6.2 shows a set of near-band-edge photoluminescence (PL) spectra obtained from GaN at various temperatures. Two sharp luminescence lines dominate the emission spectra. The intensity of the strongest emission line marked by BX was found to decrease with increasing temperature much faster than that of FX^A. It became hardly resolvable when the temperature was raised above 100 K. Such effects of temperature on the luminescence intensity indicate that the emission line results from the radiative recombination of bound excitons. The second strongest luminescence line (FX^A), together with the weak emission features such as FX^B at higher energies, can be attributed to the emissions from intrinsic free excitons associated with various interband transitions in GaN. Since as-grown GaN is always n-type, neutral donors are expected to be the most common extrinsic centers in the crystals. The bound-exciton emission line is from the radiative annihilation of excitons bound to neutral donors. One of the reasons for the predominance of the bound exciton emission line in the low-temperature luminescence spectra is that even a very small population of donor-like centers can have very large capture probability for excitons and make the bound exciton states very efficient recombination channels.

A critical parameter in determining the PL efficiency of photoexcited carriers is the effective lifetime:

$$1/\tau_{eff} = 1/\tau_R + 1/\tau_{NR}, \qquad (6.9)$$

where τ_R is the radiative lifetime and τ_{NR} the nonradiative lifetime. In addition to spontaneous radiative recombinations, other nonradiative processes rule over the decay of photoexcited carriers. The nonradiative processes which contribute to τ_{NR} include multiphonon emission, capture and recombination at impurities and defects, Auger recombination effect, and surface recombination, as well as

6.2 Optical transitions associated with excitons

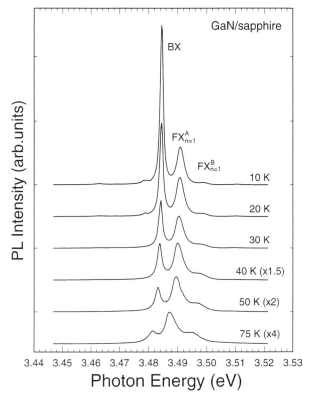

Fig. 6.2 Near-band-edge exciton luminescence spectra taken from a 7.2 μm GaN epifilm grown on sapphire by MOCVD.

diffusion of carriers away from the region of observation. These processes give rise to fast relaxation of the excited carriers down to lower states from which they decay radiatively or relax nonradiatively. Under low excitation conditions, τ_{NR} is limited by the presence of nonradiative recombination centers with the energy levels locating at or near the mid-gap, which cause Schockley–Read–Hall recombinations (Pavesi, 1994). There have been a number of time-resolved photoluminescence studies in GaN. The measured decay times for exciton emissions in GaN were found to be generally in the range of several tens of picoseconds to a few hundreds of picoseconds (Harris, 1995; Smith, 1995; Chen, 1995a; Shan, 1995a; Eckey, 1996). Shown in Fig. 6.3 is the temporal evolution of spectrally integrated exciton luminescence for both free-exciton and bound-exciton emissions observed in a GaN sample at 10 K. The time evolution for both free-exciton and bound-exciton luminescence is dominated by exponential decay. The lifetime of the main PL decay was found to be ~35 ps for the free-exciton emissions and ~55 ps for the bound-exciton emissions for the GaN sample at 10 K. These are much shorter than the theoretically estimated values of radiative lifetime for both free-exciton and bound-exciton recombination in GaN, which are on the order of nanoseconds (Shan, 1995a). The fast decay

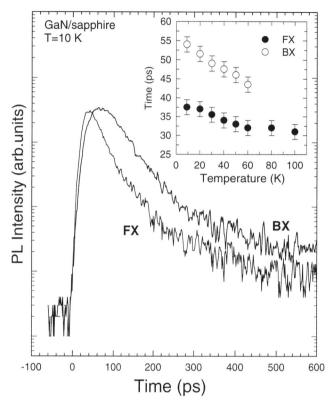

Fig. 6.3 Temporal evolution of spectrally integrated PL for intrinsic free-exciton and bound-exciton emissions in a GaN sample at 10 K. The inset plots the measured decay times for the emissions as a function of temperature.

behavior of the PL intensity indicates that the capture of excitons and trapping of carriers at defects and impurities through nonradiative recombinations dominate the decay of the exciton population. The inset shows the decay times for free-exciton and bound-exciton emission of the sample measured as a function of temperature. The bound-exciton PL decay time decreases slightly with increasing temperature before the emission thermally quenched. The free-exciton PL decay time exhibited slow decreasing with temperature for this sample. It is known that for a radiative recombination dominant system, an increase in the radiative lifetime with temperature is expected for the free excitons since their average kinetic energy is increased, while the time-integrated PL intensity remains almost unchanged ('t Hooft, 1987; Feldmann, 1987). The thermal redistribution results in a decreasing number of the excitons close enough to the Brillouin zone center for radiative recombination. Therefore, the variations of the measured PL decay time for free excitons with temperature are a result of competition between the nonradiative capture of free excitons at defects or impurities and thermally enhanced exciton–exciton and exciton–phonon scatterings. Also the lifetime is expected to be independent of temperature for bound excitons, and only the emission intensity is expected to decrease because of the thermal ionization of the bound excitons. Such phenomena were observed in a number of III–V and II–VI quantum well systems in the low-temperature regime (Feldmann, 1987; Martinez-Pastor, 1993). Although recombination from excitons bound to extrinsic states such as defects or impurities can often be very efficient at low temperatures, the measured decay time is still determined by detailed decay kinetics. The observed decrease of PL decay time with temperature for both free and bound excitons in GaN samples suggests that the incremental stronger nonradiative relaxations occur as the temperature is raised resulting in much faster decay of the exciton population.

In addition to the temperature dependent studies on the exciton luminescence in GaN, the dependence of the excitonic recombinations in GaN on pressure has also been studied to verify the theoretical estimates on the pressure coefficients for the bandgaps of III nitrides (Perlin, 1992; Gorczyca, 1993; Christensen, 1994). Pressure-dependent PL measurements were performed using the diamond-anvil pressure-cell technique to study the effects of the application of hydrostatic pressure on bound-exciton and free-exciton emissions (Shan, 1995b; Kim, 1995). The emission lines associated with the radiative decay of a free exciton or a shallow bound exciton are expected to shift with the host semiconductor bandgap under hydrostatic pressure at the same rate. This is because the excitonic electron stays in the conduction-band edge state or in the orbit of the shallow donor state associated with the conduction-band edge, and the excitonic hole bound in the Coulomb field retains the symmetry of the

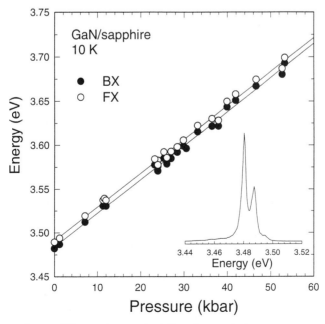

Fig. 6.4 Dependence of the peak energies of exciton emissions on pressure for GaN at 10 K. The solid lines are the least squares fits to the experimental data.

valence-band edges. The change of the intense and sharp emission lines of excitonic radiative transitions with pressure therefore provides an unmistakable signature of the direct Γ bandgap dependence for wurtzite GaN. Plotted in Fig. 6.4 are the peak energies of observed exciton emission structures as a function of applied pressure (Shan, 1995b). The solid lines in the figure are the least squares fits to the experimental data using the quadratic-fit function

$$E(P) = E(0) + \alpha P + \beta P^2, \tag{6.10}$$

where $E(P)$ is the peak energy of exciton emissions under pressure, and P is the applied pressure. The best fits to the data yield a linear slope of $\alpha = 3.86 \times 10^{-3}$ eV/kbar for the BX transition with an extremely small sublinear term of $\beta = -8 \times 10^{-7}$ eV/kbar^2. Similar results were obtained from fitting the FX transition as well. The pressure coefficient for the FX emission energies derived from the fittings is $\alpha = 3.90 \times 10^{-3}$ eV/kbar and $\beta = -1.8 \times 10^{-6}$ eV/kbar^2. The uncertainty of the linear pressure coefficient is $\sim 5 \times 10^{-5}$ eV/kbar and comes primarily from the error bar of pressure calibration (within ± 0.2 kbar), as well as the possible pressure-induced change of the exciton binding energy, which arises from an increase in the electron effective mass and a decrease in the dielectric constant with the bandgap increase (Samara, 1983; Wolford, 1985; Perlin, 1992). The pressure coefficient of the GaN bandgap determined by exciton luminescence measurements is smaller than those reported in other studies involving absorption measurements under pressure reported by Camphausen (1971) and Perlin (1992), as given in Table 6.1. The discrepancy between the PL results and the absorption results may come from the difficulty in accurately determining the transition energies in broad optical absorption curves.

Table 6.1 Pressure coefficients for the exciton emission structures and absorption edge observed in wurtzite GaN.

	$\alpha = dE/dp$ (10^{-3} eV/kbar)	$\beta = \frac{1}{2}d^2E/dp^2$ (10^{-5} eV/kbar^2)	References and methods
Free exciton	3.90 (10 K)	−0.18	Shan (PL)[a]
Bound exciton	3.86 (10 K)	−0.08	
	4.4 ± 0.1 (9 K)	−1.1 ± 0.2	Kim (PL)[b]
	4.7 ± 0.1 (300 K)	−1.6 ± 0.2	
Absorption edge	4.2 (77 K)	−	Camphausen (Absorption)[c]
	4.7 (300 K)	1.8	Perlin (Absorption)[d]

[a] W. Shan, T. J. Schmidt, R. J. Hauenstein, J. J. Song, and B. Goldenberg (1995), *Appl. Phys. Lett.* **66**, 3495.
[b] S. Kim, I. P. Herman, J. A. Tuchman, K. Doverspike, L. B. Rowland, and D. K. Gaskill (1995), *Appl. Phys. Lett.* **67**, 380.
[c] D. L. Camphausen and G. A. N. Connell (1971), *J. Appl. Phys.* **42**, 4438.
[d] P. Perlin, I. Gorczyca, N. E. Christensen, I. Grzegory, H. Teisseyre, and T. Suski (1992), *Phys. Rev.* **B45**, 13307.

Table 6.2 Comparison of III Nitride material properties with 6H–SiC and sapphire[a].

Material	Lattice parameter at room temp. (Å)	In-plane mismatch with GaN (%)	Coefficients of thermal expansion (10^{-6}/K)
GaN	$a = 3.1891$[c]	–	5.59; 3.1, 6.2[e]
	$c = 5.1855$[c]		7.75[d]; 2.8, 6.1[e]
AlN	$a = 3.112$[b]	2.5	4.15
	$c = 4.982$[b]		5.27
6H–SiC	$a = 3.08$[b]	3.5	4.2; 3.2, 4.2[e]
	$c = 15.12$[b]		4.68; 3.2, 4.0[c]
Sapphire	$a = 4.758$[b]	16.1	7.5; 4.3, 9.2[e]
	$c = 12.99$[b]		8.5; 3.9, 9.3[e]

[a] W. G. Perry, T. Zheleva, M. B. Bremser, R. F. Davis, W. Shan, and J. J. Song *J. Electronic. Mat.*(1997) **26**, 224.
[b] *Landolt–Börnstein* ed. O. Madelung, Springer, New York, 1982, Vol. 17.
[c] C. M. Balkas, C. Basceri, and R. F. Davis, *Power Diffraction* (1995), **10** 266.
[d] *Properties of Group III nitrides*, ed. J. H. Edgar, INSPEC, London, 1994.
[e] M. Leszczynski, T. Suski, P. Perlin, H. Teisseyre, I. Grzegory, M. Bockowski, J. Jun, S. Porowski, and J. Major (1995), *J. Phys. D: Appl. Phys.* **28** A149; for $T = 300$–$350\,K$ and 700–750 K, respectively.

6.2.3 Strain effects on excitonic transitions

Early work on the epitaxially grown GaN on sapphire and 6H–SiC substrates showed a relatively large difference in the energy positions of excitonic transitions reported by various groups using different spectroscopic techniques (Dingle, 1971a, b; Monemar, 1974; Shan, 1995c, 1996a; Gil, 1995; Smith, 1996; Chen, 1996), which caused a significant amount of confusion. Only recently has the importance of the effects of residual strain in the epilayers due to the mismatch of lattice parameters and coefficients of thermal expansion between GaN and the substrate materials been realized (Gil, 1995; Rieger, 1996; Shan, 1996a; Volm, 1996). Listed in Table 6.2 are the lattice parameters and coefficients of thermal expansion for GaN, AlN, sapphire, and 6H–SiC (Perry, 1997). Despite the inevitable occurrence of strain relaxation by the formation of a large density of dislocations, the residual strain has a relatively strong influence on the excitonic transition energies. Shown in Fig. 6.5 is a comparison of reflection spectra taken from two GaN epilayers grown on sapphire and a GaN epilayer on SiC at 10 K. As clearly shown by the figure, the energy positions of the exciton resonances associated with *A*-, *B*-, and *C*-exciton transitions are sample dependent. The effects of strain can be further evidenced by the variations of the lattice parameters of GaN relative to that of the virtually strain-free bulk GaN as shown in the inset of the figure, where the lattice constants of the *a*-axis measured from a few GaN epilayers grown on sapphire and SiC were plotted against those of the *c*-axis. It is generally difficult to separate the strain effects

192 *Optical properties and lasing in GaN*

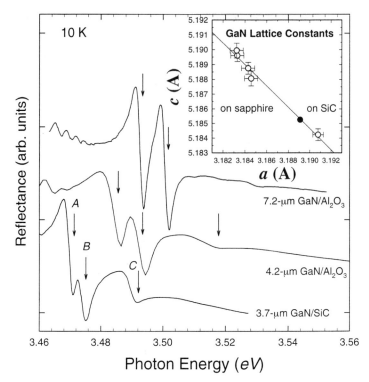

Fig. 6.5 Comparison of reflection spectra of near-band-edge excitonic transitions in one GaN/SiC and two GaN/sapphire samples at 10 K. The curves are vertically displaced for clarity. The inset plots the measured in-plane GaN lattice constants (a-axis) versus the lattice parameters along the growth direction (c-axis) by high-precision X-ray diffraction. The lattice constants of strain-free GaN (solid dot) are shown in the inset for reference.

caused by lattice parameter mismatch from the ones involving thermal-expansion mismatch so as to exactly determine their influences on the optical properties of GaN epitaxial layers. However, the figure clearly indicates that the overall effect of residual strain generated in GaN on sapphire is compressive, which results in an increased bandgap, while the stress induced in GaN on SiC is tensile, which leads to a decrease in measured exciton transition energies. The residual strain induced by thermal expansion mismatch in GaN-based epitaxial layers has the prevailing influence on the energy variations of exciton transitions, since lattice mismatch-induced strain has a completely opposite effect on the variation of GaN bandgap (Amano, 1988; Volm, 1996; Shan, 1996a).

The introduction of strain changes the lattice parameters and in some cases the symmetry of a material, which, in turn, generates variations of electronic band structure. The significance of strain effects is often directly reflected by the variation of the fundamental bandgap of the material. The strain Hamiltonian of a material having a wurtzite structure is given by Pikus (1962). Under the assumption of a strain-independent and isotropic spin–orbit interaction, the energies of the three free excitons A, B, and C can be described as (Sandomirskii, 1964; Gavini 1970)

$$E_A = E_A(0) + D_1\epsilon_{zz} + D_2(\epsilon_{xx} + \epsilon_{yy}) + D_3\epsilon_{zz} + D_4(\epsilon_{xx} + \epsilon_{yy}); \quad (6.11)$$

$$E_B = E_B(0) + D_1\epsilon_{zz} + D_2(\epsilon_{xx} + \epsilon_{yy}) + \Delta_+\left[D_3\epsilon_{zz} + D_4(\epsilon_{xx} + \epsilon_{yy})\right]; \quad (6.12)$$

$$E_C = E_C(0) + D_1\epsilon_{zz} + D_2(\epsilon_{xx} + \epsilon_{yy}) + \Delta_-\left[D_3\epsilon_{zz} + D_4(\epsilon_{xx} + \epsilon_{yy})\right], \quad (6.13)$$

where $E_i(0)$ represents strain-free exciton transition energy, D_i are deformation potentials, and the ϵ_{ii} are components of the strain tensor. Since the energy variation given in eqns (6.11)–(6.13) by D_1 and D_2 is analogous to the hydrostatic shift of a cubic semiconductor, D_1 and D_2 are combined hydrostatic deformation potentials for transitions between the conduction and the valence bands, while D_3 and D_4 are uniaxial deformation potentials characterizing the further splitting of the three topmost valence band edges for tension or compression along and perpendicular to (0001), respectively. The coefficients $\Delta_\pm$ represent the mixing of valence band orbital states by the spin–orbit interaction and are given by (Sandomirskii, 1964; Gavini, 1970)

$$\Delta_\pm = \tfrac{1}{2}\left\{1 \pm \left[1 + 8(\Delta_3/(\Delta_1 - \Delta_2))^2\right]^{-1/2}\right\}, \quad (6.14)$$

where Δ_1 is the crystal–field splitting of the Γ_9 and Γ_7 orbital states, and Δ_2 and Δ_3 are parameters, which describe the spin–orbit coupling (Fig. 6.1). With the usual definition of the strain tensor, $\epsilon_{ij} = \tfrac{1}{2}(\partial u_i/\partial x_j + \partial u_j/\partial x_i)$, where u is the displacement vector, the elastic properties of hexagonal materials can be described by the simple matrical equation:

$$\begin{bmatrix} \sigma_{xx} \\ \sigma_{yy} \\ \sigma_{zz} \\ \sigma_{xy} \\ \sigma_{yz} \\ \sigma_{zx} \end{bmatrix} = \begin{bmatrix} C_{11} & C_{12} & C_{13} & 0 & 0 & 0 \\ C_{12} & C_{11} & C_{13} & 0 & 0 & 0 \\ C_{13} & C_{13} & C_{33} & 0 & 0 & 0 \\ 0 & 0 & 0 & C_{44} & 0 & 0 \\ 0 & 0 & 0 & 0 & C_{44} & 0 \\ 0 & 0 & 0 & 0 & 0 & C_{66} \end{bmatrix} \cdot \begin{bmatrix} \epsilon_{xx} \\ \epsilon_{yy} \\ \epsilon_{zz} \\ 2\epsilon_{xy} \\ 2\epsilon_{yz} \\ 2\epsilon_{zx} \end{bmatrix}, \quad (6.15)$$

where σ_{ij} are the components of the stress tensor and C_{ij} are the elastic stiffness constants. Since the growth direction of most GaN epilayers is along the (0001) direction (z-axis), the corresponding strain keeps the wurtzite symmetry of GaN (Gil, 1995). The strain components ϵ_{ij} can then be described by

$$\epsilon_{xx} = \epsilon_{yy} = \epsilon_\parallel = (a_s - a_0)/a_0; \quad (6.16)$$

$$\epsilon_{zz} = \epsilon_\perp = (c_s - c_0)/c_0, \quad (6.17)$$

where a_0 and c_0 are lattice parameters for strain-free bulk GaN, and a_s and c_s are those for the strained GaN epilayer. Under biaxial stress conditions, the

Table 6.3 Elastic stiffness constants for GaN.

C_{11} (GPa)	C_{12} (GPa)	C_{13} (GPa)	C_{33} (GPa)	C_{44} (GPa)	C_{66} (Gpa)	References
296 ± 18	130 ± 11	158 ± 6	267 ± 18	241 ± 2	–	Savestenko[a]
390 ± 15	145 ± 20	106 ± 20	398 ± 20	105 ± 10	123 ± 10	Polian[b]
365	135	114	381	–	–	Chichibu[c]
411	149	99	389	125	131	Kim[d]

[a] V. A. Savastenko and A. U. Shelag (1978), *Phys. Stat. Sol. (a)* **48**, K135.
[b] A. Polian, M. Grimsditch, and I. Grzegory (1996), *J. Appl. Phys.* **79**, 3343.
[c] S. Chichibu, A. Shikanai, T. Azuhata, T. Sota, A. Kuramata, K. Horino, and S. Nakamura (1996), *Appl. Phys. Lett.* **68**, 3766.
[d] K. Kim, W. R. L. Lambrecht, and B. Segall (1994), *Phys. Rev. B* **50**, 1502.

components of ϵ_{xx}, ϵ_{yy}, and ϵ_{zz} are related through the elastic stiffness constants as $\epsilon_\perp = -2C_{13}/C_{33}\epsilon_\parallel$. Listed in Table 6.3 are the elastic stiffness constants for GaN.

A number of investigations of the strain effects on the exciton transition have recently been reported. The observed shifts in excitonic transition energies relative to the values of strain-free GaN resulting from an overall effect of strain on the bandgap were used to derive numerical values of the band structure parameters Δ_1, Δ_2, and Δ_3, as well as the uniaxial and hydrostatic deformation potentials of wurtzite GaN (Gil, 1995; Chichibu, 1996; Shan, 1996b; Tchounkeu,

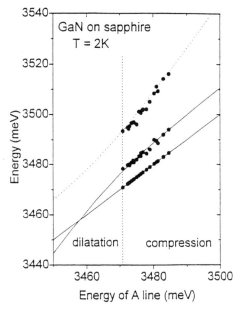

Fig. 6.6 The measured transition energies of various free excitons versus the energy position of *A*-exciton. The solid and dotted lines are fits to the data which lead to the estimates of the crystal–field splitting and the spin–orbit splitting parameters, Δ_1, Δ_2, and Δ_3, for wurtzite GaN. (From Tchounkeu, 1996.)

Table 6.4 The crystal–field splitting and the spin–orbit splitting parameters, Δ_1, Δ_2 and Δ_3, for wurtzite GaN.[a]

Δ_1 (meV)	Δ_2 (meV)	Δ_3 (meV)	References
22	3.7	3.7	Dingle[b]
72.9	5.2	5.2	Suzuki[c]
10 ± 0.1	6.2 ± 0.1	5.5 ± 0.1	Gil[d]
37.5	4	4	Chen[e]
21	5.3	5.3	Chichibu[f]
24.7	5.8	5.8	Reynolds[g]
16	4	4	Chuang[h]
22	5	5	Shikanai[i]

[a] $\Delta_C = \Delta_1$, $\Delta_{SO} = 3\Delta_2 = 3\Delta_3$
[b] R. Dingle, D. D. Sell, S. E. Stokowski, and M. Ilegems (1971), *Phys. Rev. B* **4**, 1211.
[c] M. Suzuki and T. Uenoyama (1995), *Appl. Phys. Lett.* **67**, 2527.
[d] B. Gil, O. Briot, and R.-L. Aulombard (1995), *Phys. Rev. B* **52**, R17028.
[e] G. D. Chen, M. Smith, J. Y. Lin, H. X. Jiang, S. H. Wei, M. Asif Kahn, and C. J. Sun (1996), *Appl. Phys. Lett.* **68**, 2784.
[f] S. Chichibu, A. Shikanai, T. Azuhata, T. Sota, A. Kuramata, K. Horino, and S. Nakamura (1996), *Appl. Phys. Lett.* **68**, 3766.
[g] D. C. Reynolds, D. C. Look, W. Kim, O. Aktas, A. Botchkarev, A. Salvador, H. Morkoç, and D. N. Talwar (1996), *J. Appl. Phys.* **80**, 594 (1996).
[h] S. L. Chuang and C. S. Chang (1996), *Phys. Rev. B* **54**, 2491.
[i] A. Shikanai, T. Azuhata, T. Sota, S. Chichibu, A. Kuramata, K. Horino, and S. Nakamura (1997), *J. Appl. Phys.* **81**, 417.

1996; Shikanai, 1997). Figure 6.6 shows an example of the fit of the experimentally observed exciton transition energies versus the A-exciton transition by Tchounkeu (1996), which gives an estimate of Δ_1, Δ_2, and Δ_3, as well as the deformation potentials. There have also been some theoretical studies discussing the same topic (Dingle, 1971b; Suzuki, 1995; Gil, 1995; Chen, 1996; Chichibu, 1996; Reynolds, 1996; Chuang, 1996; Shikanai, 1997). Listed in Table 6.4 is a summary of the most published values for Δ_1, Δ_2, and Δ_3. The numerical values of these three band-structure parameters, in principle, have to be used to make estimates of the deformation potentials from experimentally observed strain shifts using eqns (6.11)–(6.13) above. The individual numerical values of deformation potentials for GaN reported in recent publications are summarized in Table 6.5.

6.2.4 Exciton binding energy

Strong, sharp resonances associated with the formation of excitons could often be observed near the band edge of GaN by the various spectroscopic methods, including photoluminescence, reflectance, and absorption measurements. In fact, the exciton resonances were observed in absorption spectra even with the

Table 6.5 The interband hydrostatic deformation potentials (D_1 and D_2) of E_0 and uniaxial deformation potentials (D_3 and D_4) of the valence band for wurtzite GaN.

D_1 (eV)	D_2 (eV)	D_3 (eV)	D_4 (eV)	References
−8.16	−8.16	3.71	3.71	Gil[a]
$D_1 - D_2 =$ −5.73		5.73	−2.86	Chichibu[b]
−8.16	−8.16	−1.44	0.72	Tchounkeu[c]
−6.5	−11.8	−5.3	2.7	Shan[d]
−	−	8.82	−4.41	Shikanai[e]
0.7	2.1	1.4	−0.7	Chuang[f]
−15.33	−12.32	−3.03	1.52	Suzuki[g]

[a] B. Gil, O. Briot, and R.-L. Aulombard (1995), *Phys. Rev. B* **52** R17028.
[b] S. Chichibu, A. Shikanai, T. Azuhata, T. Sota, A. Kuramata, K. Horino, and S. Nakamura (1996), *Appl. Phys. Lett.* **68**, 3766.
[c] M. Tchounkeu, O. Briot, B. Gil, J. P. Alexis, and R.-L. Aulombard (1996), *J. Appl. Phys.* **80**, 5352.
[d] W. Shan, B. D. Little, A. J. Fischer, J. J. Song, B. Goldenberg, W. G. Perry, M. D. Bremser, and R. F. Davis (1996), *Phys. Rev. B* **54**, 16369.
[e] A. Shikanai, T. Azuhata, T. Sota, S. Chichibu, A. Kuramata, K. Horino, and S. Nakamura (1997), *J. Appl. Phys.* **81**, 417.
[f] S. L. Chuang and C. S. Chang (1996), *Phys. Rev. B* **54**, 2491.
[g] M. Suzuki and T. Uenoyama *Proc. Int. Symp. Blue Laser and LEDs*, Chiba, Mar. 5–7 1996, ed. Yoshikawa *et al.* Omsha, Tokyo, p. 368.

sample temperature well above 300 K (Song, 1996). Figure 6.7 shows optical absorption spectra taken from a 0.4 μm GaN epitaxial layer grown on sapphire in the vicinity of the band edge at selected temperatures. Three fine exciton resonance spectral features associated with the A-, B-, and C-exciton transitions between the bottom of the conduction band (Γ_7^C) and the three topmost valence band edges ($\Gamma_9^V + \Gamma_7^V + \Gamma_7^V$) can be well resolved as marked in the inset of the figure. The excitonic absorption resonances could also be clearly observed in the 300 K absorption curve, although they are thermally broadened and overlapped. These resonances persist well above room temperature up to higher than 400 K, indicating that the free excitons associated with the fundamental band edges of GaN have a substantially large binding energy. Note that, under the experimental conditions with the photon wavevector along the c-axis of GaN, the intensity of the A-exciton transition, in principle, should be stronger than or at least comparable to the B-exciton, whereas the 10 K absorption spectrum presented in the inset of Fig. 6.7 shows that the B-exciton transition appeared stronger than that of the A-exciton. The observed stronger B-exciton absorption resonance is not, however, inconsistent with what is expected. It results primarily from an overlapping of the B-exciton transition with the higher energy side of the A-exciton feature, and the absorption background of the band tail mainly associated with the $\Gamma_9^V - \Gamma_7^C$ transition, due to the small energy separation between the A- and B-exciton transitions (~ 8 meV) and relatively broad absorption line widths (~ 10 meV) associated with the respective excitonic

6.2 Optical transitions associated with excitons

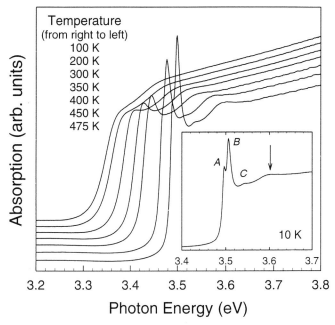

Fig. 6.7 Absorption spectra of a 0.4 μm GaN epitaxial layer in the vicinity of the fundamental absorption edge at various temperatures. The curves are vertically displaced for clarity. The inset details the near-band-edge exciton resonances. The spectral structure indicated by a vertical arrow is due to the contribution of the indirect exciton formation process which creates an exciton and an LO photon simultaneously.

transitions observed from this sample, as well as a strain-induced enhancement of oscillator strength for the B-exciton transition under compression conditions (Gil, 1997). A full treatment of excitonic absorption has been given by Elliott (1957), in which the transitions involving discrete excitonic states and band-to-band continuum are considered. By following Elliott's approach, in principle, one should be able to determine the exciton binding energy by fitting the experimental data shown in Fig. 6.7. However, the complicated fundamental bands of wurtzite GaN make it rather difficult to derive the binding energy unambiguously from the absorption spectra, mainly because of the mutual overlapping of the spectral features associated with A-, B-, and C-excitons and the structures related to the band-to-band absorption continuum.

Earlier results of photoluminescence excitation by Monemar (1974) suggested a 28 meV exciton binding energy. Low-temperature photoluminescence (PL) measurements were also reported to provide some information regarding the exciton binding energy in GaN: for instance, the observations of a small PL spectral feature on the higher energy side of the main exciton emission peaks led to the assignment of the feature to the first excited state (2s) of the A-exciton, and consequently the estimate of the binding energy of 20 meV for the A-exciton and 22 meV for the B-exciton by Reynolds (1996) and 22.7 meV for the A-exciton by Freitas (1996). Fitting to the Arrhenius plot of temperature

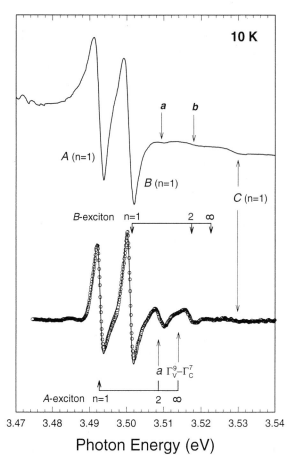

Fig. 6.8 Comparison of conventional and photoreflectance spectra taken from a 7.2 μm GaN/sapphire at 10 K. Open circles are experimental data and solid lines represent the best result of the least squares fit to the PR data. The identifications of the various spectral features are given by the notations.

dependence of integrated intensity of the free-exciton emission yielded a 26.7 meV thermal activation energy by Kovalev (1996). On the other hand, recent results obtained from reflectivity measurements, especially, photoreflectance (PR) measurements, have clearly demonstrated the signatures of transitions related to the 1s and 2s states of A-, B-, and C-exciton transitions, as well as the fundamental band-to-band ($\Gamma_V^9 - \Gamma_C^7$) transition (Shan, 1996c; Chichibu, 1996; Tchounkeu, 1996). The unambiguous observations of those transitions allow a precise determination of their energy positions and a straightforward estimate of the binding energy for the excitons using the hydrogenic model described by eqn (6.3). Figure 6.8 shows a PR spectrum along with a conventional reflection curve taken from a GaN sample at 10 K. While the spectral features A ($n = 1$) and B ($n = 1$) can be unmistakably identified to be associated with intrinsic free-exciton transitions in the conventional reflectance spectrum, the nature of the weak spectral features a and b is not immediately clear.

Fortunately, as discussed in the previous chapter, photomodulation spectroscopy is a powerful method capable of detecting weak signals so as to accurately determine their transition energies and make a positive identification for the nature of the transitions. As illustrated in the figure, the PR spectra not only consist of a series of sharp structures corresponding to most of the observed spectral features in the reflection spectra, but also exhibit, more strikingly, a pronounced enhancement of the barely observed weak spectral structures on the reflection curves such as the *a* and *b* features. Theoretical fits using the Lorentzian functional form to the PR spectra result in an energy separation of 0.008 eV between the *a* and *b* features marked in the figure, which is almost identical to the *A–B* separation, and a 0.016 eV difference for both *A–a* and *B–b* separations within experimental error (≤ 0.001 eV). Similar properties were also observed in the other GaN samples; while the absolute energy position for the main *A*- and *B*-exciton transitions varies slightly from sample to sample due to the influence of residual strain, the energy differences of the *A–a* and *B–b* features were found to be ~0.016 eV for all the samples, and the energy separation between the *a* and *b* features was found to follow closely that between the main *A*- and *B*-exciton features at lower energies for each individual sample. These observations indicate that the *a* and *b* features are indeed associated with *A*- and *B*-exciton transitions. Therefore, *a* and *b* can be attributed to the $n = 2$ excited states (2s) of the excitons. Such identifications permit a direct estimate of the binding energy for the *A*- and *B*-excitons from

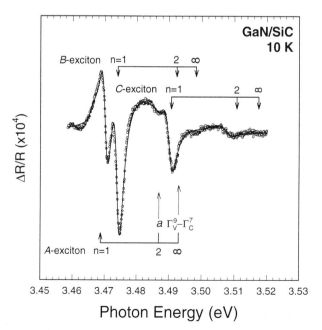

Fig. 6.9 Photoreflectance spectrum of a GaN/SiC sample. The sample exhibits much stronger transition signals associated with the 1s and 2s states of the *C*-exciton in the sample. A 0.017.2 eV energy separation of these two states is obtained by the best fit to the experimental data.

the separation between the $n = 1$ and $n = 2$ states for excitons, assuming that the hydrogenic model based on the effective mass approximation is applicable. A binding energy of $E_b \approx 0.021 \pm 0.001$ eV is obtained using $E_n = -E_b/n^2$ for the A- and B-excitons (Shan, 1996). The binding energy for the C-exciton can also be derived from the theoretical fitting to the PR spectrum taken from a GaN sample grown on SiC as shown in Fig. 6.9 using the same approach. A much stronger C-exciton transition signal could be observed from the sample. In addition to the derivative-like spectral features arising from the transitions associated with the $n = 1$ exciton states of the A-, B-, and C-excitons, the spectral features related to the transitions involving the $n = 2$ exciton states of the A- and C-excitons could be clearly observed in the PR spectrum. The best fit yields a 0.0172 eV energy separation between the 1s and 2s states of the C-exciton, which corresponds to a binding energy of ~0.023 eV, and retains a ~0.016 eV difference between the A-exciton ($n = 1$) and a feature ($n = 2$) as

Table 6.6 The values of binding energies for the intrinsic free A-, B-, and C-excitons in GaN determined by various experimental methods.

A (meV)	B (meV)	C (meV)	References and methods
28	–	–	Monemar (PLE)[a]
27	–	–	Ogino[b]
18.5; 20	–	–	Smith[c]
20	22	–	Reynolds[d]
26.1 ± 0.7	–	–	Volm[e]
22.7	–	–	Freitas[f]
25.3	–	–	Chichibu[g]
21 ± 1	21 ± 1	23 ± 1	Shan (Reflection, PR)[h]
20	18.5	–	Tchounkeu (PR)[i]
26	–	–	Shikanai (PR)[j]

[a] B. Monemar (1974), *Phys. Rev. B* **10**, 676.
[b] Ogino, T. and M. Aoki (1980), *Jpn. J. Appl. Phys.* **19**, 2395.
[c] M. Smith, G. D. Chen, J. Z. Li, J. Y. Lin, H. X. Jiang, A. Salvador, W. K. Kim, O. Aktas, A. Botchkarev, and H. Morkoç (1995), *Appl. Phys. Lett.* **67**, 3387.
[d] D. C. Reynolds, D. C. Look, W. Kim, O. Aktas, A. Botchkarev, A. Salvador, H. Morkoç, and D. N. Talwar (1996), *J. Appl. Phys.* **80**. 594.
[e] D. Volm, K. Oettinger, T. Streibl, D. Kovalev, M. Ben-Chorin, J. Diener, B. K. Meyer, J. Majewski, L. Eckey, A. Hoffmann, H, Amano, I. Akasaki, K. Hiramatsu, and T. Detchprohm (1996), *Phys. Rev. B* **53**, 16543.
[f] J. A. Freitas, Jr. K. Doverspike, and A. E. Wickenden (1996), *Mat. Res. Symp. Proc.* **395**, 485.
[g] S. Chichibu, A. Shikanai, T. Azuhata, T. Sota, A. Kuramata, K. Horino, and S. Nakamura (1996), *Appl. Phys. Lett.* **68**, 3766.
[h] W. Shan, B. D. Little. A. J. Fischer, J. J. Song, B. Goldenberg, W. G. Perry, M. D. Bremser, and R. F. Davis (1996), *Phys. Rev. B* **54**, 16369.
[i] M. Tchounkeu, O. Briot, B. Gil. J. P. Alexis, and R.-L. Aulombard (1996), *J. Appl. Phys.* **80**, 5352.
[j] A. Shikanai, T. Azuhata, T. Sota, S. Chichibu, A. Kuramata, K. Horino, and S. Nakamura (1997), *J. Appl. Phys.* **81**, 417.

6.2.5 Exciton–phonon interactions

As shown in Fig. 6.7, a weak absorption spectral feature could be clearly visible in the energy region of about 100 meV above the excitonic resonance. This is a real structure although it is weaker and broader than the main exciton absorption structures. Its temperature dependence follows the main absorption edge, as shown in Fig. 6.7. A similar feature was first observed in reflectance spectra by Dingle (1971b). They suggested that the feature is related to transitions arising from an indirect phonon-assisted absorption process. In such a process, an incident photon simultaneously creates a free exciton in its $n = 1$ band and an LO phonon, resulting in an absorption coefficient showing a maximum at the corresponding photon energy. A review of this subject is given by Planel (1973). The indirect absorption process with the simultaneous creation of a nonequilibrium exciton in its $n = 1$ band and an LO phonon followed by a cascade kinetic relaxation process with LO-phonon emissions were observed in a number of polar materials such as CdS and ZnO by excitation spectroscopy (Liang, 1968; Planel, 1973; Permogorov, 1975; Ueta, 1986). There have been few reports, however, on the observation of such phonon structures of excitons directly in the absorption spectra taken from semiconductor materials. Thus the absorption results shown in Fig. 6.7 indicate that phonon-assisted exciton formation is very efficient in GaN, primarily due to strong exciton–phonon coupling in this polar material.

In an indirect phonon-assisted exciton formation process, the most probable relaxation process in the exciton band is the phonon emission process (Ueta, 1986). When the energy of the incident light is just

$$E = E_0 + n\hbar\omega_{LO}, \tag{6.18}$$

where E_0 is the exciton energy at $K = 0$ and n is an integer, excitons created by indirect absorption quickly lose their kinetic energy by emitting LO phonons with an appropriate wavevector K before reaching the minimum at $K = 0$ of the exciton band, where they annihilate and emit light with an energy of E_0. When the excitation photon energy is

$$E = E_0 + n\hbar\omega_{LO} + \Delta E, \tag{6.19}$$

where $\Delta E < \hbar\omega_{LO}$, excitons descend toward the bottom by emitting LO phonons until they reach $E_0 + \Delta E$, and then they can relax down only by emitting acoustic phonons to dissipate the excess energy since $\Delta E < \hbar\omega_{LO}$, and virtually become *slow excitons* due to the fact that the relaxation time of the LO-phonon emission ($\tau_o \sim 10^{-13}$ s) is four orders of magnitudes shorter than that of the acoustic phonon emission ($\tau_a \sim 10^{-9}$ s) (Ueta, 1986). Such indirect phonon-assisted exciton formation processes can be verified by photoluminescence

202 *Optical properties and lasing in GaN*

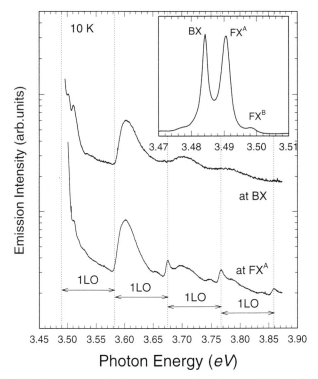

Fig. 6.10 Photoluminescence excitation spectra taken from a 2.5 μm GaN epilayer on sapphire with the detection position at the A-exciton and the bound-exciton emission resonances, respectively. The inset shows the 10 K exciton luminescence spectrum of the sample.

excitation (PLE) measurements. If the lifetime of the excitons due to other nonradiative processes is shorter than the relaxation time of the acoustic phonon emission, a change in the PLE spectrum with a period of $\hbar\omega_{LO}$ is expected to be observed when the emission intensity is measured at the free-exciton resonance while varying the excitation photon energy $E > E_0$. This is because of the fact that whatever exciton–acoustic phonon interaction occurs, the number of excitons which reach the $K = 0$ state will be small and the intensity of the resonance emission will be weak. The first observation of such PLE spectra due to nonequilibrium excitons (hot excitons) in GaN and detailed discussions were reported by Kovalev and co-workers (1996).

Shown in Fig. 6.10 are the results of PLE measurements from a 2.5 μm GaN on sapphire with detection energies at the A-exciton (FX^A) and bound-exciton (BX) emission peak positions, where the inset is the 10 K PL spectrum of the sample (Shan, 1997). The PLE spectra exhibit a series of emission resonances with 91.7 ± 0.5 meV energy spacing, which corresponds to the excitation energies of $E_0 + n\hbar\omega_{LO}$ with the LO-phonon energy $\hbar\omega_{LO} = 740\,\text{cm}^{-1}$. The PLE spectra can be clearly distinguished from the spectral structures associated with the formation of electron–hole pairs via thermalization processes by emitting

LO phonons. Otherwise, an oscillatory series beginning from the bandgap energy E_g, rather than from the exciton energy E_0, with an energy separation of $\hbar\omega_{LO}(1 + m_e/m_h)$, where m_e and m_h are the effective masses for electrons and holes, respectively, would be observed (Permogorov, 1975; Kovalev, 1996). The PLE spectrum detected at the free-exciton (FX^A) resonance consisted of two sets of spectral features: the broad features very similar to those exhibited in the PLE spectrum obtained by detecting the PL at bound-exciton (BX) resonance, and the relatively narrower photon lines highlighted by the dotted lines in Fig. 6.10. While the equidistant phonon lines exhibited in the PLE spectrum can be attributed to the radiative annihilation of those free excitons generated with initial energy at and in the vicinity of $E_0 + n\hbar\omega_{LO}$ completing the thermalization process during their lifetime reaching to the $K = 0$ minimum of the exciton band, as discussed above, the nature of the broad spectral features is not immediately clear. It is very interesting to see the broad spectral features with minima approximately at the exciting energies of $E_{BX} + n\hbar\omega_{LO}$, where E_{BX} is the bound-exciton resonance energy, in the PLE spectra measured at both bound-exciton (BX) and free-exciton (FX^A) resonances. The observation at BX resonance suggests that the process of bound-exciton formation is a fast and efficient nonradiative process of trapping slow excitons at impurities and/or defects during the period of excitons interacting with acoustic phonons, rather than a process of resonantly trapping hot excitons with LO phonon emissions, which would otherwise lead to the appearance of a series of equidistant resonant features with maxima at the excitation energies of $E_{BX} + n\hbar\omega_{LO}$. The probability of exciton trapping at impurities and defects decreases with an increase in the kinetic energy of excitons, as demonstrated by the gradual decrease of the PLE signal level with excitation photon energy. Fig. 6.11 illustrates the kinetic energy relaxation scheme for excitons within the free-exciton band observed from the GaN samples used in this work. The upper portion of the figure depicts the scheme measured at the A-exciton (FX^A) emission peak position. The equidistant phonon lines observed in the PLE spectrum detected at FX^A resonance correspond exactly to the case of free excitons with initial energy of $E_0 + n\hbar\omega_{LO}$ reaching the $K = 0$ minimum of the exciton band via thermalization process. The lower portion of the figure describes the scheme detected at bound-exciton (BX) resonance. Note that the energy separation between BX and FX^A resonances is only slightly larger than 6 meV, and therefore, a process of repopulation of the $n = 1$ free-exciton band via thermal excitation is likely to occur at a finite temperature right after the free excitons nonradiatively relaxed down to the bound exciton states. Although the occupation ratio of FX^A to BX at 10 K is roughly about 10^{-3} given by detailed balance, the density of the state of the free exciton band should be at least a few orders of magnitude larger than that of BX (Chang, 1997). The overall effect of thermal repopulation of the free-exciton band could be very significant, giving rise to the appearance of the broad spectral features at the PLE spectrum detected at FX^A, reminiscent of those measured at BX resonance. From the PLE results shown in Fig. 6.10, one can further infer that the lifetime of both free and bound excitons (τ_{ex}) in the GaN samples is very short

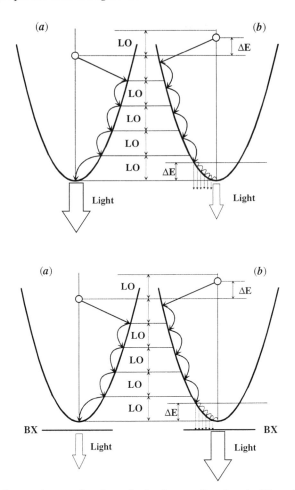

Fig. 6.11 The decay scheme of excitons in the free-exciton band with an incident photon energy (a) $E = E_0 + n\hbar\omega_{LO}$ and (b) $E = E_0 + n\hbar\omega_{LO} + \Delta E$. The upper portion depicts the scheme measured at free-exciton emission peak position (FX^A), and the lower portion describes that detected at bound-exciton resonance (BX).

compared with the relaxation time of acoustic phonon scattering, which is the condition ($\tau_0 < \tau_{ex} < \tau_a$) necessary for observing a hot-exciton oscillatory spectrum. This is consistent with the results from time-resolved exciton luminescence measurements: the measured PL decay time for free-excitons and bound-excitons was found to be on the order of a few tens to hundreds of picoseconds (Shan, 1995a; Harris, 1995; Smith, 1995). This is a clear indication that the lifetime of excitons in the GaN samples is governed by nonradiative processes. The capture of excitons and trapping of carriers at defects and/or impurities through nonradiative relaxation processes dominate the decay of the exciton population. This type of kinetic phenomena of the nonequilibrium hot excitons is also the reason why the excitation spectrum, as shown in Fig. 6.10, does not reproduce the shape of the absorption spectrum shown in Fig. 6.7.

6.3 Radiative recombinations associated with impurities and defect states

The radiative recombination process in semiconductors is primarily determined by the transitions involving band extreme or bound states (excitons, impurities, and defects). The fundamental properties of transitions associated with band extrema and excitons have been discussed in the previous section. Therefore, this section will focus mainly on the phenomena related to radiative recombination processes involving impurities and defects in GaN, which are most frequently studied by photoluminescence measurements. The nondestructive spectroscopic technique of photoluminescence concerns the radiation emitted after optical excitation and consequently reveals the intrinsic and extrinsic properties of the material under study.

6.3.1 Properties of nominally undoped GaN

Photoluminescence spectra taken from nominally undoped (as-grown) GaN samples often, if not always, exhibit a series of emission structures in the energy range approximately 3.27–2.95 eV, and a broad emission band in the yellow spectral region with the peak position around 2.2 eV, in addition to the near-band-edge exciton emissions as shown in Fig. 6.2. The intensities of these two emission bands relative to that of the exciton emissions vary from sample to

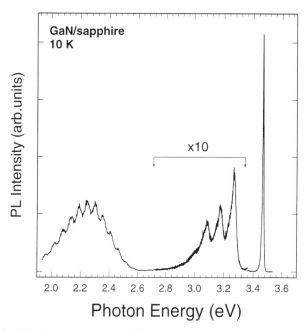

Fig. 6.12 Typical luminescence spectral signatures of nominally undoped epitaxial GaN samples.

sample depending on the crystal quality. Figure 6.12 shows typical spectral signatures of these two emission bands.

The emission structures with a characteristic spectral feature in the energy range of 3.27–2.95 eV have been identified as the radiative recombination processes arising from donor-to-acceptor pair (DAP) transitions and their phonon replicas (Dingle, 1971a). The emission energy of DAP transition is a function of the distance (r) through the initial state and final state interactions between the donor and acceptor. Assuming a simple electrostatic interaction potential between the charged donor and acceptor through a Coulomb interaction term $e^2/\epsilon_0 r$, the emission photon energy in a DAP transition is given by

$$\hbar\omega = E_g - (E_D + E_A) + e^2/\epsilon_0 r, \tag{6.20}$$

where E_D and E_A are ionization energies for the donor and acceptor, respectively. A detailed review on the subject is given by Dean (1973). The relatively broad DAP spectral structures suggest the emissions result mainly from distant pair transitions. The DAP emission structure will evolve to the free-to-bound (FB) recombination with increasing temperature as a result of the thermal ionization of the donor and the increase in the carrier concentration in the conduction band. A characteristic shift of the DAP emission toward higher energies is expected at moderate temperature range. This phenomenon was first observed by Dingle *et al.* in 1971, and the observation, combined with time-resolved spectroscopic measurements, was used to estimate the binding energies for the donor and acceptor, as well as the Coulomb interaction in GaN (Dingle, 1971a).

The broad-band emission in the yellow spectral region known as yellow luminescence has been extensively studied for its adverse effects on carrier lifetime and blue emission efficiency. It could be commonly observed regardless of the growth techniques (Ogino, 1980; Singh, 1994: Wetzel, 1994; Fertitta, 1994; Ohba, 1994; Shan, 1995b). More strikingly, this band was observed in samples implanted with a variety of atomic species (Pankove, 1976). These observations have led to the general belief that this yellow luminescence band involves the electronic states associated with defects. Recent theoretical studies on the electronic structures of impurities and native defects in GaN suggested that point defects, such as nitrogen or gallium vacancies, play important roles (Jenkins, 1989; Tansley, 1992, 1993; Neugebauer, 1994, 1996; Boguslawski, 1995). A few mechanisms were proposed regarding the electronic states involved in the radiative recombination process, which give rise to yellow luminescence. Among them, a model proposed by Glaser (1993, 1995), based on the results of optically detected magnetic resonance (ODMR) from undoped GaN epifilms, as depicted in Fig. 6.13(a), suggests a two-stage process involving three distinct states: a process of nonradiative, spin-dependent electron capture from a neutral effective-mass shallow-donor state to a singly ionized, deep-double-donor state, followed by a radiative recombination between the deep donor state and a

6.3 Radiative recombinations associated with impurities and defect states

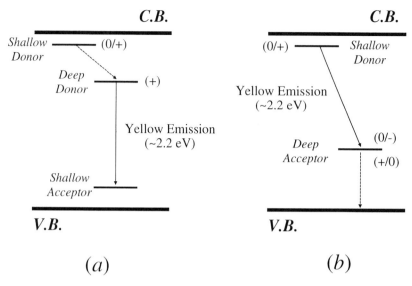

Fig. 6.13 Schematic diagrams of two models for interpreting the broadband yellow luminescence. (a) Radiative recombination involving transitions from deep donor to shallow acceptor as proposed by Glaser (1995); (b) recombination process associated with shallow donor and deep acceptors.

shallow effective-mass acceptor state with peak energy at 2.2 eV. In the meantime, the recombination from the shallow donor state to the acceptor state, which weakly competes with the recombination through the deep donor state, can account for the DAP emission band. However, another model, first proposed by Ogino (1980), attributes the yellow luminescence to the radiative recombination from a shallow-donor state to a deep localized acceptor state (Fig. 6.13(b)). Recent high-pressure PL studies found that dependence of the yellow luminescence on the applied pressure follows the bandgap of GaN (Perlin, 1995; Shan, 1995b; Kim, 1995). These results are consistent with the model of the recombination resulting from the transition between the shallow-donor state and deep-acceptor state, since a shallow-donor state shifts with the host semiconductor bandgap under hydrostatic pressure at the same rate and the effects of hydrostatic pressure on acceptor state is negligible (Wolford, 1985). On the other hand, if the yellow luminescence originates from a transition from a deep-donor state to a shallow-acceptor state, its pressure coefficient is expected to be smaller than that of the bandgap of GaN. This is because the composition of the wavefunction of a donor-like deep center consists of contributions from the entire Brillouin zone and from many bands and its pressure coefficient will be primarily determined by the pressure dependence of the average conduction band energy, since it is no longer possible to associate a deep donor level to one particular conduction band valley (Chadi, 1988).

Shown in Fig. 6.14 is the pressure dependence of the peak energy of yellow luminescence observed in GaN by Perlin (1995). A pressure coefficient of

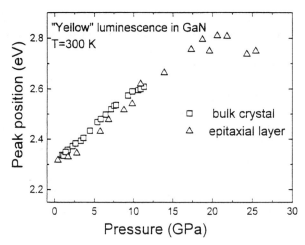

Fig. 6.14 Pressure dependence of the energy of the yellow luminescence for a bulk sample and an epitaxial film of GaN. (From Perlin, 1995).

3.0 meV/kbar was derived from the experimental data obtained at pressures below 180 kbar, which is very close to that of the bandgap of GaN. At higher pressures, the peak energy of the yellow luminescence was found to lose its linear dependence and become insensitive to the applied pressure. This experimental observation, together with the observation of free carrier freeze-out at the same pressure range by Raman and infrared absorption measurements, was explained by a pressure-induced resonant-to-deep transition; the applied pressure increases the bandgap of GaN and brings a resonant electronic level of a deep center into the forbidden gap, resulting in the capture of effective mass electrons and, consequently, the change in the pressure dependence. The *ab initio* pseudopotential calculations by Perlin *et al.* found that the native defects of the nitrogen vacancy and the interstitial gallium are shallow donors at ambient pressure, and each of them introduces a resonance approximately 0.8 eV above the bottom of the conduction band at atmospheric pressure. Based on the results of experimental and theoretical work, they further suggested that the nitrogen vacancy and the interstitial gallium result in autodoping and are responsible for the n-type conductivity commonly observed in nominally undoped GaN. And moreover, the nitrogen vacancy is energetically favorable to be the dominant donor due to its significantly lower formation energy compared to that of interstitial gallium.

It should be pointed out that whether or not the nitrogen vacancy is mainly responsible for the n-type conducting in as-grown GaN is still a subject of debate. Recent theoretical calculations by Neugebauer and Van de Walle (1994, 1996) suggested that the isolated nitrogen vacancy cannot be a source for n-type conductivity because under thermodynamic equilibrium conditions, nitrogen vacancy has a very high formation energy, about 4 eV, which is not likely to produce appreciable concentrations. On the other hand, they concluded that it is the Ga vacancy, rather than the N vacancy, which is responsible for the broadband yellow luminescence. This is based on the facts that the formation energy of the Ga vacancy depends on the position of the Fermi level and is much lower than that for the N vacancy in n-type GaN, and the electronic energy level of this point defect is a deep acceptor state, consistent with the high-pressure experimental results. Some experimental observations did show that the yellow luminescence band was noticeably suppressed by Mg- and Ge-doping or under Ga-rich growth conditions (Nakamura, 1992a; Zhung, 1996a). These results are favorably consistent with the theory that the yellow luminescence is closely related to the Ga vacancy. In addition, a number of recent investigations of introducing oxygen into GaN during growth have indicated that O acts as a donor (Seifert, 1983; Chung, 1992; Sato, 1994; Zolper, 1996). The proximity of O to N suggests that it is an effective-mass type defect (Wetzel, 1996). Therefore, O has been considered as one of the most probable candidates to account for the n-type conductivity in as-grown GaN, and consequently is involved in the yellow luminescence. In their theoretical work, Neugebauer and Van de Walle (1996) found that the triple acceptor Ga vacancy and the single substitutional donor O can form a very stable V_{Ga}–O_N complex acting as double acceptors with a relevant transition energy of 1.1 eV. Thus, the model of V_{Ga}–O_N complex, along with point defect of Ga vacancies, is also in agreement with the experimental observations in terms of the electronic properties and the energetic features.

6.3.2 Properties of doped GaN

Doping p-type has been a major challenge to the device applications of wide-bandgap semiconductors including II–VI compounds and III nitrides. Various potential p-type dopants were incorporated into GaN since the early 1970s. A detailed review on this subject is given by Strite and Morkoç (1992). Currently, p-type doping of epitaxial grown GaN typically uses Mg as the acceptor dopant, thanks to the breakthroughs of converting compensated Mg-doped GaN into p-type conducting using low-energy electron beam irradiation (LEEBI) discovered by Akasaki and coworkers (Amano, 1989) and postgrowth thermal annealing pioneered by Nakamura (1992b). Currently, free hole concentration around the above 10^{18} cm^{-3} at room temperature can be achieved for fabrications of p–n junction light-emitting diodes (Fischer, 1995). The luminescence properties of Mg-doped GaN grown by MOCVD were first studied by Amano (1990a). They found that, at low temperatures (4.2 K), samples with low Mg concentration show DAP emission with the peak position at 3.27 eV and its LO-phonon replicas, indicating that Mg acts as an acceptor. The PL spectra of highly Mg-doped GaN were dominated by broad deep-level emission with the peak at about 2.95 eV. The room-temperature luminescence was found to be doping-concentration dependent, as shown in Fig. 6.15. When Mg concentration is lower than 2×10^{20} cm^{-3} and higher than 5×10^{19} cm^{-3}, blue emission with the peak at 2.7 eV could be observed, together with a much weaker yellow luminescence band. Once Mg concentration is over 2×10^{20} cm^{-3}, only the yellow band was observed. Based on these experimental results, they concluded

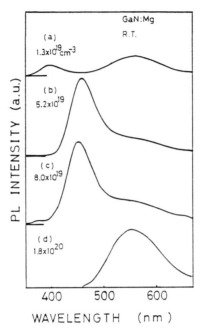

Fig. 6.15 PL spectra at room temperature of Mg-doped GaN layers. The Mg concentrations are (a) $1.3 \times 10^{19}\,\text{cm}^{-3}$; (b) $5.2 \times 10^{19}\,\text{cm}^{-3}$; (c) $8.0 \times 10^{19}\,\text{cm}^{-3}$; and (d) $1.8 \times 10^{20}\,\text{cm}^{-3}$. (From Amano, 1990a.)

that it is important to control the Mg concentration in GaN for efficient blue emission. Similar results have also been observed from Mg-doped GaN prepared with other techniques (Ilegems, 1973; Yang, 1996; Kim, 1996).

It is well known that all Mg-doped GaN samples grown with techniques that use NH_3 as a nitrogen source or which furnish a hydrogen-rich environment, such as MOCVD, require an additional, deliberate processing step to electrically activate the acceptor dopant in order to dissociate the electrically inactive Mg–H complex. These unpassivating processes are carried out either by low-energy electron beam irradiation (LEEBI) (Amano, 1989) or N_2-ambient thermal annealing above 600°C (Nakamura, 1992b). Götz and co-workers reported a systematic study on the activation kinetics of Mg acceptors as a function of postgrowth isochronal rapid thermal annealing temperature (1996a). The thermal treatment results in a reduction in resistivity by six orders of magnitude, and the p-type conductivity was found to be dominated by the acceptor of the substitutional Mg with an activation energy of about 170 meV. They found that PL emission intensities and peak positions of the spectral features, as well as the overall line shapes of as-grown Mg-doped GaN, vary with annealing temperatures (600–775 °C), as shown in Fig. 6.16. These observations suggest a correlation between the near-band-edge PL lines and activated Mg. Studies of the effects of hydrogen passivation on MBE grown Mg-doped GaN using ammonia as the nitrogen source have also been reported (Yang, 1996; Kim, 1996). While the postgrowth thermal annealing treatment significantly alters the PL emission

6.3 *Radiative recombinations associated with impurities and defect states* 211

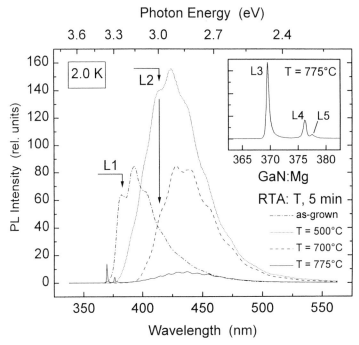

Fig. 6.16 PL spectra of Mg-doped GaN in the as-grown state and after annealing at incremental temperatures T. Zero phonon lines (ZPL) are labeled as L1 (382 nm) and L2 (414 nm). The inset shows the near-band-edge region of the spectrum after annealing at 775 °C. The lines are labeled L3 (369.55 nm), L4 (376.25 nm), and L5 (377.55 nm). (From Götz, 1996a.)

in MOCVD Mg-doped GaN, it was found to have marginal effects on p-type resistivity and PL emission in MBE Mg-doped GaN, indicating that fewer Mg–H complexes exist in MBE-grown GaN. Nevertheless, a high beam flux ratio of ammonia over Ga was found to affect the PL emissions as well, although the amount of Mg–H complexes in the Mg-doped GaN grown by MBE is not as large as that grown by MOCVD (Yang, 1996).

The incorporation of other common column II acceptors, such as Zn, Cd, and Be, into GaN to achieve p-type doping has been attempted, and the related optical properties have been investigated (Ilegems, 1973; Pankove, 1973, 1974, 1976; Boulou, 1979; Monenar, 1980; Bergman, 1987). Unfortunately, the results obtained from PL measurements suggested that the doping introduces deep states within the gap and only produces a broad mid-gap luminescence band. The binding energy for Zn ($\sim$550 meV) and Cd ($\sim$350 meV) was also found to be much higher than that for Mg. Therefore, the effectiveness of these two impurities as p-type dopant species will be very limited, even though the accurate thermal ionization energy for them is not definitive (Strite, 1992). Although p-type doping of GaN using Be was investigated previously without much success (Pankove, 1974; Ilegems, 1973), recent work on Be-doped GaN grown by reactive MBE using ammonia reported the achievement of p-type conduction without involving any post-growth annealing and suggested that Be can form acceptor states about 250 meV above the valence band of GaN based on the observation of near UV luminescence at room temperature (Salvador, 1996). Zolper *et al.* recently reported ion implantation doping of GaN using Ca as a p-type dopant to test a theoretical argument for Ca to be a shallower acceptor than Mg based on the d-state electron relaxation effects in GaN and the lack of d-state electrons in Ca proposed by Strite (1994). An estimated ionization level of ~169 meV for Ca was derived from their results, which is very close to that reported for Mg (Zolper, 1996).

Controlled n-type doping of GaN is commonly achieved by the incorporation of Si. Ge is another dopant used for n-doping. Both Si and Ge act as a single donor by substituting Ga in the GaN lattice. Si-doped GaN was primarily characterized by electrical methods such as variable Hall effect measurements. The recently reported values of the donor activation energies are ~17–27 meV (Hacke, 1994; Gaskill, 1995; Götz, 1996b). Typical PL spectra taken from Si-doped GaN exhibit features which are usually observed for n-type GaN. The most distinguishable part of the PL spectra of Si-doped GaN, compared to that from undoped GaN, should probably be the near-band-edge exciton emission features: the former only exhibit a very strong emission line of excitons bound to neutral shallow donors with the free-exciton emission on the higher energy side hardly observable at low temperatures, whereas the latter show both emission signatures (Freitas, 1996; Götz, 1996b). The intensity of near-band-edge emission relative to that of the broadband yellow emission in Si-doped GaN was also found to increase with the dopant concentration (Nakamura, 1992a; Hacke, 1994). An optical Si donor level relative to the bottom of the conduction band of GaN was found to be ~22 ± 4 meV by Götz *et al.*, based on their observations of PL features associated with radiative transitions of free carriers in the conduction band to acceptors, as well as donor–acceptor transitions involving the Si donor level as the initial state (Götz, 1996b).

6.4 Stimulated emission and lasing

The quantum system of a direct gap material described by single electron band structures predicts that the excitation of electrons from the valence band to the conduction band can result in a filling of these bands to levels corresponding to the carrier quasi-Fermi levels. A photon with energy $E_g < h\nu < \Delta E_f$, where ΔE_f is the difference in electron and hole quasi-Fermi levels (Bernard, 1961), can stimulate recombination and experience net amplification, or gain, providing the cross-section for this process exceeds that for all other absorption processes, such as intervalence band transitions or intraband free carrier absorption (Voos, 1980). The very large probability for radiative recombination in direct gap materials provides a tremendous possibility of stimulated emission resulting from the radiative recombination of carriers at reasonable excitation densities.

Group III nitrides (AlN, GaN and InN and their alloys) have direct energy gaps spanning the wavelength range of ~200 nm to ~650 nm at room temperature and are currently considered to be the most promising UV–blue laser diode materials. The understanding of stimulated emission and laser action in GaN and related heterostructures is of great importance in terms of both scientific and technological aspects. The recent demonstration of pulsed InGaN LDs by Nichia was rapidly followed by continuous-wave (CW) LD demonstration a year later, and now the lifetime of the continuous-wave (CW) LD is evidently being augmented at an even more impressive pace by the same group.

This section is divided into two subsections: (i) optically pumped stimulated emission (SE) and laser action and (ii) current injection laser diodes. In the first subsection, we first discuss widely used experimental configurations for optical pumping and then present experimentally observed SE threshold values reported in the literature and factors affecting the threshold level in GaN and related structures, including the difficulties forming an efficient resonator cavity with commonly used substrate materials. We also mention some of the discrepancies in observed Fabry–Perot interference modes and sample dimensions reported. The gain values reported in the literature are presented, along with the mechanisms proposed, including the self-formed quantum dots playing a role in InGaN lasing. Definitive experimental verification of possible mechanisms is still lacking. Finally, a brief discussion of surface emitting laser action follows. It should be kept in mind that the quantitative values such as SE and lasing threshold and gain can be very much dependent on the samples. And even the relatively good nitride samples that current technology offers still suffer from inhomogeneity and imperfections, which, in part, are related to lattice-mismatched substrates and the difficulty of incorporating aluminum and indium uniformly in the alloy samples. Additionally, estimating the pumping power density and hence the threshold value can vary with different laser systems and the method of pumping density estimates employed by different groups. In the second subsection, a brief chronological description of the development of the current injection laser diodes is given with the emphasis on their optical performance and spectral output.

6.4.1 Optically pumped stimulated emission and laser action

The optical pumping technique is the method capable of providing very high excitation densities necessary for the occurrence of stimulated emission and lasing. The advantage of the optical pumping technique is that the study of stimulated emission and lasing phenomena can be conducted without electrical contacts on samples so that somewhat complicated device-processing procedures can be avoided. Usually a pulsed laser has to be used for optical pumping in order to generate sufficient numbers of carriers in the conduction band. Examples of those pulsed lasers used for optical pumping in Group III nitrides include a nitrogen laser, and the third and fourth harmonics of an Nd:YAG laser. Tuning of the wavelengths of pump lasers can generally be achieved by doubling dye lasers or Ti:sapphire laser beams to the UV range. The most

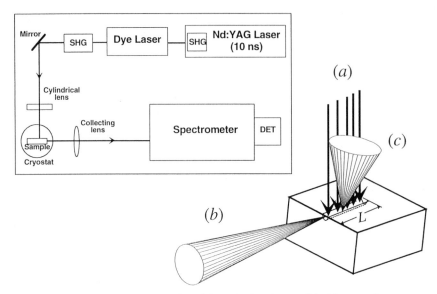

Fig. 6.17 Schemes of optical pumping: (a) excitation beam; (b) side-pumping geometry: a laser beam is focused on a variable width slit into the sample front surface to form a rectangular excitation spot. The length of excitation stripe (L) can be varied by changing the slit size. The emission signal is collected from sample edge; (c) vertical-pumping configuration: emission can be collected from sample front surface in a backscattering geometry. The inset shows an example of an optical pumping experimental setup for wide-gap Group III nitrides. The primary excitation source is a tunable dye laser pumped by a doubled ND:YAG laser at 532 nm. Its output is then frequency-doubled into UV for high-power excitation in either side-pumping or vertical-pumping geometry.

commonly used pumping schemes are side-pumping and backscattering configurations. Figure 6.17 is the schematic depicting these two experimental configurations. In the side-pumping geometry, samples are cleaved or cut into bar-like specimens, and the pumping laser beam is focused on the surface of a bar-like sample under study using a cylindrical lens to form a rectangular excitation spot. The two edges of the sample act as reflective mirrors naturally forming a resonant cavity. The emission signal is collected from the edge perpendicular to the illuminated surface of the sample and then fed into a detection system including a spectrometer, a photodetector such as a PMT or CCD array, and a data acquisition system. In the backscattering configuration, the emission signal is measured from the same side of the sample as the excitation laser beam direction. In the backscattering geometry, the sample inhomogeneity in the surface normal to the growth direction can be probed with high spatial resolution by using a short focal length lens such as a microscope objective lens (Bidnyk, 1997). The backscattering geometry is also useful in the feasibility study of surface emitting laser action. With transparent substrate samples such as GaN/sapphire, a forward scattering, as opposed to a backscattering configuration, can be employed. In this case, the emission can be adversely affected by the inhomogeneity of the sample along the growth direction and the poor quality of the sample in the interface region, since the emission passes through the entire sample thickness, as well as the sapphire substrates. This geometry, of course, is not suitable for opaque substrate materials such as GaN/6H–SiC.

The first observation of optically pumped lasing in GaN was made by Dingle *et al.* using needle-like GaN single crystals at the sample temperature of 2 K (1971c). Since then, there have been a number of investigations on optically pumped stimulated emission and laser action in GaN and related heterostructures in recent years (Amano, 1990, 1991, 1994; Yung, 1994; Khan, 1991, 1996; Kim, 1994; X. Yang, 1995; Zubrilov, 1995; Tanaka, 1996; Schmidt, 1996; Frankowsky, 1996; Aggarwal, 1996; Maki, 1996; Redwing, 1996; Wiesmann, 1996; Zhang, 1996; Shmagin, 1997; Hofmann, 1997; Loeber, 1997). Optically pumped UV lasing was observed even at 475 K from a bar-like sample of MOCVD GaN grown on sapphire, and furthermore, the lasing threshold was found not to be very sensitive to the temperature change (X. Yang, 1995). Figure 6.18 summarizes the reported values of the pumping threshold for stimulated emission and lasing in GaN and related InGaN/GaN and GaN/AlGaN heterostructures prepared by different growth techniques using various substrate materials. The measured threshold values from double-heterostructure (DH) or separate-confinement-heterostructure (SCH) samples are generally lower than the values for bulk-like GaN epifilms by approximately one order of magnitude. This reduction in threshold is due mainly to the effects of carrier confinement and waveguiding associated with the DH and SCH structures (Schmidt, 1996). It is known that in order to achieve lasing in a semiconductor laser structure, a recombination region to produce optical gain by stimulated emission, at least partially overlaps with the optical waveguide region, to confine the optical field (Kressel, 1993). In addition, an optical cavity is required to provide the feedback necessary to produce a self-consistent optical mode with the structure. In the case of the DH structure, the photoexcited carriers are localized in an active region, and an efficient optical waveguide that is contiguous to the recombination region is readily formed. In the case of SCH, a larger optical waveguide is incorporated to symmetrically locate around the recombination region. SCH is an optimal structure for low-threshold operation because the peak of the optical field is at the recombination region, the optical-mode size is determined by the spacing and refractive-index step of the outer waveguide layers, and the inner active region can be structured independently to obtain the maximum carrier confinement leading to the minimum lasing threshold (Kressel, 1993).

Shown in Fig. 6.19 are emission spectra taken from a GaN/AlGaN SCH sample growth by MBE at different pumping power densities (Schmidt, 1996). Under the conditions of low-excitation power densities, the observed spectra exhibit a relatively weak and broad emission feature with the peak position around 365 nm, and the emission intensities linearly increase with the excitation power density, showing the typical characteristic of spontaneous emission. As the excitation power density increases, a sharp, narrow emission feature appears on the higher energy side of the spontaneous emission peak. The position of the maximum of this new emission feature is ~361.5 nm. The emission is polarized

216 *Optical properties and lasing in GaN*

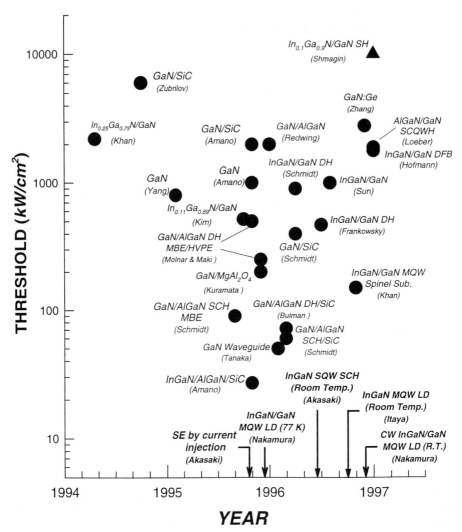

Fig. 6.18 A brief summary of the recently reported stimulated emission and lasing thresholds in GaN and related materials. The circles represent optical pumping results obtained using ns pulsed lasers, the triangles fs pulsed laser pumping. The arrows mark important events regarding electroluminescence and current-injection laser diodes. If not specified, the substrate material is sapphire, and the Group III nitride sample growth method is MOCVD/MOVPE.

along the edge of the sample, and its emission intensity increases superlinearly with the excitation power. This new emission structure becomes the dominant feature as the pumping power density is further increased. Under the conditions of high-excitation power densities, the output of the emission from the GaN SCH samples was very intense. The observations of spectral narrowing and a superlinear increase in intensity with the excitation power density, as well as the complete suppression of the broad emission background, are typical characteristics of the occurrence of stimulated emission. The onset of the steep rise of the

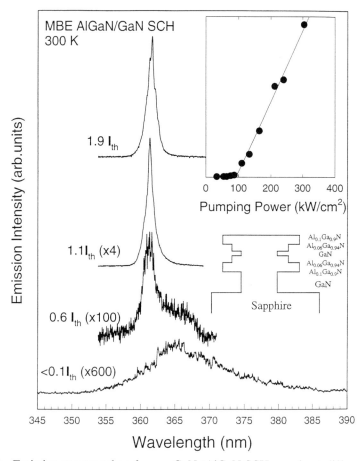

Fig. 6.19 Emission spectra taken from a GaN/AlGaN SCH sample at different pumping power densities. The spectra were vertically displaced for the clarification. I_{th} (~ 90 kW/cm²) represents the pumping threshold for stimulated emission. The change of emission intensity as a function of pumping power density is shown in the inset. The onset of the superlinear increase of emission intensity is defined as the pumping threshold. (From Schmidt, 1996.)

emission intensity plotted as a function of pumping power density, as illustrated in the inset of the figure, is defined as the threshold for stimulated emission. The pumping threshold for stimulated emission was determined to be ~ 90 kW/cm² for these GaN/AlGaN SCH samples. The measured threshold value is approximately one order of magnitude less than that for MOCVD-grown GaN bulk-like epifilms reported by the same group (X. Yang, 1995). Note that the SCH sample shown in Fig. 6.19 was not cut into a bar-like shape to form a cavity, and therefore, the observation of the interference modes cannot be expected in this case. With the proper preparation of sample edges, such as cutting and polishing, further reduction in threshold can be achieved.

Generally, the threshold value for stimulated emission and lasing under optical pumping conditions can be affected by parameters which are dependent

218 *Optical properties and lasing in GaN*

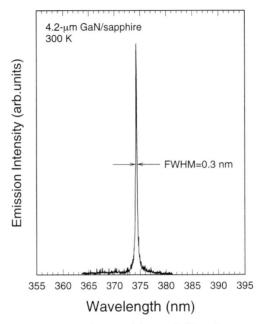

Fig. 6.20 Lasing spectrum taken from a 4.2 μm GaN epilayer at room temperature (300 K). The sample edges were fine polished, and external mirrors were imposed to form a resonant cavity. The full width at half maximum of the laser emission line is 3 Å.

on the pumping source and sample. For a given optical pumping source, the most important influence on the threshold is from the sample itself, with the threshold varying from sample to sample. This can be classified into two groups: one is associated with material properties, such as impurities, crystallinity, and defects; the other is related to the sample preparation, such as, in the case of lasing studies, laser cavity length and the quality of sample edge facets. The difficulties associated with forming high-quality facets by cleaving wurtzite GaN grown on sapphire are well known. Usually the specimens used in optical pumping experiments were just small pieces simply cut off from the large wafers without too much success in finessing the cut surfaces. Schmidt *et al.* demonstrated the improvement of the surface by mechanical polishing of the edge surface of GaN/sapphire, which, however, is a time-consuming process. By attaching external dielectric high-reflectivity mirrors to the sample, these authors observed high contrast Fabry–Perot fringes and, in some cases, nearly a single mode-like lasing peak with FWHM of ∼0.3 nm as shown in Fig. 6.20 (Schmidt, 1996). Kuramata *et al.* reported the formation of the optical cavity by cleaving GaN grown on $MgAl_2O_4$ (1996). The use of such spinel substrates allows the cleavage of GaN epifilms on top in some direction to obtain smooth cleaved facets. Stimulated emission was observed from those samples, and the threshold of ∼200 kW/cm^2 was determined. This value is indeed significantly lower than the reported threshold values of ∼800 kW/cm^2 for GaN samples without cleaved edge facets (X. Yang, 1995). Another issue associated with the threshold is the inhomogeneity of the crystal quality for a single piece of GaN. It is known that there is a very large dislocation density ($\sim 10^8$–10^{10} cm^{-2}) in GaN epilayers grown on commonly used substrate materials of sapphire and 6H–SiC. Some microstructures, such as grains and pillars, are often observed to be co-existing with dislocations (Liliental-Weber, 1996). In addition, cracks are easily generated in the GaN layers grown on 6H–SiC. Therefore, the emission spectral lineshape and the pumping threshold values are sometimes found to depend on the excitation spots on a single piece of sample (Song, 1997). In some literature, observations of Fabry–Perot cavity modes in stimulated emission and lasing spectra were reported (Aggarwal, 1996; Zubrilov, 1995; X. Yang, 1995; Redwing, 1996; Tanaka, 1996), but it is rather difficult to associate the observed mode spacings with sample dimensions. One of the reasons for these difficulties is associated with the fact that the different groups use different indices of refraction n, and its dispersion $dn/d\lambda$ in calculating the Fabry–Perot interference mode spacings using

$$\Delta \lambda = \lambda_0^2 / (2 n_g L), \tag{6.21}$$

where λ_0 is the emission wavelength, $n_g = n - \lambda_0 (dn/d\lambda)$, and L is the cavity length. Another difficulty is the pumping laser shot-to-shot noise, together with poor sample edge surfaces, yielding different mode structures at each laser shot. Additionally, mode fringes can also be affected by parallel microcracks in GaN forming a cavity. Zubrilov and co-workers (1995) attributed the high-quality fringes they observed in GaN on 6H–SiC to such internal microcavities of the order of tens of microns in size.

Knowledge of the dependence of optical gain on optical pumping density or injection current is important for understanding the stimulated recombination processes and developing practical devices such as laser diodes. There have been a number of experimental and theoretical optical-gain studies for GaN and related materials including InGaN/GaN and GaN/AlGaN heterostructures (Fang, 1995; Meney, 1995; Chow, 1995, 1996; Suzuki, 1996; Domen, 1996; Ahn, 1996; Shmagin, 1997). Some theoretical calculations predict that the maximum gain value in GaN could be very high (Chow, 1995, 1996; Domen, 1996). In particular, the calculation results of Domen *et al.* suggested that GaN has a large gain of over 10^4 cm^{-1} and no gain saturation up to 25 000 cm^{-1}, although the transparent carrier density is quite high at 5×10^{10} cm^{-3}. However, the vast majority of experimentally determined values using the variable-stripe-length (VSL) excitation method are found to be at least one order of magnitude smaller in various samples with different structures (Kim, 1994; Song, 1996; Frankowsky, 1996; Wiesmann, 1996; Shmagin, 1997). In the VSL experiment, the excited stripe on the sample surface is varied, and the emission is collected from the sample edge. Then the intensity of emission spectra at every wavelength or selected wavelengths is analyzed as a function of the length of the excited stripe l:

$$I_E = I_0 \{\exp(g \times l) - 1\}, \tag{6.22}$$

220 Optical properties and lasing in GaN

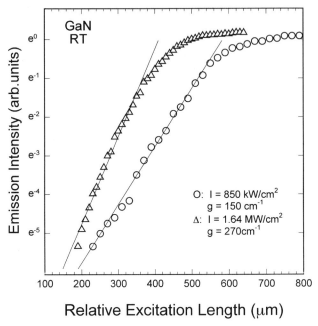

Fig. 6.21 The variation of the peak emission intensity with the length of the excitation stripe on an GaN sample surface at selected power levels at room temperature. The estimated optical gain values are $g = 150\,\text{cm}^{-1}$ and $g = 270\,\text{cm}^{-1}$ for the pumping densities of $850\,\text{kW/cm}^2$ (○) and $1.64\,\text{MW/cm}^2$ (△), respectily. The pumping threshold $I_{\text{th}} = 800\,\text{kW/cm}^2$.

where g is the optical gain coefficient, at different excitation power densities. Shown in Fig. 6.21 are the variations of emission intensity with the length of the excitation stripe on a GaN sample surface at two different pumping power levels at room temperature. The optical gain values for the GaN sample given in the figure were estimated using eqn (6.22) (Song, 1996). Figure 6.22 shows the measured optical gain spectra from an InGaN/GaN double-heterostructure at various pumping power levels reported by Frankowsky and co-workers (1996). The solid lines in the figure are the calculated spectra with a model based on band-to-band transitions. Although the good agreement between the calculated and measured gain spectra led Frankowsky *et al.* to the conclusion that the band-to-band transitions are mostly responsible for the optical gain in this structure, they also suggest that excitonic enhancement of the gain may have to be taken into account in order to explain the magnitude of gain they observed. In fact, Chow (1995) predicted an excitonic enhancement of the optical gain by a factor of two or so in GaN. Further study is needed to verify the origin of the gain in GaN or InGaN.

The investigation of surface emitting laser action by optical pumping is an important step in the development of the vertical-cavity surface-emitting laser (VCSEL). It can also be used for relatively simple testing of lasing since the emission does not depend on the sample size or the edge conditions. There are

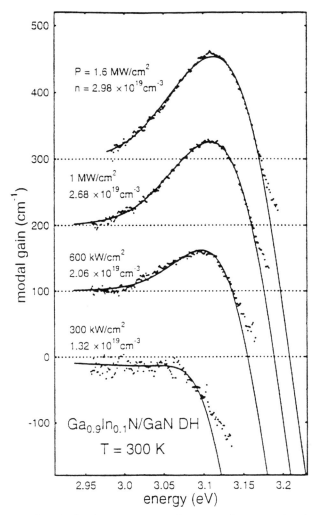

Fig. 6.22 Measured optical gain spectra at various pumping power levels and comparison with calculated spectra (solid lines). The carrier density is determined by fitting the spectra. (From Frankowsky, 1996.)

several reports of the observation of surface emitting stimulated emission in GaN and InGaN films by optical pumping with a backscattering configuration (Khan, 1991, 1994; Amano, 1994; Yung, 1994; Zhang, 1996; Shmagin, 1997). It should be pointed out that caution needs to be taken to block the SE signal emitted from sample edges, which could sometimes be unintentionally collected in the backscattering direction (Bagnall, 1996, Wiesmann, 1996, Bidnyk, 1997). Recently, Redwing and co-workers reported the observation of optically pumped vertical cavity lasing from GaN (1996). In this case, the 10 μm thick GaN lasing medium was sandwiched between MOCVD-grown $Al_{0.4}Ga_{0.6}N/Al_{0.12}Ga_{0.88}N$ Bragg reflectors, which formed a vertical cavity.

Recently, Narukawa and co-workers (1997a,b) reported the observation of InGaN quantum dot-like features self-formed, associated with indium concentration inhomogeneity in the well region of purple–blue laser diode structures. They attributed the main radiative recombination in these quantum wells to the excitons localized at deep traps probably originating from the indium-rich region of the quantum wells acting as quantum dots. It was suggested that the process of self-formation of quantum dots may be a result of the intrinsic nature of InGaN ternary alloys associated with the indium compositional modulation due to phase separation (Ho, 1996; Narukawa, 1997). These quantum dots are considered to play an important role in the lasing processes of InGaN MQWs at room temperature (Chichibu, 1996; Narukawa, 1997). It is likely that the high quantum efficiency of InGaN-based LDs is due mostly to the large localization of excitons and hence substantially reduced nonradiative recombination channels. However, the origin of the trapping centers of these localized excitons has not been clarified yet. The idea of self-formed quantum dots contributing significantly to InGaN MQW lasing processes is still rather new and currently under extensive investigations. Further work is needed to understand better the formation processes of quantum dot-like features and their roles in radiative recombination and lasing processes in InGaN-based laser diodes. The determination of the nature of the lasing transition in InGaN MQWs will continue to be of considerable scientific interest and technological importance and requires further detailed study.

6.4.2 Current injection laser diodes

The most significant and important application of stimulated emission is the generation of coherent radiation in a laser. Tremendous efforts have been directed toward achieving practical current-injection laser diodes using GaN and its related materials. Akasaki (1995) first reported the observation of room-temperature stimulated emission in an InGaN/GaN quantum well structure using pulsed current injection. The device used is actually an SCH structure consisting of an InGaN active layer about 7.5 nm thick, GaN waveguiding layers with total thickness of about $0.4\,\mu$m thick, and AlGaN cladding layers about $0.5\,\mu$m thick for both p- and n-type sides. Strong and narrow stimulated emission with an FWHM of 3 nm was clearly observed, in addition to the broad near-band-edge spontaneous emission at a current density of $1\,\text{kA/cm}^2$ for the device.

In December 1995, fabrication of laser diodes (LDs) using wide-bandgap III–V nitride materials was announced for the first time by Nichia (Nakamura, 1996a). The LDs emit coherent light at 417 nm from an InGaN based MQW structure under pulsed current injection at room temperature. The first InGaN MQW LD device consisted of a 300 Å GaN buffer grown at a low temperature of 550 °C, a 3 μm n-type GaN:Si layer, a 0.1 μm n-type $\text{In}_{0.1}\text{Ga}_{0.9}\text{N}$:Si layer, a 0.4 μm n-type $\text{Al}_{0.15}\text{Ga}_{0.85}\text{N}$:Si layer, a 0.1 μm n-type GaN:Si layer, a 26-period $\text{In}_{0.2}\text{Ga}_{0.8}\text{N}/\text{In}_{0.05}\text{Ga}_{0.95}\text{N}$ multiple quantum well (MQW) structure with a well layer thickness of 25 Å and a barrier layer thickness of 50 Å, a 200 Å p-type

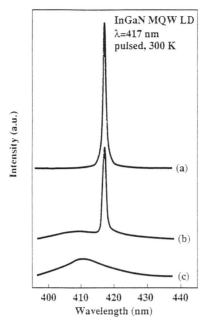

Fig. 6.23 The optical spectra for the first Nichia InGaN MQW LD: (a) at a current of 2.1 A; (b) at 1.7 A; and (c) at 1.3 A. The intensities of the spectra scales are in arbitrary units and are different. (From Nakamura, 1996a.)

$Al_{0.2}Ga_{0.8}N$ layer, a 0.1 μm p-type GaN : Mg layer, a 0.4 μm p-type $Al_{0.15}Ga_{0.85}N$: Mg layer, and a 0.5 μm p-type GaN : Mg layer. Sapphire with (0001) orientation was used as the substrate. Reactive ion etching (RIE) was employed to form mirror-like cavity facets. The roughness of the obtained facets was ~500 Å. High reflective coating (60–70%) was used to reduce the threshold current. The device with a stripe geometry was operated at room temperature under pulsed current-biased conditions with a pulse width of 2 μs and a pulse period of 2 ms. The stimulated emission was observed to occur at a current of 1.7 A, which corresponds to a threshold current density of 4 kA/cm². Figure 6.23 shows the emission spectra of the first reported current-injection InGaN MQW LD. Below the threshold, the emission is characteristic of spontaneous emission with a peak at 410 nm and an FWHM of 20 nm. Above the threshold current, strong stimulated emission at 417 nm with an FWHM of 1.6 nm became dominant. An elliptical far-field pattern could also be observed once the stimulated emission occurred.

Shortly after their first report on the achievement of InGaN LDs, Nakamura (1996b, 1996c) reported the fabrication of LDs with the same device structure using spinel (111) $MgAl_2O_4$ substrate. Polished facets with a roughness of about 50 Å were used as the resonant cavity end mirrors for the LDs. Laser action was achieved under similar operating conditions for the first LDs. The threshold current density and differential quantum efficiency per facet were found to be almost the same as the devices grown on sapphire substrates with much rougher facet surfaces. The performance of the LDs was considerably improved after the device geometry was changed from the stripe geometry to the ridge geometry and the substrates were switched from the (0001) C-face sapphire to (11$\bar{2}$0) A-face sapphire (Nakamura, 1996d). The number of the active QWs of the new LDs was noticeably reduced, and the layer thickness of the QWs was slightly increased. The threshold current density of the ridge-geometry LDs was found to be half that of the stripe-geometry LDs. The differential quantum efficiency of the former is also much higher than the latter. In addition, the ridge-geometry LDs showed a much higher operation tolerance to the duty ratio of a pulsed current at room temperature. These differences were attributed to the high lateral confinement of light propagation and to the absence of etching damage in the gain region of the active layer due to the ridge geometry. The Nichia group also studied the external differential quantum efficiency as a function of the cavity length of an LD with an $In_{0.2}Ga_{0.8}N/In_{0.05}Ga_{0.95}N$ MQW structure consisting of seven 25 Å thick undoped $In_{0.2}Ga_{0.8}N$ layers as gain medium (Nakamura, 1996e). An internal quantum efficiency of 86% and an intrinsic loss of $54\,cm^{-1}$, as well as a threshold gain of $110\,cm^{-1}$, were obtained. A minority carrier lifetime of 2.5 ns was derived from the measurement of the pulse response of the LD device and, consequently, a carrier density of $1.3 \times 10^{19}\,cm^{-3}$ at lasing threshold was calculated. Very recently, the same group again reported continuous-wave (CW) operation of InGaN LDs (Nakamura, 1996f,g, 1997b). Compared with the earlier performance of their LDs, the threshold voltage was significantly reduced from the previous 20–30 V to 11 V for 233 K operation (Nakamura, 1996f) and 5.5 V for room temperature operation (Nakamura, 1997d), respectively, by adjusting growth, ohmic contacts, and doping profile conditions. The lifetime of LDs for CW operation at room temperature was first reported to be only one second and then improved to be 24–40 minutes. These relatively short lifetimes were due to large heat effects caused by relatively high operating currents and voltages (Nakamura, 1997b). However, the CW operation lifetime since then has been improved significantly, to over 1,000 hours (Nakamura, 1997c, 1997c–e).

Akasaki (1996) reported the achievement of the shortest wavelength semiconductor LDs. UV lasing at 376 nm was demonstrated by a nitride-based SCH single quantum well (SQW) structure with current injection at room temperature. The SCH LD uses a 1c–.5 nm thick $In_{0.1}Ga_{0.9}N$ SQW as the active layer without facet coatings. The device was operated at room temperature under pulse-biased conditions. Strong and narrow stimulated emission at 376 nm with the FWHM of 1.5 Å was clearly observed when the injection current density exceeded $2.9\,kA/cm^2$. Room temperature pulsed operation of nitride-based MQW LDs with cleaved mirror facets on conventional C-face sapphire substrates was reported by Itaya et al. of Toshiba (1996). The cleaved facets were carried out by partially scribing along the $\langle 11\bar{2}0 \rangle$ orientation of the sapphire substrate, after lapping the substrate, to form a $(1\bar{1}00)$ facet plane. The roughness of the cleaved facets, caused primarily by a small tilted angle ($\sim 2°$) between the cleaved surface of sapphire and nitride layers, was determined to be within 50 Å. The lasing emission with a peak position at 417 nm and FWHM of

1.5 Å was observed under high current injection conditions at room temperature. The threshold current density was found to be around $50\,kA/cm^2$ with a 20 V operating voltage for the LD with cleaved facets.

Nakamura et al. reported several interesting features observed in the CW lasing emission spectra from their InGaN-based LDs (Nakamura, 1997d). Spectrally very clean, high-finesse longitudinal laser modes at ~405.83 nm were observed with mode spacings $\Delta\lambda = 0.42$ Å from the LDs with the cavity length of 550 μm, consistent with the theoretical estimate. This is the smallest mode spacing observation reported so far in nitride lasing. These longitudinal modes appeared to be modulated by a somewhat periodic envelope with larger spacings of 1.3 ~ 9.2 Å. A series of these emission bands with the large irregular spacings were attributed to the possible mode hopping associated with transitions between adjacent quantum well or quantum dot-like confined energy levels. Similar mode hopping was also observed under the pulsed operation conditions (Nakamura, 1996g, 1997b). In addition, Nakamura et al. also observed large mode hopping with temperature. The peak shifts were due to the temperature dependence of the gain profile associated with various factors, including the bandgap change with temperature (Nakamura, 1996g, 1997d). Another interesting feature in the CW LD operation was that at threshold current, many sharp longitudinal modes appear. With a 20% increase of the CW injection current, however, only a single, sharp line dominates, a phenomenon not yet clearly understood.

Concerning the energy position of the lasing emission with respect to that of the spontaneous emission, the former is usually red-shifted from the latter in semiconductors. The red shift of SE and lasing observed from various semiconductors is often attributed to the renormalization of the semiconductor bandgap in the presence of high excitation densities (Voos, 1980). Nakamura et al. reported a blue shift of the lasing peak compared with the spontaneous emission from some of their InGaN LDs under the pulsed current injection at room temperature (Nakamura, 1996d), while their earlier results exhibited a red shift (Nakamura, 1996a–c). A blue shift of SE in comparison with the spontaneous peak position was also reported by Schmidt et al. in a GaN/AlGaN SCH under optical pumping conditions (Schmidt, 1996). In the SCH samples, the active lasing medium is Si-doped GaN, and the blue shift was attributed to the saturation of the available density of impurity levels, leading to the blue shift arising from the radiative recombination processes to be dominated by the band-to-band transitions. In the case of the blue shift exhibited by undoped InGaN MQWs used in Nakamura's LDs, transition processes involving impurity levels can be ruled out. A further blue shift was observed above the threshold by Nakamura et al. (1996d), when the pulsed current injection density is increased. For example, at an injection current of 199 mA, a single peak at 411.3 nm appeared, whereas at 200 mA another peak showed up at 411.0 nm. With a further increase of current to 203 mA, three subgroups of emission features were observed at 407.7 nm, 409.5 nm, and 410.9 nm, respectively. The origin of these emission subbands is not clear as yet. One possible origin of these subbands is associated with the transitions between quantum confined energy levels in quantum wells or quantum dot-like structures (Nakamura, 1996g). The blue shift recently observed in the electroluminescence spectra of the InGaN MQW with the increase of the driving current by Chichibu *et al.* was explained by the combined effects of the quantum confinement Stark effects and the band filling of the localized exciton states (Chichibu, 1997). A clear understanding of the origin and the blue shift observed at a higher current in the pulsed LD operation requires further detailed investigations of the InGaN MQWs from the structural and optical points of view.

In general, tracing the origin of a blue or red shift in MQWs can be quite complicated due to the inevitable imperfection in MQWs, such as interface roughness, well thickness fluctuations, and alloy disordering effects. Growth of high-quality homogeneous nitride alloy layers, such as AlGaN and InGaN, is still a technical challenge. Especially in the case of MQWs containing InGaN layers, further complications arise from the difficulty of incorporating indium in the first place and the intrinsic nature of the current InGaN growth processes. As mentioned in the previous section, it has been suggested by several groups that indium phase separation occurs in InGaN alloys, leading to indium compositional modulation (Ho, 1996; Narukawa, 1997). Especially, the quantum dot-like features suggested to be present in the MQWs of the Nichia LDs invoke the picture of localized excitons playing a role in Nichia's current injection LDs (Narukawa, 1997; Kawakami, in press). To optimize the current injection InGaN LDs, therefore, a much better understanding is needed of the growth processes of InGaN, the structural properties of InGaN, and its quantum wells on the atomic scale, as well as their optical properties.

6.5 Optical nonlinearities of GaN

Nonlinear optical properties are an important aspect of any materials to be used for photonics and optoelectronics. Stimulated emission and lasing processes are, in fact, nonlinear optical phenomena, and as discussed in the previous sections, are being extensively investigated for the development of practical current-injection LDs operating at the UV–blue spectral region. Although only very limited work has so far been reported in the study of different types of optical nonlinearities of GaN, it has been found that the material exhibits several interesting nonlinear optical effects, which provide important information for us to better understand its nonlinear properties and to exploit the intriguing possibilites of its device applications.

6.5.1 Harmonic generations and multiphoton spectroscopy

Optical third harmonic generation spectroscopy was carried out by Miragliotta and co-workers (1994) in MOCVD-grown wurtzite GaN on sapphire to measure dispersion and magnitude of the third order susceptibility tensor element $\chi^{(3)}_{xxxx}$, where the x-axis is perpendicular to GaN growth direction. A resonant enhancement was observed in $\chi^{(3)}_{xxxx}$, near the fundamental energy gap, E_g, of GaN when

6.5 Optical nonlinearities of GaN

tuning the photon energy $\hbar\omega$ in the vicinity of $3\hbar\omega \sim E_g$. The nanosecond photons in the range of $1.0 \sim 1.35\,\text{eV}$ were generated by stimulated Stokes–Raman scattering of tunable dye laser radiation using a pressurized H_2 gas cell. The peak value of $\chi^{(3)}_{xxxx}$ was 2.7×10^{-11} esu at the absorption edge. This is approximately twice as large as the predicted values in ZnSe in the same photon energy region, despite the lower bandgap energy of the latter. Miragliotta et al. attribute this to higher ionicity, hence the lower polarizability of ZnSe compared with GaN (Miragliotta, 1994).

In second-harmonic (SH) generation experiments, the second-order susceptibility tensor elements were measured in GaN at 532 nm. The magnitude of $\chi^{(2)}_{zzz}(-2\omega, \omega, \omega)$ was measured to be nearly twice as large as $\chi^{(2)}_{xxz}$ in the same sample, and approximately 22 times as large as $\chi^{(3)}_{xxx}$ in quartz ($\sim 1.2 \times 10^{-9}$ esu) (Miragliotta, 1993). Recent calculations of the second harmonic susceptibility by Chen (1995) based on Kohn–Sham local density approximation closely agree with the experiments for wurtzite GaN.

More recently, Miragliotta (1996) investigated the optical second-harmonic response of a reverse-biased GaN film grown by MOCVD. In this work, the nanosecond tunable dye laser source was used so as to tune the SH photon energy through the fundamental absorption edge. With the application of a d.c. field, the nonlinear response can result from coupling two optical fields of frequency ω and one d.c. field via third-order nonlinear processes. In this electric field-induced second harmonic (EFISH) generation, the second harmonic polarization at 2ω is linearly dependent on the applied d.c. field and quadratically dependent on the incident laser field $E(\omega)$. It was demonstrated that the EFISH response from n-type GaN can be resonantly enhanced when the two-photon energy approaches the bandgap energy. Also the magnitude of the EFISH signal was found to exceed the 2ω contribution from the intrinsic second-order nonlinearity when a field of the order of $100\,\text{kV/cm}$ was applied to the GaN surface. Figure 6.24 shows the SH reflected intensity from the GaN/electrolyte interface for various reverse bias potentials. With the application of a reverse bias d.c. field, a resonance appeared at 3.43 eV with a linewidth of $\sim 50\,\text{meV}$. This narrow resonance was considered to be due to the resonant behavior of $\chi^{(3)}$. This sharp resonance observed at the band edge also suggests that SH spectroscopy may be used to probe the electronic band structure of GaN (Miragliotta, 1996).

Harmonic generation, as well as multiphoton absorption-induced photoluminescence spectroscopy, can also be used to probe the mid-gap states using tunable laser sources. Recently, Yang and co-workers (1997) observed extensive spectral structures below the bandgap in the second-harmonic generation spectra taken from GaN on sapphire. A strong resonance was observed at two photon energies of $2\hbar\omega \sim 2.85\,\text{eV}$, but its origin is not yet well understood. The mid-gap defect states on sapphire were also observed by Libbon (1997) in GaN on sapphire by using the two-color, two-photon excitation spectroscopy technique. The excitation source was a tunable picosecond optical parametric generator/amplifier pumped by the third harmonic of a picosecond pulsed Nd:YAG laser. The photoluminescence in the UV region was monitored while

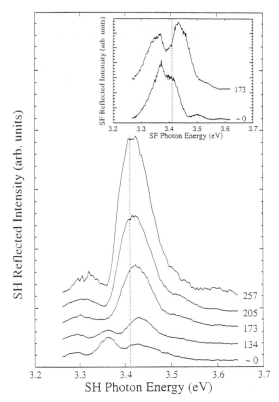

Fig. 6.24 The second harmonic reflected intensity from the GaN/electrolyte interface for various reverse bias potentials. The dashed line represents the position of the band edge in the unbiased sample. The number to the right of each spectrum is the value of the dc field at the GaN surface in kV/cm. The inset shows sum frequency generation spectra taken from the same GaN sample using a fixed 532 nm and a tunable IR source. (From Miragliotta, 1996.)

tuning one of the infrared picosecond pulses, and the increase in the UV PL intensity was observed when $\hbar \omega$ was resonant with states at ~ 1.0 eV above the valence band (Kim, 1997). Although the origin of these states is not clear as yet, this work, together with the SH generation spectroscopy work mentioned above, demonstrates the usefulness of the multiphoton spectroscopy technique in detecting subband gap states.

6.5.2 Picosecond four-wave mixing: below bandgap excitation

Taheri *et al.* reported the performance of degenerate four-wave mixing (FWM) experiments in GaN grown on sapphire by MOCVD (Taheri, 1996). Fig. 6.25 depicts the schematics of the experimental setup. The degenerate wave mixing experiments were performed using frequency-doubled pulses from a mode-locked, Q-switched Nd:YAG laser operating with a repetition rate of 10 Hz.

6.5 Optical nonlinearities of GaN

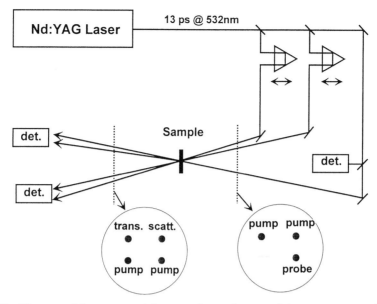

Fig. 6.25 Picosecond four-wave mixing experimental setup. A boxcar geometry is used. The insets show the cross-section view of beam spots in front of and behind the sample.

The laser pulse has a quasi-Gaussian spatial and temporal profile with an e^{-1} of 13 picoseconds. The second harmonic output of the laser at a wavelength of 532 nm was equally split in energy into two pump beams and one probe beam. These beams were spatially and temporally recombined in the samples in a forward propagating boxcar geometry (Eichler, 1986). The advantage of this geometry over the conventional counter propagating technique is that the scattered signals are spatially separated from input beams. This enhances the signal-to-noise ratio and allows direct measurement of the absolute scattering efficiency. The two pump beams were σ-polarized, while the probe beam was π-polarized in order to increase the signal-to-noise ratio and minimize interaction between the pump and probe beams. At the front surface of the sample, the two pump beams had a beam radius of 500 μm and a small crossing angle, 2θ, of 2°. The probe beam was incident on the interference region at an angle of 1.7° with respect to the pump beams. Its beam radius was slightly smaller at 400 μm. The smaller radius allows the central portion of the pump beams to be monitored, reducing the error that can occur because of the finite radial extent of the pump beams. In such a wave-mixing geometry, the pump beams set up an interference pattern in their overlap region with fringe spacing $\Lambda = \lambda/2\sin\theta$, where λ is the wavelength of the beams in free space. Intensity dependence causes the index of refraction in the light regions of the interference pattern to differ from that in the dark region. The overlap region, therefore, behaves as an index grating with spacing of Λ. An incident probe beam will be diffracted from this grating in the directions satisfied by the phase matching conditions. An energy meter preceded by a polarizer was used to detect the π-polarized

230 Optical properties and lasing in GaN

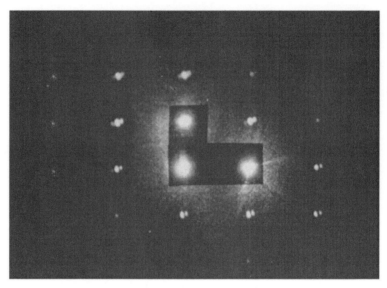

Fig. 6.26 Photograph of the wave mixing signals showing three attenuated pump and probe beams and higher order nonlinear diffracted signal spots. The laser photon energy (2.33 eV) is well below the room-temperature GaN bandgap (3.42 eV).

diffracted signal. Figure 6.26 shows a photograph of the diffracted signals, together with the three attenuated incident beams when they are temporally overlapped in a thick GaN sample. The picture was taken at room temperature with a total incident irradiance of 1.7 GW/cm². The clearly visible patterned spots are diffracted higher order wave-mixing signals in the phase-matching directions.

By delaying the arrival of the probe beam, the time response of these nonlinear optical changes was measured. Fig 6.27 shows the scattering efficiency as a function of the probe-beam delay. The scattering efficiency is defined as the ratio of the scattered to the transmitted probe beam:

$$\eta(t) \equiv \frac{I_{\text{scattered}}}{I_{\text{transmitted}}} = \sin^2\left(\frac{k\Delta n(t)d}{2}\right) \simeq \left(\frac{k\Delta n(t)d}{2}\right)^2, \qquad (6.23)$$

where k is the probe wavevector, $\Delta n(t)$ is the time-dependent index change, and d is the sample thickness. At zero delay, a maximum scattering efficiency 4×10^{-5} corresponding to an index change $\Delta n(0)$ of 1.2×10^{-3} was obtained. This change can be related to the pump beam intensity through an effective nonlinear refractive coefficient n_2, defined as

$$\Delta n(0) = n_2 I_0. \qquad (6.24)$$

With the pump beam irradiance of 1.2 GW/cm², a value of 1×10^{-3} cm²/GW was derived for the effective nonlinear refractive index n_2 from eqn (6.23). Such a value is much larger than the expected range for semiconductors. For example, ZnSe has a value of 6.7×10^{-5} cm²/GW at 532 nm (Sheik-Bahae, 1990, 1991). A large value of n_2 in GaN indicates the potential of its technological importance

6.5 *Optical nonlinearities of GaN* 231

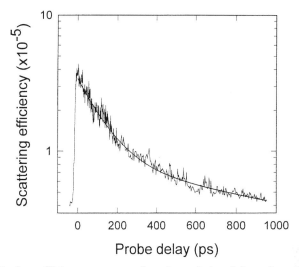

Fig. 6.27 Scattering efficiency, η, as a function of the delay of probe beam. Two characteristic time constants of $100 \pm 10\,\text{ps}$ and $1.1 \pm 0.2\,\text{ns}$ can be derived from the nonexponential decay. (From Taheri, 1996.)

for photonic and optoelectronic device applications, since for practical applications, it is advantageous to have large nonlinear refractive indices or the nonlinear absorption coefficients that have fast time responses (Chemla, 1987).

Based on the fact that the measured scattering efficiency against the probe beam delay exhibits a nonexponential decay with two characteristic time constants of $100 \pm 10\,\text{ps}$ and $1.1 \pm 0.2\,\text{ns}$, as shown in Fig. 6.27, Taheri *et al.* suggested that the change in the refractive index is caused primarily by the generation of free carriers in the conduction band, and the lifetime of the index grating is predominantly governed by the decay of the free-carrier population through various recombination processes rather than the diffusion of the carriers from the light regions of the grating to the dark regions, diminishing the index modulation. The bound electron contribution can also be ruled out, since their time response is shorter than the picosecond pulses used here.

Since the third-order nonlinear optical susceptibility components can exhibit resonance enhancement when the photon energy approaches the energy bandgap or excitonic transitions (Levenson, 1987), an even larger diffraction efficiency, and hence a larger nonlinear index of refraction than that observed at 532 nm, can be expected in GaN at shorter wavelengths. Particularly interesting will be four-wave mixing work in the vicinity of excitons in GaN-based quantum well structures, as in the case of GaAs-based quantum wells and superlattices (Chemla, 1987).

6.5.3 Femtosecond four-wave mixing: exciton resonance

Recently, there has been intense research activity in subpicosecond coherent spectroscopy of semiconductors and their quantum wells, stimulated in part by the ready availability of tunable femtosecond laser sources. Extensive studies

232 *Optical properties and lasing in GaN*

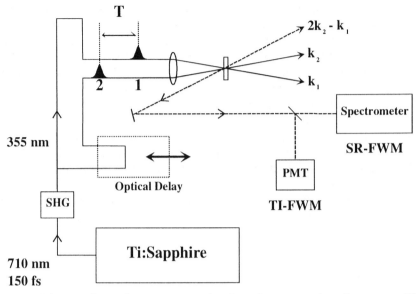

Fig. 6.28 The experimental arrangements for the coherent-transient degenerate FWM measurement.

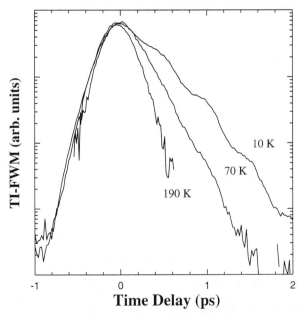

Fig. 6.29 Time-integrated four-wave mixing (TI-FWM) signal in the reflection geometry near the *B*-exciton resonance at 10, 70, and 190 K for a 7.2 μm epilayer of GaN grown on sapphire.

have been performed on GaAs, ZnSe, and related materials and their quantum wells in the coherent regime (Leo, 1991; Fischer, 1994). Fortunately, GaN intrinsic free excitons can be resonantly excited as well using the second harmonic of a self-mode-locked Ti:sapphire laser in the high-energy region of the tuning curve (~700 nm). Using this laser system, Fischer and co-workers (1997) performed femtosecond four-wave mixing (FWM) experiments on the A- and B-exciton transitions in GaN (Song, 1997). Fig. 6.28 illustrates the experimental arrangements for the coherent-transient degenerate FWM measurement. The second harmonic of 710 nm, 150 fs pulses from a mode-locked Ti:sapphire is used to resonantly excite the intrinsic free excitons in GaN. The FWM signal is detected in the reflection geometry in the phase-matched direction and measured as a function of optical delay between the two pulses in order to obtain the time-integrated (TI) FWM signal. Alternatively, the energy spectrum of the signal at zero delay can be measured by sending the signal into a spectrometer to take the spectrally resolved (SR) FWM signal. Time-integrated (TI) FWM data are shown in Fig. 6.29 at three temperatures (T = 10, 70, and 190 K). The delay time at each temperature was fitted to a single exponential decay to determine the decay rate as a function of temperature. The homogeneous linewidth can then be determined from the relation $\Gamma_{hom} = 2\hbar/T_2$, where T_2 is the pure dephasing time (Fischer, 1994, 1997). By performing

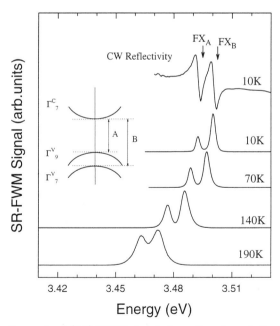

Fig. 6.30 Spectrally-resolved (SR) FWM signals from the A- and B-exciton resonances near zero delay at 10, 70, 140, and 190 K plotted together with CW reflectivity data from the same sample. The A(FX$_A$) and B(FX$_B$) intrinsic free excitons are labeled, and the inset shows the excitonic band structure. From the temperature dependent exciton linewidth changes, the exciton-LO phonon coupling coefficient can be derived (Fischer, 1997).

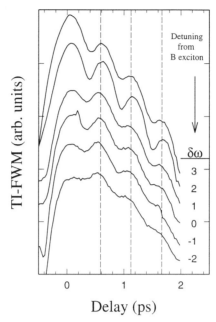

Fig. 6.31 TI-FWM data at 10 K at different detunings $\delta\omega = \hbar(\omega_d - \omega_2)$ around the B-exciton. (ω_d = detection frequency). The laser is set to excite both the A- and B-exciton resonances. The fact that no phase shift was observed as a function of energy position across the B-exciton resonance indicates that the observed beats are true quantum beats. The curves taken at different detunings are offset for clarity.

spectrally resolved (SR)-FWM, as shown in Fig. 6.30, Fischer et al. determined that the excitons are nearly homogeneously broadened even at low temperatures, and therefore, the FWM decay time τ_{FWM} is taken as equal to $T_2/2$. In Fig. 6.30, the lack of background signals clearly indicates that the dominant contributions to FWM signals come from excitons rather than the band continuum. Similar observations were also made in both ZnSe and GaAs quantum wells (Fischer, 1994; Kim, 1992). The exciton line can therefore be accurately measured using the SR-FWM data, such as those shown in Fig. 6.30, without any background, which plagues the analysis of absorption measurement results. In addition, for relatively thick samples, it is virtually impossible to perform absorption measurements or experiments. The SR-FWM data at zero delay shown in Fig. 6.30 represent a clear demonstration of A- and B-excitons without the interference of bound excitons that can complicate the interpretation of photoluminescence spectra at low temperature.

Figure 6.31 shows the quantum beats between the A- and B-excitons observed by TI-FWM. When the laser is tuned roughly in the middle of the A- and B-excitons, well-defined beating, whose period coincides with $\Delta E/h$, where ΔE is the energy separation between the A- and B-excitons measured in SR-FWM, can be clearly observed. Different curves in Fig. 6.31 correspond to TI-FWM taken as a function of decay for several different positions across the B-exciton

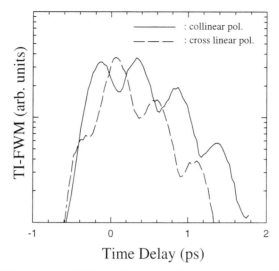

Fig. 6.32 TI-FWM signal at 10 K for co-linear (solid) and cross-linear (dashed) polarization geometries.

resonance. The phases are completely in sychronization at all the energies, which indicates that the observed beating is due to the true quantum beat associated with the coherent superposition of the *A*- and *B*-excitons, rather than to the so-called polarization beat (Lyssenko, 1993). In Fig. 6.32, TI-FWM in both the co-linear (solid lines) and the cross-linear (dotted lines) polarization geometries is shown. The maxima at one geometry correspond to the minima at the other, which indicates the existence of a 180° phase shift between the quantum beats. Such observations are very similar to the HH-LH quantum beat observed in GaAs quantum wells (Schmitt-Rink, 1992).

With the advance of Group III nitride growth technology, sophisticated structures such as quantum wells will become readily available, and coherent femtosecond spectroscopy can be extensively applied to the excitons in quantum well-type structures. However, one of the obstacles in femtosecond spectroscopy in GaN at low temperatures or GaN/AlGaN quantum wells is the difficulty in generating tunable and stable femtosecond UV laser beams in their excitonic region. Finding a way to extend the tunability well beyond 345 nm with substantial power levels, which is not readily available with commercial femtosecond lasers, is highly desirable in femtosecond laser technology.

6.6 Acknowledgments

We are particularly grateful to Professor Y. C. Chang for illuminating discussions on electron–LO-phonon interactions. Special thanks to Judy Nicholas for her patient and expert editorial assistance. We have tried to provide as many references as possible for those who seek more detailed information concerning these topics. However, it is impossible to include all relevant references in the current environment, where new references appear nearly every week through conferences, on Internet Web sites, and in journal publications. Any omission is unintentional.

References

Aggarwal, R. L., P. A. Maki, R. J. Molnar, Z.-L. Liau, and I. Melngailis (1996), *J. Appl. Phys.* **79**, 2148.
Ahn, D. and S. H. Park (1996), *Appl. Phys. Lett.* **69**, 3303.
Akasaki, I. H. Amano, S. Sota, H. Sakai, T. Tanaka, and M. Koike (1995), *Jpn. J. Appl. Phys.* **34**, L1517.
Akasaki, I., S. Sota, H, Sakai, T. Tanaka, M. Koike, and H. Amano (1996), *Electronics Lett.* **32**, 1105.
Amano, H., K. Hiramatsu, and I. Akasaki (1988), *Jpn, J. Appl. Phys.* **27**, L1384.
Amano, H., M. Kito, K. Hiramatsu, and I. Akasaki (1989), *Jpn. J. Appl. Phys.* **28**, L2112.
Amano, H., M. Kito, K. Hiramatsu, and I. Akasaki (1990), *J. Electrochem. Soc.* **137**, 1639.
Amano, H., T. Asahi, and I. Akasaki (1990), *Jpn. J. Appl. Phys.* **29**, L205.
Amano, H., T. Asahi, M. Kito, and I. Akasaki (1991), *J. Lum.* **48/49**, 889.
Amano, H., T. Tanaka, Y. Kunii, K. Kato, S. T. Kim, and I. Akasaki (1994) *Appl. Phys. Lett.* **64**, 1377.
Bagnall, D. M., and K. P. O, Donnell (1996), *Appl. Phys. Lett.* **68**, 3197.
Bergman, P., G. Ying, B. Monemar, and P. O. Holtz (1987), *J. Appl. Phys.* **61**, 4589.
Bernard, M. G. A. and G. Duraffourg (1961), *Phys. Stat. Sol.* **1**, 699.
Bidnyk, S., T. J. Schmidt, G. H. Park, and J. J. Song (1997), *Appl. Phys. Lett.* **71**, 729.
Boguslawski, P., E. Briggs, and J. Bernholc (1995), *Phys. Rev. B* **51**, 17255.
Boulou, M., M. Furtado, G. Jacob, and D. Bois (1979), *J. Lumin.* **18/19**, 767.
Camphausen, D. L. and G. A. N. Connell (1971), *J. Appl. Phys.* **42**, 4438.
Capaz, R. B., H. Kim, and J. D. Joannopoulos (1995), *Phys. Rev. B* **51**, 17755.
Chadi, J. and K. J. Chang (1988), *Phys. Rev. Lett.* **61**, 873.
Chang, Y. C. (1997), personal communication.
Chemla, D. S., D. A. B. Miller, and P. W. Smith (1987), in *Semiconductors and Semimetals*, Vol. **24**, ed. R. Dingle, Academic Press, San Diego, Ch. 5.
Chen, G. D., M. Smith, J. Y. Lin, H. X. Jiang, M. A. Khan and C. J. Sun (1995), *Appl. Phys. Lett.* **67**, 1653.
Chen, G. D., M. Smith, J. Y. Lin, and H. X. Jiang, S. H. Wei, M. Asif Kahn, and C. J. Sun (1996), *Appl. Phys. Lett.* **68**, 2784.
Chen, J., Z. H. Levine, and J. W. Wilkins (1995), *Appl. Phys. Lett.* **67**, 1129.
Chichibu, S., T. Azuhata, T. Sota, and S. Nakamura (1997), *Mat. Res. Symp. Proc.* **449**, 653.
Chichibu, S., A. Shikanai, T. Azuhata, T. Sota, A. Kuramata, K. Horino, and S. Nakamura (1996), *Appl. Phys. Lett.* **68**, 3766.
Chow, W. W., A. Knorr, and S. W. Koch (1995), *Appl. Phys. Lett.* **67**, 754.
Chow, W. W., A. F. Wright, and J. S. Nelson (1996), *Appl. Phys. Lett.* **68**, 296.
Christensen, N. E. and I. Gorczyca (1994), *Phys. Rev. B* **50**, 4397.
Chuang, S. L. and C. S. Chang (1996), *Phys. Rev. B.* **54**, 2491.
Chung, B.-C. and M. Gershenzon (1992), *J. Appl. Phys.* **72**, 651.
Dean, J. P. (1982), in *Prog. Cryst. Growth Charact.*, Pergamon, Oxford, 1982, Vol. 5, p.89.
Dean, P. J. (1973), in *Progress in Solid State Chemistry*, ed. J. O. McCaldina and G. Somorjai, Pergamon, Oxford, Vol. 8, p.1.

Dingle, R. and M. Ilegems (1971a), *Solid State Commun.* **9**, 175.
Dingle, R., D. D. Sell, S. E. Stokowski, and M. Ilegems (1971b), *Phys. Rev. B* **4**, 1211.
Dingle, R., K. L. Shaklee, R. F. Leheny, and R. B. Zetterstrom (1971c), *Appl. Phys. Lett.* **19**, 5.
Domen, K., K. Kondo, A. Kuramata, and T. Tanahashi (1996), *Appl. Phys. Lett.* **69**, 94.
Eckey, L., J.-Ch. Holst, P. Maxim, R. Heitz, A. Hoffmann, I. Broser, B. K. Meyer, C. Wetzel, E. N. Mokhov, and P. G. Baranov (1996), *Appl. Phys. Lett.* **68**, 415.
Eichler, H. J., P. Gunter, and D. W. Pohl. (1986), *Laser-Induced Dynamic Grating*, Springer-Verlag, Berlin, p. 47.
Elliott, R. J. (1957), *Phys. Rev.* **108**, 1384.
Fang, W. and S. L. Chuang (1995), *Appl. Phys. Lett.* **67**, 751.
Feldmann, J., G. Peter, E. O. Göbel, P. Dawson, K. Moore, C. Foxon, and R. J. Elliott (1987), *Phys. Rev. Lett.* **59**, 2337.
Fertitta, K. G., A. L. Holmes, J. G. Neff, F. J. Ciuba, and R. D. Dupuis (1994), *Appl. Phys. Lett.* **65**, 1823.
Fischer, A. J., D. S. Kim, J. Hays, W. Shan, J. J. Song, D. B. Eason, J. Ren, J. F. Schetzina, H. Luo, J. K. Furdyna, Z. Q. Zhu, T. Yao, J. F. Klem, and W. Schafer (1994), *Phys. Rev. Lett.* **73**, 2368.
Fischer, A. J., G. H. Park, J. J. Song, D. S. Kim, D. S. Yee, R. Horning and B. Goldenberg (1997), *Phys. Rev. B* **56**, 1077.
Fischer, S., C. Wetzel, E. E. Haller, and B. K. Meyer (1995), *Appl. Phys. Lett.* **67**, 1298.
Frankowsky, G., F. Steuber, V. Harle, F. Scholz, and A. Hangleiter (1996), *Appl. Phys. Lett.* **68**, 3746.
Freitas, Jr. J. A., K. Doverspike, and A. E. Wickenden (1996), *Mat. Res. Symp. Proc.* **395**, 485.
Gaskill, D. K., A. E. Wickenden, K. Doverspike, B. Tadayon, and L. B. Rowland (1995), *J. Electron. Mater.* **24**, 1525.
Gavini, A. and M. Cardona (1970), *Phys. Rev. B* **1**, 672.
Gil, B., O. Briot, and R.-L. Aulombard (1995), *Phys. Rev. B.* **52**, R17028.
Gil, B. and O. Briot (1997), *Phys. Rev. B* **55**, 2530.
Glaser, E. R., T. A. Kennedy, H. C. Crookham, J. A. Freitas, Jr., M. Asif Khan, D. T. Olson, and J. N. Kuznia (1993), *Appl. Phys. Lett.* **63**, 2673.
Glaser, E. R., T. A. Kennedy, K. Doverspike, L. B. Rowland, D. K. Gaskill, J. A. Freitas, Jr., M. Asif Khan, D. T. Olson, J. N. Kuznia, and D. K. Wickenden (1995), *Phys. Rev. B.* **51**, 13326.
Gorczyca, I. and N. E. Christensen (1991), *Solid State Commun.* **80**, 335.
Gorczyca, I. and N. E. Christensen (1993), *Physica B* **185**, 410.
Götz, W., N. M. Johnson, C. Chen, H. Liu, C. Kuo, and W. Imer (1996), *Appl. Phys. Lett.* **68**, 3144.
Götz, W., N. M. Johnson, J. Walker, D. P. Bour, and R. A. Street (1996), *Appl. Phys. Lett.* **68**, 669.
Hacke, P., A. Maekawa, N. Koide, K. Hiramatsu, and N. Sawaki (1994), *Jpn. J. Appl. Phys.* **33**, 6443.
Harris, C. I., B. Monemar, H. Amano, and I. Akasaki (1995), *Appl. Phys. Lett.* **67**, 840.
Haynes, J. R. (1960), *Phys. Rev. Lett.* **4**, 361.
Ho, I. and G. B. Stringfellow (1996), *Appl. Phys. Lett.* **69**, 2701.
Hofmann, R., V. Wagner, H.-P. Gauggel, F. Adler, P. Ernst, A. Sohmer, H. Bolay, F. Scholz, and H. Schweizer (1977), *Mat. Res. Soc. Symp. Proc.* **449**, 1197.
Ilegems, M. and R. Dingle (1973), *J. Appl. Phys.* **44**, 4234.
Itaya, K., M. Onomura, J. Nishio, L. Sugiura, S. Saito, M. Suzuki, J. Rennie, S. Y. Nunoue, M. Yamamoto, H. Fujimoto, Y. Kokubun, Y. Ohba, G. I. Hatakoshi, and M. Ishakawa (1996), *Jpn. J. Appl. Phys.* **35**, L1315.

Jenkins, D. W. and J. D. Dow (1989), *Phys. Rev. B*. **39**, 3317.
Kawakami, Y., Y. Narukawa, S. Fujita, and S. Fujita, in *Proc. Int. Symp. Quantum Struct. Photonic Appl.*, Mar. 6–8, Sendai, Japan, in press.
Khan, M. A., D. T. Olson, J. M. Van Hove, and J. N. Kuznia (1991), *Appl. Phys. Lett.* **58**, 1515.
Khan, M. A., S. Krishnankutty, R. A. Skogman, J. N. Kuznia, D. T. Olson, and T. George (1994), *Appl. Phys. Lett.* **65**, 520.
Khan, M. A., C. J. Sun, J. W. Yang, Q. Chen, B. W. Lim, M. Zubair Anwar, A. Osinsky, and H. Temkin (1996), *Appl. Phys. Lett.* **69**, 2418.
Kim, D., I. H. Libon, C. Voelkmann, Y. R. Shen, and V. Petrova-Koch (1997), *Phys. Rev. B* **55**, R4907.
Kim, D. S., J. Shah, J. E. Cunningham, T. C. Damen, W. Schafer, M. Hartmann, and S. Schmidt-Rink (1992), *Phys. Rev. Lett.* **68**, 1006.
Kim, K., W. R. L. Lambrecht, and B. Segall (1994), *Phys. Rev. B* **50**, 1502.
Kim, S., I. P. Herman, J. A. Tuchman, K. Doverspike, L. B. Rowland, and D. K. Gaskill (1995), *Appl. Phys. Lett.* **67**, 380.
Kim, S. T., H. Amano, I. Akasaki, and N. Koide (1994), *Appl. Phys. Lett.* **64**, 1535.
Kim, W., A. Salvador, A. E. Botchkarev, O. Aktas, and H. Morkoç (1996), *Appl. Phys. Lett.* **69**, 559.
Knox, R. S. (1963), in *Solid State Physics*, Suppl. 5, eds. Seitz and Turnbull, Academic Press, New York.
Kovalev, D., B. Averboukh, D. Volm, B. K. Meyer, H. Amano, and I. Akasaki (1996), *Phys. Rev. B* **54**, 2518.
Kressel, H. and D. E. Ackley (1993), in *Device Physics*, ed. C. Hilsum, North-Holland, Amsterdam, Ch. 8.
Kuramata, A., K. Horino, K. Domen, R. Soejima, H. Sudo, and T. Tanahashi (1996), *Mat. Res. Symp. Proc.* **395**, 61.
Lambrecht, W. R. L., B. Segall, J. Rife, W. R. Hunter, and D. K. Wickenden (1995), *Phys. Rev. B* **51**, 13516.
Lampert, M. A. (1958), *Phys. Rev. Lett.* **1**, 450.
Leo, K., E. O. Gobel, T. C. Damen, J. Shah, S. Schmitt-Rink, W. Schafer, J. F. Muller, K. Kohler, and P. Ganser (1991), *Phys. Rev. B* **44**, 5726.
Levenson, M. D. and S. S. Kano (1987), *Introduction to Nonlinear Laser Spectroscopy*, Academic Press, New York, Ch. 2.
Liang, W. Y. and A. D. Yoffe (1968), *Phys. Rev. Lett.* **20**, 59.
Libbon, I. H., C. Voelkmann, D. Kim, V. Petrova-Koch, and Y. R. Shen (1997), *Mat. Res. Symp. Proc.* **449**, 621.
Liliental-Weber, Z., S. Ruvimov, Ch. Kisielowski, Y. Chen, W. Swider, J. Washburn, N. Newman, A. Gassmann, X. Liu, L. Schloss, E. R. Weber, I. Grzegory, M. Bockowski, J. Jun, T. Suski, K. Pakula, J. Baranowski, S. Porowski, H. Amano, and I. Akasaki (1996), *Mat. Res. Symp. Proc.* **395**, 351.
Loeber, D. A. S., N. G. Anderson, J. M. Redwing, J. S. Flynn, G. M. Smith, and M. A. Tischler (1997), *Mat. Res. Soc. Symp. Proc.* **449**, 1203.
Lyssenko, V. G., J. Erland, I. Balslev, K.-H. Pantke, B. S. Razbirin, and J. M. Hvam (1993), *Phys. Rev. B* **48**, 5720.
Maki, P. A., R. J. Molnar, R. L. Aggarwal, Z.-L. Liau, and I. Melngailis (1996), *Mat. Res. Symp. Proc.* **395**, 919.
Martinez-Pastor, J., A. Vinattieri, L. Cazzaresi, M. Colocci, Ph. Roussignol, and G. Weimann (1993), *Phys. Rev. B* **47**, 10456.
Meney, A. T. and E. P. O. O'Reilly (1995), *Appl. Phys. Lett.* **67**, 3013.
Merz, J. L., H. Kukimoto, K. Nassau, and J. W. Shiever (1972), *Phys. Rev. B* **6**, 545.
Min, B. J., C. T. Chan, and K. H. Ho (1992), *Phys. Rev. B* **45**, 1159.
Miragliotta, J. and D. K. Wickenden (1994), *Phys. Rev. B* **50**, 14960.

Miragliotta, J. and D. K. Wickenden (1996), *Phys. Rev. B* **53**, 1388.
Miragliotta, J., D. K. Wickenden, T. J. Kistenmacher, and W. A. Bryden (1993), *J. Opt. Soc. Am. B* **10**, 1447.
Monemar, B. (1974), *Phys. Rev. B* **10**, 676.
Monemar, B., O. Lagerstedt, and H. P. Gislason (1980), *J. Appl. Phys.* **51**, 625.
Nakamura, S., T. Mukai, M. Senoh, and N. Iwasa (1992a), *Jpn. J. Appl. Phys.* **31**, 2883.
Nakamura, S. T. Mukai, M. Senoh, and N. Iwasa (1992b), *Jpn. J. Appl. Phys.* **31**, L139.
Nakamura, S., M. Senoh, S. Nagahama, N. Iwasa, T. Yamada, T. Matsushita, H. Kiyoku, and Y. Sugimoto (1996a), *Jpn. J. Appl. Phys.* **35**, L74.
Nakamura, S., M. Senoh, S. Nagahama, N. Iwasa, T. Yamada, T. Matsushita, H. Kiyoku, and Y. Sugimoto (1996b), *Appl. Phys. Lett.* **68**, 2105.
Nakamura, S., M. Senoh, A. Nagahama, N. Iwasa, T. Yamada, T. Matsushita, H. Kiyoku, and Y. Sugimoto (1996c), *Appl. Phys. Lett.* **68**, 3270.
Nakamura, S., M. Senoh, S. Nagahama, N. Iwasa, T. Yamada, T. Matsushita, H. Kiyoku, and Y. Sugimoto (1996d), *Appl. Phys. Lett.* **69**, 1477.
Nakamura, S., M. Senoh, S. Nagahama, N. Iwasa, T. Yamada, T. Matsushita, Y. Sugimoto, and H. Kiyoku (1996e), *Appl. Phys. Lett.* **69**, 1568.
Nakamura, S., M. Senoh, S. Nagahama, N. Iwasa, T. Yamada, T. Matsushita, Y. Sugimoto, and H. Kiyoku (1996f), *Appl. Phys. Lett.* **69**, 3034.
Nakamura, S., M. Senoh, S. Nagahama, N. Iwasa, T. Yamada, T. Matsushita, Y. Sugimoto, and H. Kiyoku (1996g), *Appl. Phys. Lett.* **69**, 4056.
Nakamura, S., M. Senoh, S. Nagahama, N. Iwasa, T. Yamada, T. Matsushita, Y. Sugimoto, and H. Kiyoku (1997a), *Appl. Phys. Lett.* **70**, 616.
Nakamura, S., M. Senoh, S. Nagahama, N. Iwasa, T. Yamada, T. Matsushita, Y. Sugimoto, and H. Kiyoku (1997b), *Appl. Phys. Lett.* **70**, 868.
Nakamura, S., M. Senoh, S. Nagahama, N. Iwasa, T. Yamada, T. Matsushita, Y. Sugimoto, and H. Kiyoku (1997c). *Appl. Phys. Lett.* **70**, 1417.
Nakamura, S. (1997d), *Mat. Res. Symp. Proc.* **449**, 1135.
Nakamura, S. (1997e), *24th International Symposium on Compound Semiconductors*, San Diego.
Narukawa, Y., Y. Kawakami, M. Funato, Sz. Fujita, Sg. Fujita, and S. Nakamura, (1997), *Appl. Phys. Lett.* **70**, 981.
Neugebauer, J. and C. G. Van de Walle (1994), *Phys. Rev. B* **50**, 8067.
Neugebauer, J. and C. G. Van de Walle (1996), *Appl. Phys. Lett.* **69**, 503.
Ogino, T. and M. Akoki (1980), *Jpn. J. Appl. Phys.* **19**, 2395.
Ohba, Y. and A. Hatano (1994), *Jpn. J. Appl. Phys.* **33**, L1367.
Palummo, M., C. M. Bertoni, L. Reining, and F. Finocchi (1993), *Physica B* **185**, 404.
Pankove, J. I. and J. E. Berkeyheiser (1974), *J. Appl. Phys.* **45**, 3892.
Pankove, J. I. and J. A. Hutchby (1976), *J. Appl. Phys.* **47**, 5387.
Pankove, J. I., M. T. Duffy, E. A. Miller, and J. E. Berkeyheiser (1973), *J. Lumin.* **8**, 89.
Pavesi, L. and M. Guzzi (1994), *J. Appl. Phys.* **75**, 4779.
Perlin, P., I. Gorczyca, N. E. Christensen, I. Grzegory, H. Teisseyre, and T. Suski (1992), *Phys. Rev. B* **45**, 13307.
Perlin, P., T. Suski, H. Teisseyre, M. Leszcynski, I. Grzegory, J. Jun, S. Porowski, P. Boguslawski, J. Bernholc, J. C. Chervin, A. Polian, and T. D. Moustakas (1995), *Phys. Rev. Lett.* **75**, 296.
Permogorov, S. (1975), *Phys. Stat. Sol.* **68**, 9.
Perry, W. G., T. Zheleva, M. B. Bremser, R. F. Davis, W. Shan, and J. J. Song (1997), *J. Electronic. Mats.* **26**, 224.
Pikus, G. (1962), *Sov. Phys.-JETP* **14**, 1075.
Planel, R., A. Bonnot, and C. Benoit á la Guillaume (1973), *Phys. Stat. Sol. (b)* **58**, 251.
Redwing, J. M., D. A. S. Loeber, N. G. Anderson, M. A. Tischler, and J. S. Flynn (1996), *Appl. Phys. Lett.* **69**, 1.

Reynolds, D. C., D. C. Look, W. Kim, O. Aktas, A. Botchkarev, A. Salvador, H. Morkoç, and D. N. Talwar (1996), *J. Appl. Phys.* **80**, 594.
Rieger, W., T. Metzger, H. Angerer, R. Dimitrov, O. Ambacher, and M. Stutzmann (1996), *Appl. Phys. Lett.* **68**, 970.
Rubio, A., J. L. Corkill, M. L. Cohen, E. L. Shirley, and S. G. Louie (1993), *Phys. Rev. B* **48**, 11810.
Rubio, A. and M. L. Cohen (1995), *Phys. Rev. B* **51**, 4343.
Salvador, A., W. Kim, O. Atkas, A. Botchkarev, Z. Fan, and H. Morkoç (1996), *Appl. Phys. Lett.* **69**, 2692.
Samara, G. A. (1983), *Phys. Rev. B* **27**, 3494.
Sandomirskii, V. B. (1964), *Sov. Phys. Solid State* **6**, 261.
Sato, H., T. Minami, E. Yamada, M. Ishii, and S. Takata (1994), *J. Appl. Phys.* **75**, 1045.
Schmidt, T. J., X. H. Yang, W. Shan, J. J. Song, A. Salvador, W. Kim, O. Aktas, A. Botchkarev, and H. Morkoç (1996), *Appl. Phys. Lett.* **68**, 1820.
Schmitt-Rink, S., D. Bennhardt, V. Heuckeroth, P. Thomas, P. Aaring, G. Maidorn, H. Bakker, K. Leo, D. S. Kim, J. Shah, and K. Köhler (1992), *Phys. Rev. B* **46**, 10460.
Seifert, W., R. Franzheld, E. Butter, H. Sobotta, and V. Riede (1983), *Cryst. Res. Technol.* **18**, 383.
Shan, W., A. J. Fischer, S. J. Hwang, B. D. Little, R. J. Hauenstein, X. C. Xie, and J. J. Song (1997), unpublished.
Shan, W., A. J. Fischer, J. J. Song, G. E. Bulman, H. S. Kong, M. T. Leonard, W. G. Perry, M. D. Bremser, and R. F. Davis (1996a), *Appl. Phys. Lett.* **69**, 740.
Shan, W., R. J. Hauenstein, A. J. Fischer, and J. J. Song, W. G. Perry, M. D. Bremser, R. F. Davis, and B. Goldenberg (1996b), *Phys. Rev. B* **54**, 13460.
Shan, W., B. D. Little, A. J. Fischer, and J. J. Song, B. Goldenberg, W. G. Perry, M. D. Bremser, and R. F. Davis (1996c). *Phys. Rev. B* **54**, 16369.
Shan, W., T. J. Schmidt, R. J. Hauenstein, and J. J. Song, and B. Goldenberg (1995c), *Appl. Phys. Lett.* **66**, 3495.
Shan, W., T. J. Schmidt, X. H. Yang, S. J. Hwang, J. J. Song, and B. Goldenberg (1995b), *Appl. Phys. Lett.* **66**, 985.
Shan, W., X. C. Xie, J. J. Song, and B. Goldenberg (1995a), *Appl. Phys. Lett.* **67**, 2512.
Sheik-Bahae, M., D. C. Hutchings, D. J. Hagan, and E. W. Van Stryland (1991), *IEEE J. Quantum Electron.* **27**, 1296.
Sheik-Bahae, M., A. A. Said, T.-H. Wei, D. H. Hagan, and E. W. Van Stryland (1990), *IEEE J. Quantum Electron.* **26**, 760.
Shikanai, A., T. Azuhata, T. Sota, S. Chichibu, A. Kuramata, K. Horino, and S. Nakamura (1997), *J. Appl. Phys.* **81**, 417.
Shmagin, I. K., J. F. Muth, S. Krishnankutty, R. M. Kolbas, S. Keller, U. K. Mishra, S. P. DenBaars (1997), *Mat. Res. Symp. Proc.* **449**, 1209.
Singh, R., R. J. Molnar, M. S. Ünlü, and T. D. Moustakas (1994), *Appl. Phys. Lett.* **64**, 336.
Sirenko, Yu . M., J. B. Jeon, K. W. Kim, M. A. Littlejohn, and M. A. Stroscio (1996), *Phys. Rev. B* **53**, 1997.
Smith, M., G. D. Chen, J. Y. Lin, H. X. Jiang, A. Salvador, B. N. Sverdlov, A. Botchkarev, and H. Morkoç (1995), *Appl. Phys. Lett.* **66**, 3474.
Smith, M., G. D. Chen, J. Y. Lin, H. X. Jiang, M. Asif Khan, C. J. Sun, Q. Chen, and J. W. Yang (1996), *J. Appl. Phys.* **76**, 7001.
Song, J. J., W. Shan, T. Schmidt, X. H. Yang, A. Fischer, S. J. Hwang, B. Taheri, B. Goldenberg, R. Horning, A. Salvador, W. Kim, O. Atkas, A. Botchkarev, and H. Morkoç (1996), *SPIE Proc.* **2693**, 86.
Song, J. J., A. J. Fischer, W. Shan, B. Goldenberg, and G. E. Bulman (1997), *Inst. Phys. Conf. Ser.* No. 155, 355.
Strite, S. (1994), *Jpn. J. Appl. Phys.* **33**, L699.

Strite, S. and H. Morkoç (1992), *J. Vac. Sci. Technol. B* **10**, 1237.
Suzuki, M. and T. Uenoyama (1995), *Phys. Rev. B* **52**, 8132.
Suzuki, M. and T. Uenoyama (1996a), *Jpn. J. Appl. Phys.* **35**, 1420.
Suzuki, M. and T. Uenoyama (1996b), *Appl. Phys. Lett.* **69**, 3378.
Taheri, B., J. Hays, J. J. Song, and B. Goldenberg (1996), *Appl. Phys. Lett.* **68**, 587.
Tanaka, T., K. Uchida, A. Watanabe, and S. Minagawa (1996), *Electronics Lett.* **32**, 35.
Tansley, T. L. and R. J. Egan (1992), *Phys. Rev. B* **45**, 10942.
Tansley, T. L. and R. J. Egan (1993), *Physica. B* **185**, 190.
Tchounkeu, M., O. Briot, B. Gil, J. P. Alexis, and R. L. Aulombard (1996), *J. Appl. Phys.* **80**, 5352.
't Hooft, G. W., W. A. J. A. van der Poel, and L. W. Molenkamp (1987), *Phys. Rev. B* **35**, 8281.
Ueta, M., H. Kanzaki, K. Kobayashi, Y. Toyozawa, and E. Hanamura (1986), *Excitonic Processes in Solids*, Springer-Verlag, Berlin.
Volm, D., K. Oettinger, T. Streibl, D. Kovalev, M. Ben-Chorin, J. Diener, B. K. Meyer, J. Majewski, L. Eckey, A. Koffmann, H. Amano, I. Akasaki, K. Hiramatsu, and T. Detchprohm (1996), *Phys. Rev. B* **53**, 16543.
Voos, M., R. F. Leheny, and J. Shah (1980), in *Optical Properties of Solid*, ed. M. Balkanski, North-Holland, Amsterdam, Ch. 6.
Wetzel, C., T. Suski, J. W. Ager III, W. Walukiewicz, S. Fischer, B. K. Meyer, I. Grzegory, and S. Porowski (1996), *Proc. 23rd Int. Conf. Phys. Semicond.*, eds. M. Scheffler and R. Zimmerman, World Scientific, Singapore, p. 2929.
Wetzel, C., D. Volm, B. K. Meyer, K. Pressel, S. Nilsson, E. N. Moknov, and P. G. Baranov (1994), *Appl. Phys. Lett.* **65**, 1033.
Wiesmann, D., I. Brener, L. Pfeiffer, M. A. Khan, C. J. Sun (1996), *Appl. Phys. Lett.* **69**, 3384.
Wolford, D. J. and J. A. Bradley (1985), *Solid State Commun.* **53**, 1069.
Yang, N., A. G. Yodh, M. A. Khan, and C. J. Sun (1997), *Bull. Am. Phys.* **42**, 319.
Yang, T., S. Nakajima, and S. Sakai (1995), *Jpn. J. Appl. Phys.* **34**, 5912.
Yang, X. H., T. Schmidt, W. Shan, J. J. Song, and B. Goldenberg (1995), *Appl. Phys. Lett.* **66**, 1.
Yang, Z., L. K. Li, and W. I. Wang (1996), *Mat. Res. Soc. Symp. Proc.* **395**, 169.
Yung, K., J. Yee, J. Koo, M. Rubin, N. Newman, and J. Ross (1994), *Appl. Phys. Lett.* **64**, 1135.
Zhang, X., P. Kung, D. Walker, A. Saxler, and M. Razeghi (1996), *Mat. Res. Symp. Proc.* **395**, 625.
Zhang, X., P. Kung, A. Saxler, D. Walker, and M. Razeghi (1996), *J. Appl. Phys.* **80**, 6544.
Zolper, J. C., R. G. Wilson, S. J. Pearton, and R. A. Stall (1996), *Appl. Phys. Lett.* **68**, 1945.
Zubrilov, A. S., V. I. Nikovlaev, V. A. Dmitriev, K. G. Irvine, J. A. Edmond, and C. H. Carter, Jr. (1995), *Inst. Phys. Conf. Ser.* **141**, 525.
Zubrilov, A. S., V. I. Nikovlaev, V. A. Dmitriev, K. G. Irvine, J. A. Edmond, and C. H. Carter, Jr. (1995), *Appl. Phys. Lett.* **67**, 533.

7 Defect spectroscopy in the nitrides

Bruno K. Meyer, Axel Hoffmann, and Patrick Thurian

7.1 Introduction

Nitride-based semiconductors (GaN, InN, AlN) and their ternary alloys are now the open and very attractive playground of physicists and engineers towards the implementation of optoelectronic devices in the blue and ultraviolet (UV) spectral range. In the last three years tremendous efforts have been undertaken and remarkable breakthroughs have been achieved allowing for the realization of blue and green light-emitting diodes and laser structures [1–3]. Specially the improvement in heteroepitaxial growth is made possible by the use of sophisticated low-temperature buffer layer growth.

The quality of GaN films is commonly judged by three different criteria: (i) the structural properties as obtained by X-ray diffraction and the half-width of the rocking curves, (ii) the electrical properties as given by the electron and hole mobilities, and (iii) the luminescent properties as measured by the ratio of near-bandgap excitonic recombination with respect to deep level emission at low temperatures and the small half-width of the excitonic recombination lines. Current state-of-the-art undoped GaN films grown on sapphire or 6H–SiC exhibit free carrier concentrations below $10^{17}\,\text{cm}^{-3}$, mobilities around $900\,\text{cm}^2/\text{Vs}$ and recombination dominated by free-exciton decay at room temperature. Much progress has been achieved towards the identification of the shallow residual donors and acceptors involved in excitonic recombination. The margins of error with respect to binding energies of free and bound excitons have been considerably reduced, and the influence of residual strain in the films on the line positions of excitonic recombination is now well established.

The combination of different optical methods, like photoluminescence (PL), far infrared absorption spectroscopy (FIR) and Raman spectroscopy, is a very powerful tool for the detection and identification of defects because of the sensitivity and the nondestructive character of these methods. Most of the luminescence processes, except free-exciton recombination, are directly related to energy levels of the defects present in the gap. The optical properties of the defect and its identification give valuable feedback for the optimization of growth procedures in device fabrication.

This chapter is organized in three parts. In the first part (7.1–7.2) a summary of the excitonic properties and shallow donors and acceptors is presented. Knowledge of PL and infrared absorption spectroscopy experiments is combined to reach conclusions about the localization energies of neutral donor-bound excitons and donor binding energies. We briefly summarize the status of acceptor-bound excitons and acceptor doping in GaN. Excitons bound to extended defects are discussed, and we emphasize the role of the interface between sapphire and GaN in spatially and time-resolved photoluminescence experiments.

In the second part (7.3), deep defects and their influence on optical properties are investigated. The properties of the notorious yellow luminescence band recombination mechanisms and its relation to defect microstructure will be discussed.

Finally (7.4–7.5), a state-of-the-art overview of transition metal (TM) centers in nitride compounds is given. The TMs are often present as unintentional contamination in the growth process. They act as minority lifetime killers and open up additional nonradiative recombination channels for excited carriers, leading to a decrease of the near-bandgap emission intensity. In addition, they can be used for the fabrication of semi-insulating substrates. The knowledge of the energy position of the TM in the bandgap of GaN and AlN allows some prediction of the band offsets according to the Langer and Heinrich rule [4, 5].

7.2 Shallow impurities in GaN

7.2.1 Excitonic and impurity-related recombination in undoped GaN epitaxial films

Excitonic recombination in GaN has been a well-established feature for several decades, but it was only recently, with the development of modern epitaxy methods, that free-exciton emission (FX) could dominate the optical spectra. However, undoped films are often of n-type conduction, and therefore in most cases the excitonic luminescence arises from the annihilation of excitons bound to neutral shallow donors ((D^0, X) or I_2). In order to distinguish between free- and bound-exciton emissions, a systematic study of the temperature dependence of luminescence intensity, line positions, and lineshapes has been done in the last three years. It was found that line positions depend critically on the differences between the thermal expansion coefficients of the substrate materials and the epitaxial layers (thermal mismatch). Thermal mismatch is responsible for the UV- and red-shifted line positions of free and bound excitons in GaN (there is certainly an influence from the buffer layer, too). Depending on the nature of the donor present in the layers and its binding energy, the energetic position of the (D^0, X) line can also vary. This is the case if Haynes' rule is valid, i.e. the localization energy of the exciton bound to a neutral defect is proportional to its respective binding energy. GaN epitaxial films are commonly grown by metalorganic vapor phase epitaxy (MOVPE) or by molecular beam epitaxy (MBE). The best MOVPE samples have exciton linewidths between 2 and 3 meV, and for MBE-grown layers linewidths around 6 meV are reported [6] (values at low temperatures). However, in epitaxial films grown by hydride vapor phase epitaxy (HVPE) the neutral donor-bound exciton linewidth was as narrow

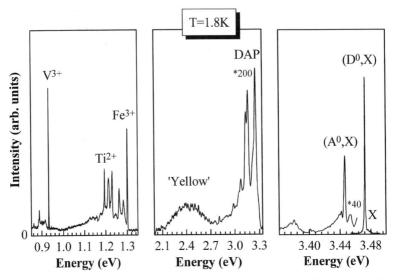

Fig. 7.1 Luminescence transitions commonly seen in undoped GaN epitaxial films.

as 0.9 meV. A recent report using homoepitaxial growth (MOVPE GaN on GaN) also reported values around 1 meV [7]. This improvement in layer quality made it possible to distinguish between intrinsic and extrinsic recombination.

Before coming to the fine structure of free and bound excitons (emitting in the energy range from 3.43 eV to 3.49 eV) we give a short overview of the luminescence transitions commonly seen in undoped GaN films (Fig. 7.1). In the spectral range below 3.4 eV (if we neglect the longitudinal optical (LO) phonon replicas from FX and (D^0,X) in some samples deeply bound excitons are observable. The recombination lines appear in pairs with a constant intensity ratio. They are located at 3.36 eV and 3.31 eV. At around 3.27 eV a donor–acceptor (D–A) pair band appears, close to the position where also the Mg acceptor related D–A pair band shows up. The acceptor involved in the recombination is still unidentified. The bands centered around 3 eV (violet band) and at 2.2 eV (yellow band) have a Gaussian lineshape, indicating strong vibronic coupling (see Section 7.3.2). The transition metal impurities in GaN emit in the spectral range from 1.3 to 0.8 eV (see Section 7.4).

Figures 7.2 and 7.3 give an overview of the luminescence close to the energy gap at low temperatures. Using a logarithmic scale for the intensity (Fig. 7.2) spectrally resolved free (A-exciton) and bound exciton transitions followed by a sequence of LO phonon replicas having the periodicity of exactly 92 meV can be observed. We note that for GaN on sapphire free and bound exciton lines are resolved on the no-phonon line but not on the 1LO and 2LO replicas. This is in contrast to a 3 μm thick GaN film grown on 6H–SiC (Fig. 7.3). The luminescence is dominated by the neutral donor-bound exciton line at 3.460 eV. At higher energies free-exciton transitions involving the A-valence band (3.467 eV) and the C-valence band (3.4865 eV) are seen. For the 1LO replica the intensity ratio of the free to the bound exciton line has changed, which also holds for

7.2 Shallow impurities in GaN

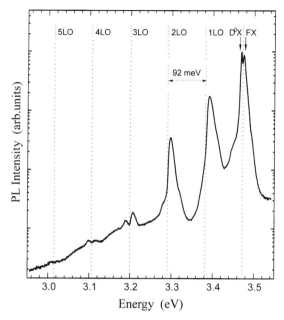

Fig. 7.2 Luminescence spectrum of GaN grown by MOVPE on sapphire showing free and bound exciton emission and the corresponding longitudinal optical phonon replicas of 92 meV periodicity.

the 2LO line. The donor-bound exciton is 6.2 meV below the A-exciton. The line that is approximately 30 meV below the free-exciton line arises from an acceptor-bound exciton. Note the different energetic positions of the free A-exciton lines of GaN films on sapphire and on SiC.

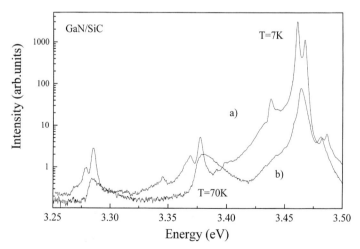

Fig. 7.3 Luminescence spectrum of GaN grown by MOVPE on 6H–SiC measured at (a) 7 K and (b) 70 K. Free and bound exciton transitions are observed.

The assignment of these lines becomes clear when comparing with a measurement at elevated temperatures (see Fig. 7.3(b). The free-exciton binding energy is of the order of 22 meV, hence free-exciton luminescence can be observed up to room temperature. Bound excitons have only small localization energies and rapidly lose intensity with increasing temperatures. This effect is most pronounced for the acceptor-bound exciton at 3.4378 eV. The free exciton, as well as its phonon replicas, shows considerable broadening on the high-energy site of the line with a Maxwellian distribution [8]. Donor-bound excitons show a symmetrical broadening of the line with increasing temperature (see Section 7.2.4).

7.2.2 Donor-bound excitons

The defects causing n-type conductivity in undoped GaN epitaxial films are still partly unidentified. The free carrier concentrations can vary over a wide range, from 10^{17} up to 10^{19} cm^{-3} [9]. Following recent theoretical arguments [10, 11], native defects as well as extrinsic impurities are believed to be responsible. The nitrogen vacancies or the gallium interstitials are the prime candidates for intrinsic defects; they should induce a shallow effective-mass-type state in GaN. The extrinsic defects producing shallow donor levels could come from the Group VI elements on Group V sites. Oxygen would be one candidate. From the Group IV elements, Si on a Ga site is known to be a very efficient donor dopant. Götz *et al.* [12] and Wickenden *et al.* [13] reported on Hall effect measurements on Si doped GaN films. Two donor-electron activation energies are found, ranging from 12 to 17 meV and 32 to 37 meV. Götz *et al.* [12] attribute the donor with the lower activation energy to Si and results from secondary ion mass spectroscopy seem to support this conclusion.

The HVPE grown film on sapphire had a layer thickness of 400 μm and hence corresponds to the strain free case. The exciton linewidth was as narrow as 0.8 meV at 1.5 K and made it possible to resolve even more details on the free-exciton recombination. Considering a comparison with calorimetric absorption and reflection measurements the transitions of the A-, B-, and C-excitons were attributed (for details see ref. [14]). However, the origin of some of these lines remained unclear. At high resolution and using polarization spectroscopy additional features become visible (Fig. 7.4). In the following we will only address the localization energies of the neutral donor-bound excitons. As seen in Fig. 7.4, the dominant transition occurs at 6–7 meV below the free A-exciton. It is a doublet with an energy separation of 0.8 meV. This splitting originates from the presence of two shallow donors having different localization and hence donor binding energies. There is an additional more pronounced transition for the polarization vector parallel to the *c*-axis. This is 3.5–4.5 meV below the free A-exciton and mentioned also in a report of Merz *et al.* [15]. From its behavior with temperature and in an electric field we presented strong evidence that it is a donor-bound exciton [6]. With respect to the B-exciton at 3.484 eV it has the same localization energy as the A-exciton, i.e. between 7 and 8 meV. It exhibits the same fine structure as the dominant transition, and it gives strong evidence

7.2 Shallow impurities in GaN 247

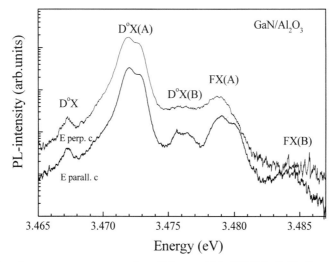

Fig. 7.4 Luminescence spectrum of a 400 μm thick HVPE GaN film on sapphire showing free exciton recombinations with the A- and B-valence bands. Additional fine stucture is due to donor bound exciton states.

that it is the same donor-bound exciton but bound to the B-valence band (see also ref. [15]). A third transition appears at 3.4673 eV and has an energy separation with respect to the A-exciton of 11.4 ± 0.2 meV. It is assigned by Baranowski et al. [16] to an acceptor-bound exciton (see Section 7.2.7). Summarizing these results, we can now accurately determine the localization energies and the energetic distances to the free exciton. We obtain 6.2 and 7.0 meV. We note that there is excellent agreement with data on homoepitaxial GaN films, where the corresponding values are 6.0 meV and 6.9 meV [16].

7.2.3 Activation energies and binding energies of donors and donor-bound excitons

The bound exciton lines rapidly decrease in intensity with increasing temperature, whereas free-exciton emission is observable up to room temperature. This reflects the fact that the bound-exciton lines have small localization energies significantly smaller than the free-exciton binding energy. In most cases it was possible to follow the intensity decrease of the (D^0, X) complex lines over only two decades, and one can hence only deduce the localization energy. However, for a sample with slightly higher free carrier concentration (free-exciton emission could not be observed) the donor-bound exciton emission was strong enough that it could be followed up to room temperature. The Arrhenius plot now goes over six orders of magnitude and exhibits two activation energies (Fig. 7.5). One is the localization energy and the higher one represents the donor binding energy. The thermally activated dissociation of bound excitons must involve two activation energies. By exciting with photon energies above the bandgap we create free excitons. They try to localize at the neutral donors

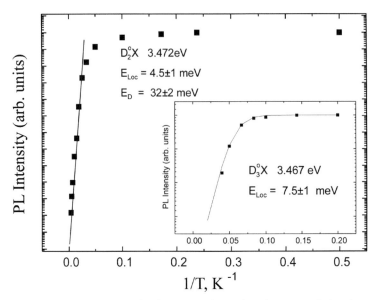

Fig. 7.5 Arrhenius plot for the luminescence intensity decrease of the donor bound exciton of an MOVPE grown film GaN on sapphire and a fit with two activation energies. The inset shows a similar analysis for a deeper bound exciton.

(localization energy). At elevated temperatures we have an interplay between the localization at and ionization of the neutral donor, reducing the number of available neutral donors.

The data are analyzed using the following expression:

$$I(0)/I(K) = 1/[1 + c_1 \exp(-\Delta E_1/kT) + c_2 \exp(-\Delta E_2/kT)], \quad (7.1)$$

where c_1 and c_2 are prefactors and ΔE_1 and ΔE_2 are the activation energies, respectively.

On the MOVPE film one obtains activation energies of 4.5 ± 1 meV and 32 ± 2 meV, respectively.

In a MOVPE sample GaN on GaN Pakula et al. [7] reported on the observation of two strong lines at 3.4666 eV and 3.4719 eV being attributed to donor- and acceptor-bound excitons. This assignment is supported by temperature-dependent luminescence experiments on an HVPE film (Fig. 7.5). We noticed a transition lying 5.1 ± 0.2 meV below the main donor-bound exciton line (Fig. 7.4). Pakula et al. [7] obtained 5.3 meV. Its activation energy amounts to 7.5 ± 1 meV, i.e. the activation energy is connected with the localization energy. Its intensity was too low to be seen at higher temperatures; therefore an estimate of the respective donor binding energy is difficult.

7.2.4 The bound-exciton linewidth

An important aspect of the material quality of GaN epitaxial films is the free/bound-exciton linewidth measured at low temperatures. Together with the

7.2 Shallow impurities in GaN

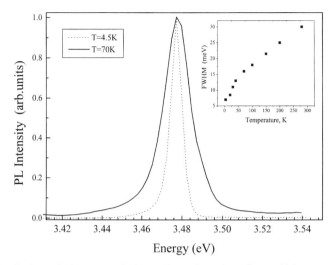

Fig. 7.6 Evolution of the neutral donor bound exciton line width as a function of temperature.

X-ray rocking curves, these two parameters are used for quality control. Here, we show the result for the (D^0, X) exciton complex. Its linewidth depends on the homogeneous broadening mechanism described with Γ_1 (related to the radiative lifetime by the uncertainty relation), just as from the inhomogeneous broadening parameter Γ_2 caused by residual strain, crystal imperfections etc. and from a parameter Γ_3 which describes the exciton–phonon interaction. In general the linewidth is thus a sum of all individual contributions

$$\Gamma(T) = \Gamma_1 + \Gamma_2 + \Gamma_3 \tag{7.2}$$

We assume here that the temperature dependence of the linewidth arises from the exciton–phonon interaction. The neutral donor-bound exciton decay in GaN is less than 60 ps (corresponding to a homogeneous width of 0.1 meV) at low temperatures, and shortens with rising temperatures, i.e. the linewidth is not influenced by the lifetime. The low-temperature value of the linewidth (full width at half maximum (FWHM)) was 0.9 meV for the HVPE layer, and for the MOVPE layers typically 2–3 meV. In Fig 7.6 we show the evolution of the FWHM of the donor-bound exciton line as a function of temperature. At around 280 K the linewidth has increased approximately by a factor of four. We ascribe this behavior due to exciton–phonon interactions with acoustical and optical phonons. The dispersion relation ω vs. $\mathbf{k}$ of the phonons in GaN is determined [17] based on a comparison of the experimental Raman scattering data with calculated phonon dispersion curves as well as the group-theoretically derived selection rules.

7.2.5 Residual donor binding energies

The shallow defects also give rise to Rydberg excited states commonly seen in

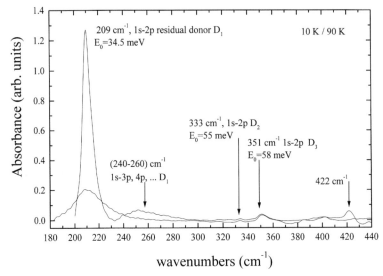

Fig. 7.7 Fourier transform infrared absorption spectrum of a free standing GaN film grown by HVPE at two different temperatures. Transition energies and assignments are given.

far-infrared conductivity or absorption experiments. From the transition series the ground state binding energy can be calculated within the effective-mass theory (EMT). Recently, Fourier transform infrared (FTIR) absorption [18] on GaN thick films were reported, where the absorption of a residual donor was observed. There is an additional report that residual donors have binding energies of 54 and 57 meV based on the analysis of distant donor–acceptor pair recombination lines [15].

These FTIR absorption experiments have been extended on a 400 μm substrate-free HVPE GaN films. Five temperature-dependent and hence electronic absorption bands at 209 cm^{-1}, between 240 and 280 cm^{-1}, at 333 and 351 cm^{-1} and a hot line at 422 cm^{-1} (Fig. 7.7) were identified. The line at 209 cm^{-1} has been reported before [18] in experiments on similar films, however, growing on a sapphire substrate. This prominent line is now at slightly lower energies; the shift of 6 wavenumbers might be caused by the release of strain due to the removal of the substrate. The second broader band around 250 cm^{-1} is lower in intensity; the small modulations on top of it are due to interference fringes. From an analysis of the data using the effective mass approach and attributing the line at 209 cm^{-1} with the 1s ground state of 2p excited state transition of a shallow donor we obtained from the series limit a binding energy of 34.5 meV. Hence the 1s to 3p transition should be at $8/9 \times 34.5$ meV (247.7 cm^{-1}) and the 1s to 4p transition is expected to be found at 261 cm^{-1}. This range of transition energies is marked in Fig. 7.7 with an arrow. The transitions to higher excited states are expected to be considerably lower in intensity and the broad absorption band between 240 and 280 cm^{-1} can be understood as the overlap of these transitions.

Two additional lines at 333 cm^{-1} and 351 cm^{-1} are also of electronic origin, as demonstrated by temperature-dependent measurements. Using the same approach (i.e. attributing them to 1s to 2p transitions) the corresponding binding energies are determined as 55 and 58 meV. This is in remarkable agreement with the binding energies deduced from luminescence experiments by Kaufmann et al. [15] (54 and 57 meV).

7.2.6 Localization and donor/acceptor binding energies

An estimate of the shallow donor binding energy E_D can also be made by using the effective-mass theory provided the electron effective mass is known with high precision. Cyclotron resonance experiments [19] on a high-quality GaN epitaxial film gave a precise value for the electron effective mass (polaron mass), $m^* = 0.220 \pm 0.005 m_0$. Using 9.7 for the dielectric constant one calculates $E_D = 13.58 m^*/\varepsilon^2$ (eV) = 31.7 meV.

Although Si is used as an efficient n-type dopant its binding energy is still debated. In undoped and Si-doped GaN films Hall effect measurements are derived with the conclusions that two donors are electrically active: the one with a binding energy around 17 meV is attributed to Si whereas for the second one, with a binding energy around 32 meV, no identification was possible. The conclusions from Götz et al. [12] and Wickenden et al. [13] are, however, in conflict with recent magneto-optical studies on Si doped GaN [20]. Wang et al. [20] monitored the behavior of the 1s to 2p$^+$ transition in the magnetic field range from 15 to 27 T. From a linear fit they deduced a transition energy at zero magnetic field of 21.7 meV (3/4 of E_D) and hence calculated for the binding energy of Si a value of 29 meV. This linear fit slightly underestimates the binding energy. In Fig. 7.8, we show the splitting of the 2p excited states as a function of the magnetic field [20]. The nonlinear behavior at fields below 10 T

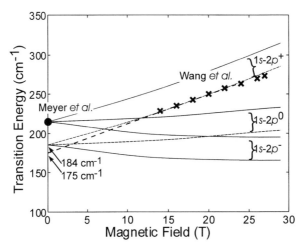

Fig. 7.8 The 1s to 2p interimpurity transition in an external magnetic field (after Ref. [21]).

is evident and the zero magnetic field value is 184 cm^{-1} instead of 175 cm^{-1}. The Si donor binding energy is hence 30.4 meV. Also shown in Fig. 7.8 is the zero-field value of the donor obtained from the FTIR investigations mentioned above.

In many semiconductors, there is a well-defined proportionality to the respective donor binding energy (Haynes' rule), and a unique description of the impurity bound excitons is usually searched in terms of Haynes' rule [22]. It simply means that for the localization energy of the exciton with respect to the binding energy of the impurity (E_D for the donors, E_A for the acceptors) a constant ratio is found. In a semi-empirical theory, Halstedt and Aven derived for the ratios the following result $\alpha = E_L/E_D = 0.2$ and $\beta = E_L/E_A = 0.1$ in line with the work from Hopfield. Haynes' rule is, however, not always applicable. CdTe is an example where it holds for the donors, $\alpha = 0.245$, but for the neutral acceptor-bound excitons β ranges between 0.07 and 0.1 [23].

The localization energies found for the two donor-bound excitons are 6.2 ± 0.2 meV and 7.0 ± 0.2 meV. The two donor bonding energies are 30.4 meV and 34.5 meV. The ratio α is 0.2 ± 0.01, assuming that Haynes' rule is valid.

The bound exciton line at 11.4 meV below the free exciton would hence correspond to a donor binding energy of 56 ± 2 meV. In the FTIR experiments a donor with a binding of 58 meV would be needed to explain the transition at 350 cm^{-1}.

It is difficult to make a decisive identification as to the chemical nature of the second slightly deeper donor. Substituting for gallium sites and being donors are possibly elements from the fourth group of the periodic table: C, Si, Ge, etc. Carbon is believed to act as an acceptor, and some experiments indeed provide evidence for p-type conduction due to carbon [24]. SIMS experiments give no indication of Ge as a trace impurity. On nitrogen sites the Group VII elements can act as donors: O, S, Se, Te, etc. GaN films can have high oxygen concentrations and the free electron carrier concentrations scale with the oxygen content in the films. Oxygen doping experiments concluded, however, that O in GaN introduces a deeper donor state with a binding energy of around 80 meV. HVPE uses quartz reactors and the reactivity of NH_3 and H radicals can easily introduce Si into the films. In MOVPE films using AlN buffer layers, O and Si (the high affinity of oxygen to Al is known and Si is a contaminant of Al-related organometallic compounds) are most likely to be incorporated. Whether the 34.5 meV donor is due to oxygen has to be clarified in future experiments.

In the most recent work on Zn and Mg acceptor doped thin GaN films [15] the neutral acceptor-bound excitons were studied, and it was concluded that Haynes' rule holds for the acceptors; the value for β is 0.1. The corresponding localization energies with respect to the free A-exciton line are: $E(Mg^0, X) = 19 \pm 4$ meV and $E(Zn^0, X) = 34 \pm 4$ meV. Leroux et al. [25] made a detailed comparative optical characterization study of undoped and Mg-doped GaN films grown by epitaxial techniques. They could clearly distinguish between the D–A pair band caused by the residual acceptor with $E_A = 220$ meV and the Mg acceptor with $E_A = 265$ meV. On increasing the Mg doping the D–A pair band broadens and eventually becomes the 'blue band' at 2.85 eV. The intensity decrease of the Mg-related D–A pair band with increasing temperature gives a thermal activation energy of 160 ± 10 meV, consistent with the measured Hall depth, and should not be mixed up with the optical binding energy of the Mg acceptor of 265 meV. In Fig. 7.3 a line at 30 meV below the free-exciton line is seen, and based on Kaufmann's results [15] it is now clear that Zn was a residual trace impurity.

One comment on the exciton bound to the neutral acceptor Cd: Illegems *et al.* [26] as well as Lagerfeld *et al.* [27] reported on the Cd bound exciton. Lagerfeld obtained a localization energy of 21 ± 1 meV; this value has a higher precision than those from [26] (here the corresponding value was 19 meV), since the free and neutral donor-bound excitons are also seen in the same sample. If, for the (Cd^0, X) complex, Haynes' rule is valid, the Cd acceptor binding energy would be close to Mg: around 230 meV. Cd doping, however, strongly enhances a donor–acceptor pair band, with a zero phonon line at 3 eV. According to the zero phonon line position the acceptor binding energy is close to 500 meV, and judged by this result Cd is ineffective as an acceptor dopant. However, the correlation found for Mg and Zn should be a motivation to reconsider Cd as a shallow acceptor in GaN.

Coming to the role of carbon as dopant, electrical measurements indicate that C behaves as an acceptor [24]. No optical data are available at present. The exciton line at 11.4 meV below the free-exciton line is, according to Ref. [28], an acceptor-bound exciton. Additional experimental facts seem to support this identification (Section 7.2.7). Using Haynes' rule, the acceptor binding energy would be 120 ± 10 meV. Such a shallow acceptor would certainly be of great benefit for GaN and hopefully an efficient p-type dopant. There is currently no experimental evidence for such a shallow acceptor level.

7.2.7 Magneto-optical investigations of the bound exciton complexes in GaN

The valence band is split by the combined interaction of the hexagonal crystal-field and spin–orbit interactions. The degeneracy of the heavy hole–light hole band is lifted and the heavy hole band $J = 3/2$, $m_J = \pm 3/2$, is on top of the valence band. The valence band structure is of considerable importance for the identification of the effective mass type acceptors in GaN. They should reflect the properties of the valence band structure, i.e. involve the heavy hole band. This situation is similar to hexagonal CdS or (4H, 5H) SiC. In optically detected magnetic resonance (ODMR) experiments on CdS and SiC, shallow effective mass type acceptors have been observed [29, 30]. The angular dependencies are a one-to-one image of the heavy hole valence band. The resonance positions follow the $g_h \cos \theta$ behavior, where θ is the angle between the static magnetic field and the c-axis of the crystal. For the Zeeman experiments the sample orientation must be varied from B parallel to B perpendicular, with respect to the c-axis of the crystal. For the donor-bound exciton the respective splitting in the magnetic field occurs in the neutral donor ground state and the bound exciton state. The neutral donor ground state splits by

$$\Delta E = g_e \mu_B B_0, \quad (7.3)$$

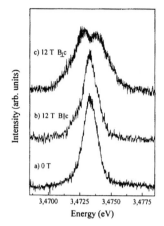

Fig. 7.9 Photoluminescence of the neutral donor-bound exciton for different magnetic fields and orientations of the sample (after Ref. [37]).

where g_e is the electron g-value in GaN, μ_B is Bohr's magneton, and B_0 is the applied magnetic field. The donor g-value and its anisotropy were determined from electron paramagnetic resonance (EPR) and optically detected magnetic resonance (ODMR) experiments [31–33]. They are $g = 1.951$ and $g = 1.9483$ for parallel and perpendicular to the c-axis, respectively. The neutral donor-bound exciton is formed by the coupling of two angular momenta $J_1 = J_2 = 1/2$ of the electrons to the angular momentum of the hole $J_3 = 3/2$, $m_J = \pm 3/2$. The total angular momentum is hence $J = 3/2$ ($J_1 + J_2 = 0$) and the magnetic properties are solely determined by the hole. The Zeeman splitting is given by

$$\Delta E = g_h \, \mu_B B_0 \cos\theta \tag{7.4}$$

where θ is the angle between the applied magnetic field and the c-axis of the crystal. For $\theta = 90°$, i.e. perpendicular to the c-axis, the contribution of the hole vanishes and only the donor Zeeman splitting remains (for neutral acceptor-bound excitons initial and final state splitting are exchanged but give the same information as from (D^0, X)).

For B parallel to the c-axis neither a splitting nor a broadening (error bar smaller than ± 0.1 meV) occurs (see Fig. 7.9). The same result was found in CdS [34]. It was explained by an accidental cancellation of the Zeeman contributions from the electron and hole Zeeman interactions — the hole g-value equals the electron g-value. For B parallel to the c-axis, one observes a broadening of the neutral donor-bound exciton line with increasing magnetic field and a partially resolved splitting at the highest available magnetic field. The spectra were deconvoluted into the sum of two Gaussian curves with the same half-width as obtained at zero magnetic field. A least squares fit to the experimental data points is drawn as a solid line in Fig. 7.10. The slope is $1.1 \times 10^{-4} \times B$ (eV T), i.e. (1.1 ± 0.1) meV at 10 T. For the corresponding g-value one calculates 1.9 ± 0.1, very close to the electron/donor g-value in GaN of 1.95 [31–33].

In III–V and II–VI semiconductors the hole g-values range between 0.7 and

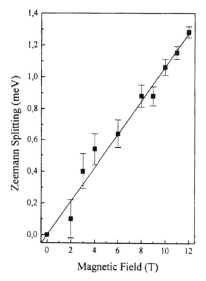

Fig. 7.10 Magnetic field dependence of the Zeeman splitting of D^0X; squares are the experimental data, the drawn line gives a best fit (after Ref. [37]).

0.8 [35]. For a $J = 3/2 \pm 3/2$ state the energy separation is given by $\Delta E = 3g_h \mu_B B_0$, and since it equals the Zeeman splitting of the neutral donor one obtains $3g_h = g_e$. With $g_e = 1.95$ one calculates $g_h = 0.65 \pm 0.1$, quite consistent with values from other compound semiconductors. In the framework of the Kohn–Luttinger formalism [36], $3g_h = 6\kappa$, neglecting cubic terms. Hence κ is 0.325 ± 0.03.

Stepniewski et al. [28] recently reported on Zeeman experiments on homoepitaxial GaN films. The advantage of lattice matching with the GaN substrate resulted in very narrow linewidths of 1.2 meV and 0.6 meV for the transitions at 3.472 eV (D^0, X) and 3.466 eV (A^0, X) respectively. A clear splitting could already be seen at 7.5 T (field direction perpendicular to the c-axis). No splitting was observed for the magnetic field parallel to the c-axis, confirming the results of Volm et al. [37]. Based on the intensity change of the field split lines the authors favor an acceptor-bound exciton as the origin of the 3.466 eV line. The g-value was 2.04, whereas for the (D^0, X) complex a value of 1.708 was obtained.

7.3 Deep structural defects in GaN

7.3.1 Structural defects in GaN

This section will demonstrate that structural defects strongly influence the optical transitions in GaN. Two examples will be given. First, structural minority phases in GaN will be investigated. Then, we will show that the dominating recombination processes near the interface between GaN and its lattice-mismatched substrate differ strongly from those far away from the interface [38],

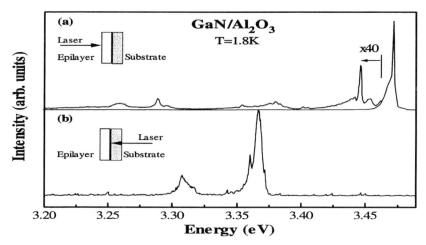

Fig. 7.11 Low-temperature photoluminescence spectra of a 400 μm epitaxial HVPE GaN films on sapphire. (a) Excitation is from the epilayer side; (b) excitation is from the substrate side.

i.e. near the surface of the epilayer. We will demonstrate that in structurally perturbed regions, such as in the vicinity of the interface, the radiative recombination of excitons is quenched and deeply bound dislocation excitons [39] and often yellow luminescence appear in the spectrum.

In Fig. 7.11 we show a comparison of near bandgap luminescence when exciting the layer (a) from the epilayer side and (b) through the substrate. An HVPE-grown 400 μm thick single-crystalline GaN film was used for this investigation to separate high-quality regions from strongly perturbed regions close to the interface between sapphire and the GaN film. Under excitation from the epilayer side, strong neutral donor-bound exciton recombination dominates. Magnifying the portion of the spectrum below I_2 makes the neutral acceptor-bound exciton line a 3.446 eV visible as well as phonon replicas from I_2. A line at 3.26 eV might arise from the cubic phase of GaN, which could be present in a small amount in hexagonal films. The spectrum obtained after excitation through the substrate is substantially different. The bound exciton transitions are no longer observed, and instead one sees a complex structure of lines between 3.37 eV and 3.31 eV. These lines have been reported by different groups (see e.g. Ref. 40 and references therein), and they are commonly found in layers grown by different techniques and on different substrates. Wetzel et al. [40] reported on detailed luminescence investigations, which are briefly summarized. Despite their energetic location approximately 130 meV and 190 meV below bandgap, the two lines (or line groups if not resolved spectroscopically) exhibit small thermal activation energies in the temperature-dependent PL measurements, viz. 14 meV (3.31 eV) and 27 meV (3.36 eV). They are of excitonic origin, as demonstrated by the time-resolved PL experiments of Eckey et al. [39] (see Fig. 7.12). They reveal very fast dynamics dominated by monoexponential decays in the picosecond range. Lifetimes vary between 45 ps and 650 ps, depending on

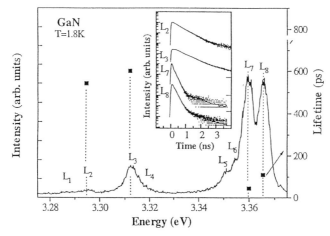

Fig. 7.12 Luminescence spectrum of deeply bound excitons. The filled squares give the decay constants (right scale) at the respective energy positions. In the inset the respective luminescence transients are shown (after Ref. [39]).

which spectral position is detected. Furthermore, they show a weak coupling to the lattice, as judged from the intensities of the respective phonon replicas. Under hydrostatic pressure, a very small (if any) shift is found, characteristic of strongly localized defects. The lines always appear together and certainly all these experimental findings remain one of the properties of strongly localized excitons in CdS [41 and references therein], ZnSe, and ZnTe [42]. In ZnTe [42] a strong correlation with high densities of dislocations within the interface region of the ZnTe films and the GaAs substrate has been found. It is therefore believed and supported by theoretical estimates [43] that these pairs of excitonic lines originate from dislocations. Shreter *et al.* [43] made the statement more specific, arguing that they stem from bound excitons on *c*-axis screw dislocations. The authors concluded that in spite of the high recombination activity of 90° dislocations in GaN they do not significantly affect the efficiency of optoelectronic devices, since the carrier lifetime is comparable with the carrier diffusion time to the dislocation. Detailed transmission electron microscopy and cathodoluminescence (under wavelength imaging) as is done for CdS [41] are needed to verify this suggestion.

7.3.2 Deep center yellow luminescence

The quality of epitaxial nitride films is very often judged by the ratio of bandgap luminescence to deep center luminescence. The key role here is played by the defect-induced recombination centered on 2.2 eV (yellow luminescence). To complete the picture we mention the ion implantation-induced luminescence bands associated with Mg, Zn, Cd, C, As, Hg, and Ag, reported in 1976 by Pankove and Hutchby [44]. Ogino and Aoki [45] made a detailed investigation of the mechanism of the yellow luminescence [see also 46, 47]. From the variation of the half-width as a function of temperature a characteristic phonon energy of 40 ± 5 meV was obtained. The luminescence has a Gaussian lineshape with a half-width at low temperatures of 430 meV. The luminescence intensity decreased, with an activation energy of 860 ± 40 meV at temperatures above 480 K. It was concluded that this corresponds to the thermal release of a hole, since the GaN material studied was of n-type conduction. The recombination should involve a shallow 25 meV deep donor and a deep acceptor located 860 meV above the valence band. The occurrence of this particular luminescence in bulk crystallities as well as in epitaxial films independent of the growth technique gave the motivation for recent investigations to unravel the microscopic nature of the defects involved in the recombination. Using optically detected magnetic resonance, Glaser et al. [48] found that the signals from the shallow donor as well as from a deep donor. Contrary to Ogino and Aoki [45], they proposed a recombination mechanism from the deep donor located approximately 1 eV below the conduction band to a shallow effective mass-type acceptor present as a residual impurity. There are, however, two other reports which are in favor of the original Ogino [45] interpretation. Suski et al. [49] studied the behavior of the luminescence upon hydrostatic pressure. The band shifts to the blue linearly for pressures up to 18 GPa. The shift of 30 meV/GPa is very close to the pressure coefficient of the bandgap, and hence evidence for a shallow to deep center recombination [50]. For pressures above 20 GPa there is hardly any change of the peak position. The behavior was identical for a GaN bulk sample and for an MBE-grown GaN epitaxial film. Hofmann et al. [51] reported on ODMR, luminescence excitation, and time-resolved photoluminescence experiments. The decay dynamics—the decay times extend from nanoseconds to milliseconds—were analyzed and the distribution of lifetimes is consistent with the participation of the shallow donor. Time-delayed luminescence experiments showed no energy shift of the yellow band and the authors concluded there is almost no or only a very small Coulomb interaction between the two recombination partners of the 2.2 eV band and hence of the shallow donor to deep donor type. Luminescence excitation and two-color stimulation experiments [52] are consistent with this interpretation, but also indicate a more complex excitation behavior (see Fig. 7.13). The luminescence could be excited with photon energies well below the bandgap energy of 3.5 eV, starting around 2.5 eV. The dominant excitation is via free-exciton absorption, but there are also contributions from neutral donor- and acceptor-bound excitons. Whereas in luminescence the neutral donor-bound exciton is the dominating recombination expected for an n-type direct bandgap semiconductor in the PLE spectrum, the absorption due to acceptor-bound excitons is more pronounced than the donor-bound exciton features. A shallow donor (D) to deep donor (DD) recombination in any case requires free carriers, i.e. free electrons and holes according to the recombination scheme

$$D^+ + DD^0 + h\nu \text{ (exc.)} \rightarrow D^0 + DD^+ + h\nu \text{ (2.2 eV)} \quad (7.5)$$

where an electron is captured at the shallow donor and a hole at the deep

7.3 Deep structural defects in GaN   259

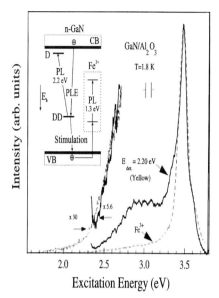

Fig. 7.13 Photoluminescence excitation spectra of the yellow emission and of the Fe^{3+} luminescence. The inset shows the level scheme of the yellow luminescence and Fe^{3+} recombination and excitation processes. The observed transitions are given as hole transitions to account for the n-type character of the samples (after Ref. [52]).

donor, respectively. The participation of free and bound excitons points to an Auger-type excitation and a hole transfer from the bound exciton state to the deep state of the yellow emission. The involvement of a hole can also be inferred from the two-color stimulation experiments reported by Hoffmann et al. [52]. They noted that the luminescence excitation of the 2.2 eV band and Fe^{3+}, a common residual contaminant in GaN, showed striking similarities. Again the excitation via free and bound exciton adsorption dominates. Between 2.0 and 3.0 eV the IR luminescence of Fe^{3+} at 1.29 eV is excited by the $Fe^{3+/2+}$ charge transfer band. Through this charge transfer process free holes are generated. These free holes can be captured by the deep donor and will hence stimulate the yellow emission intensity. Indeed, the luminescence intensity of the Fe^{3+} line and the yellow band can be influenced by additional light illumination. With a photon energy of 1.5 eV the luminescence of Fe^{3+} is enhanced, whereas the yellow band is reduced in intensity. The spectral dependence starts at 1.2 eV with a maximum at 1.5 eV and disappears at around 1.9 eV. From these two-color stimulation experiments the energy level of the deep center was determined to be 1.2 eV above the valence band.

Cathodoluminescence has been used to map the spatial variation of the near bandgap and yellow luminescence. Based on a correlation with transmission electron microscopy images, Ponce et al. [53] suggested that the yellow band is associated with the presence of extended defects, such as dislocation at low-angle grain boundaries, or point defects which nucleate at the dislocation. These findings are supported by spatially resolved Raman and PL experiments [38], which indicated that the 2.2 eV band is particularly strong at or close to the interface between sapphire and the GaN layer, most certainly the area with a very high number of extended defects.

Further information about the defects involved in the recombination comes from growth studies. There is no clear correlation between growth conditions and the appearance and intensity of the yellow band. It is, however, suppressed in p-type GaN:Mg, i.e. the deep state should be a defect with a low (high) formation energy under n-type (p-type) conditions. Neugebauer et al. [54], based on all these experimental findings, used first principles calculations to investigate native defects and complexes of them as possible candidates for the deep center. They concluded that the gallium vacancy, an acceptor, could be the most likely candidate. In n-type material it induces an energy level at approximately 1.1 eV above valence band (2−/3− level). Other possible defect structures could be pair defects consisting of V_{Ga} and Si_{Ga} or V_{Ga} and O_N acceptor–donor pairs, hence similar to the famous self-activated centers in II–VI compounds. They also rule out the original Ogino and Aoki [45] suggestion: a V_{Ga}–C_N pair defect.

7.4 Transition metals in GaN

Despite this enormous progress in growth and device technology there is only a little information about deep defects in these materials. Transition metals (TMs) form deep defects and can be expected to be common contaminations of Group III nitrides grown at very high temperatures. Even though their technological relevance for growing high-resistivity material was demonstrated for Fe- and Cr-doped GaN layers two decades ago [55], the first detailed information on TM defects was reported only recently [56–65].

Usually, the TMs are incorporated substitutionally on the Ga cation site in GaN. Thus, the neighborhood of the TM is given by the nearest four nitrogen atoms. Compared with other III–V compounds, the electronegativity of N is much larger higher than for P and As, leading to a more ionic bond. Because of the localized wavefunction of the deep impurity and the similar lattice constants of GaN and AlN, together with their large bandgaps, it is expected that internal transitions of the same transition metal element appear at about the same energy.

Due to the fact that the chemical identification of TM-related luminescence bands in GaN and AlN is still a matter of discussion, we consider the different zero-phonon line (ZPL) energies of the transition metal elements in GaN (Section 7.4) and AlN (Section 7.5) separately.

Finally, we summarize the results and discuss the relative energy level positions of each TM in GaN and AlN compared with GaAs, which are important for the prediction of band offsets according to the Langer and Heinrich rule [4, 5].

Up to now, the most coherent understanding has been obtained for the 1.3 eV luminescence band in GaN, which is generaly attributed to Fe^{3+}. Therefore we

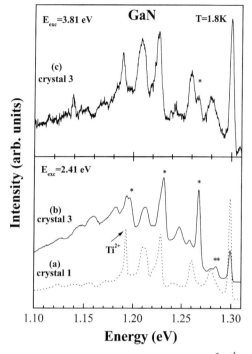

Fig. 7.14 Low-temperature photoluminescence of the $Fe^{3+}(^4T_1(G)-{}^6A_1(S))$ center. Shown are the n-type crystal 1 (a) and the semi-insulating sample 1 (b, c). Spectra (a) and (b) were recorded using the green Ar laser line at 2.412 eV for excitation. For spectrum (c) crystal 2 was excited at 3.81 eV. On exciting 2.412 eV, new ZPLs at 1.268 eV and 1.286 eV are observed in the semi-insulating sample and are tentatively attributed to Fe-related defect complexes. These lines and the corresponding phonon replica are labeled by * and **, respectively (after Ref. [68]).

start with a summary of results obtained for this luminescence and also describe the experimental methods. These methods were also applied for the investigations of other TM-related luminescence bands.

7.4.1 The 1.3 eV luminescence

Maier et al. observed a structured near-infrared luminescence band in hexagonal GaN [57], which is dominated by a zero-phonon line (ZPL) at 1.299 eV. This luminescence is also observed in unintentionaly doped GaN crystals grown on SiC and Al_2O_3 substrates by several other groups [60, 66]. A typical luminescence spectrum of an iron-doped 38 μm thick sample grown on Al_2O_3 is shown in Fig. 7.14. At lower photon energies a rich phonon sideband structure is apparent. The respective phonon energies are close to those of intrinsic lattice phonons in GaN. A detailed discussion of the local vibrational properties of this defect is given later on after identification of the defect center. The main keys to the identification of the corresponding defect are magnetic resonance experiments in combination with magneto-optical data.

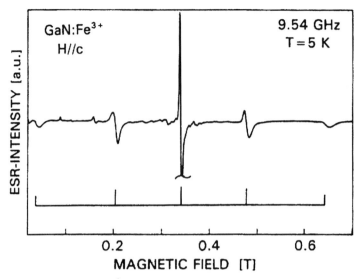

Fig. 7.15 ESR-spectrum of Fe^{3+} in a 40 μm thick GaN layer on sapphire substrate for $H \parallel c$ and at 5 K. The position of the expected five fine-structure transitions is indicated. Additional weaker signals arise from forbidden transitions and from Cr^{3+} trace impurities in the sapphire substrate (after Ref. [57]).

7.4.1.1 Electron spin resonance

Maier et al. [57] reported on electron spin resonance (ESR) of Fe^{3+}. The ESR spectrum is shown in Fig. 7.15. Five fine-structure transitions with line positions typical for an $S = 5/2$ ion in axial (C_{3V}) symmetry are observed. Because of random strain broadening, the linewidths of the outer pairs of fine-structure transitions exceed that of the central line. The spin-Hamiltonian appropriate for substitutional Fe^{3+} in the wurtzite (C_{6V}) host reads:

$$H = g\mu_B H S + a/6\left(S_\xi^4 + S_\eta^4 + S_\zeta^4 - 707/16\right) + D(S_z^2 - 35/12)$$
$$+ 7F/36(S_z^4 - 95/14 S_z^2 + 81/16) \tag{7.6}$$

assuming an isotropic g-factor. Here, the z-axis lies along the hexagonal c-axis of the crystal, whereas ξ, η, and ζ refer to the cubic axes at the two cation lattice sites in the wurtzite structure. Although crystallographically equivalent, these two sites can be distinguished magnetically. This is manifested by a doubling of the five fine-structure lines, apparent for a general orientation of the magnetic field H with respect to the crystalline c-axis. It should be pointed out that the above Hamiltonian has been used before to analyze the ESR spectrum of Fe^{3+} in zinc oxide (ZnO) [ref. mentioned in 57], which also crystallizes in the hexagonal wurtzite structure. The parameters of the above spin-Hamiltonian for GaN:Fe^{3+} were extracted from the ESR spectra taken under various orientations. The results are compiled in Table 7.1, which also quotes corresponding values previously obtained for ZnO:Fe^{3+}.

7.4 Transition metals in GaN

Table 7.1 Spin Hamiltonian parameters of Fe^{3+} in GaN and ZnO at room temperature (after ref. [57]).

	GaN	ZnO
$g_\parallel$	1.990 ± 0.005	2.0062 ± 0.0002
$g_\perp$	1.997 ± 0.005	
$D(10^{-4}\,cm^{-1})$	-713 ± 5	-595 ± 2
$a - F(10^{-4}\,cm^{-1})$	$+52 \pm 6$	$+37 \pm 2$
$a(10^{-4}\,cm^{-1})$	$+48 \pm 5$	$+41 \pm 3$

Using these parameters, the angular dependence of the ESR spectrum under a rotation of the crystal from $H \parallel c$ to $H \perp c$ was calculated (Fig. 7.16). Good agreement with the experimental points is seen to exist. We also note the predicted splittings for arbitrary orientation of H with respect to the crystalline c-axis.

The ESR spectrum can also be observed at 300 K. Because of the high Debye temperature of GaN (600 K), spin–lattice relaxation does not apparently lead to excessive ESR line broadening at room temperature.

7.4.1.2 Optically detected magnetic resonance

In order to establish a firm correlation between the ESR data and the photoluminescence data, optically detected magnetic resonance (ODMR) experiments are the most powerful tool. The ODMR and ESR experiments described here were done by Maier *et al.* with the same crystal [57]. The ODMR setup was built around a 4T split-coil superconducting magnet and operated at K-band (18–26 GHz) microwave frequencies. The total luminescence (800 nm–1650 nm) was monitored with a Ge detector. ODMR spectra recorded on this luminescence band at 1.3 eV reveal the fingerprint of the $^6A_1(S)$ state of an electronic d^5

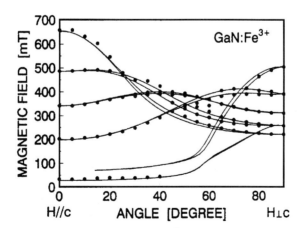

Fig. 7.16 Angular dependence of the GaN:Fe^{3+} ESR spectrum for a rotation of the crystal from $H \parallel c$ to $H \perp c$. The solid lines have been calculated using the parameters quoted in Table 7.1 (after Ref. [57]).

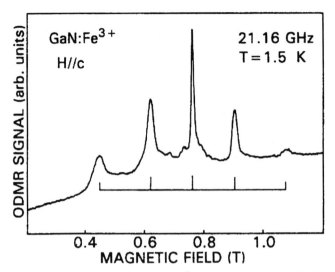

Fig. 7.17 ODMR spectrum of Fe^{3+} in GaN (after Ref. [57]).

configuration. The ODMR spectrum, shown in Fig. 7.17, was essentially identical to the ESR spectrum of Fe^{3+}, shown in Fig. 7.15.

Because of the lower temperature and higher magnetic fields used in the ODMR experiment, compared with the ESR experiment, thermal depopulation

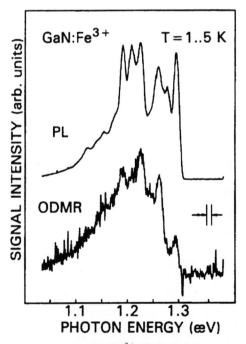

Fig. 7.18 ODMR excitation spectrum of Fe^{3+} in GaN. For comparison, the infrared PL spectrum of this sample is shown in the upper trace. Both spectra have been recorded under identical spectral resolution (after Ref. [57]).

of the electronic Zeeman levels is more prominent in the ODMR spectrum. This explains the unequal intensity of the outer fine-structure transitions in Fig. 7.17 and also allows a determination of the signs of the spin-Hamiltonian parameters. The parameters that were obtained are $g_\parallel = 1.998 \pm 0.005$, $D = -(707 \pm 8)10^{-4}$ cm^{-1}, $a - F = (85 \pm 25)10^{-4}$ cm^{-1}.

The intensity of the central $-\frac{1}{2} \leftrightarrow \frac{1}{2}$ Fe^{3+} ODMR transition as a function of the photon energy was also measured. Such an ODMR excitation spectrum is shown in the lower part of Fig. 7.18; it is seen to be almost identical to the infrared PL spectrum, which is shown for comparison in the upper part of Fig. 7.18. The above ODMR data thus strongly support the assumption that the 1.3 eV infrared luminescence of GaN arises from Fe^{3+} (3d^5).

7.4.1.3 Zeeman investigations of the luminescence

ODMR results show only the $^6A_1$ ground state resonance. The ODMR signal results from the nonresonant excitation process, allowing for no unambiguous identification of the luminescence center. An unambiguous identification of the defect center was done by Heitz et al. [61] using Zeeman measurements for the ZPL of the 1.3 eV luminescence band, giving direct evidence for the $^6A_1$ ground state. The Zeeman data were recorded for n-type samples grown by the sublimation sandwich technique on 6H–SiC substrates which contain Fe as unintentional impurity. The luminescence is excited by a 488 nm line Ar$^+$ laser,

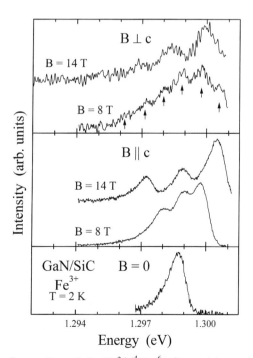

Fig. 7.19 The zero-phonon line of the Fe^{3+}($^4T_1$–$^6A_1$) transition at $T = 2$ K for $B = 0$ T, $B \parallel c$, and $B \perp c$ (after Ref. [61]).

dispersed by a 0.75 m double grating monochromator and detected with a cooled Ge photodiode. The Zeeman experiments are carried out using a 15 T split-coil superconducting magnet. The sample investigated in the Zeeman experiments shows the ZPL at 1.2988 eV with a full width at half-maximum (FWHM) of 850 μeV (see Fig. 7.19(a)). We attribute the asymmetric lineshape to inhomogeneous strain in the GaN/6H–SiC samples. Figures 7.19(b) and (c) show the zero-phonon region at 8 and 14 T for magnetic field orientations $B \parallel c$ and $B \perp c$, respectively.

The Zeeman behavior is anisotropic, as anticipated, for a trigonally coordinated point defect in hexagonal GaN. The 8 T spectrum (Fig. 7.19(c)) shows six equally spaced Zeeman components. The relative intensity of the components does not change for sample temperatures up to 10 K, demonstrating the observed splitting to orginate in the ground state. The sixfold splitting is clear evidence for a $^6A_1$ ground state of the luminescence and therefore an electronic ^5d configuration of the luminescence center. All six components are resolved only for magnetic fields between 6 and 10 T. At higher fields the outer components become too weak, and at lower fields the FWHM of the ZPL exceeds the Zeeman splitting. The relative intensity of the Zeeman components depends critically on the magnetic field orientation. For $B \perp c$ up to six components could be resolved, whereas for $B \parallel c$ only three lines are observed. Qualitatively the same Zeeman behavior has been reported for the $Fe^{3+}(^4T_1-^6A_1)$ transition in ZnO, which is a good reference material for hexagonal GaN. Both semiconductors have a wurtzite structure with almost the same lattice parameters and similar bandgaps. From the comparison with Fig. 3 of [74] it becomes clear that the three components observed for $B \parallel c$ in GaN correspond to transitions to the three lowest components of the $^6A_1$ ground state. The lowest Zeeman component of the excited $^4T_1(G)$ multiplet corresponds to $S_z = -3/2$ and, thus, transitions into the $S_z = -5/2, -3/2,$ and $-1/2$ components are expected to dominate the spectra. Figure 7.20 compiles the energies of the Zeeman components observed for the ZPL at 1.2988 eV. For $B \parallel c$ a linear Zeeman behavior is observed, with the center of the sixfold splitting shifting towards lower energies, whereas for $B \perp c$ an additional nonlinear shift towards lower energies occurs.

The fit (see Fig. 7.20) was done with the ESR parameters of Fe^{3+} in GaN grown on SiC. The parameters are in agreement with those for GaN grown on Al_2O_3, but indicate a larger trigonal crystal field in GaN grown on 6H–SiC.

At zero magnetic field the $^6A_1$ ground state multiplet splits into three Kramers doublets with the $\pm 1/2$ and $\pm 3/2$ states lying 59 and 38 μeV above the $\pm 5/2$ ground state. This zero-field splitting is much too small to be resolved in the luminescence experiments. Thus, no detailed comparison of the fine-structure data is possible. Only the magnetic field dependence of the excited state has been taken as a variable for the fitting. Good agreement with the experimental data confirms the attribution of the 1.3 eV luminescence band to the $^4T_1-^6A_1$ transition of Fe^{3+} already proposed on the basis of ODMR data. The excited state is found to match the behavior for Fe^{3+} in ZnO [74]. For $B \parallel c$ a linear low-energy shift corresponding to a g-value of -2.81 ± 0.05 (-2.71 in the case of ZnO) is observed.

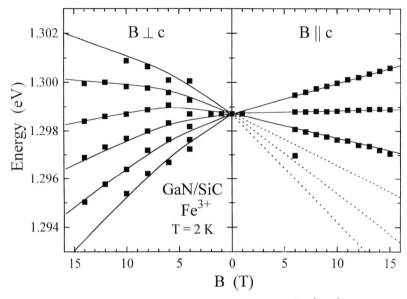

Fig. 7.20 Zeeman splitting of the zero-phonon line of the Fe^{3+} ($^4T_1$–$^6A_1$) transition for the magnetic field orientations $B \parallel c$ and $B \perp c$ (after Ref. [61]).

Heitz et al. [61] measured a luminescence decay time of 8 ms at $T = 1.8$ K, which is about twice as long as that observed at 77 K. Millisecond lifetimes are typical for the spin- and symmetry-forbidden (neglecting fine-structure interactions) $^4T_1$–$^6A_1$ transition of Fe^{3+} in semiconductors, indicating a high quantum efficiency of the relaxation process.

In conclusion, it is unambiguously shown that the 1.299 eV ZPL is related to the $^4T_1$–$^6A_1$ transition of isolated Fe^{3+} on a substitutional Ga site.

7.4.1.4 Photoluminescence excitation and time-resolved photoluminescence

However, the Fe-doped crystal shows additional ZPLs under different excitation conditions. Figure 7.14 represents typical PL spectra of the Fe^{3+} ($^4T_1$–$^6A_1$) luminescence for different excitation energies. On exciting the sample at 2.412 eV, new ZPLs at 1.268 eV and 1.286 eV are observed (indicated by * and **), superimposed on the luminescence of isolated Fe^{3+} with its ZPL at 1.299 eV. The full width at half-maximum (FWHM) of the 1.268 eV ZPL is 3 meV, whereas for the 1.299 eV line the FWHM is only 1 meV. A lifetime of 7.8 ± 0.5 ms and 2.6 ± 0.5 ms is detected for the 1.299 eV and 1.268 eV ZPL, respectively (see Fig. 7.21). Temperature-dependent PL experiments reveal the detection of additional ZPLs on the high-energy side of the respective ZPL due to the thermal population of higher excited fine-structure states. For the $^4T_1$ state of isolated Fe^{3+} these three states are 1.8, 2.6, and 3.8 meV above the lowest component at 1.2988 eV. For the 1.2687 eV ZPL these states are 2.0 and 3.5 meV above the lowest component.

The presence of PL from Fe complexes renders the spectral position and the

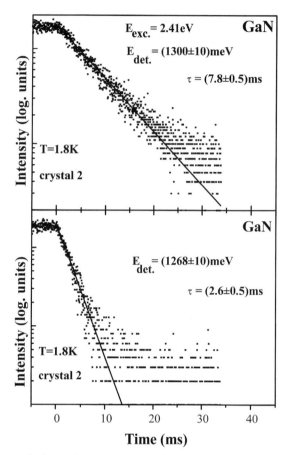

Fig. 7.21 Decay of the 1.299 eV ZPL and of the 1.268 eV ZPL at $T = 1.8$ K (after Ref. [124]).

width of the detection window critical for the reliability of the PLE experiments. Spectrum (a) in Fig. 7.22 presents the PLE of isolated Fe^{3+}, detecting luminescence only in a 10 meV window around 1.299 eV. The excitation behavior of the Fe^{3+} luminescence depends critically on the stable charge state of iron. The spectra of semi-insulating samples show structured absorption bands not observed for the n-type samples (c).

For the semi-insulating sample, fine structure is resolved around 2.0 eV and 2.8 eV and a broad excitation band appears in the UV spectral region (Fig. 7.22(a)). Heitz et al. [68] performed high-resolution laser PLE of these fine structures. The occurrence of sharp ZPLs in the PLE spectrum of the Fe^{3+} luminescence together with the Fe^{3+} EPR signal in the crystal directly demonstrates the presence of the neutral 3+ charge state in the unexcited samples.

Figure 7.23 shows a high-resolution spectrum of the intracenter excitation band at 2.01 eV of crystal 2, which was excited by a tunable dye laser.

The authors were not able to resolve these PLE resonances for the n-type

7.4 Transition metals in GaN 269

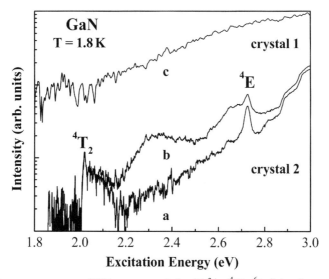

Fig. 7.22 Low-temperature PLE spectra of the Fe^{3+} ($^4T_1$–$^6A_1$) luminescence for the two samples. Crystal 1 is n-type containing Fe^{2+} (spectrum c). Crystal 2 contains practically only Fe^{3+} (spectra a and b). Luminescence in a 10 meV window around the ZPL at 1.299 eV was detected for spectra a and c, whereas the detection energy is 1.268 eV for spectrum b (after Ref. [125]).

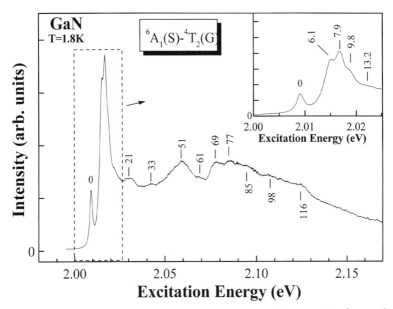

Fig. 7.23 Laser PLE spectrum of sample 2 in the region of the Fe^{3+} ($^6A_1(S)$–$^4T_2(G)$) transition at 1.8 K. The inset gives the ZPL region on an enlarged energy scale. Luminescence in a 10 meV window around the ZPL position of 1.299 eV was detected. Energy shifts with respect to the ZPL at 2.0091 eV are given in meV (after Ref. [68]).

sample 1. Luminescence spectra excited at the various PLE fine structures show the typical Fe^{3+} luminescence, which proves that all belong to the Fe^{3+} center. The set of ZPLs around 2.01 eV is followed by a vibronic sideband. The labels in Fig. 7.23 give the respective enegy differences (in meV) between peaks in the sideband and the ZPL at 2.0091 eV. This vibronic sideband shows no distinct replica due to optical phonon modes of hexagonal GaN [69]. Only a weak step around 69 meV almost coincides in energy with the E_2(high) mode which dominates the sideband of the $^4T_1(G)-^6A_1(S)$ luminescence (Fig. 7.14(a)). The inset of Fig. 7.23 shows the ZPL region over an enlarged energy scale. At least four ZPLs (at 2.0091 eV, 2.0152 eV, 2.0170 eV, and 2.0188 eV) are resolved with full widths at half-maximum (FWHM) down to 1.2 meV. Transitions into higher excited quartet states are known to lead to structured bands in PLE spectra of the $^4T_1(G)-^6A_1(S)$ luminescence [127]. Thus, we attribute this 2.01 eV absorption band to the $^6A_1(S)-^4T_2(G)$ transition of Fe^{3+}. $^4T_2(G)$ is the quartet state next to the luminescent $^4T_1(G)$ state. The series of weak and broad peaks in the sideband of the 2.01 eV absorption indicates a dynamical Jahn–Teller effect in the excited state.

For the narrow bandgap III–V semiconductors no optical transitions involving higher excited quartet and doublet states of Fe^{3+} have been reported [70–72], indicating that they are degenerate with the valence band. Obviously, for GaN the $Fe^{3+/2+}$ acceptor level lies high enough in the bandgap to allow the observation of higher excited crystal-field states in the bandgap. The $^4T_2(G)$ state is expected to show a strong Jahn–Teller coupling to ε-type models, which can reduce the fine structure to a doublet. This was verified experimentally for Mn^{2+} in ZnS [127]. In hexaonal host crystals, however, the lower defect symmetry stabilizes the Fe^{3+} center against the Jahn–Teller coupling, as recently established for the $^4T_1(G)$ state in ZnS, ZnO, and also GaN [73, 74, 61]. A similar effect on the $^4T_2(G)$ state in GaN would explain the richer fine structure spectrum observed for the 2.01 eV absorption (inset of Fig. 7.23). A detailed assignment of the observed fine structure, however, requires more detailed experimental data.

Figure 7.24 shows the fine structure observed for the semi-insulating sample 2 around 2.8 eV on an enlarged energy scale. The observed linewidths are not limited by the experimental resolution even for lamp excitation, but rather result from the limited crystal quality. The structure starts with a single ZPL at 2.731 eV with a FWHM of 16 meV, which is followed by a series of step-like resonances. The step period of approximately 75 meV almost corresponds to TO modes of hexagonal GaN ($E_{TO}(\Gamma) = 70.5$ meV) [69]. However, it is more likely that local vibrational modes of the Fe center are involved. This would explain the energy difference between the phonon energy and the step period [67]. Additionally, it is questionable whether the step-like structure can be interpreted as a phonon sideband of the intense ZPL at 2.731 eV. The energy separation between this ZPL and the first step amounts to 157 meV and, thus, corresponds to neither the step period of 75 meV nor the typical phonon modes of hexagonal GaN [69]. Moreover, the strong intensity of the 2.73 eV line makes it an unlikely candidate for the ZPL of the step-like PLE structure. Therefore,

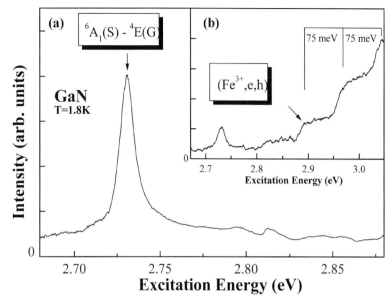

Fig. 7.24 High-resolution PLE spectrum of crystal 2 recorded at 1.8 K showing the Fe^{3+} ($^6A_1(S)$–$^4E(G)$) transition and the (Fe^{3+}, e, h) complex in the low-energy onset of the $Fe^{2+/3+}$ charge-transfer band. Luminescence in a 10 meV window around the ZPL position of 1.299 eV was detected (after Ref. [68]).

we treat the 2.731 eV absorption as a separate feature and tentatively attribute it to an intracenter transition involving the next excited quartet state, the $^6A_1(S)$–$^4E(G)$ transition. This assignment is supported by the weak phonon sideband typical for this transition [75].

The spectral appearance of the step-like structure showing at least four replicas with increasing intensity is unusual for an intracenter transition of a TM. However, it is rather typical for deeply bound excitons [76] or deeply bound electron–hole pairs [77–79]. The steps are located on the low-energy onset of the broad and efficient UV excitation band which we assign to the $Fe^{3+/2+}$ charge-transfer transition. We therefore attribute this structure to the formation of a shallow bound state at the Fe center.

Shallow bound states are now well established for the near-midgap acceptor Fe in III–V semiconductors [79–81]. The Coulomb interaction between the negatively charged Fe^{2+} ion and a hole in the valence band results in the formation of an (Fe^{2+}, h) complex. In a first approximation the bound-hole state can be described by effective-mass theory and thus as a (transient) shallow acceptor state. This complex is in principle an excited state of Fe^{3+} and relaxes nonradiatively to the excited $^4T_1(G)$ state. The exchange interaction between the loosely bound hole and the core holes influences the fine structure. In GaAs, GaP, and InP Fe forms a deep acceptor level near the middle of the bandgap with the core wavefunctions well localized at the Fe ion. The binding energy of the (Fe^{2+}, h) complex amounts to only a few tens of meV. As a consequence the overlap between the hole and the core wavefunctions is small and thus the exchange interaction is weak. This fact is demonstrated by the fivefold fine structure observed at the low-energy onset of the $Fe^{3+/2+}$ charge-transfer band in InP, GaP, and GaAs [79–81], which reflects the Fe^{2+} term scheme. The exact location of the $Fe^{3+/2+}$ acceptor level in GaN is not clear yet. The observation of the charge-transfer band in PLE should provide this information. However, the superposition of shallow bound states of TMs to the low-energy slope of a charge-transfer band makes a fit of the ionization band impossible in order to determine the onset energy [77, 78]. Nevertheless, it is reasonable to identify the energy (3.17 ± 0.10 eV) at which the step-like structure vanishes and the broad structureless charge-transfer band begins to dominate with the energy position of the deep $Fe^{3+/2+}$ acceptor level. This assignment yields a binding energy of (280 ± 100) meV for the shallow bound state of the Fe center in GaN.

Therefore, the deep $Fe^{3+/2+}$ acceptor level in GaN is close to the conduction-band minimum. This makes the hybridization of the core states important. Additionally, the binding energy of 280 meV of the shallow complex is much higher than in the other III–V semiconductors. Both effects add up to a strong exchange interaction. Therefore, a deeply bound electron–hole complex (Fe^{3+}, e, h) is the appropriate description of this state. Nevertheless, the ionization products are still Fe^{2+} and a free hole. A very similar situation was demonstrated for shallow bound states of Ni comparing cubic ZnS and hexagonal CdS [78]. The strong phonon coupling causing the step-like structure, as seen in Fig. 7.22, is typical for such a deeply bound electron–hole pair [79]. Recent calculations show that the hybridization of the TM ground state with valence band states is favored in hexagonal host crystals [82]. Indeed, all known shallow states of TMs in wurtzite crystals have the character of a (TM, e, h) complex [77, 78].

The PLE spectra of the Fe^{3+} ($^4T_1(G)$–$^6A_1(S)$) luminescence in semi-insulating GaN samples (Fig. 7.22) allow us to deduce a comprehensive term scheme of the Fe^{3+} center in GaN (Fig. 7.25).

Four crystal field states of Fe^{3+} are identified. Their energy positions give a wealth of information for detailed ligand-field calculations, which are beyond the scope of this chapter. In principle, optical transitions of a d^5 configuration involving the $^5A_1(S)$ ground state have a low transition probability because of their spin-flip character. Only the luminescent $^4T_1(G)$ state exhibiting an ms lifetime is observed in narrow bandgap III–V compounds such as InP, GaP, and GaAs. The higher excited states are expected to be degenerate with the conduction band. However, detailed information on higher excited states was derived by PLE investigations for Mn^{2+} in the wide bandgap II–VI semiconductors [75, 127]. From our results we conclude that the crystal field splitting of the ^4G multiplet of Fe^{3+} in GaN is approximately three times larger than that of Mn^{2+} in II–VI compounds. This strong crystal field may be the result of the comparatively small lattice constant of GaN, the 3+ charge state of iron, and the high electronegativity of nitrogen. An increased crystal-field strength has been reported for other TMs in GaN too [62, 63].

The possible excitation processes of the Fe^{3+} luminescence in the semi-insulating samples containing iron in the neutral 3+ charge state are obvious

7.4 Transition metals in GaN 273

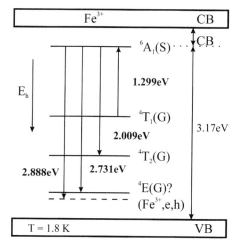

Fig. 7.25 Energy levels of the Fe^{3+} center in hexagonal GaN in the hole picture. The observed transition energies are given for a crystal temperature of $T = 1.8\,\mathrm{K}$ (after Ref. [68]).

from the term scheme in Fig. 7.25. The sharp lines at 2.01 eV and 2.73 eV result from intracenter absorption into the excited crystal field states and the (Fe^{3+}, e, h) complex. The broad UV band is attributed to the acceptor-type charge-transfer process

$$Fe^{3+}\left({}^6A_1(S)\right) + h\nu \to Fe^{2+} + h_{VB} \to (Fe^{3+})^*. \tag{7.7}$$

The recombination of holes with Fe^{2+} centers provides the most efficient excitation process of the Fe^{3+} luminescence in both III–V [79–81] and II–VI [73, 74] semiconductors. For n-type GaN samples the direct excitation of Fe^{3+} is more complicated. Fe^{3+} has to be generated in an excited state starting from Fe^{2+} in a photoionization process:

$$Fe^{2+}({}^5E) + h\nu \to (Fe^{3+})^* + e_{CB}. \tag{7.8}$$

The probability of this transition is expected to be comparatively low due to the s–d character of the electron transition. However, the low-energy thresholds for excitation into the ${}^4T_1(G)$ or ${}^4T_2(G)$ states are expected at 1.64 eV and 2.35 eV, respectively, whereas the experimental threshold is found at 2.2 eV; crystal 1 in Fig. 7.22. Thus, it is questionable if the observed PLE can be attributed to the $Fe^{2+/3+}$ charge-transfer excitation process (eqn 7.9)). The efficient visible excitation of the Fe^{3+} luminescence observed in n-type samples is more likely caused by the simultaneous presence of other defects in GaN acting as activator centers. In n-type II–VI semiconductors excitation of the Fe^{3+} luminescence is provided by the capture of holes generated in charge-transfer processes of other TM defects [73, 74]. Two-color stimulation experiments indicate that there is a similar excitation mechanism in n-type GaN with the deep defect of the yellow luminescence acting as a hole source [39]. Thus, the PLE spectra of the Fe^{3+} luminescence in n-type GaN yield information on other defects rather than on the Fe center itself. The PLE spectrum of crystal 2 represents a superposition of the PLE spectra observed in the n-type sample 1 and semi-insulating sample 2, showing that the Fermi level is pinned at the Fe level and that both Fe^{3+} and Fe^{2+} are present in the unexcited crystal.

The PLE spectra presented in the paper of Heitz et al. [68] differ considerably from those reported by Baur et al. [58, 59]. They assume that the better defined detection window enabled them to separate different luminescence processes and, thus, to obtain more reliable data. The ionization energy of 3.17 eV determined in the work of Heitz et al. [68] is considerably higher than the value of 2.5 eV found by Baur et al. [58, 59]. It should be noted that the observation of the formation of (Fe^{3+}, e, h) complex on the low-energy onset of the $Fe^{3+/2+}$ charge-transfer band does not allow for much uncertainty in the deep acceptor level position.

The energy level of the acceptor close to the conduction band allows us to understand an otherwise very puzzling fact. Up to now, it was not possible to detect the $Fe^{2+}(^5E-^5T_2)$ transition in absorption or luminescence in n-type or semi-insulating samples, even though it is very prominent in other semiconductors. The lack of luminescence may be the result of efficient nonradiative relaxation processes resulting from the large phonon energies [83, 84]. Taking into account the position of the $Fe^{3+/2+}$ level, only 340 meV below the conduction band of GaN, it is now very likely that the excited $^5T_2$ state of Fe^{2+} is degenerate with the conduction band, explaining the absence of the intra-center transitions of Fe^{2+} in optical spectra.

Compared with the Fe^{3+} PLE in n-type and semi-insulating samples, two additional features peaking at 2.3 eV and 2.65 eV are observed in the PLE spectrum of the 1.268 eV ZPL. (Fig. 7.22, spectrum b). The 2.65 eV resonance is rather broad in comparison with the 2.73 eV PLE resonance of Fe^{3+}. This broadening can be caused by the larger energy separation within the ^4E state of the Fe^{3+} complex due to the larger crystal field. A similar observation was made for Mn^{2+} centers in cubic ZnS [85]. Here, additional axial crystal fields due to polytype effects of nearby impurity atoms cause a larger energy separation within the ^4E-state of the d^5 configuration of Mn^{2+}. The other additional feature is a broad PLE structure peaking at 2.3 eV. The peak energy of 2.3 eV agrees well with the expected $Fe^{2+/3+}$ low energy threshold of 2.36 eV within the $^4T_2$ state of isolated Fe^{3+}. This indicates that the other atom within the Fe complex might be a nearby donor, resulting in an (Fe^{2+}, D^+) complex in the unexcited crystal. The 1.268 eV PL of the Fe complex is excited via a charge-transfer process of the Fe^{2+}:

$$(Fe^{2+}, D^+) + h\nu \ (2.3\,eV) \rightarrow (Fe^{3+*}, D^+) + e_{CB} \tag{7.9}$$

followed by the radiative relaxation of the excited (Fe^{3+*}, D^+) center causing the 1.268 eV luminescence and the (very slow) capture of the electron.

7.4.1.5 Zeeman investigation of the iron-related 1.268 eV emission

In order to get an insight into the electronic structure of the defect, we

7.4 Transition metals in GaN

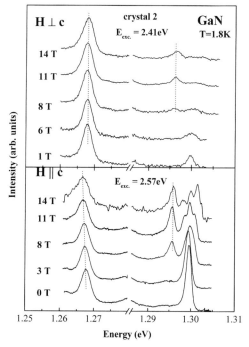

Fig. 7.26 Zeeman spectra of the ZPL region in the configuration $H \parallel c$ and $H \perp c$ for $T = 1.8$ K. For $H \parallel c$ the crystal is excited at 2.57 eV, whereas for $H \perp c$ the crystal is excited at 2.4 eV. No Zeeman splitting is observed for the 1.268 eV ZPL. With increasing magnetic field, a new ZPL at 1.296 eV is observed for both configurations. The relative intensity change between the 1.299 eV ZPL and the 1.268 eV ZPL in both Zeeman configurations is due to the different excitation behavior of the ZPLs (after Ref. [124]).

performed Zeeman spectroscopy of the ZPL region (Fig. 7.26). For $H \parallel c$, the three components [61] of the Fe^{3+} ZPL are well resolved above $B = 11$ T. In contrast to the Fe^{3+} behavior, no Zeeman-splitting is observed for the 1.268 eV ZPL. Only a slight shift to lower energies of 61 ± 5 μeV/T is observed in this configuration. For $H \perp c$ the sixfold splitting of the Fe^{3+} ZPL [61] is not resolved, but a broadening with increasing magnetic field is observed. The 1.268 eV ZPL does not show any shift or broadening in this configuration. This indicates that the electronic structure of the Fe-related defect is strongly modified by the nearby impurity atom. The relative intensity change between the 1.299 eV ZPL and the 1.268 eV ZPL in both Zeeman configurations is due to the different excitation behavior of the ZPLs.

Additionally, a new ZPL at 1.296 eV is observed with increasing magnetic field for both Zeeman configurations. The excitation behavior of this line is similar to that of isolated Fe^{3+}. This line is also specific for the Fe-doped crystal 2 and was not observed for GaN epilayers grown on SiC [61]. This might indicate that the Fe doping is responsible for this line. If we explain this line by ferromagnetic coupling between two nearby iron atoms, we should expect a fast decay within the ground state, because the spin-selection rule is lifted. More investigations have to be done to clarify the origin of this new ZPL.

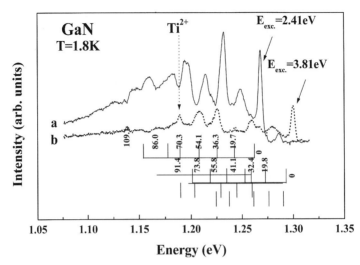

Fig. 7.27 PL-spectrum of the 1.268 eV luminescence (a) in comparison to the Fe^{3+} luminescence (b) of the semi-insulating GaN sample 2. The energy differences of the vibronic peaks to the respective ZPL are given in meV (after Ref. [125]).

7.4.1.6 Local vibrational mode (LVM) properties

The spectral shape of the phonon sidebands of the 1.299 eV luminescence as well as the 1.268 eV luminescence (Fig. 7.27) looks quite similar. However, the relative intensity and the energy difference of the vibronic peaks to each ZPL vary slightly. Thus, different local vibrational modes are involved for both defects.

In order to get an insight into the local vibrational system of iron in GaN we performed calculations using a valence–force model. The interatomic forces of cubic and hexagonal semiconductors had been parameterized in consideration of symmetry requirements by Keating [86]. The simple model for bond stretching and bond bending forces was extended by Kane [87] to include up to third nearest neighbor interactions. He demonstrated that the phonon dispersion curves of various crystals can be interpreted by a small number of parameters including an effective atomic charge to take the long-range Coulomb interaction into account. We used this model to set up the dynamical matrix, the eigenvalues of which are the squares of the vibration frequencies. In order to apply the Keating–Kane model to the calculation of LVMs at point defects in semiconductors we used a cluster of 295 vibration atoms around the defects in GaN. The vibration frequencies were obtained from a numerical diagonalization of the dynamical matrix. The localization of the vibrational mode is determined from the vibration amplitudes as obtained from the corresponding components of the eigenvector of the dynamical matrix. We selected the local vibrational modes by the condition that the sum of the squared components of the eigenvector corresponding to the atoms in the first three shells is larger than 0.3. We adopted the scaling-factor approximation (SFA) [88] to describe the change of the interatomic forces in the vicinity of the defect. The scaling factor s is defined by $p_d = p + sp$, where p is the bond-stretching or bond-bending

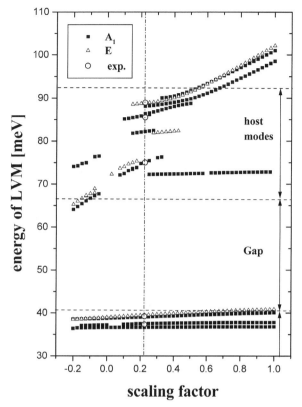

Fig. 7.28 Energy of the local vibrational modes of iron in GaN. For a scaling factor of 0.225 the best fit for the observed local vibrational modes of iron is obtained (after Ref. [126]).

valence-force parameter of the perfect crystal and p_d the corresponding parameter at the defect. The scaling factor was fitted to obtain the observed energy of the LVM.

The four LVM which were observed in the PL phonon sideband are located at 37.7 meV, 39.2 meV, 85.5 meV, and 88.9 meV. Additionally, in the Fe^{3+}–Fe^{2+} charge-transfer band a step period of 75 meV is observed in the Fe^{3+} PLE spectrum (Figs. 7.22, 7.24). These peaks were fitted with a scaling factor of 0.225. The best fit reveals LVM energies of about 88.8 meV (E), 86.0 meV (A_1), 75.1 meV (A_1), 39.5 meV (E), and 36.7 meV (A_1).

However, this excellent agreement should be compared with results obtained by first principles calculations, which are not available so far.

7.4.1.7 Conclusion on the 1.30 eV luminescence results

On the basis of electron spin resonance, optically detected magnetic resonance, and Zeeman studies of the zero phonon line it is unambiguously clear that the radiative recombination within the Fe^{3+} ($^4T_1$–$^6A_1$) transition causes the 1.3 eV luminescence.

The photoluminescence excitation results for the Fe^{3+} ($^4T_1(G)$–$^6A_1(S)$) luminescence in hexagonal GaN give rise to the observation of new crystal field transitions. The transitions $^6A_1(S)$–$^4T_2(G)$, and $^6A_1(S)$–$^4E(G)$ were resolved with ZPLs at 2.01 eV and 2.731 eV, respectively. Additionally, a deeply bound electron–hole complex (Fe^{3+}, e, h) with a binding energy of (280 ± 100) meV was observed. A strong phonon coupling leads to a step-like structure in PLE on the low-energy side of the $Fe^{3+/2+}$ charge-transfer band. The step period of 75 meV is explained by LVM of the Fe center in GaN. The $Fe^{3+/2+}$ acceptor level is located 0.34 eV below the conduction band minimum, which has important implications for the use of the $Fe^{3+/2+}$ level to determine band offsets.

The similar spectral shape of the vibronic sideband and the energy position, together with the decay time, indicate the close correlation between the isolated Fe^{3+} ($^4T_1$–$^6A_1$) transition at 1.299 eV and the 1.268 eV defect. Additionally, the PLE results can be explained in the Fe term-scheme. On the basis of these observations, the 1.268 eV is tentatively attributed to radiative recombination within an (Fe^{3+}, D^+) defect. However, the Zeeman behavior of the 1.268 eV ZPL does not show the clear fingerprint of the d^5 configuration. This indicates that the electronic structure of the 1.268 eV transition is strongly modified by the nearby impurity atom.

7.4.2 The 1.19 eV luminescence

Baur et al. [56] first reported spectra of the 1.19 eV luminescence for GaN grown on Al_2O_3. The most detailed study of this luminescence was performed by Heitz et al. [63]. The luminescence band shows a single sharp ZPL at 1.1934 eV with an FWHM of 150 μeV (inset in Fig. 7.29). The FWHM of this ZPL varies slightly from sample to sample, but is always significantly smaller than that of the 1.3 eV ZPL (800 μeV) present in the same crystals. The phonon sideband of the 1.19 eV luminescence is unusually weak for an internal d–d transition of a 3d impurity in III–V compounds.

The strongest phonon replica has about 2% of the intensity of the ZPL. Magnified spectra reveal a weak coupling to acoustical phonons with a pronounced structure near E_2 (low) (18 meV) and same sharp phonon replicas in the region of optical phonons of the hexagonal GaN host crystal. The phonon replicas with energies of 65.1 and 70.5 meV almost coincide in energy with the A_1 (TO) and the E_2 (high) observed in Raman spectra [69]. Two well-defined replicas L_1 and L_2, corresponding to phonon energies of 67.8 and 82.2 meV, are assigned to local modes of the defect center [67]. The weak phonon sideband and the observation of sharp host phonon replicas indicate that the electron–phonon coupling is weak in both the excited and the ground state of the 1.19 eV luminescence. The luminescence decays exponentially with a time constant of 65 ± 15 μs. This lifetime is typical for a symmetry-forbidden d–d transition. Figure 7.30 shows the zero-phonon region of the 1.19 eV band for various sample temperatures. The sample is excited at 2.41 eV. The 1.19 eV luminescence band is only weakly affected by temperatures up to 50 K. At higher temperatures the luminescence intensity drops and the ZPL shifts

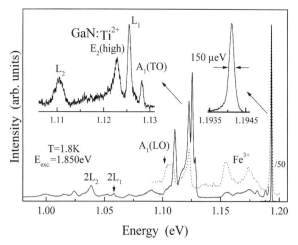

Fig. 7.29 Luminescence spectrum of the 1.19 eV emission in n-type GaN excited at 1.83 eV. The insets show on enlarged energy scales the ZPL at 1.1934 eV and the region of pronounced optical phonon replicas (after Ref. [63]).

towards lower energies and becomes broader. It must be emphasized that no hot lines occur on the high-energy side of the ZPL, indicating the lack of further fine-structure components. The observation of hot lines has been claimed in one paper [62], but later on these lines were attributed to some transition-metal complex centers [89].

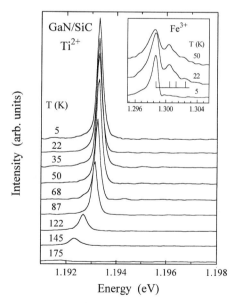

Fig. 7.30 The ZPL region of the 1.19 eV emission in GaN for various sample temperatures. The position of the hot line at 1.196 eV claimed in Ref. [62] by Baur et al. is indicated by the dotted line.

280 *Defect spectroscopy in the nitrides*

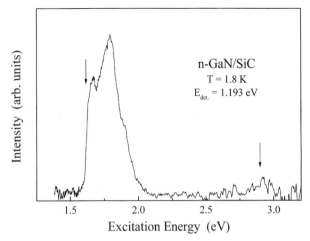

Fig. 7.31 Excitation spectrum of the 1.19 eV emission in n-type GaN ($n > 10^{17}\,\text{cm}^{-3}$) at $T = 1.8\,\text{K}$. The luminescence is detected at 1.1934 eV. The arrows indicate excitation bands corresponding to transitions into higher excited $^3T_1$ states at 1.62 and 2.8 eV (after Ref. [63]).

Figure 7.31 shows a typical excitation spectrum of the 1.19 eV luminescence in n-type GaN recorded at an excitation density below $20\,\text{mW}\,\text{cm}^{-2}$.

The excitation spectrum is dominated by a band around 1.7 eV. The absorption starts at 1.62 eV and shows a triplet structure. The shape of the 1.7 eV band is rather typical for an intracenter transition and is thus attributed to absorption ending in a higher excited multiplet. A further excitation band exists around 2.8 eV, which could either be a charge transfer or further intracenter transition. The excitation efficiency in the blue/green spectral region is at least a factor of 10 smaller than in the 1.7 eV band. When excited in the blue/green spectral region with an Ar laser, the ZPL of the 1.19 eV band is observed with a broad luminescence background originating from the Fe^{3+} transition (Fig. 7.14). 1.7 eV excitation leads to a background-free PL spectrum (Fig. 7.30).

7.4.2.1 *Zeeman experiments*

In order to identify the electronic structure of the luminescence center, Heitz *et al.* [63] have investigated the Zeeman behavior of the ZPL. Figure 7.32 shows the spectral region of the ZPL in magnetic fields oriented either parallel ($B \parallel c$) or perpendicular to the c-axis of the hexagonal GaN sample.

In the Zeeman experiments the spectral resolution was limited by the experimental setup to $250\,\mu\text{eV}$. For $B \perp c$ the ZPL splits into a symmetrical triplet. For $B \parallel c$ each component of the triplet additionally splits into a doublet. Measurements at higher temperatures show no intensity variation for the components of the triplet, which indicates that these components represent the splitting of the ground state of the luminescence. On the contrary, the high-energy component of the doublets resolved for $B \parallel c$ gains intensity with increasing temperature; hence this splitting is in an excited state.

7.4 Transition metals in GaN

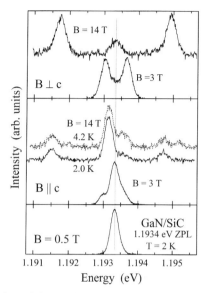

Fig. 7.32 Zeeman splitting of the 1.1934 eV ZPL with the magnetic field oriented either parallel or perpendicular to the axis of the hexagonal GaN. The spectral resolution is limited by the experimental setup (after Ref. [63]).

Figure 7.33 compiles the Zeeman components of the 1.1934 eV ZPL in dependence of the magnetic field strength. The open squares correspond to lines dominating at 2 K and open circles represent those gaining intensity at higher temperatures. Solid and dashed lines in Fig. 7.33 represent a fit of the experimental data using

$$\Delta E = g_{\text{gr}} \mu_B B m_J^{\text{gr}} + g_{\text{exc}} \mu_B B m_J^{\text{exc}} \quad (7.10)$$

with g_{gr} and g_{exc} the g values of the excited and the ground state, respectively, and μ_B Bohr's magneton. We assume a triplet for the ground state and a doublet for the excited state. The ground state g value is found to be isotropic with $g = 1.98 \pm 0.03$, whereas the g value of the excited state is anisotropic. The doublet splitting observed for $B \parallel c$ corresponds to a g value of 0.51 ± 0.05, whereas for $B \perp c$ no splitting could be resolved, giving an upper limit of 0.15. In principle, the trigonal crystal field in hexagonal GaN should result at least in a zero-field splitting of the triplet ground state. We are not able to resolve the zero-field splitting, hence yielding an upper limit of 40 μeV. It has to be noted that the Zeeman splitting of the 1.1934 eV ZPL is totally symmetric with respect to the zero-field position. In general, term interactions with nearby fine structure components lead to a superimposed nonlinear shift towards lower energies for the fine-structure transition lowest in energy. Thus the Zeeman experiments show, in agreement with the temperature-dependent data, the absence of such states for the 1.19 eV emission.

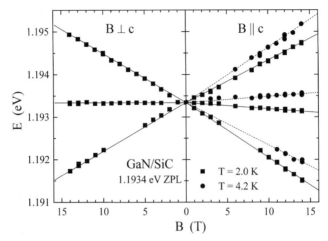

Fig. 7.33 Observed Zeeman splittings (open squares and circles) of the 1.1934 eV ZPL with the magnetic field oriented either parallel or perpendicular to the axis of the hexagonal GaN. Full and dashed lines represent a fit with eqn (7.10) (after Ref. [63]).

The electronic and chemical origin of the 1.19 eV emission, which is obviously connected with an omnipresent contaminant, is not clear. Recently, it has been attributed to the symmetry-forbidden $^3T_2(F)$–$^3A_2(F)$ transition of Cr^{4+}. The major argument was associated with the observation of a hot line 3.1 meV above the ZPL, which is claimed but not shown in Ref. 89. Such a splitting would be rather typical for a $^3T_2(F)$ state of an electronic d^2 configuration. In our samples, however, we do not observe this hot line. On the contrary, our results of both Zeeman and temperature-dependent measurements of the 1.1934 eV ZPL show the absence of such a state, which makes the assignment of the 1.19 eV emission to the $^3T_2(F)$–$^3A_2(F)$ transition of Cr^{4+} doubtful.

The results allow us some general remarks about the luminescence center. First of all, it has to be an isolated point defect. Although we were not able to resolve polarization of the optical transitions expected in a hexagonal host, the anisotropic Zeeman splitting of the excited state with respect to the c-axis of the hexagonal GaN evidences a point defect. The shape of the phonon sideband and the temperature behavior of the ZPL of the 1.19 eV luminescence are rather typical for rare earth defects. However, unintentional contamination with rare earth elements is unlikely and those rare earth centers having nearly matching transition energies should show fine structure in the ground state, which is not observed. Additionally, it would be difficult to explain the intracenter excitation bands showing a strong phonon coupling. The 4d and 5d transition metals are also unlikely contaminants and, additionally, the large mass difference relative to the replaced Ga atom would result in local modes dominating the phonon sideband in luminescence, which is not the case.

The Zeeman results yield detailed information about the electronic origin of the 1.19 eV luminescence band. The ground state is obviously a spin triplet

($S = 1$) with an isotropic g value of 1.98 and a trigonal zero-field splitting of less than 40 μeV. Only the electronic d^2 configuration offers an adequate ground state multiplet with the $^3A_2(F)$ state. The identification of the excited state is much more difficult. Depending on the cubic crystal-field strength, either the triplet $^3T_2(F)$ or the singlet $^1E(D)$ will be the first excited state, and, thus, should be observed in luminescence. For 3d defects in semiconductors with a d^2 configuration, the $^3T_2(F)$ state has been identified as the first excited state [90–93]. It is characterized by a doublet structure with a splitting of a few meV. A strong dynamical Jahn–Teller coupling to ε-type modes quenches the first-order spin–orbit coupling, leading to a doublet splitting by second-order interactions. For V^{3+} in the cubic semiconductors InP, GaAs, and GaP, the splitting amounts to 1.53, 1.27, and 1.93 meV, respectively [91–94]. This doublet structure is missing in the case of the excited state of the 1.19 eV emission in GaN. The hexagonal crystal structure of GaN will alter the fine structure of the $^3T_2(F)$ multiplet due to the additional trigonal crystal field and a stabilization of the center against Jahn–Teller distortions [61]. A comparison with ZnO shows, however, that only additional fine structure can be expected [95]. The $^3T_2(F)$ multiplet shows a characteristic Zeeman behavior in cubic III–V semiconductors [90–94]. The lower fine-structure state shows an anisotropic doublet splitting which quenches for $B \parallel [111]$. This configuration corresponds to $B \perp c$ in hexagonal GaN. In principle, this is consistent with our experimental results (Fig. 7.33), but term interaction with the excited state of the hot line causes a clearly resolvable nonlinear low-energy shift of the components of the low-temperature line, which is not observed for the 1.1934 eV ZPL in GaN. Additionally, the c_{3v} crystal field is found to strongly alter the Zeeman behavior. In ZnO an almost isotropic g value of 1.78 is observed [95] for the low-temperature line, which differs drastically from our results. From the above discussion, it must be concluded that in GaN the singlet $^1E(D)$ state is shifted below the triplet $^3T_2(F)$ state, as indicated in Fig. 7.34.

The $^1E(D)$–$^3A_2(F)$ transition explains most of the experimental fine-structure data of the 1.19 eV emission in GaN. Considering spin–orbit interaction, the ZPL corresponds to the dipole-allowed Γ_3–Γ_5 transition. The trigonal crystal field splits the threefold-degenerate Γ_5 ground state into a Γ_1 singlet and a Γ_3 doublet, leading to two polarized ZPLs. The 1.1934 eV ZPL is too broad to allow us to resolve the doublet splitting and shows no clear polarization behavior. For both the $^1E(D)$ and the $^3A_2(F)$ multiplet, any Jahn–Teller interaction is expected to be weak. This explains the weak phonon sideband in luminescence, which shows mainly replicas due to modes of the hexagonal host crystal. The lack of any fine-structure splitting in both multiplets accounts for the totally symmetric Zeeman behavior and the weak temperature dependence. It is interesting to note that the temperature behavior of the 1.1934 eV ZPL is quite similar to that of the $^2E(D)$–$^4A_2(F)$ transition, e.g. V^{2+} in MgO [96], described by 49.6 and 46.8 meV for α and β, respectively. In a Tanabe–Sugano diagram [97], the $^1E(D)$ multiplet shifts almost parallel to the $^3A_2(F)$ ground state. Thus, a variation of the cubic crystal field due to internal strain in the samples has

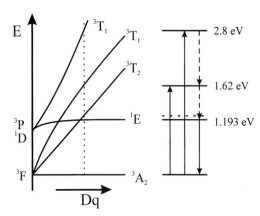

Fig. 7.34 Schematic Tanabe–Sugano diagram with the dotted line giving the proposed crystal-field strength for Ti^{2+} in GaN. The observed optical transitions are indicated by arrows in the term scheme on the right-hand side (after Ref. [63]).

little influence on the transition energy. Indeed, the FWHM of the 1.1934 eV ZPL is much smaller than that of the ZPLs of the 1.3 and 1.047 eV emissions in the investigated samples. The luminescence decay time of 65 μs is, however, unexpectedly short for a spin- and symmetry-forbidden transition like the $^1E(D)$–$^3A_2(F)$ transition [98, 61]. Competing nonradiative relaxation processes, which generally cannot be neglected for internal transitions of 3d elements, shorten the lifetime of the exciting state but simultaneously decrease the quantum efficiency of the luminescence transition [84]. However, for the 1.19 eV emission the electron–phonon coupling is weak in both the excited and ground states and the luminescence is comparatively intensive, making competing nonradiative processes unlikely. The short lifetime is rather the result of strong mixing with the triplet $^3T_2(F)$ multiplet situated at slightly higher energies by spin–orbit interaction. Obviously, the 1.19 eV center is close to the crossing point of the $^3T_2(F)$ and $^1E(D)$ multiplet in the Tanabe–Sugano diagram (Fig. 7.34). The 1.7 eV absorption band observed in excitation spectra of the 1.19 eV emission (Fig. 7.31) is characteristic for the d^2 configuration. Absorption spectra of V-doped ($3d^2$) III–V semiconductors always show more or less the same triplet structure as observed here in GaN [99]. It is attributed to the spin- and symmetry-allowed transition from the $^3A_2(F)$ ground state to the first excited $^3T_1$ state, as indicated in Fig. 7.33. The structure in the absorption band is the result of a dynamical Jahn–Teller coupling in the $^3T_1$ state [100]. The weak excitation band centered at 2.8 eV might be attributed to the transition into the second $^3T_1$ state. The proposed energy scheme of the 1.19 eV luminescence center is given on the right-hand side of Fig. 7.34.

A straightforward chemical identification of the luminescence center of the 1.19 eV band is not possible yet. Among the 3d transition metals, only three candidates offer an electronic d^2 configuration in III–V semiconductors: Ti^{2+}, V^{3+}, and Cr^{4+}. A chemical analysis of our samples shows that all three are unintentional dopants. However, the observation of efficient excitation of the $^1E(D)-^3A_2(F)$ luminescence via the intracenter $^3A_2(F)-^3T_1$ absorption bands in n-type samples (Fig. 7.31) demonstrates the presence of the luminescence center in its active charge state in n-type material. Otherwise, an efficient charge-transfer process of the luminescence center has to take place prior to excitation, leading to a two-step excitation mechanism. Such a process is rather unlikely at excitation densities below $20\,\mathrm{mW\,cm^{-2}}$ used in our excitation measurements (Fig. 7.31). Thus, we propose isolated Ti^{2+} on Ga sites as the luminescence center, since Ti is the only 3d transition-metal center expected to be stable in the d^2 configuration in n-type GaN. This assignment to Ti^{2+} was confirmed by doping experiments by Pressel *et al.* [64] where different transition metals (Ti, V, Cr) were introduced during crystal growth. Only the Ti-doped GaN sample showed the 1.19 eV emission.

However, it is claimed by Baur *et al.* [56] that the intensity of the 1.19 eV luminescence increases with increasing Mg doping, i.e. with a lower-lying Fermi level. These arguments make the assignment to V^{3+} or Cr^{4+} also possible.

A controversial discussion of the chemical nature of the defect center was given by Fermi-level pinning arguments in the framework of the internal reference rule of Langer and Heinrich [4, 5], comparing the charge-transfer levels of the transition metals in GaN, AlN, and GaAs. These charge-transfer energies have implications for the band offsets between the different host compounds. A detailed discussion will be given at the end of this chapter after the complete presentation of the various transition-metal luminescence bands in GaN and AlN.

7.4.3 The 1.047 eV luminescence

In contrast to the 1.30 eV and 1.19 eV luminescence, the 1.047 eV luminescence has not been observed as a natural contaminant in GaN samples grown by metalorganic vapor phase epitaxy or vapor phase epitaxy on Al_2O_3 substrates. It is only observed in samples grown by the sublimation sandwich technique on 6H−SiC [64, 66, 101, 102]. Figure 7.35 shows the PL spectrum of a GaN sample, which was grown by the sublimation sandwich technique on a 6H−SiC subtrate.

The emissions labeled I and II in Fig. 7.35 do not originate from this defect because in contrast to the 1.047 eV emission they were not observed in spectra excited with the 647 nm red light of a Kr ion laser. The inset of Fig. 7.35 shows three detailed PL spectra in the range of the ZPL at 1.047 eV (labeled A) detected at different temperature. With rising temperature a hot line (labeled B) shows up 8 meV higher in energy, which indicates a splitting of the excited state. The corresponding level scheme is depicted in the inset of Fig. 7.35.

The inset of Fig. 7.36 shows a PLE spectrum of the 1.047 emission. It is clearly seen that this emission can be excited with light not only in the near band-edge region at energies higher than 3.35 eV but also in the optical range between 1.5 and 1.75 eV. All these features are superimposed on a broad background signal.

286 *Defect spectroscopy in the nitrides*

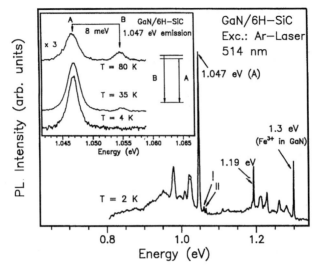

Fig. 7.35 Photoluminescence spectra of GaN/6H–SiC in the region between 0.8 and 1.35 eV. Three ZPLs peaking at 1.3 eV, 1.19 eV, and 1.047 eV (A) appear by Ar ion laser (514 nm) excitation. The inset shows three spectra detected at 4 K, 35 K, and 80 K. With rising temperature additional to the ZPL at 1.047 eV, labeled A, the hot line, labeled B, appears. The corresponding level scheme is shown in the right part (after Ref. [101]).

The spectrum of Fig. 7.36 shows the very strong enhancement of the PL intensity in the range between 1.45 eV and 1.8 eV in detail. A set of at least six peaks can be observed. The peaks labeled C (1.518 eV) and D (1.618 eV) belong to the ZPLs. The other four peaks can be explained by coupling to phonon modes [69]. A phonon with energy 71 meV (due to E_2(high)) couples to the absorptions C and D. The two peaks at 1.662 and 1.681 eV can be attributed to a two-phonon coupling to the absorption C: 71 meV (E_2(high)) + 73 meV (E_2(high)) and 71 meV (E_2(high)) + 92 meV (A_1(LO), E_1(LO)). The peaks are relatively sharp and therefore a charge-transfer transition can be most definitely excluded. The PLE measurements reveal a higher excited state of the luminescence center by which the 1.047 eV emission can be excited. The two ZPLs C and D belong to the fine-structure splitting of this excited state. The deduced level scheme of the defect with the two transitions A and B, observed in PL, and the two absorptions C and D, observed in PLE, is depicted in the left part of Fig. 7.36. We assume that the higher excited state observed in PLE is still in the bandgap because otherwise the fine-structure lines should be broader. From the PLE spectra it can be concluded that the defect is already in its luminescent charge state in the dark (without laser excitation) in the n-type material.

As we observe only one ZPL at 2 K the 1.047 eV emission can belong to an internal transition of either a d^1 (ground state ^2E), a d^2 (ground state $^3A_2$), a d^5 (ground state $^6A_1$), or a d^7 (ground state $^4A_2$) configuration, which show no splitting of the ground state in the tetrahedral environment of the four nitrogen atoms. A possible weak splitting caused by the hexagonal crystal field (reduction to c_{3V}) is not observed. A reason is the relatively large line width of about

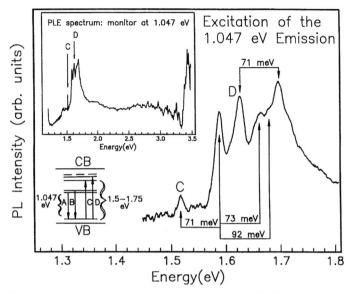

Fig. 7.36 Photoluminescence excitation spectrum of the 1.047 eV emission. The monitor energy was set to 1.047 eV. On a broad background we observe not only excitation in the near band-edge region, but also strong below-bandgap peaks in the range between 1.5 eV and 1.75 eV. We observed at least two ZPLs labeled C and D. GaN-related phonon modes couple to these transitions. The PLE data reveal that the luminescence center is already present in the dark. The suggested level scheme of this defect in the bandgap is indicated in the inset (after Ref. [101]).

1.5 meV at 2 K. We can exclude the d^1 configuration because this configuration has no additional excited state. The appearance of the excited state in the PL excitation spectrum (Fig. 7.35) cannot be explained with a d^1 configuration. Fe^{3+} ($3d^5$) can be excluded because the 1.3 eV emission is already identified as Fe^{3+}. Preliminary Zeeman measurements up to 7T show no shift of the maximum of the 1.047 eV ZPL. With increasing magnetic field the linewidth gets broader but the lineshape of the ZPL remains symmetric. Thus, we can also exclude the $^4T_1-^6A_1$ spin-flip transition of Mn^{2+} ($3d^5$). For Mn^{2+} we expect a fine-structure splitting and Zeeman behavior caused by a Jahn–Teller distortion similar to Fe^+, which is not observed. The decay time of less than a few μs supports the exclusion of the $^4T_1-^6A_1$ spin-flip transition of a d^5 configuration. The following 3d elements have a d^2 or d^7 electronic configuration that can fit to the experimenal results on the 1.047 eV emission: Sc^+ (d^2), Ti^{2+} (d^2), V^{3+} (d^2), Cr^{4+} (d^2), Fe^+ (d^7), Co^{2+} (d^7), and Ni^{3+} (d^7). Sc^+ (scandium is a very rare element) and Cr^{4+} (our samples are n-type samples and the element is in its luminescence charge state already in the dark) can be most definitely excluded. The two remaining possible contaminations with d^2 configuration are Ti^{2+} and V^{3+}. Baur et al. [89] identify a ZPL at 0.94 eV, which shows an excited state at 1.5 meV above the ground state, with V^{3+} (see next part of this chapter). For V^{3+} we expect a splitting of the excited state in the range of 1–2 meV instead of

8 meV as observed here. Also, the shape of the excitation band at about 1.6 eV is completely untypical for the $^3A_2(F)-^3T_1(F)$ transition of Ti^{2+} and V^{3+}. Our studies of the 1.19 eV emission indicate that the emission is (most) probably caused by the Ti^{2+} center [63]. For the reasons mentioned above a $3d^2$ state can be excluded. According to the experimental observations only a $3d^7$ electronic configuration can explain the 1.047 eV emission. A possible attribution candidate is nickel in its 3+ charge state. But from the PLE we know that the luminescence center is already present in the samples without illumination. We expect Ni^+ and Ni^{2+} to be the stable configurations in the dark in n-type GaN material. We cannot completely exclude the $^4T_2(F)-^4A_2(F)$ transition of Ni^{3+}, but it is improbable in n-type GaN. We know that cobalt is not a dominant natural contamination in crystal growth, but it is the best candidate that explains the present experimental results data. The 1.047 eV emission is thus attributed to the $^4T_2(F)-^4A_2(F)$ transition of Co^{2+}. The observed splitting of 8 meV for the $^4T_2(F)$ state caused by spin–orbit coupling is reasonable. The absorptions at about 1.6 eV observed in PLE spectroscopy can be attributed to the $^4A_2(F)-^4T_1(F)$ transition of Co^{2+}. However, the resulting crystal field parameter of Co^{2+} in GaN is twice as large as for other semiconductors, e.g. ZnO. In ZnO, the crystal field parameter of Ni^{3+} is of the order of 0.7 eV [103]. This indicates that the Co^{2+} attribution is still tentative.

In conclusion it can be summarized that the 1.047 eV luminescence is most probably related to a $3d^7$ configuration. The most promising 3d impurities are Co^{2+} and Ni^{3+}. The 2+ charge state of the 3d impurity is in better accordance with the Co^{2+} attribution, whereas the order of magnitude of the crystal field parameter is in better accordance with Ni^{3+}.

7.4.4 The 0.93 eV luminescence

Baur et al. studied the 0.931 eV emission in detail [89]. A PL structure consisting of a ZPL at 0.931 eV and phonon replica close in energy to intrinsic lattice phonons is shown in Fig. 7.37.

The lifetime of the ZPL is less than 0.4 ms. The inset shows the ZPL region at an elevated temperature with an additional hot ZPL separated by $13 \,cm^{-1}$ (1.6 meV) from the cold ZPL. This PL structure is ascribed to neutral vanadium impurities (V^{3+}) and is typical only for HVPE samples. So far no PLE data are reported. However, Baur et al. [89] observed the 0.94 eV ZPL best while exciting the samples with 413 nm excitation of a Kr^+ laser. Thick samples showing the V^{3+} correlated PL exhibit absorption bands at 2.54 and 2.74 eV. The assignment is mainly based on the doublet structure of the ZPL region and a comparison with well-defined V^{3+} luminescence centers in other III–V compounds. For GaAs, GaP, and InP, vanadium-correlated ZPLs between 0.73 and 0.83 eV have been assigned to the symmetry forbidden $^3T_2(F)-^3A_2(F)$ crystal field transition of cation-substitutional $V^{3+}(3d^2)$. At higher sample temperatures hot ZPLs separated by 1.2 meV, 1.9 meV, and 1.5 meV from the cold ZPL have been reported for these materials [94]. The analogy to the GaN PL data in Fig. 7.37 is obvious. The line position of the ZPL is mainly determined by the cubic crystal

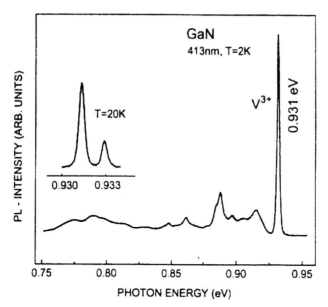

Fig. 7.37 PL spectrum observed from an undoped wurtzite GaN HVPE sample under 413 nm Kr⁺ laser excitation. The higher energy ZPL in the inset is a hot line (after Ref. [89]).

field strength (Dq), which increases with the ionicity of the host crystal. Therefore, the higher energy of the emission in GaN is not surprising. The hot ZPL can be explained by a moderately strong dynamic Jahn–Teller effect [94]: The first order spin–orbit interaction within the excited $^3T_2(F)$ state is quenched by electron–phonon coupling, which forces the four triply and a higher lying sixfold quasi-degenerate level separated by second-order spin–orbit coupling. Thermal occupation of the upper level at elevated temperatures yields an additional hot ZPL. The splitting between the ZPLs is mainly determined by the spin–orbit coupling constant λ, which is characteristic for the incorporated impurity atom. The splitting of 1.6 meV in the GaN PL is in excellent agreement with the splittings observed for V^{3+} impurities in other III–V compounds. It should be mentioned that a similar splitting has been claimed for the 1.047 eV ZPL in GaN grown on 6H–SiC [66, 102]. However, exact data are not published in [66] and [102]. Later on, Pressel et al. [101, 64] reported a splitting of 8 meV for the excited state of the 1.047 eV luminescence, which is too large for V^{3+}. Trigonal field splittings due to the axial field in wurtzite GaN are not observable in Fig. 7.37. This indicates that either the trigonal field is small or that such splittings are also quenched within the $^3T_2(F)$ state by Jahn–Teller interaction.

In order to get more information about the energy level scheme of V^{3+} ($3d^2$) in GaN, Baur et al. [89] numerically diagonalized the complete set of static crystal field energy matrices (Liehr et al. [104]). These matrices include Coulomb interaction (parameters B and C), the cubic crystal field (parameter Dq) and the spin–orbit coupling (parameter λ). These results using parameter values ($B = 810 \text{ cm}^{-1}$, $C = 3660 \text{ cm}^{-1}$, $\lambda = 89 \text{ cm}^{-1}$, $Dq = 754 \text{ cm}^{-1}$), close to those

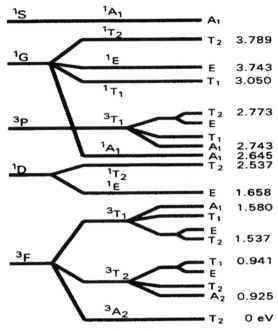

Fig. 7.38 Energy level diagram for V^{3+} ($3d^2$) on a Ga site in GaN. The energy values are evaluated by a numerical diagonalization of the static crystal field energy matrices using the parameters given in the text (after Ref. [89]).

known for the free V^{3+} ion ($B = 860\,\text{cm}^{-1}$, $C = 4128\,\text{cm}^{-1}$, $\lambda = 105\,\text{cm}^{-1}$), are shown in Fig. 7.38.

Comparing these parameters, one should keep in mind that, compared to the free ion λ is reduced in semiconductors due to covalency. Once the spin–orbit splitting of the $^3T_2(F)$ state in the static crystal field is calculated, the dynamic Jahn–Teller model for this state (Aszodi et al. [94]) can be applied. Choosing the above parameters, the line position of the 0.931 eV ZPL as well as the line splitting is correctly described, whereas λ values typical for other possible d^2 impurities produce splittings, which are either much too small (Ti^{2+}) or much too large (Cr^{4+}, Nb^{3+}, Mo^{4+}). The coincidence of the 2.54 and 2.74 eV absorption bands with the calculated energies of the $^1T_2(D)$ and the $^3T_1(P)$ states of V^{3+} suggests that the absorptions are related to these states.

7.4.5 The 0.82 eV luminescence

Recently, a further luminescence structure in GaN was observed by using ion implantation by Kaufmann et al. [105]. After ion implantation and annealing the GaN layers exhibit a new series of intense transitions at 820 meV. The spectra in Fig. 7.39 were found using an implantation dose of $10^{13}\,\text{cm}^{-2}$ producing a doping of about $10^{18}\,\text{cm}^{-3}$ vanadium within a depth of 100 nm. The upper trace of Fig. 7.39 was taken at $T = 5\,\text{K}$. The lower two spectra (Fig. 7.39(b) and (c)) show that the PL is stable up to 30 K. Even, at room temperature the main line

7.4 Transition metals in GaN 291

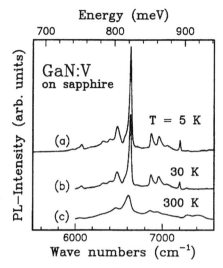

Fig. 7.39 Photoluminescence spectra recorded at several temperatures for a GaN film on sapphire. The defect luminescence is generated by vanadium implantation (10^{13} cm^{-2}) followed by a heat treatment under growth conditions (after Ref. [105]).

is reduced only by less than a factor of 10 compared to 5 K. All lines in Fig. 7.39 are broad, showing a FWHM of about 4 meV. There is no thermalization up to 30 K between the several lines. The transition at 0.82 eV depends on the implantation dose. Heat treatment is essential to activate the spectrum. Furthermore, the spectrum cannot be generated by ion implantation of another transition-metal, like chromium or iron, followed by annealing, and it cannot be activated by heat treatment alone. V^{4+} in SiC placed on the hexagonal silicon site shows a series of transitions (α lines) in the same spectral region (950 meV) [106]. The 820 meV transition observed, here, could be a candidate of the corresponding transition in wurtzite-type GaN. However, as mentioned above, the 4+ charge state seems to be unlikely in n-type GaN, but with the laser light excitation, charge transfer of the impurity is possible. But in SiC the linewidth is much smaller, and thermalization indicates that the series of lines there originate from the electronic structure of one defect. As this is not the case here, it is likely that we are dealing with several arrangements of the vanadium ion with radiation damage. It is also useful to compare these results with the properties of vanadium in ZnSe. Here, V^{3+}, V^{2+}, and V^+ can be observed in the same crystal, depending on the Fermi level position and on the excitation energy [107]. Thereby, the energy of the corresponding transition decreases from V^{3+} to V^{2+} and V^+. In this context it is also probable that the 0.82 eV PL in GaN is related to V^{2+} or V^+. Also V^{2+} with nearby radiation-induced defects seems to be a probable interpretation.

Undoped GaN crystals without any further heat treatment show the narrow bandgap PL transitions mentioned above. None of these transitions above 0.9 eV could be activated by ion implantation of chromium and iron followed by annealing. In particular, any correlation between the 0.931 eV luminescence and V doping is not observed.

In conlusion, it was shown by Kaufmann *et al.* [105] that the implantation of vanadium followed by heat treatment gives rise to an intense PL spectrum at 0.82 eV. The center occurs in several modifications, presumably a vanadium ion with a lattice vacancy. The microscopic structure of the defect is not yet clear; magnetic resonance techniques and Zeeman investigations are needed to clarify this.

7.5 Transition metals in AlN

Deep defects like transition metals are believed to be omnipresent defects in the nitride compounds. However, no quantitative information for the TM content in AlN and InN is available yet. As mentioned above, the nearest neighbors of the transition metal are nitrogen in AlN and GaN and due to the fact that the wavefunctions of the TM are strongly localized, it is expected that the energy of identical TM recombinations occurs in the same spectral range for the nitride compounds. Up to now, only four publications concerning the PL of defects in AlN exist [58, 59, 89, 108].

Baur *et al.* [58, 59, 89] reported two emission centers in AlN which have ZPLs at 1.201 eV and 1.297 eV. On the basis of a comparison of their results with two emissions at 1.19 eV and 1.3 eV in GaN they attributed these two emissions to Cr^{4+} and Fe^{3+}. The spectral shape and the excitation behavior of each PL is quite similar compared with GaN and AlN.

7.5.1 The 1.297 eV luminescence

As for the 1.3 eV in GaN, a decay time in the ms range of 4.5 ms is detected at 77 K [58] confirming the attribution of the luminescence to a spin forbidden $^4T_2$–$^6A_1$ transition. The AlN crystals investigated by Baur *et al.* [58, 89] were polycrystalline ceramics (SHAPAL, Tokuyama Soda Co. Ltd) (see Fig. 7.40).

For the 1.297 eV luminescence strong excitation is observed, exciting the AlN above 3.0 eV [58, 59]. However, nothing is known about the charge state of iron in the AlN crystals, but it seems to be reasonable to attribute the broad PLE structure to the $Fe^{3+/2+}$ charge-transfer process because of the similarity to GaN. The fact that the Fe-acceptor ionization energies are comparable is remarkable in a sense, since the bandgap energies (GaN: $E_g = 3.5$ eV; AlN: $E_g = 6.3$ eV) differ strongly. However, because of the similar nitrogen neighborhood of the iron center incorporated on the metal side in both compounds a comparable charge-transfer energy of the (FeN_4) complexes is expected. Nevertheless, it was shown that the PLE spectra of Fe^{3+} in GaN depend on the charge state of iron (see Section 7.4.1). Thus, it is necessary to investigate AlN crystals with different charge states of iron in order to determine the value of the $Fe^{3+/2+}$ charge-transfer energy precisely. These charge-transfer energies of the iron center allow a prediction of the band offsets between GaN and AlN according to the Langer and Heinrich rule [4, 5]. This will be done at the end of this chapter.

7.5 Transition metals in AlN

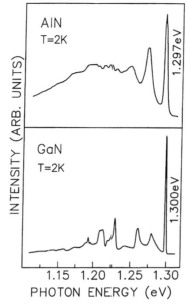

Fig. 7.40 Photoluminescence spectra of GaN and AlN at $T = 2\,\text{K}$ (after Ref. [58]).

7.5.2 The 1.20 eV luminescence

For a better comparison, some details of the 1.19 eV ZPL in GaN are repeated here to elucidate similarities between the 1.19 eV ZPL in GaN and the 1.201 eV ZPL in AlN and to discuss the chemical origin in the context of the AlN results. The upper part of Fig. 7.41 shows the ZPL region of the previously reported 1.193 eV ZPL in GaN under 676 nm Kr$^+$ ion laser excitation observed by Baur et al. [89]. The inset in Fig. 7.41 shows the ZPL region of the 1.193 eV for different samples under above-bandgap excitation. Additional lines are seen to occur, in particular a ZPL at 1.196 eV which formerly was assigned as a hot line of the 1.193 eV ZPL [62]. However, more thorough investigations of further samples revealed that the 1.196 eV ZPL does not show a typical hot line behavior [63, 89]. Possibly the additional lines in the inset of Fig. 7.41 originate from impurity pairs. The fact that the lines are broadened in the Cr-doped sample suggest that chromium atoms are involved. However, as mentioned above, Ti-doped GaN crystals give rise to the observation of the 1.19 eV ZPL too [64]. This indicates that the results obtained from codoped crystals have to be interpreted carefully.

The AlN ceramics in the lower trace of Fig. 7.41 exhibit a PL structure with a ZPL at 1.201 eV, which strongly resembles the 1.193 eV band in GaN in line position, phonon coupling, and excitation behavior. EPR measurements of the same sample reveal a sharp isotropic signal at $g = 1.9970$ and a clearly resolved hyperfine structure with a hyperfine coupling constant of $A = 1.907\,\text{mT}$ (Fig. 7.42). The number of hyperfine lines and their fractional intensity (10%) are

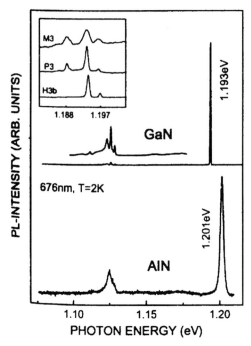

Fig. 7.41 Photoluminescence spectra from an undoped GaN sample (upper traces) and from polycrystalline AlN ceramics (lower trace). Excitation was with the 675 nm line of a krypton laser. The inset shows details of the zero phonon line for various samples (after Refs. [58, 59]).

consistent only with the isotope ^{53}Cr ($I = 3/2$, 9.5%). It can be seen in the upper trace in Fig. 7.42 that the EPR signal nearly vanishes after fast neutron irradiation. This suppression of the EPR signal is accompanied by a tremendous increase in the 1.201 eV PL intensity. Because of its similarity with the 1.193 eV PL in GaN, the 1.201 eV PL in AlN is also assigned to the $^1E(D)-^3A_2(F)$ crystal field transition. Bauer et al. [89] suggested that the most likely explanation for the EPR signal in Fig. 7.42 is a chromium impurity with a d^1 electron configuration, i.e. Cr^{5+}. Its $^2E(D)$ ground state can produce an isotropic spectrum provided that the ^2E trigonal field splitting due to the wurtzite axial field in AlN is small and comparable to the splittings induced by random strains [109]. Neutron irradiation is known to create nitrogen vacancies in AlN [110], thus leading to a change of the Fermi level. This could explain the decrease in the Cr^{5+} EPR together with the increase in the 1.201 eV PL if the equilibrium state of the chromium impurity changes from Cr^{5+} (3d^1) to Cr^{4+} (3d^2) after neutron irradiation. Baur suggests that these observations favor an assignment of the 1.201 eV ZPL in AlN and of the corresponding 1.193 eV ZPL in GaN to Cr^{4+}. However, Baur et al. [89] conclude that a correlation of these bands with Ti impurities cannot be definitely ruled out.

Now we check whether the results of Baur could also be explained by Ti. It is known that Ti^{2+} (3d^2) and Ti^{3+} (3d^1) are also possible charge states of titanium.

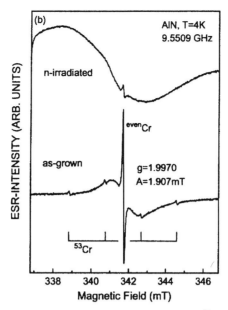

Fig. 7.42 Electron paramagnetic resonance of chromium Cr^{5+} observed from polycrystalline AlN ceramics. The lower trace shows the hyperfine splitting with the ^{53}Cr isotope having nuclear spin $I = 3/2$. The chromium-related signal is reduced in intensity after neutron irradiation (upper trace) (after Ref. [89]).

In particular, the increase of nitrogen vacancies in AlN should cause a higher-lying Fermi level, because these nitrogen vacancies are known to be donors in GaN [11]. If Ti is also present as a residual impurity in the AlN crystal investigated by Baur et al. [89], the decrease in the EPR intensity together with the increase of the 1.201 eV PL by neutron irradiation can be explained by the simultaneous recharging of chromium and titanium. The higher-lying Fermi level leads both to a simultaneous Cr^{5+}/Cr^{4+} and Ti^{3+}/Ti^{2+} charge-transfer process. This makes it possible to explain on the one hand the decrease of the Cr^{5+} EPR signal and on the other the increase of the Ti^{2+} PL. As mentioned above, it seems much more likely to have impurities in the 2+ charge state than in the 4+ charge state in n-type crystals. If Cr^{4+} and Ti^{2+} where present in the AlN crystal after neutron irradiation this might indicate that the electrons were mainly localized around the nitrogen ligands due to the large electronegativity of nitrogen. Otherwise the 4+ charge state of Cr^{4+} could not be realized in the n-type crystal, in which the 2+ charge state is most probable. It should be pointed out that Baur et al. reported also an increasing of the 1.19 eV ZPL in GaN with increasing Mg-concentration [56], i.e. a lower-lying Fermi level, which contradicts the arguments given for AlN given by the same authors.

In conclusion, it is shown that the 1.193 eV ZPL in GaN and the 1.201 eV ZPL exhibit similar PL spectra and PLE excitation behavior. Thus, it seems probable to assign the 1.201 eV emission in AlN to the $^1E-^3A_2$ transition of a d^2 configuration. The doping experiments in GaN give evidence for both Ti^{2+} on

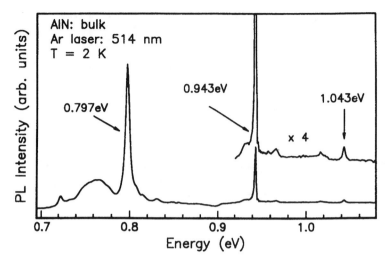

Fig. 7.43 Luminescence spectra of crystalline AlN needles excited with the 514 nm line of an argon ion laser (after Ref. [108]).

the one hand [64] and for Cr^{4+} on the other hand [56]. The experimental results shown for the 1.201 eV emission in AlN could be explained by Ti^{2+} and Cr^{4+} as well. No clear distinction is possible; further investigations are required to clarify the chemical origin of the TM impurity center in GaN and AlN, respectively.

7.5.3 The 1.043 eV luminescence

Pressel et al. [108] reported PL spectra of AlN needles in the near infrared spectral region. Figure 7.43 is a typical luminescence spectrum of these needles. One weak ZPL at 1.043 eV and two intense ZPL peaking at 0.943 eV and 0.797 eV are observed by excitation with the 514 nm line of an Ar^+ ion laser. Only, the 0.943 eV luminescence center could be detected under UV excitation.

The FWHM of the 1.043 eV emission is about 2 meV. The defect center shows coupling to phonon modes. The peaks in the phonon sideband at 86.3 meV and 112.8 meV almost coincide with the modes observed in Raman spectroscopy on these needles ($E_1(TO) = 83.1$ meV, $E_1(LO) = 113.2$ meV, $A_1(LO) = 110.7$ meV [111]). The mode at 77.9 meV is a little lower in energy than the corresponding Raman mode. Thus, these modes are probably caused by local modes of the defect. The 1.043 eV luminescence is still visible with about the same intensity at 80 K. One only observes a small shift of about 0.2 meV to lower energies, which is typical for internal electronic transitions of 3d elements [128]. On the basis of the arguments given in the discussion of the 1.047 eV luminescence in GaN, Pressel et al. [108] attribute the emission to the $^4T_2$–$^4A_2$ transition of a 3d element with a $3d^7$ electronic configuration. Co^{2+} and Ni^{3+} are discussed as luminescence centers.

7.5.4 The 0.943 eV luminescence

The emission at 0.943 eV is close to an emission at 0.931 eV in GaN reported by

7.5 *Transition metals in AlN*   297

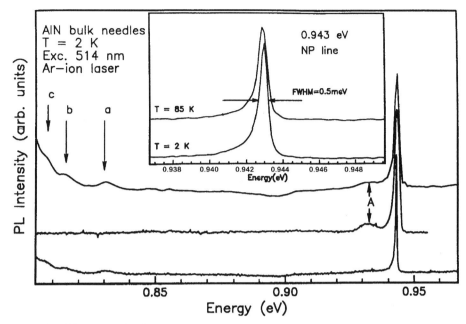

Fig. 7.44 PL spectra in the energy range of the 0.943 eV ZPL. In the inset two spectra detected at 2 K and 85 K are depicted. The FWHM of the peak is 0.6 meV (after Ref. [108]).

Baur *et al.* [89]. They observed a hot line 1.6 meV higher in energy and attribute the 0.931 eV emission to an internal electronic transition of V^{3+}. Figure 7.44 shows the 0.932 eV emission for three different samples. The inset compares two spectra recorded at different sample temperatures.

With rising temperature no hot line is observed in AlN for the 0.943 eV emission. The energy shift towards lower energy between the 2 K and the 85 K spectrum is about 0.2 meV. In some samples the emission showed a shoulder, labeled A, on the low-energy side. Possibly this is caused by internal strain or rather a disturbance in the neighborhood of the luminescence center. The FWHM is 0.5 to 0.8 meV depending on the sample. For rising temperatures up to 85 K the linewidth only slightly increases from 0.5 meV to 0.7 meV. Phonon sidebands could practically not be resolved for this emission. By chance the weak emission at 0.832 eV, labeled a, is 112.8 meV from the 0.943 eV NP line in this sample. But the emission at 0.832 eV belongs most definitely to the fine structure of the 0.797 eV luminescence center and not to the coupling of a phonon mode of AlN.

Preliminary Zeeman measurements performed on this ZPL reveal a symmetric threefold splitting in the configuration $H \perp c$ with a g value of 1.96 ± 0.07 [112]. The splitting does not depend on the rotating angle between the magnetic field H and the crystal axis. This indicates an isotropic splitting of the ground and excited states respectively. These results are in accordance with the expected behavior of V^{3+} ($3d^2$) in hexagonal hosts.

298 *Defect spectroscopy in the nitrides*

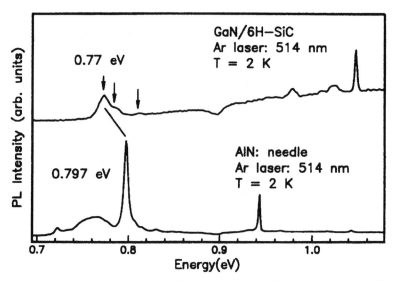

Fig. 7.45 PL spectrum of the 0.797 eV emission. In addition to the main ZPL one observes further weak satellite peaks indicated by the rake. The inset compares two spectra of the 0.797 eV emission detected at 2 K and 85 K (after Ref. [108]).

7.5.5 The 0.797 eV luminescence

In general the strongest luminescence in the AlN needles is the 0.797 eV emission. Figure 7.45 shows a detailed spectrum. Besides the strong ZPL at 0.797 eV one resolves four weak satellites at 0.832 eV (a), 0.815 eV (b), 0.811 eV (c), and 0.777 eV (e). The whole set of lines is indicated by the rake with five arrows.

The FWHM of the main peak at 0.797 eV is 4 meV in our samples. As shown in the inset of Fig. 7.45, no hot lines could be observed up to 85 K. Similar to the 0.943 eV emission, the strong 0.797 eV emission also shows nearly no increase in linewidth and only a small shift of about 0.2 meV to lower energy. On the low-energy side of the 0.797 eV emission a broad shoulder appears. The maximum of this shoulder is close to the E_2(low) (30.8 meV) phonon mode, which is observed in Raman measurements [111]. 76 meV lower in energy a further set of peaks appears. The spacing of the two main peaks at 0.722 eV (d') and 0.782 eV (c') agrees excellently with the A_1 (TO) (75.6 meV) mode, which can be observed in Raman measurements [111].

In general the spectrum of the 0.797 eV luminescence center looks similar to the well-studied 0.84 eV emission of Cr in GaAs [113]. This luminescence center is attributed to an internal electronic transition of Cr^{2+} on a Ga substitutional lattice site. The tetrahedral environment of the Cr center is disturbed by a defect atom on one of the four As nearest neighbor sites.

The 0.797 eV center in AlN may be caused by a similar defect. A $^5T_2$ ground state is a good possibility; thus possible candidates are Mn^{3+}, Cr^{2+}, and V^+. Comparing these interpretations with the results obtained by ion implantation of vanadium in GaN causing the 0.82 eV ZPL, a vanadium-related defect seems to be the most promising candidate. As mentioned above, lower ZPL energies are expected for V^{2+} and V^+ in comparison with V^{3+}. This is also supported by the presence of the 0.94 eV emission in the same samples, tentatively attributed to V^{3+}.

7.5.6 Summary of the TM emissions in GaN and AlN

As already pointed out, it seems possible to transfer the identifications of internal electronic transitions in GaN to AlN. In Table 7.2 transition metal emissions in AlN and GaN are summarized.

Table 7.2 Comparison of TM emissions in GaN and AlN. The possible identifications are also given. The detailed discussion is given in the corresponding sections of the text.

GaN Energy (eV)	Possible identification	AlN Energy (eV)	Possible identification
1.299	$3d^5 Fe^{3+}(^4T_1-^6A_1)$	1.297	$3d^5 Fe^{3+}(^4T_1-^6A_1)$
1.193	$3d^2(^1E-^3A_2)Ti^{2+}$, Cr^{4+}	1.201	$3d^2(^1E-^3A_2)Ti^{2+}$, Cr^{4+}
1.047	$3d^7(^4T_2-^4A_2)Co^{2+}$, Ni^{3+}	1.043	$3d^7(^4T_2-^4A_2)Co^{2+}$, Ni^{3+}
0.931	$3d^2(^3T_2-^3A_2)V^{3+}$	0.943	$3d^2 V^{3+}$
0.82	TM complex with V	0.797	TM complex with V

7.5.7 Band offsets between GaN and AlN

Energy levels of transition metal impurities in different isovalent semiconductors are almost constant, if referred to the vacuum energy [114, 115]. This can be

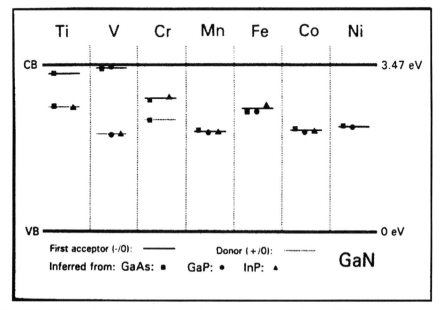

Fig. 7.46 Energy levels of 3d transition metal ions in GaN in different charge states inferred from GaAs, GaP, and InP (after Ref. [89]).

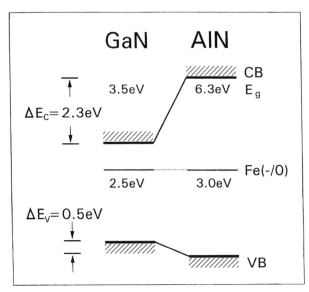

Fig. 7.47 Band offsets in the conduction and valence bands of the GaN/AlN system, as predicted from the common $(-/0)$ acceptor reference level of iron (after Refs. [58, 59]).

used to estimate band discontinuities if the energy levels of at least one transition metal are known in two isovalent semiconductor compounds [4, 5, 116], or, vice versa, to estimate energy levels from known band discontinuities. Baur et al. [89] applied this procedure to derive the levels of 3d transition metals in GaN from known levels in InP, GaP, and GaAs [117], assuming the band discontinuities given in [59]. They started their procedure by comparing especially the Fe energy levels in these III–V compounds (see Fig. 7.46).

One predicts the first acceptor level and the first donor level of vanadium in GaN to be at 2.0 and 3.4 eV above the valence band maximum, respectively. However, other values for the position of $Fe^{3+/2+}$ acceptor level than those reported (2.5 eV above the valence band) have been recently determined by PLE of Fe^{3+} using GaN in n-type and semi-insulating GaN crystals by Heitz et al. [68] (see Section 7.4.1). In contrast to the 2.5 eV (see Fig. 7.47) they determined a value of 3.17 eV for the energy distance between the $Fe^{3+/2+}$ charge-transfer level and the valence band. With this value and the assumption that the relative energy positions of the 3d elements in Fig. 7.26, taken from GaAs, are also valid in GaN, the acceptor levels of Ti, V, and Cr were degenerate with the conduction band. Thus, it seems unlikely that ZPLs with an FWHM of 0.2 meV of Ti^{2+} or Cr^{4+} (attributed to the 1.19 eV PL) or the V^{3+} ZPL (attributed to the 0.94 eV ZPL) will be observed.

This means that either the energy level position of the $Fe^{3+/2+}$ level is not correctly determined or the relative positions of the impurity levels known for GaAs, GaP, and InP could not be transferred directly to the more ionic semiconductor GaN. In order to check this argument we compare the validity of the internal reference rule [4, 5] for other materials with different bonding.

Even though the internal reference rule has been validated for many semiconductor/TM combinations it is not clear if it can be used unambiguously to predict band offsets. Following the internal reference rule [4, 5], the position of the $Fe^{3+/2+}$ acceptor level yields a type-II band alignment for the GaAs/GaN heterojunction with valence and conduction band offsets of 2.63 eV and 0.65 eV, respectively. These offsets are in disagreement with previous results of electrical measurements. These favored a type-I configuration with significant offsets in both the valence and conduction bands [118, 119]. Obviously, the internal reference rule does not hold for GaN in conjunction with GaAs, GaP, InP. The reason may be the large difference between the electronegativity of nitrogen (3.00) on one side and that of phosphorus (1.64) as well as arsenic (1.57) on the other [120]. This argument can be supported considering ZnO which is comparable to GaN with respect to electronic and lattice properties. Like nitrogen among the Group V elements, oxygen has a much higher electronegativity than the other Group VI elements. With these similarities in mind it is interesting to note that for ZnO the internal reference rule does not hold with respect to the other II–VI semiconductors. Both the $Cu^{2+/+}$ [121] and the $Fe^{3+/2+}$ [74] levels are found directly below the conduction band minimum, which is not consistent with the internal reference rule. The theories presently available do not account for these deviations. Among the nitrides the internal reference rule is assumed to give reasonable values for band offsets. With a valence band offset of 0.8 eV for GaN/AlN [122, 123] the $Fe^{3+/2+}$ acceptor level is expected to be 4 eV above the valence band of AlN. This energy is much larger than the charge-transfer energy of 2.97 eV derived from PLE spectra [58, 59].

7.6 Conclusions

Defect identification using optical and magnetic resonance spectroscopy has a long tradition in III–V/II–VI semiconductors. Transfering the knowledge to the Group III nitride semiconductors helped to establish common characteristics for transition metals as well as for shallow residual impurities. A conclusive finite picture had not emerged in all cases, but remains as a challenge for defect spectroscopists in the next years.

Acknowledgements

The authors acknowledge financial support from the German Research Foundation (DFG). The authors wish to thank their collaborators and colleagues at various institutions who contributed to that work. Specifically they are grateful to Professor I. Akasaki, Dr H. Amano, Professor K. Hiramatsu, Dr Th. Detchprohm, Dr R. Heitz, L. Eckey, Dr K. Pressel, Professor H. C. Alt, Professor J. Schneider, Dr U. Kaufmann and our students—the workhorses in this research program.

References

1. S. Nakamura, T. Mukai, and M. Senoh, *Appl. Phys. Lett.* **64**, 1678 (1994).
2. S. Nakamura, *Jpn. J. Appl. Phys.* **35**, L74 (1996).
3. S. Nakamura, *Proc. International Symposium on Blue Laser and Light Emitting Diodes (ISBLLED)*, Chiba, Japan, p. 119 (1996).
4. J. M. Langer and H. Heinrich, *Phys. Rev. Lett.* **55**, 1414 (1985).
5. J. M. Langer, C. Delerue, M. Lannoo, and H. Heinrich, *Phys. Rev. B* **38**, 7723 (1988).
6. D. Volm, K. Oettinger, T. Streib, M. Ben-Chorin, J. Diener, B. K. Meyer, J. Majewski, L. Eckey, A. Hoffmann, H. Amano, I. Akasaki, K. Hiramatsu, and T. Detchprohm, *Phys. Rev. B* **53**, 165433 (1996).
7. K. Pakula, A. Wysmolek, K. P. Korona, J. M. Baranowski, R. Stepniewski, I. Grzegory, M. Bockowski, J. Jun, S. Krukowski, M. Wroblewski, and S. Porowski, *Solid State Commun.* **97**, 919 (1996).
8. D. Kovalev, B. Averboukh, D. Volm, B. K. Meyer, H. Amano, and I. Akasaki, *Phys. Rev. B* **54** 2518 (1996).
9. S. Strite and H. Morkoc, *J. Vac. Sci. Technol. B* **10**, 1237 (1992).
10. P. Boguslawski, E. Briggs, T. A. White, M. G. Wensell, and J. Bernholc, *Mat. Res. Soc. Symp. Proc.* **339**, 693 (1994); P. Boguslawski, E. Briggs, and J. Bernholc, *Phys. Rev. B* **51**, 17255 (1995).
11. J. Neugebauer and C. G. Van de Walle, *Phys. Rev. B* **50**, 8067 (1994).
12. W. Götz, N. M. Johnson, C. Chen, C. Kuo, and W. Imler, *Appl. Phys. Lett.* **68**, 3144 (1996).
13. S. E. Wickenden, D. K. Gaskill, D. D. Koleske, K. Doverspike, D. S. Simons, and P. H. Chi, *Mat. Res. Proc.* **395**, 679 (1996).
14. L. Eckey, I. Podlowski, A. Göldner, A. Hoffmann, I. Broser, B. K. Meyer, D. Volm, T. Streibl, K. Hiramatsu, T. Detchprohm, H. Amano, and I. Akasaki, *Inst. Phys. Conf. Ser.* **142**, 943 (1996).
15. C. Merz, M. Kunzer, U. Kaufmann, H. Amano, and I. Akasaki, *Semicond. Sci. Technol.* **11**, 712 (1996); U. Kaufmann, M. Kunzer, C. Merz, I. Akasaki, and H. Amano, *Mat. Res. Proc.* **395**, 633 (1996).
16. J. M. Baranowski and S. Porowski, *The Physics of Semiconductors*, ed. M. Scheffler and R. Zimmermann, World Scientific, p. 497 (1996).
17. H. Siegle, L. Filippidis, G. Kaczmarczyk, A. P. Litvinchuk, A. Hoffmann, and C. Thomsen, *The Physics of Semiconductors*, ed. M. Scheffler and R. Zimmermann, World Scientific, p. 537 (1996).
18. B. K. Meyer, D. Volm, A. Graber, H. C. Alt, T. Detchprohm, H. Amano, and I. Akasaki, *Solid State Commun.* **95**, 597 (1995).
19. M. Drechsler, B. K. Meyer, T. Detchprohm, H. Amano, and I. Akasaki, *Jpn. J. Appl. Phys.* **34**, L1178 (1995).
20. Y. J. Wang, H. K. Ng. K. Doverspike, D. K. Gaskill, T. Ikedo, I. Akasaki, and H. Amano, *J. Appl. Phys.* **79**, 8007 (1996).
21. C. Wetzel, personal communication.
22. J. R. Haynes, *Phys. Rev. Lett.* **4**, 351 (1960).
23. Th. Kuhn, *Diplomarbeit*, Würzburg, Germany (1989).
24. S. J. Pearton, C. R. Abernathy, and F. Ren. *Electron. Lett.* **30**, 527 (1994).
25. M. Leroux, B. Beaumont, N. Grandjean, C. Golivet, P. Gibart, J. Massies, J. Leymarie, A. Vasson, and A. M. Vasson, *Mat. Sci. Eng. B* **43**, 237 (1997).
26. M. Ilegems, R. Dingle, and R. A. Logan, *J. Appl. Phys.* **43**, 3797 (1972).
27. O. Lagerstedt and B. Monemar, *J. Appl. Phys.* **45**, 2266 (1974).

28. R. Stepniewski, A. Wysmolek, K. Pakula, and J. M. Baranowski, M. Potemski, G. Martinez, I. Grzegory, M. Wroblewski, and S. Porowski, *The Physics of Semiconductors*, ed. M. Scheffler and R. Zimmermann, World Scientific, p. 549 (1996).
29. J. L. Patel, E. Nicholls, and J. J. Davies, *J. Phys. C* **14**, 1339 (1981).
30. G. Baranov and N. G. Romanov, *Appl. Mag. Res.* **2**, 361 (1991).
31. E. R. Glaser, T. A. Kennedy, H. C. Crookham, J. A. Freitas, Jr., M. Asif Khan, D. T. Olson, and J. N. Kuznia, *Appl. Phys. Lett.* **63**, 2673 (1993).
32. K. Maier, M. Kunzer, U. Kaufmann, J. Schneider, B. Monemar, I. Akasaki, and H. Amano, *Mat. Sci. Forum* **87**, 143–147 (1994).
33. W. E. Carlos, J. A. Freitas, Jr., M. Asif Khan, D. T. Olson, and J. N. Kuznia, *Phys. Rev. B* **48**, 17878 (1993).
34. D. G. Thomas, J. J. Hopefield, and W. M. Augustvniak, *Phys. Rev.* **140**, 202 (1965).
35. O. Madelung, M. Schultz, and H. Weiss (eds), *Zahlenwerte und Funktionen, Halbleiter*, Landholt-Björnstein, Neue Serie, Band III/17a, 17b, Springer-Verlag, Berlin, 1982.
36. J. M. Luttinger and W. Kohn, *Phys. Rev.* **97**, 869 (1955).
37. D. Volm, T. Streibl, B. K. Meyer, T. Detchprohm, H. Amano, and I. Akasaki, *Solid State Commun.* **96**(2), 53–56 (1995).
38. H. Siegle, P. Thurian, L. Eckey, A. Hoffmann, C. Thomsen, B. K. Meyer, H. Amano, I. Akasaki, D. Detchprohm, and K. Hiramatsu, *Appl. Phys. Lett.* **68**, 1265 (1996).
39. L. Eckey, J.-Ch. Holst, P. Maxim, R. Heitz, A. Hoffmann, I. Broser, B. K. Meyer, C. Wetzel, E. N. Mokhov, and P. G. Baranov, *Appl. Phys. Lett.* **68**, 415 (1996).
40. C. Wetzel, S. Fischer, J. Krüger, E. E. Haller, R. J. Molnar, T. D. Moustakas, E. N. Mokhov, and P. G. Baranov, *Appl. Phys. Lett.* **68**, 18 (1996).
41. A. Hoffmann, J. Christen, and J. Gutowski, *Advanced Materials for Optics and Electronics*, **1**, 25–28 (1992).
42. A. Naumov, K. Wolf, T. Reisinger, H. Stanzl, and W. Gebhardt, *J. Appl. Phys.* **73**, 2581 (1993).
43. Y. G. Shreter and Y. T. Rebane, *Int. Conference Extended Defects in Semiconductors 96*, Giens, France, 7–12 September, 1996.
44. J. I. Pankove and J. A. Hutchby, *J. Appl. Phys.* **47**(76), 5387 (1976).
45. T. Ogino and M. Aoki, *Jpn. J. Appl. Phys.* **19**, 2395 (1980).
46. R. J. Molnar and T. D. Moustakas, *J. Appl. Phys.* **76**, 4587 (1996).
47. R. Singh, R. J. Molnar, M. S. Ünlü, and T. D. Moustakas, *Appl. Phys. Lett.* **64**, 336 (1994).
48. E. R. Glaser, T. A. Kennedy, K. Doverspike, L. B. Rowland, D. K. Gaskill, J. A. Freitas, Jr., M. Asif Khan, D. T. Olson, J. N. Kuznia, and D. K. Wickenden, *Phys. Rev. B* **51**, 13326 (1995).
49. T. Suski, P. Perlin, H. Tesseire, M. Leszczynski, I. Grzegory, J. Jun, M. Bockowski, S. Porowski, and T. D. Mouestakas, *Appl. Phys. Lett.* **67**, 2188 (1995).
50. P. Perlin, T. Suski, H. Teyssieire, M. Leszczynski, I. Grzegory, J. Jun, S. Porowski, P. Boguslawski, J. Bernholc, J. C. Chervin, A. Polian, and T. D. Mouestakas, *Phys. Rev. Lett.* **75**, 296 (1995).
51. D. M. Hofmann, D. Kovalev, G. Steude, B. K. Meyer, A. Hoffmann, L. Eckey, R. Heitz, T. Detchprohm, H. Amano, and I. Akasaki, *Phys. Rev. B* **52**, 16702 (1995).
52. A. Hoffmann, L. Eckey, P. Maxim, J.-Chr. Holst, R. Heitz, D. M. Hofmann, D. Kovalev, G. Steude, D. Volm, B. K. Meyer, T. Detchprohm, H. Amano, and I. Akasaki, *Proc. Topical Workshop on III–V Nitrides, TWN '95*, Nagoya, 1995, *Solid State Electron.* **41**(2), 275 (1997).
53. F. A. Ponce, D. P. Bour, W. Götz, and P. J. Wright, *Appl. Phys. Lett.* **68**(1), 917 (1996).
54. J. Neugebauer and C. G. Van de Walle, *Appl. Phys. Lett.* **69**(4), 503 (1996).

55. B. Monemar and O. Lagerstedt, *J. Appl. Phys.* **50**, 6480 (1979).
56. J. Baur, K. Maier, M. Kunzer, U. Kaufmann, J. Schneider, H. Amano, I. Akasaki, T. Detchprohm, and K. Hiramatsu, *Appl. Phys. Lett.* **64**, 857 (1994).
57. K. Maier, M. Kunzer, U. Kaufmann, J. Schneider, B. Monemar, I. Akasaki, and H. Amano, *Mater. Sci. Forum* **143–147**, 93 (1994).
58. J. Baur, K. Maier, M. Kunzer, U. Kaufmann, and J. Schneider, *Appl. Phys. Lett.* **65**, 2211 (1994).
59. J. Baur, M. Kunzer, K. Maier, U. Kaufmann, and J. Schneider, *Mater. Sci. Eng. B* **29**, 61 (1995).
60. K. Pressel, S. Nilsson, C. Wetzel, D. Volm, B. K. Meyer, I. Loa, P. Thurian, R. Heitz, A. Hoffmann, E. N. Mokhov, and P. G. Baranov, *Mater. Sci. Technol.* **12**, 90 (1996).
61. R. Heitz, P. Thurian, I. Loa, L. Eckey, A. Hoffmann, I. Broser, K. Pressel, B. K. Meyer, and E. N. Mokhov, *Appl. Phys. Lett.* **67**, 2822 (1995).
62. J. Baur, U. Kaufmann, M. Kunzer, J. Schneider, H. Amano, I. Akasaki, T. Detchprohm, and K. Hiramatsu, *Appl. Phys. Lett.* **67**, 1140 (1995).
63. R. Heitz, P. Thurian, K. Pressel, I. Loa, L. Eckey, A. Hoffmann, I. Broser, B. K. Meyer, and E. N. Mokhov, *Phys. Rev.* **52**, 16508 (1995).
64. K. Pressel, R. Heitz, L. Eckey, I. Loa, P. Thurian, A. Hoffmann, B. K. Meyer, S. Fischer, C. Wetzel, and E. E. Haller, *MRS Symp. Proc.* **395**, 491 (1995).
65. P. Thurian, L. Eckey, H. Siegle, J.-C. Holst, P. Maxim, R. Heitz, A. Hoffmann, C. Thomsen, I. Broser, K. Pressel, I. Akasaki, H. Amano, K. Hiramatsu, T. Detchprohm, D. Schikora, M. Hankeln, and K. Lischka, *Proc. ISBLLED*, Chiba, Japan, We-14, 180 (1996).
66. C. Wetzel, D. Volm, B. K. Meyer, K. Pressel, S. Nilsson, E. N. Mokhov, and P. G. Baranonov. *Appl. Phys. Lett.* **65**, 1033 (1994).
67. P. Thurian, G. Kaczmarczyk, H. Siegle, R. Heitz, A. Hoffmann, I. Broser, B. K. Meyr, R. Hoffbauer, and U. Scherz, *Mater. Sci. Forum* **196–201**, 1571 (1995).
68. R. Heitz, P. Maxim, L. Eckey, P. Thurian, A. Hoffmann, I. Broser, and B.-K. Meyer, *Phys. Rev. B* **55**, 4382 (1997).
69. H. Siegle, L. Eckey, A. Hoffmann, C. Thomsen, B. K. Meyer, D. Schikora, M. Hankeln, and K. Lischka, *Solid State Commun.* **96**, 943 (1995).
70. K. Pressel, G. Bohnert, A. Dörnen, B. Kaufmann, I. Denzel, and K. Thonke, *Phys. Rev. B* **47**, 9411 (1993).
71. B. Kaufmann, D. Haase, A. Dörnen, C. Hiller, and K. Pressel, *Mater. Sci. Forum* **143–147**, 797 (1994).
72. K. Pressel, G. Bohnert, G. Rückert, A. Dörnen, and K. Thonke, *J. Appl. Phys.* **71**(11), 5703 (1992).
73. A. Hoffmann, R. Heitz, and I. Broser, *Phys. Rev. B* **41**, 5806 (1990).
74. R. Heitz, A. Hoffmann, and I. Broser, *Phys. Rev. B* **45**, 8977 (1992).
75. U. W. Pohl, H.-E. Gumlich, and W. Busse, *Phys. Stat. Sol. (b)* **125**, 773 (1984).
76. R. E. Dietz, D. G. Thomas, and J. J. Hopfield, *Phys. Rev. Lett.* **8**, 391 (1962).
77. R. Heitz, A. Hoffmann, P. Thurian, and I. Broser, *J. Phys. C: Condens. Matter* **4**, 157 (1992).
78. R. Heitz, A. Hoffmann, and I. Broser, *Phys. Rev. B* **48**, 8672 (1993).
79. L. Podlowski, R. Heitz, T. Wolf, A. Hoffmann, D. Bimberg, I. Broser, and W. Ulrici, *Mater. Sci. Forum* **143–147**, 311 (1994).
80. K. Pressel, G. Rückert, A. Dörnen, and K. Thonke, *Phys. Rev. B* **46**, 13171 (1992).
81. K. Pressel, A. Dörnen, G. Rückert, and K. Thonke, *Phys. Rev. B* **47**, 16267 (1993).
82. P. Dahan, V. Fleurov, and K. A. Kikoin, *Mater. Sci. Forum* **196–201**, 755 (1995).
83. L. Podlowski, R. Heitz, P. Thurian, A. Hoffmann, and I. Broser, *J. Lumin.* **58**, 252 (1994).

84. R. Heitz, L. Podlowski, P. Thurian, A. Hoffmann, and I. Broser, *Proc. 22nd ICPS*, World Scientific Publishing, p. 2379 (1994).
85. U. W. Pohl and H. E. Gumlich, *Phys. Rev. B* **40**, 1194 (1989).
86. P. N. Keating, *Phys. Rev.* **145**, 637 (1966).
87. E. O. Kane, *Phys. Rev. B* **31**, 7865 (1985).
88. R. M. Feenstra, R. J. Hauenstein, and T. C. McGill, *Phys. Rev. B* **28**, 5793 (1983).
89. J. Baur, U. Kaufmann, M. Kunzer, J. Schneider, H. Amano, I. Akasaki, T. Detchprohm, and K. Hiramatsu, *Mater. Sci. Forum* **196–201**, 55 (1995).
90. D. Buhmann, H.-J. Schulz, and M. Thiede, *Phys. Rev. B* **19**, 5360 (1979).
91. U. Kaufmann, H. Ennen, J. Schneider, R. Wörner, J. Weber, and F. Köhl, *Phys. Rev. B* **25**, 5598 (1982).
92. M. J. Kane, M. S. Skolnick, P. J. Dean, W. Hayes, B. Cockayne, and W. R. MacEwan, *J. Phys. C* **17**, 6455 (1984).
93. G. Armelles, J. Barrau, D. Thebault, and M. Brousseau, *J. Phys. (Paris)* **45**, 1795 (1984).
94. G. Aszodi and U. Kaufmann, *Phys. Rev. B* **32**, 7108 (1985).
95. R. Heitz, A. Hoffmann, B. Hausmann, and I. Broser, *J. Lumin.* **48/49**, 689 (1991).
96. G. F. Imbush, W. M. Yen, A. L. Schawlow, D. E. McCumber, and M. D. Sturge, *Phys. Rev.* **133**, A1029 (1964).
97. Y. Tanabe and S. Sugano, *J. Phys. Soc. Jpn.* **9**, 766 (1954).
98. G. Guillot, C. Benjeddou, P. Leyral, and A. Nouailhat, *J. Lumin.* **31/32**, 439 (1984).
99. B. Clerjaud, C. Naud, B. Deveaud, B. Lambert, B. Plot, G. Bremont, C. Benjeddou, G. Guillot, and A. Nouhailat, *J. Appl. Phys.* **58**, 4207 (1985).
100. C. A. Bates, J. L. Dunn, and W. Ulrici, *J. Phys. Condensed Matter* **2**, 607 (1990).
101. K. Pressel, S. Nilsson, R. Heitz, A. Hoffmann, and B. K. Meyer, *J. Appl. Phys.* **79**, 3214 (1996).
102. C. Wetzel, D. Volm, B. K. Meyer, K. Pressel, S. Nilsson, E. N. Mokhov, and P. G. Baranow, *Mat. Res. Soc. Symp. Proc.* **339**, 453 (1994).
103. P. Thurian, R. Heitz, A. Hoffmann, and I. Broser, *J. Crystal Growth* **117**, 727 (1992).
104. A. D. Liehr and C. J. Ballhausen, *Ann. Phys. (NY)* **2**, 134 (1959).
105. B. Kaufmann, A. Dörnen, V. Härle, H. Bolay, F. Scholz, and G. Pensl, *Appl. Phys. Lett.* **68**(2), 203 (1996).
106. J. Schneider, H. D. Müller, K. Maier, W. Wilkening, F. Fuchs, A. Dörnen, S. Leibenzeder, and R. Stein, *Appl. Phys. Lett.* **56**, 1184 (1990).
107. G. Goetz, U. W. Pohl, and H.-J. Schulz, *J. Phys. C* **4**, 8253 (1992); G. Goetz, U. W. Pohl, H.-J. Schulz, and M. Thiede, *J. Lumin.* **60/61**, 16 (1994).
108. K. Pressel, R. Heitz, S. Nilsson, P. Thurian, A. Hoffmann, and B. K. Meyer, *MRS Symp. Proc.* **395**, 613 (1995).
109. K. Maier, J. Schneider, W. Wilkening, S. Leibenzeder, and R. Stein, *Mater. Sci. Eng. B* **11**, 27 (1992).
110. K. Atobe, M. Honda, N. Fukuoka, M. Okada, and M. Nakagawa, *Jpn. J. Appl. Phys.* **29**, 150 (1990).
111. L. Filippidis, H. Siegle, A. Hoffmann, C. Thomsen, K. Karch, and F. Bechstedt, *Phys. Stat. Sol. (b)* **198**, 621 (1996).
112. P. Thurian, I. Loa, P. Maxim, K. Pressel, A. Hoffmann, and C. Thomsen, *Appl. Phys. Lett.*, in press.
113. Ch. Uihlein and L. Eaves, *Phys. Rev. B* **26**, 4473 (1982).
114. L. A. Lebedo and B. K. Ridley, *J. Phys. C: Solid State Physics*, **15** L961 (1982).
115. M. J. Caldas, A. Fazzio, and A. Zunger, *Appl. Phys. Lett.* **45**, 671 (1984).
116. A. Zunger, *Phys. Rev. Lett.* **54**, 849 (1985).
117. B. Clerjaud, in *Semiconductor Physics*, ed. A. M. Stoneham, Adam Hilger, Bristol and Boston, p. 117 (1986).

118. G. Martin, S. Strite, J. Thornton, and H. Morkoc, *Appl. Phys. Lett.* **58**, 2375 (1991).
119. X. Huang, T. S. Cheng, S. E. Hooper, T. J. Foster, L. C. Jenkins, J. Wang, C. T. Foxon, J. W. Orton, L. Eaves, and P. C. Main, *J. Vac. Sci. Technol. B* **13**, 1582 (1995).
120. J. C. Phillips. *Phys. Rev.* **42**, 317 (1970).
121. E. Mollwo, G. Müller, and P. Wagner, *Solid State Commun.* **13**, 1283 (1973).
122. G. A. Martin, S. C. Strite, A. Botchkarev, A. Agarwal, A. Rockett, W. R. L. Lambrecht, B. Segall, and H. Morkoc, *Appl. Phys. Lett.* **65**, 610 (1994).
123. E. A. Albanesi, W. R. Lambrecht, and B. Segall, *J. Vac. Sci. Technol. B* **12**, 2470 (1994).
124. P. Thurian, R. Heitz, L. Eckey, P. Maxim, V. Kutzer, A. Hoffmann, I. Broser, K. Pressel, and B. K. Meyer, *Proc. 23rd ICPS*, Berlin, 2897 (1996).
125. P. Thurian, A. Hoffmann, L. Eckey, P. Maxim, R. Heitz, I. Broser, K. Pressel, B. K. Meyer, J. Schneider, J. Baur, and M. Kunzer, Proc. MRS Meeting, Boston, 1996, in press.
126. C. Göbel, C. Schrepel, U. Scherz, P. Thurian, G. Kaczmarczyk, and A. Hoffmann, ICDS S7, Aveiro, Portugal, to be published in *Mat. Science Forum*.
127. R. Parrot, A. Geoffroy, C. Naud, W. Busse, and H.-E. Gumlich, *Phys. Rev. B* **23**, 5288 (1981).
128. B. DiBartolo, *Optical Interactions in Solids*, Wiley, New York (1968).

8 Electronic and optical properties of GaN-based quantum wells

Masakatsu Suzuki and Takeshi Uenoyama

8.1 Introduction

There is increasing interest in the use of GaN and related materials for opto-electronic devices in the short wavelength region, due to their wide direct bandgap and physical hardness. During the past few years, remarkable progress has been made in the development of optical devices based on wurtzite (WZ) GaN. At present, high-brightness blue light-emitting diodes (LEDs), based on WZ InGaN/AlGaN doublehetero (DH) (Nakamura *et al.* 1994) or single quantum well (SQW) (Nakamura *et al.* 1995) structures, are commercially available. Very recently the pulse-lasing action of multi-quantum well (MQW) laser diodes (LDs) by current injection at room temperature was reported for the first time (Nakamura *et al.* 1996). Furthermore, the recent development of crystal growth techniques has made it possible to obtain high-quality zincblende (ZB) crystals as well as WZ ones (Mizuta *et al.* 1986). However, many fundamental material and device characteristics of the III–V nitrides are less understood than conventional ZB compounds.

There are many problems to be clarified in the electronic and optical properties of the III–V nitrides and their QW structures. Some of them are as follows:
- What is the difference between the III–V nitrides and the other ZB III–V semiconductors?
- What are the physical parameters of the III–V nitrides?
- What are the characteristics of subband and optical gain of WZ QWs?
- What is the difference between WZ and ZB nitrides?
- What are the strain effects on WZ and ZB nitrides and their QWs?

In order to clarify these problems, we have studied electronic and optical properties of bulk GaN, AlN, and GaN/AlGaN QW structures and discussed them from a fundamental point of view, on the basis of the first-principles electronic band calculation and $k \cdot p$ theory (Suzuki and Uenoyama 1995; Kamiyama *et al.* 1995; Suzuki *et al.* 1995, 1997; Uenoyama and Suzuki 1995; Suzuki and Uenoyama 1996a,b,c,d,e). The practical approach of linking them provided the unknown physical parameters and made it possible to analyze the III–V nitrides and their QW structures. As a result a fundamental understanding of them was obtained, and it led to guidelines for material and device design with the III–V nitrides.

308 *Electronic and optical properties of GaN-based quantum wells*

This chapter is organized as follows. In Section 8.2, the method and results of the first-principles electronic band calculations are described. In Section 8.3, the $\mathbf{k} \cdot \mathbf{p}$ theory with strain effects for the WZ structure is explained, and the numerical derived physical parameters are presented for WZ and ZB nitrides. In Section 8.4, the method and results of the subband calculations for GaN/AlGaN QWs are described. In Section 8.5, the optical gain calculations for bulk GaN and GaN/AlGaN QWs are shown and discussed. Section 8.6 summarizes the chapter.

8.2 Electronic band structures of bulk GaN and AlN

8.2.1 Method of first-principles band calculations

GaN and AlN take not only hexagonal WZ structure (space group C_{6v}^4:P6$_3$mc) but also cubic ZB structure (space group T_d^2:F$\bar{4}$3m). Figure 8.1 shows the first Brillouin zones (BZs) of (a) the WZ and (b) the ZB structures. The first-principles electronic band calculations for both WZ and ZB structures have been performed by using the full-potential linearized augmented plane wave (FLAPW) method (Wimmer *et al.* 1981) within the local density functional approximation (LDA) (Gunnarson and Lundquvist 1976). In general, the nitrogen atom has such a strong electron affinity that the valence charge densities of the III−V nitrides tend to be well localized. In such a case, a conventional pseudo-potential approach seems to be inadequate. Therefore, the FLAPW method, which is well-known to be very powerful in such cases, is adopted as the method of the first-principles band calculation.

In the FLAPW calculations for the WZ (ZB) structure, we use the criteria that $l_{\max} = 7$ (7) and $|\mathbf{k} + \mathbf{G}|_{\max} = (2\pi/a) \times 3.5$ (5.5), with $\mathbf{k}$ being a wavevector in the first BZ and $\mathbf{G}$ being a reciprocal lattice vector, for the construction of the basis functions. The expansion of the potential and the charge density is up to $l_{\max} = 6$ (6) and $|\mathbf{k} + \mathbf{G}|_{\max} = (2\pi/a) \times 7.0$ (10.0). The charge density have

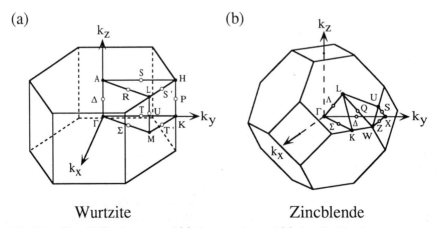

Wurtzite   Zincblende

Fig. 8.1 First Brillouin zones of (a) the wurtzite and (b) the zincblende structures.

been self-consistently determined using 28 (19) meshed k points in an irreducible wedge of the first BZ. The iteration process has been repeated until the calculating total energy of the crystal converges into less than 1 mRyd. As for the lattice constants, the experimentally observed values (our theoretically optimized values) have been used: $a = 6.0263$ a.u., $c = 9.7982$ a.u ($a = 8.5941$ a.u.) for GaN (Maruska and Tietjen 1969) and $a = 5.8808$ a.u., $c = 9.4146$ a.u. ($a = 8.3426$ a.u.) for AlN (Yim et al. 1973). The internal parameter u in the WZ structure was fixed at the ideal value of $(3/8)c$. As the muffin-tin sphere radii we adopted the same values of $0.29a$ ($0.21a$) for all atoms, since the use of the full-potential ensures that the calculation is completely independent of the choice of muffin-tin sphere radii. In our calculations, the Ga 3d states are treated as a part of the valence band states, since they are relatively high in energy even though they may constitute the well-localized narrow band. In fact, according to our calculation, the Ga 3d states are considerably hybridized with the N 2s states. Thus, the Ga 3d states should not be treated as core states. The importance of self-consistent treatment of the Ga 3d states has been independently pointed out by the other researchers as well (Fiorentini et al. 1993; Wenchang et al. 1993).

8.2.2 Results of first-principles band calculations

The calculated band structures for WZ GaN and AlN are in good agreement with the previous theoretical results within the LDA (Fiorentini et al. 1993; Miwa and Fukumoto 1993; Rubio et al. 1993; Wenchang et al. 1993). It is well known that the LDA yields quite reliable ground state properties, but seriously underestimates the bandgap for all usual semiconductors. A more accurate band structure can be obtained by the quasi-particle correction using the GW approximation (Hedin 1965; Hybertsen and Louie 1985). The GW calculations have been already applied to these nitrides as well (Rubio et al. 1993; Palummo et al. 1994). The results show that the quasi-particle correction has a little k-dependence. In general, however, the LDA wavefunctions are very close to the exact ones, and the correction causes almost a rigid shift of the conduction band against the valence band and the broadening of both bands. Now, we are mainly interested in the band structure in the close vicinity of the VBM and the CBM, which are the Γ points for GaN and AlN in the WZ and the ZB structures. Therefore, the k-dependence on the GW correction is negligible, and the LDA calculation seems to be sufficiently accurate for our purposes.

Figures 8.2(a) and (b) show the electronic band structures of WZ and ZB GaN, respectively. The spin–orbit interaction is included, and the energy dispersions are along high-symmetry lines of the first BZ. As an example, the characteristics of WZ GaN are described. The energy bands consist of four parts, which are three occupied bands and an unoccupied band. The lowest energy bands, which are localized between the -0.70 Ryd and -0.65 Ryd, mostly originate from the N 2s states. The upper sides of these states are hybridized with the Ga 3d states. The second lowest energy bands are located between about -0.55 Ryd and -0.38 Ryd, where the lower energy side almost originates

310 *Electronic and optical properties of GaN-based quantum wells*

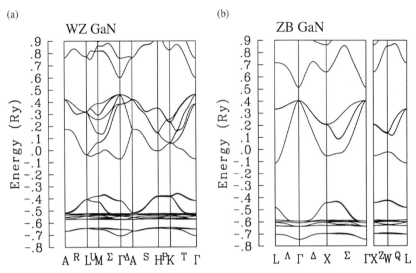

Fig. 8.2 Electronic band structures of (a) wurtzite and (b) zincblende GaN.

from the Ga 3d well-localized states, and the upper one consists of the hybridized states between the Ga 3d and the N 2s states. Due to these hybridization, the Ga 3d states must be included in the valence band states. In case of AlN, there is no hybridization between the cation d and the N 2s states, since the Al atom has no d electrons in the ground state. Consequently, the lowest energy band is combined with the upper side of the second lowest one, and these two bands become one narrow band. In the intermediate energy region, there are broad energy bands whose width is about 0.53 Ryd. The VBM is located at the top of these bands and at the Γ point. These energy bands arise from N 2s, 2p and the Ga 4s, 4p states, however, the vicinity of the VBM almost consists of the N 2p states. The highest energy region separated by about 0.14 Ryd above the valence band is the conduction bands, and the CBM is also located at the Γ point. The vicinity of the CBM originates from the s states of N and Ga. Comparing the result of AlN with that of GaN, we see that the main features in these two energy regions are very similar, except for the order of energy levels around the VBM.

Next, we focus on the features of electronic structures near the VBM. We note that the electronic structures around the VBM of GaN in the WZ structure, even in the ZB structure, are diffeent from conventional ZB III–V semiconductors. Figure 8.3(a) shows the splitting at the top of the valence bands of GaN under the influence of the crystal–field and spin–orbit interaction. Figures 8.3(b) and (c) show the schematic band structures around the VBM of WZ and ZB GaN, respectively. In the WZ structure, the VBM is split into two-fold and non-degenerate states even in the absence of the spin–orbit interaction. Using the irreducible representations of group theory, the former state is labeled Γ_6, whose wave functions transform like x and y, and the latter one is labeled Γ_1, whose wavefunction transforms like z. The energy splitting

8.2 Electronic band structures of bulk GaN and AlN

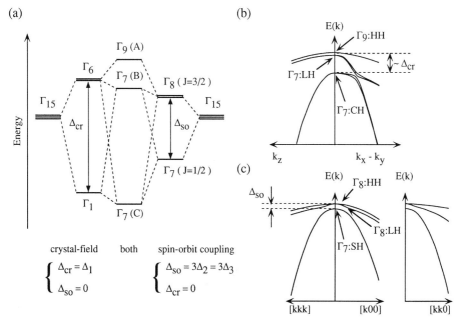

Fig. 8.3 (a) Splitting at the top of valence bands of GaN under the influence of the crystal–field and spin–orbit coupling. (b) and (c) show the schematic band structures around the top of the valence bands of wurtzite and zincblende GaN, respectively. The figures are not drawn to scale.

between these two levels is induced by the hexagonal symmetry of the WZ structure; therefore this splitting is called the crystal–field splitting Δ_{cr}. The order of the two levels depends on the kind of materials, the ratio c/a of lattice constants, and the internal parameter u. In case of GaN, the Γ_6 level is higher than the Γ_1 level, while in the case of AlN, the Γ_1 level is higher. This difference between AlN and GaN is in agreement with the result by Rubio et al. (1993). Considering the spin–orbit interaction, the two-fold degenerate Γ_6 state is split into the Γ_9 and Γ_7 levels, and the non-degenerate Γ_1 state is labeled Γ_7 as well. Then the two Γ_7 levels can be mixed by the spin–orbit interaction. The energy splitting among these three levels by the spin–orbit interaction is called a spin–orbit splitting. According to the FLAPW calculations, the spin–orbit splittings are much smaller than the crystal–field splitting in the III–V nitrides, though the crystal–field splitting itself is small.

The hole effective masses of the three levels Γ_9, Γ_7, and Γ_7 have a large k-dependence, as shown in Fig. 8.3(b), where we have labeled three hole bands as HH (heavy), LH (light), and CH (crystal–field split-off), based on the feature in the k_x–k_y plane. The mass of the Γ_9 band is heavy along any $\boldsymbol{k}$ direction. On the other hand, that of the upper Γ_7 band is light in the k_x–k_y plane but is heavy along the k_z direction, and that of the lower Γ_7 band is the very reverse. The lower Γ_7 band is split off from the Γ_9 and the upper Γ_7 bands, even without the spin–orbit coupling. These features show that the $\boldsymbol{k}$-dependence on the hole masses is not negligible when we carry out the material design or the characteristic analysis of the GaN-based QW devices like LDs.

In the ZB structure, on the other hand, the threefold degenerate Γ_{15} states are not split into the two-fold Γ_8 state and the non-degenerate Γ_7 state unless the spin–orbit interaction is taken into account. In Fig. 8.3(c), we have labeled three hole bands as HH (heavy), LH (light), and SH (spin–orbit split-off), conventionally. However, the LH mass is as heavy as the HH one along any k direction, and it is quantitatively different from ZB GaAs. This difference is very significant, and it originates from very small spin–orbit splitting energies compared with conventional III–V compounds. Therefore we must pay attention to it: we must treat equivalently the upper six valence bands in the analysis of GaN-based QW devices.

8.2.3 Strain effect on electronic band structures
8.2.3.1 *Biaxial and uniaxial strains in the (0001) plane of wurtzite*

Figure 8.4 shows the schematic band structure in the k_x–k_y plane around the VBM of bulk WZ GaN, (a) without strain, (b) with a biaxial strain, and (c) with a uniaxial strain in the (0001) plane (*c*-plane). It is found that each energy dispersion of HH, LH, and CH bands is almost unchanged even under stress because the weak spin–orbit coupling makes the orbital character, such as $|X\rangle$, $|Y\rangle$, and $|Z\rangle$, dominant. Under the condition of the compressive (tensile) biaxial strain, only the crystal–field splitting energy becomes effectively large (small) since the strained crystal has still the C_{6v} symmetry, as well as the unstrained one. Then the HH mass is still heavy, and the density of states (DOS) at the VBM is not so reduced. This is very different from ZB crystals, where the HH mass becomes light and the DOS decreases considerably since the T_d symmetry is changed to the D_{2d} one and the degeneracy is removed. On the other hand, the uniaxial strain in the *c*-plane gives the anisotropic energy splitting in the k_x–k_y plane since it causes symmetry lowering from C_{6v} to C_{2v}. When the compressive (tensile) uniaxial strain along the *y*-direction or the tensile (compressive) one along the *x*-direction are induced, the HH band along the k_x direction, as well as the LH band along the k_y direction, moves to the higher (lower) energy side. Thus, the DOS at the VBM is largely reduced, compared with the conditions without strain or with a biaxial strain.

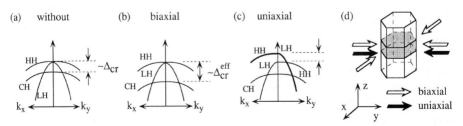

Fig. 8.4 Schematic band structure in the k_x–k_y plane around the top of valence bands of the wurtzite GaN, (a) without strain, (b) with a biaxial strain, and (c) with a uniaxial strain in the *c*-plane. (d) shows the direction of each strain.

8.3 k·p theory with strains for wurtzite 313

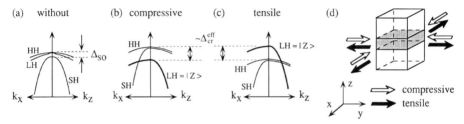

Fig. 8.5 Schematic band structure in the k_x–k_y plane around the top of the valence bands of the zincblende GaN, (a) without strain, (b) with a compressive biaxial strain, and (c) with a tensile biaxial strain in the x–y plane. (d) shows the direction of each strain.

8.2.3.2 Biaxial strain in the (001) plane of zincblende

Figure 8.5 shows the schematic band structure along the k_x and k_z directions around the VBM of bulk ZB GaN, (a) without strain, (b) with a compressive biaxial strain, and (c) with a tensile biaxial strain in the (001) plane (x–y plane). It is found that the strain effect on the HH and LH bands is almost the same as in conventional III–V compounds, where the heavy hole mass becomes light and the light hole mass becomes heavy due to the removal of the degeneracy of the Γ_8 state. However, the heavy hole mass of the strained ZB GaN is still quite heavy because both the heavy and light hole masses of the unstrained ZB GaN are much heavier than those of conventional III–V compounds. Thus, the biaxial strain is not so effective on the reduction of the DOS at the VBM as in conventional ZB III–V compounds. Furthermore, the spin–orbit interaction is so weak that the orbital character, such as $|X\rangle$, $|Y\rangle$, and $|Z\rangle$, is dominant. As a result, it looks as if biaxial strained ZB GaN is quasi-WZ GaN. In other words, under the biaxial stress, the crystal-field splitting between the $|Z\rangle$ state and the other states is induced due to symmetry lowering from T_d to D_{2d}. It is the same situation as the case of WZ GaN, though the crystal symmetry is not D_{2d} but C_{6v}. Then the compressive biaxial strain lifts the $|X\rangle$ and $|Y\rangle$ states above the $|Z\rangle$ state, and the tensile one lifts the $|Z\rangle$ state above the $|X\rangle$ and $|Y\rangle$ states. The former situation is similar to WZ GaN, and the latter is similar to WZ AlN.

8.3 k·p theory with strains for wurtzite

Generally speaking, optical and transport properties are governed by the electronic band structures in the close vicinity of the VBM and the CBM. Thus, the **k·p** theory is not only an appropriate method to describe the electronic structures around the VBM and the CBM, but also a convenient method to analyze the QW devices on the basis of the effective mass approximation. In the analyses of conventional ZB semiconductors, we frequently assume a parabolic band for the conduction bands, and the 4×4 and/or 6×6 Luttinger–Kohn Hamiltonians are used to describe the upper valence bands (Luttinger and Kohn 1955). In treating the valence bands together with the conduction bands, we often make use of the 8×8 Kane Hamiltonian (Kane 1957). For WZ materials, however, we must consider the hexagonal symmetry because the Luttinger–Kohn

314   Electronic and optical properties of GaN-based quantum wells

and Kane Hamiltonians are constructed under the condition of cubic symmetry. Let us show the $\mathbf{k} \cdot \mathbf{p}$ and strain Hamiltonians for the WZ structure.

8.3.1 8×8 $\mathbf{k} \cdot \mathbf{p}$ Hamiltonian

We take the following basis functions in that order:

$$|S,\uparrow\rangle, |S,\downarrow\rangle, \frac{1}{\sqrt{2}}|(X+iY),\uparrow\rangle, \frac{1}{\sqrt{2}}|(X+iY),\downarrow\rangle,$$

$$|Z,\uparrow\rangle, |Z,\downarrow\rangle, \frac{1}{\sqrt{2}}|(X-iY),\uparrow\rangle, \frac{1}{\sqrt{2}}|(X-iY),\downarrow\rangle,$$

where $|S\rangle, |X\rangle, |Y\rangle$, and $|Z\rangle$ are the Bloch functions at the Γ point, transforming like atomic s, p_x, p_y, and p_z-like functions, respectively, under the operations of the C_{6v} symmetry group. $|\uparrow\rangle$ and $|\downarrow\rangle$ are spin functions corresponding to spin-up and -down states, respectively. Then, the Hamiltonian has the following form:

$$\mathbf{H}(k) = \begin{pmatrix} \mathbf{H}'_{cc} & \mathbf{H}'_{cv} \\ \mathbf{H}'^{\dagger}_{cv} & \mathbf{H}'_{vv} \end{pmatrix}, \quad (8.3.1)$$

where $\mathbf{H}'_{cc}$ and $\mathbf{H}'_{vv}$ are 2×2 and 6×6 matrices for the conduction bands and the valence bands, respectively, without the interaction between them. The interaction with the other bands is treated as a second-order perturbation. The interaction between the conduction bands and the valence bands is described by $\mathbf{H}'_{cv}$, whose matrix elements are written as

$$\mathbf{H}'_{cv} = \begin{pmatrix} Q & 0 & R & 0 & Q^* & 0 \\ 0 & Q & 0 & R & 0 & Q^* \end{pmatrix}, \quad (8.3.2)$$

where

$$Q = \frac{1}{\sqrt{2}} \frac{\hbar}{m_0} \langle S|p_x|X\rangle(k_x + ik_y) = \frac{1}{\sqrt{2}} P_\perp (k_x + ik_y),$$

$$R = \frac{\hbar}{m_0} \langle S|p_z|Z\rangle k_z = P_\parallel k_z.$$

Note that the momentum matrix elements, $P_\parallel$ and $P_\perp$, are different from each other in the WZ structure due to the anisotropy between the directions parallel and perpendicular to the c-axis. $P_\parallel^2$ and $P_\perp^2$ can be analytially derived from the WZ 8×8 Hamiltonian as follows:

$$P_\parallel^2 = \frac{\hbar^2}{2m_0}\left(\frac{m_0}{m_e^\parallel} - \frac{m_0}{m_c^\parallel}\right)\frac{(E_g + \Delta_1 + \Delta_2)(E_g + 2\Delta_2) - 2\Delta_3^2}{E_g + 2\Delta_2}, \quad (8.3.3)$$

$$P_\perp^2 = \frac{\hbar^2}{2m_0}\left(\frac{m_0}{m_e^\perp} - \frac{m_0}{m_c^\perp}\right)\frac{\{(E_g + \Delta_1 + \Delta_2)(E_g + 2\Delta_2) - 2\Delta_3^2\}E_g}{(E_g + \Delta_2)(E_g + \Delta_1 + \Delta_2) - \Delta_3^2}, \quad (8.3.4)$$

where $m_e^\parallel$ and $m_e^\perp$ denote the electron effective masses along the k_z direction and in the k_x-k_y plane, respectively. Here, note that $m_c^\parallel$ and $m_c^\perp$ stand for the sum of a free electron mass and the influence of higher energy bands. If we neglect the influence of higher energy bands ($m_c^\parallel = m_c^\perp = m_0$) and assume the quasi-cubic approximation ($\Delta_{cr} = \Delta_1$, $\Delta_{so} = 3\Delta_2 = 3\Delta_3$) (Bir and Pikus 1974), the expressions consisting of only several observable parameters are obtained as follows:

$$P_\parallel^2 \simeq \frac{\hbar^2}{2m_0}\left(\frac{m_0}{m_e^\parallel} - 1\right) \frac{(E_g + \Delta_{cr} + \Delta_{so})E_g + \frac{2}{3}\Delta_{cr}\Delta_{so}}{E_g + \frac{2}{3}\Delta_{so}}, \quad (8.3.5)$$

$$P_\perp^2 \simeq \frac{\hbar^2}{2m_0}\left(\frac{m_0}{m_e^\perp} - 1\right) \frac{(E_g + \Delta_{cr} + \Delta_{so})E_g + \frac{2}{3}\Delta_{cr}\Delta_{so}}{E_g + \Delta_{cr} + \frac{2}{3}\Delta_{so} + (\Delta_{cr}\Delta_{so}/3E_g)}. \quad (8.3.6)$$

Furthermore, assuming that $\Delta_{cr} = \Delta_1 = 0$, the expression for the ZB structure is obtained:

$$P_{ZB}^2 = \frac{\hbar^2}{2m_0}\left(\frac{m_0}{m_e^*} - 1\right) \frac{(E_g + \Delta_{so})E_g}{E_g + \frac{2}{3}\Delta_{so}}, \quad (8.3.7)$$

where m_e^* stands for the electron effective mass averaged all over the *k*-directions. This expression is in agreement with the expression used for conventional ZB semiconductors.

8.3.2 2 × 2 and 6 × 6 *k* · *p* and the strain Hamiltonian

Owing to the large bandgaps of GaN and AlN, we can treat the interaction $\mathbf{H}'_{cv}$ within the second-order perturbation. Then the 8 × 8 Hamiltonian can be split into the 2 × 2 Hamiltonian $\mathbf{H}_{cc}$ for the CBM and 6 × 6 Hamiltonian $\mathbf{H}_{vv}$ for the VBM. The strain effect can be easily included by the same symmetry consideration as the *k* · *p* terms and by the straightforward addition of the corresponding terms. The matrix elements of $\mathbf{H}_{cc}$ and $\mathbf{H}_{vv}$ are given by

$$\mathbf{H}'_{cc} = \begin{pmatrix} E_c & 0 \\ 0 & E_c \end{pmatrix}, \quad (8.3.8)$$

$$\mathbf{H}'_{vv} = \begin{pmatrix} F & 0 & -H^* & 0 & K^* & 0 \\ 0 & G & \sqrt{2}\Delta_3 & -H^* & 0 & K^* \\ -H & \sqrt{2}\Delta_3 & \lambda & 0 & I^* & 0 \\ 0 & -H & 0 & \lambda & \sqrt{2}\Delta_3 & I^* \\ K & 0 & I & \sqrt{2}\Delta_3 & G & 0 \\ 0 & K & 0 & I & 0 & F \end{pmatrix}, \quad (8.3.9)$$

where

$$E_c = E_s^0 + \frac{\hbar^2 k_z^2}{2m_e^{\parallel}} + \frac{\hbar^2(k_x^2 + k_y^2)}{2m_e^{\perp}} + a_c(\varepsilon_{xx} + \varepsilon_{yy} + \varepsilon_{zz}),$$

$$F = \Delta_1 + \Delta_2 + \lambda + \theta,$$

$$G = \Delta_1 - \Delta_2 + \lambda + \theta,$$

$$H = \mathrm{i}(A_6 k_z + \mathrm{i} A_7)(k_x + \mathrm{i} k_y) + \mathrm{i} D_6(\varepsilon_{xx} + \mathrm{i} \varepsilon_{yz}),$$

$$I = \mathrm{i}(A_6 k_z - \mathrm{i} A_7)(k_x + \mathrm{i} k_y) + \mathrm{i} D_6(\varepsilon_{xx} + \mathrm{i} \varepsilon_{yz}),$$

$$K = A_5(k_x^2 - k_y^2 + 2\mathrm{i} k_x k_y) + D_5(\varepsilon_{xx} - \varepsilon_{yy} + 2\mathrm{i}\varepsilon_{xy}),$$

$$\lambda = E_p^0 + A_1 k_z^2 + A_2(k_x^2 + k_y^2) + D_1 \varepsilon_{zz} + D_2(\varepsilon_{xx} + \varepsilon_{yy}),$$

$$\theta = A_3 k_z^2 + A_4(k_x^2 + k_y^2) + D_3 \varepsilon_{zz} + D_4(\varepsilon_{xx} + \varepsilon_{yy}).$$

E_s^0 and E_p^0 represent the energy level of s-like and p-like functions at $k = 0$, respectively. Δ_1 and $\Delta_{2,3}$ correspond to the crystal–field and spin–orbit splitting energies, respectively. A_i are the valence band parameters, corresponding to the Luttinger parameters γ_i in ZB crystals, a_c and D_i are the deformation potentials for the conduction and the valence bands, respectively. ε_{ij} ($i, j = x, y, z$) are the ij-components of a strain tensor, where the diagonal terms ε_{ii} are positive for tension. According to the FLAPW calculations and group theory, the lowest conduction band state is an almost s-like state for the WZ structure as well as for the ZB structure, and it is isotropic, in contrast to the valence band states which show a strong anisotropy. Therefore, if the small anistropy of the energy dispersions is neglected, namely, if the parabolic band is assumed, the electronic structure around the CBM is given by

$$E_c(k) \simeq E_s^0 + \frac{\hbar^2 k^2}{2m_e^*} + a_c \varepsilon. \tag{8.3.10}$$

Generally, the strain–stress relations for hexagonal crystals with the C_{6v} symmetry are expressed as

$$\begin{pmatrix} \sigma_{xx} \\ \sigma_{yy} \\ \sigma_{zz} \\ \sigma_{xy} \\ \sigma_{yz} \\ \sigma_{zx} \end{pmatrix} = \begin{pmatrix} C_{11} & C_{12} & C_{13} & 0 & 0 & 0 \\ C_{12} & C_{11} & C_{13} & 0 & 0 & 0 \\ C_{13} & C_{13} & C_{33} & 0 & 0 & 0 \\ 0 & 0 & 0 & C_{44} & 0 & 0 \\ 0 & 0 & 0 & 0 & C_{44} & 0 \\ 0 & 0 & 0 & 0 & 0 & C_{66} \end{pmatrix} \begin{pmatrix} \varepsilon_{xx} \\ \varepsilon_{yy} \\ \varepsilon_{zz} \\ \varepsilon_{xy} \\ \varepsilon_{yz} \\ \varepsilon_{zx} \end{pmatrix}, \tag{8.3.11}$$

where C_{ij} are the stiffness constants, and $C_{66} = (C_{11} - C_{12})/2$. When the WZ crystal is strained in the (0001) plane and free along the [0001] direction

($\sigma_{xx} \neq 0$, $\sigma_{yy} \neq 0$, $\sigma_{zz} = 0$ and $\sigma_{xy} = \sigma_{yz} = \sigma_{zx} = 0$), the strain tensor has only three nonvanishing elements:

$$\varepsilon_{xx} = \frac{a - a_0}{a_0}, \quad \varepsilon_{yy} = \frac{b - a_0}{a_0},$$

$$\varepsilon_{zz} = \frac{c - c_0}{c_0} = -\frac{C_{13}}{C_{33}}(\varepsilon_{xx} + \varepsilon_{yy}), \quad (8.3.12)$$

where a_0 and c_0 are the lattice constants of the unstrained WZ crystal, and a, b, and c are those of the strained crystal. If the crystal is pseudomorphically grown along the [0001] direction on a hexagonal substrate ($\sigma_{xx} = \sigma_{yy}$, $\sigma_{zz} = 0$, and $\sigma_{xy} = \sigma_{yz} = \sigma_{zx} = 0$), the strain is biaxial, namely symmetric in the (0001) plane. Then the lattice constants a and b are equal to that of the substrate material a_{sub}, and the nonvanishing elements of the strain tensor are satisfied by the relations

$$\varepsilon_{xx} = \varepsilon_{yy} = \frac{a_{\text{sub}} - a_0}{a_0}, \quad \varepsilon_{zz} = -\frac{2C_{13}}{C_{33}}\varepsilon_{xx}. \quad (8.3.13)$$

When the WZ crystal is uniaxially strained in the (0001) plane and free along the other direction ($\sigma_{xx} \neq 0$, $\sigma_{yy} = \sigma_{zz} = 0$ and $\sigma_{xy} = \sigma_{yz} = \sigma_{xx} = 0$), the strain tensor has only two nonvanishing elements:

$$\begin{pmatrix} \varepsilon_{yy} \\ \varepsilon_{zz} \end{pmatrix} = \frac{1}{C_{13}^2 - C_{11}C_{33}} \begin{pmatrix} C_{12}C_{33} - C_{13}^2 \\ C_{11}C_{13} - C_{12}C_{13} \end{pmatrix} \varepsilon_{xx}. \quad (8.3.14)$$

8.3.3 Quasi-cubic approximation

From the point of view of the local symmetry, the relations among the Luttinger-like parameters A_i, the splitting energies Δ_i and the deformation potentials D_i for the WZ structure are discussed. The relations between A_i and the Luttinger parameters γ_i for the ZB structure are also discussed. Note that the local coordination on the atomic position of the WZ structure is the same as that of the ZB structure. The two structures are different mainly at the relative positions of the third neighbors and beyond, and it is obvious that the layer stacking along the [0001] direction in the WZ structure is equivalent to that along the [111] direction in the ZB structure. Thus, what is called quasi-cubic approximation to the WZ structure is introduced (Bir and Pikus 1974). First, let us transform the Luttinger–Kohn Hamiltonian for the ZB structure to another coordinate system, in which the z'-axis is along the [111] direction and the x'- and y'-axes along [11$\bar{2}$] and [$\bar{1}$10] directions, respectively. Next, comparing this transformed Hamiltonian with the WZ Hamiltonian, we can establish the relations among the parameters. They are satisfied by

$$\Delta_2 = \Delta_3, \quad A_1 = A_2 + 2A_4,$$
$$A_3 = -2A_4 = \sqrt{2}A_6 - 4A_5, \quad A_7 = 0, \quad (8.3.15)$$
$$D_1 = D_2 + 2D_4, \quad D_3 = -2D_4 = \sqrt{2}D_6 - 4D_5.$$

Moreover, the following relations between A_i and γ_i can be established:

$$A_1 = -(\gamma_1 + 4\gamma_3),\ A_2 = -(\gamma_1 - 2\gamma_3),\ A_3 = 6\gamma_3,$$
$$A_4 = -3\gamma_3,\ A_5 = -(\gamma_2 + 2\gamma_3),\ A_6 = -\sqrt{2}(2\gamma_2 + \gamma_3). \quad (8.3.16)$$

8.3.4 Numerical derivation of physical parameters

8.3.4.1 Method of numerical derivation

Thanks to the first-principles FLAPW band calculations, we can obtain the energy levels for arbitrary $\boldsymbol{k}$ points. On the other hand, on the effective mass approximation, the electronic structure around the CBM is given by

$$E_c(\boldsymbol{k}, \varepsilon) = E_s^0 + \frac{\hbar^2 k^2}{2m_e^*} + a_c \varepsilon, \quad (8.3.17)$$

and the electronic structure around the VBM are obtained by solving the equation

$$D(\boldsymbol{k}, \varepsilon) = \det|\mathbf{H}_{6\times6}(\boldsymbol{k}, \varepsilon) - E_v(\boldsymbol{k}, \varepsilon)\mathbf{I}_{6\times6}| = 0. \quad (8.3.18)$$

However, the required physical parameters, such as effective masses and deformation potentials, are mostly unknown for the III–V nitrides. Thus, in order to derive these parameters, we link the first-principles band calculations with the effective mass approximation. The electron and hole effective masses, the Luttinger-like parameters, the crystal–field and spin–orbit splitting energies, deformation potentials, and the momentum matrix elements have been derived from reproducing the calculated first-principles band structures near the band edges by using the effective mass theory. In order to clarify the quantitative reliability of our procedure, it was applied to ZB GaAs at first. As a result, we obtained $m_e^* = 0.062m_0$ ($0.067m_0$), $\gamma_1 = 7.1$ (6.85, 6.95), $\gamma_2 = 2.2$ (2.10, 2.25) and $\gamma_3 = 2.4$ (2.90, 2.86) in units of $\hbar^2/2m_0$. The calculated values are in good agreement with the observed values in round brackets (Stillman *et al.* 1971; Hess *et al.* 1976; Skolnick *et al.* 1976). For the WZ structure, here, it is difficult to accurately determine such many parameters at the same time. Thus we adopted the quasi-cubic approximation for simplicity.

8.3.4.2 Crystal–field and spin–orbit splitting energies

The calculated results of the crystal–field and spin–orbit splitting energies are shown in Table 8.1. The calculated values of Δ_{so} for WZ and ZB GaN are in good agreement with the experimentally observed values (Dingle *et al.* 1971[a]; Chichibu *et al.* 1996[b]; Flores *et al.* 1994[c]). On the other hand, the calculated value of Δ_{cr} for WZ GaN is in disagreement with the observed values, though the sign ($\Delta_{cr} > 0$) and the relative size ($|\Delta_{cr}| > |\Delta_{so}|$) are correct. However, the electronic structures around the VBM are so sensitive to strains that further accurate measurements using homogeneous strain-free crystals are desirous for more detailed comparison. Moreover, we must perform the first-principles

Table 8.1 Crystal–field and spin–orbit splitting energies (meV) of GaN and AlN.

			$\Delta_{cr(1)}$	Δ_{so}	Δ_2	Δ_3
Wurtzite	GaN	calculated	72.9	15.6	5.2	5.2
		observed[a]	22(±2)	11(+5, −2)	–	–
		observed[b]	21	16	–	–
	AlN	calculated	−58.5	20.4	6.8	6.8
Zincblende	GaN	calculated		20		
		observed[c]		17(±1)		
	AlN	calculated		20		

calculations with lattice parameters free and investigate the dependence on the ratio c/a and the internal parameters u.

8.3.4.3 Luttinger-like parameters

The calculated results of the Luttinger-like parameters are shown in Table 8.2 All the parameters are obtained within the quasi-cubic approximation. The values of A_i transformed from γ_i, with the help of the quasi-cubic approximation, are also listed. We note that the differences between the direct and the transformed results are not so large, although the latter are obtained by simply transforming without taking the relaxation of c/a and u into account. Therefore the quasi-cubic approximation for the WZ structure is fairly good and useful. The similarity between the WZ and the ZB structures has been also investigated by other researchers (Yeh *et al.* 1992), where it is reported that the trend in the difference of bandgaps can be explained by band folding and symmetry changing. From our results and this result, it is clarified that the quasi-cubic approximation for the WZ structure is valid in the III–V nitride system.

8.3.4.4 Electron effective mass

The calculated electron effective masses are summarized in Table 8.3. The

Table 8.2 The Luttinger-like valence band parameters of GaN and AlN. All values, except A_7 (Ryd cm), are in units of $\hbar^2/2m_0$.

Wurtzite		A_1	A_2	A_3	A_4	A_5	A_6	A_7
GaN	calculated	−6.56	−0.91	5.65	−2.83	−3.13	−4.85	0.00
	$\gamma_i \Rightarrow A_i$	−6.98	−0.56	6.42	−3.21	−2.90	−3.66	0.00
AlN	calculated	−3.95	−0.27	3.68	−1.84	−1.95	−2.92	0.00
	$\gamma_i \Rightarrow A_i$	−3.98	−0.26	3.72	−1.86	−1.63	−1.98	0.00

Zincblende		γ_1	γ_2	γ_3				
GaN	calculated	2.70	0.76	1.07				
AlN	calculated	1.50	0.39	0.62				

Table 8.3 Electron effective masses (m_0) of GaN and AlN. m_e^* denotes the average electron effective mass, and the other superscripts stand for the k-dependency.

Wurtzite		m_e^*	$m_e^{\perp(\Sigma)}$	$m_e^{\perp(T)}$	$m_e^{\parallel(\Delta)}$
GaN	calculated	0.18	0.18	0.18	0.20
	observed[a]	0.20 ± 0.02	0.20 ± 0.02	0.20 ± 0.02	0.20 ± 0.06
	observed[b]	≈ 0.19	–	–	–
	observed[c]	0.25	–	–	–
	observed[d]	$0.20 \sim 0.30$	–	–	–
	observed[e]	0.27 ± 0.06	–	–	–
	observed[f]	0.22 ± 0.03	–	–	–
	observed[g]	0.236 ± 0.005	–	–	–
	observed[h]	0.20 ± 0.005	–	–	–
AlN	calculated	0.27	0.25	0.25	0.33

Zincblende		m_e^*	$m_e^{[100](\Delta)}$	$m_e^{[110](\Sigma)}$	$m_e^{[111](\Lambda)}$
GaN	calculated	0.17	0.17	0.17	0.18
AlN	calculated	0.30	0.30	0.30	0.30

experimentally observed values for WZ GaN are also listed, compared with the theoretically derived values. The calculated values for WZ GaN are in good agreement with the observed values (Barker and Ilegems 1973[a]; Kosicki et al. 1970[b]; Dingle and Ilegems 1971[c]; Cunningham et al. 1972[d]; Rheinländer and Neumann 1974[e]; Pankove et al. 1975[f]; Meyer et al. 1995[g]; Drechsler et al. 1995[h]). Comparing the electron effective mass of WZ GaN with that of ZB GaN, the former is slightly heavier than the latter. Generally speaking, the electron effective mass is proportional to the bandgap and inversely proportional to the square of the momentum matrix element between the CBM and the VBM. According to the first principles calculations as well as the observed results, the bandgap of WZ GaN is nearly equal to or slightly larger than that of ZB GaN. The momentum matrix elements in both WZ and ZB structures are almost the same due to the same local symmetry (see Table 8.5). Therefore, the above simple rule is consistent with our calculated result.

8.3.4.5 Hole effective masses

The Luttinger-like parameters are merely fitting parameters and not so physically meaningful compared with the hole effective masses. Thus, the feature of three hole bands around the VBM are discussed. The calculated hole masses are summarized in Table 8.4. According to the calculated results, the hole masses have a considerable k-dependence, and the lowest conduction band is strongly coupled with only one hole band in both structures. This is caused by the small spin–orbit interaction in the III–V nitrides. Furthermore, the coupled hole mass becomes much lighter than the other two hole masses. In Table 8.4, the experimentally observed values (Kosicki et al. 1970[a]; Dingle and Ilegems 1971[b]; Pankove et al. 1975[c]) are also listed. Experimentally, the k-dependence of

Table 8.4 Hole effective masses (m_0) of GaN and AlN. m_{hh}, m_{lh} and m_{ch} (m_{sh}) denote heavy, light, and crystal–field (spin–orbit) split-off hole masses, respectively, for the wurtzite (zincblende) structure.

	Wurtzite	m_{hh}	m_{lh}	m_{ch}	Zincblende	m_{hh}	m_{lh}	m_{sh}
GaN	$k_x (\Sigma)$	1.61	0.14	1.04	$[k00](\Delta)$	0.86	0.86	0.17
	$k_y (T)$	1.44	0.15	1.03	$[kk0](\Sigma)$	4.89	0.84	0.15
	$k_z (\Delta)$	1.76	1.76	0.16	$[kkk](\Lambda)$	1.74	1.74	0.15
					observed[a] $m_h^* \geq 0.6$			
					observed[b] $m_h^* = 1 \sim 2$			
					observed[c] $m_h^* = 0.8 \pm 0.2$			
AlN	$k_x (\Sigma)$	10.42	0.24	3.81	$[k00](\Delta)$	1.40	1.40	0.32
	$k_y (T)$	5.02	0.25	3.61	$[kk0](\Sigma)$	$\gg 10$	1.41	0.27
	$k_z (\Delta)$	3.53	3.53	0.25	$[kkk](\Lambda)$	3.72	3.72	0.25

hole masses has not yet been reported in detail. The calculated hole mass of the heavy hole band, which is the lowest in hole energy among the three hole bands, seems to be slightly heavier than observed one. However, both the crystal–field and the spin–orbit splittings are so small that the observed hole mass may be the averaged value among them, namely the density of states mass. Thus we estimated the typical hole mass by averaging the k-dependent masses with the weight of the star of k. In averaging the hole masses, two (heavy and light) or three (heavy, light, and crystal–field split-off) hole bands are considered. In both cases, the average hole mass m_h^* is about $0.95 \sim 1.10 m_0$. This is consistent with the experimentally observed values.

8.3.4.6 Momentum matrix elements

The momentum matrix elements of WZ GaN have been determined by fitting the energy dispersions with the analytical expression. The obtained values are shown in Table 8.5. The result shows that the anisotropy is not so large ($P_\parallel \simeq P_\perp$). Next, we examine whether the second forms of the analytical expressions can yield reasonable momentum matrix elements by giving the observed parameters. As the observed parameters, we adopted the following values: $E_g = 3.5$ eV (Monemar 1974), $m_e^*/m_0 = 0.20 \pm 0.02$ (Barker and Ilegems 1973), $\Delta_{cr} = 22$ meV, and $\Delta_{so} = 11$ meV (Dingle et al. 1971). The obtained result, which is also shown in Table 8.5, is in good agreement with the FLAPW results. This

Table 8.5 Momentum matrix elements of GaN and GaAs. All values are in units of $(\hbar^2/2m_0)$ Ryd.

		Wurtzite		Zincblende
		$P_\perp^2$	$P_\parallel^2$	P^2
GaN	FLAPW	0.96	1.13	–
	$k \cdot p$ (exp.)	1.04 ± 0.14	1.04 ± 0.14	1.03 ± 0.14
GaAs	$k \cdot p$ (exp.)	–	–	1.55

Table 8.6 Deformation potentials (eV) of GaN and AlN. The values obtained within the quasi-cubic approximation are also listed in parentheses.

Wurtzite	D_1	D_2	D_3	D_4	D_5	D_6
GaN	−13.87	−13.71	2.92	−1.56	−2.05	–
	(−15.35)	(−12.32)	(3.03)	(−1.52)	(−2.05)	(−3.66)
AlN	−12.34	−9.36	4.76	−2.08	−2.57	–
	(−12.92)	(−8.46)	(4.46)	(−2.23)	(−2.57)	(−4.12)

Zincblende	a	b	d
GaN	(−13.33)	(−2.09)	(−1.75)
AlN	(−9.95)	(−2.17)	(−2.57)

agreement shows that the influence of higher energy bands is negligible ($m_c^\parallel = m_c^\perp \simeq m_0$) under the condition of the experimental observations. In other words, if we use the observed bandgap and effective masses, then the second forms of the analytical expressions give the reasonable momentum matrix elements. Taking this approach, the momentum matrix elements of ZB GaN and GaAs can be also calculated. As the observed parameters, we used the following values for ZB GaAs: $E_g = 1.42$ eV (Stillman et al. 1971), $m_e^*/m_0 = 0.067$ (Stillmann et al. 1970) and $\Delta_{so} = 340$ meV (Williams and Rehn 1968). For ZB GaN, the same values that we used for WZ GaN were adopted. It is found that the momentum matrix element of ZB GaN is almost the same as that of ZB GaN because of the same local symmetry and also that that of GaN is about 30% smaller than that of ZB GaAs.

8.3.4.7 Deformation potentials

In order to derive the deformation potentials of WZ GaN and AlN, we have performed further FLAPW band calculations under conditions with some kind of strain, which are biaxial and unaxial strains in the (0001) plane and a uniaxial strain along the [0001] direction within a few per cent in all cases. The deformation potentials have been determined by a linear fitting of the energy shift at the Γ point. The obtained values are shown in Table 8.6, compared with the values obtained within the quasi-cubic approximation. It is found that the quasi-cubic approximation is considerably good for the deformation potentials as well. This means that we can use the quasi-cubic approximation in deriving the deformation potentials of ZB nitrides for simplicity. The deformation potentials of ZB GaN and AlN have been obtained by transforming D_i for WZ nitrides. The results obtained are listed in Table 8.6.

8.4 Subband structures of GaN / AlGaN quantum wells

8.4.1 Method of subband calculations

In this section, the subband structures in the QWs are treated. They are

8.4 Subband structures of GaN / AlGaN quantum wells

described by the direct product between the rapidly varying Bloch wavefunction and the slowly varying envelope function. In general, the latter is calculated within the effective mass approimation (Luttinger and Kohn 1955). In treating the QW structures, a confining potential $V(z)$ along the growth direction z is added and k_z becomes an operator in the $\mathbf{k} \cdot \mathbf{p}$ Hamiltonian. For the conduction band states, it is enough that only the lowest Γ_7 state is considered. For the valence band states, the upper six states must be explicitly treated in both the WZ and ZB structures. Because they are very close to one another, due to the small spin–orbit and crystal–field splitting energies, in order to consider the mixing between these valence band states the WZ and ZB 6×6 Hamiltonians are used.

The wavefunctions at the valence (conduction) bands can be expressed by

$$\Psi_n^{v(c)}(\mathbf{k}_\perp, z) = \sum_\nu \psi_{n,\nu}^{v(c)}(\mathbf{k}_\perp, z) u_\nu^{v(c)} e^{i\mathbf{k}_\perp \cdot \mathbf{r}_\perp}, \qquad (8.4.1)$$

where $u_\nu^{v(c)}$ are the Bloch functions at the VBM (CBM), $\psi_{n,\nu}^{v(c)}(\mathbf{k}_\perp, z)$ are the envelope functions, and n is the subband index. In the analyses of the WZ nitrides, we take the following basis functions:

$$u_1^c = |S, \uparrow\rangle, \qquad u_2^c = |S, \downarrow\rangle, \qquad (8.4.2)$$

$$u_1^v = \frac{1}{\sqrt{2}} |X + iY, \uparrow\rangle, \; u_2^v = \frac{1}{\sqrt{2}} |X + iY, \downarrow\rangle, \qquad (8.4.3)$$

$$u_3^v = |Z, \uparrow\rangle, \qquad u_4^v = |Z, \downarrow\rangle, \qquad (8.4.4)$$

$$u_5^v = \frac{1}{\sqrt{2}} |X - iY, \uparrow\rangle, \; u_6^v = \frac{1}{\sqrt{2}} |X - iY, \downarrow\rangle. \qquad (8.4.5)$$

These envelope functions satisfy

$$\sum_{\nu'} \left\{ H_{\nu\nu'}^{v(c)}\left(\mathbf{k}_\perp, \frac{\hbar}{i}\frac{\partial}{\partial z}\right) + V_\nu^{v(c)}(z)\delta_{\nu,\nu'} + H_{\nu\nu'}^{v(c)}(\varepsilon) \right\} \psi_{n,\nu'}^{v(c)}(\mathbf{k}_\perp, z)$$

$$= E_n^{v(c)}(\mathbf{k}_\perp)\psi_{n,\nu}^{v(c)}(\mathbf{k}_\perp, z), \qquad (8.4.6)$$

where $\nu = 1, 2, \ldots, 6$ ($\nu = 1, 2$) for the valence (conduction) band states, and

$$V_\nu^v(z) = \begin{cases} V_{hh}(z) & \text{(for } \nu = 1, 6) \\ V_{lh}(z) & \text{(for } \nu = 2, 5) \\ V_{ch}(z) & \text{(for } \nu = 3, 4). \end{cases} \qquad (8.4.7)$$

The Hamiltonian H^c is already diagonal. Furthermore, since the CBM mostly consists of s-like states, the conduction subband dispersions are nearly parabolic. Thus, the following discussion is limited to the valence subband structure.

For the valence subband states, the envelope functions can be described by

$$\psi_{n,v}^{v}(\mathbf{k}_\perp, z) = \sum_v \sum_m C_{n,v}^m(\mathbf{k}_\perp) \phi_m^\mu(z), \qquad (8.4.8)$$

where ϕ_m^μ ($\mu = $ hh, lh, ch) are the eigenfunctions for the diagonal element operators, which satisfy the following equations:

$$\left\{ (A_1 + A_3) \frac{\partial^2}{\partial z^2} + \Delta_1 + \Delta_2 + V_{\text{hh}}(z) \right\} \phi_m^{\text{hh}}(z) = \varepsilon_m^{\text{hh}} \phi_m^{\text{hh}}(z), \qquad (8.4.9)$$

$$\left\{ (A_1 + A_3) \frac{\partial^2}{\partial z^2} + \Delta_1 - \Delta_2 + V_{\text{lh}}(z) \right\} \phi_m^{\text{lh}}(z) = \varepsilon_m^{\text{lh}} \phi_m^{\text{lh}}(z), \qquad (8.4.10)$$

$$\left\{ A_1 \frac{\partial^2}{\partial z^2} + V_{\text{ch}}(z) \right\} \phi_m^{\text{ch}}(z) = \varepsilon_m^{\text{ch}} \phi_m^{\text{ch}}(z). \qquad (8.4.11)$$

When $V_\mu(z)$ are symmetric in the z direction, each subband m has two-fold degeneracy. Note that ϕ_m^μ are not the eigenstates at $\mathbf{k} = 0$, since the off-diagonal element $\sqrt{2}\Delta_3$, which is the spin–orbit coupling between LH and CH bands, is not considered. However, the boundary conditions on ϕ_m^μ at the heterojunctions are consistent with those of the eigenstates at $\mathbf{k} = 0$, since the matrix element $\sqrt{2}\Delta_3$ is even with respect to k_z. Then the differential eigenvalue problem is transformed to an algebraic one with the coefficients of $C_{n,v}^m(\mathbf{k}_\perp)$. This procedure has been already introduced in the study of the valence subband structure of GaAs/AlGaAs QWs (Broido and Sham 1985). Therefore the subband structure can be obtained by solving a matrix differential eigenalue problem.

As for the required physical parameters, we used our theoretically derived values (see Section 8.3) except for the elastic stiffness constants, at which the experimentally observed values for WZ GaN are adopted (Sheleg and Savastenko 1979). As for the elastic stiffness constants of ZB GaN, they can be obtained by transforming the elastic stiffness tensor of WZ GaN. The band offsets have been also determined by the first-principles FLAPW calculations. The values of the band offsets in the $GaN/Al_{0.2}Ga_{0.8}N$ interface and the elastic stiffness constants of GaN are listed in Table 8.7. Using these parameters, the subband structure in WZ and ZB $GaN/Al_{0.2}Ga_{0.8}N$ SQWs were calculated.

Table 8.7 Elastic stiffness constants (10^{11} dyn/cm^2) of GaN and band offsets (eV) of $GaN/Al_{0.2}Ga_{0.8}N$ interface.

	GaN					$GaN/Al_{0.2}Ga_{0.8}N$	
	C_{11}	C_{12}	C_{13}	C_{33}	C_{44}	ΔE_v	ΔE_c
Wurtzite	29.6	13.0	15.8	26.7	2.41	0.11	0.43
Zincblende	25.3	16.5	–	–	6.05	0.07	0.44

8.4 Subband structures of GaN/AlGaN quantum wells

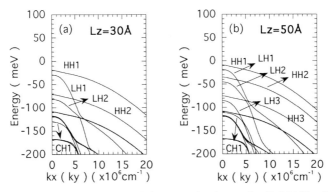

Fig. 8.6 Valence subband structures of the unstrained wurtzite GaN/Al$_{0.2}$Ga$_{0.8}$N single quantum wells, with the well lengths L_z being (a) 30 Å and (b) 50 Å.

8.4.2 Subband in wurtzite quantum wells

The valence subband structures in WZ GaN/Al$_{0.2}$Ga$_{0.8}$N QWs, with the well lengths being (a) 30 Å and (b) 50 Å, are shown in Fig. 8.6, where the strain effect by the lattice mismatch between GaN and Al$_{0.2}$Ga$_{0.8}$N is neglected. As long as the $\boldsymbol{k} \cdot \boldsymbol{p}$ Hamiltonian is expanded within the second order of $\boldsymbol{k}$, the energy dispersions are isotropic in the k_x–k_y plane. This is very different from conventional ZB QWs, and one of the remarkable features in WZ QWs. In the subband structure of WZ QWs, the energy bands can be labeled as HH$_i$, LH$_i$, and CH$_i$, as well as in WZ bulk GaN (see Fig. 8.3(b)). For any well length, the coupling between HH$_i$ bands and the other kinds of bands is weak, and the HH$_i$ bands are nearly parabolic. On the other hand, the LH$_i$ bands are strongly coupled with the CH$_i$ bands with a different parity, through the matrix elements H or I including $A_6 k_z k_\pm$. However, the feature of hole bands typically shown by the above three labels is almost unchanged, even in the QWs, because the crystal symmetry is still C$_{6v}$ as well as in the bulk. This character originates from the hexagonal crystal symmetry and the weak spin–orbit coupling of the nitrogen atom. Thus, it also seems to be intrinsic in WZ nitride QWs and to be very different from conventional ZB GaAs/AlGaAs QWs. In the case of conventional ZB QWs, the heavy hole mass becomes light and the light hole mass becomes heavy, compared with the bulk, due to the symmetry lowering.

Next, the well length dependence of the subband structure is discussed. Roughly speaking, the subband energies at the Γ point are inversely proportional to the bulk's hole masses along the confinement direction and to the square of the well length. This is generally called the quantum size effect. As for the bulk's masses along the k_z direction, the LH band is as heavy as the HH band and much heavier than the CH band. This is the reason why only CH$_i$ bands are more split off from the HH$_i$ and LH$_i$ bands in the QWs than in the bulk. In other words, the crystal-field splitting energy Δ_{cr} becomes effectively large due to the quantum size effect. For $L_z = 30$ Å, the CH$_1$ band is the only bound state among the CH$_i$ bands, and the LH$_2$–CH$_1$ coupling causes the non-parabolicity of the LH$_2$ band. For $L_z = 50$ Å, the interval between the same

326 *Electronic and optical properties of GaN-based quantum wells*

labeled subbands becomes shorter than for $L_z = 30$ Å due to the quantum size effect; then the DOS around the VBM becomes a little larger. In WZ nitride QWs, thus, there is little well length dependence of the subband structure at the top of the valence band due to very weak spin–orbit coupling and the unchanged symmetry. Therefore the difference between the QWs with different well lengths is dependent on the number of bound states alone.

Here, once again, we pay attention to the fact that the subband energy dispersions have similar characteristics to the bulk ones. The diagonal operators F and G in the $\mathbf{k}\cdot\mathbf{p}$ Hamiltonian have the same parameters $A_1 + A_3$ and Δ_1, and $\Delta_2 \sim 5$ meV is much smaller than the valence band offset $\Delta E_v \sim 100$ meV, so that ϕ_m^{hh} and ϕ_m^{lh} are almost same eigenfunctions, i.e. $\langle \phi_m^{hh} | \phi_{m'}^{lh} \rangle \sim \delta_{m,m'}$. Here, the mixing between HH_i and LH_i with the same indices through $A_5 k_{\pm}^2$ is very strong but meaningless, because its effect is excluded by transforming the basis function to $|X\rangle$, $|Y\rangle$, and $|Z\rangle$, and any mixing other than with these is almost zero. Therefore the non-parabolicity of the energy dispersions is very weak. Then the hole effective masses in the k_x–k_y plane are approximately given by $m_{hh} \sim -(A_2 + A_4 - A_5)^{-1} = 1.64 m_0$, and $m_{lh} \sim -(A_2 + A_4 + A_5)^{-1} = 0.15 m_0$. These are the same as the bulk's hole masses. Generally, this property would be observed for the other III–V nitrides as well, since the spin–orbit coupling of the nitrogen atom is negligible. As for the CH bands, CH_1 is so weakly coupled with LH_2 through $A_6 k_z k_{\pm}$ that m_{ch} is about equal to A_2^{-1} and still heavy. This weak coupling modifies only the LH_2 band at $k \sim 6 \times 10^6$ cm^{-1} compared with the LH_1 and LH_3 bands, due to the same parity of the envelope functions of the LH_1 and LH_3 band as that of the CH_1 band with respect to the $k_{\pm}$.

On the other hand, in the extreme case where Δ_2 is infinity, the hole effective masses in the plane of the quantum well are given by $m_{hh} \sim -(A_2 + A_4)^{-1} = 0.27 m_0$, and $m_{lh} \sim -(A_2 + A_4/3)^{-1} = 0.54 m_0$. Note that m_{hh} becomes lighter than m_{lh}. This property has been observed in conventional ZB GaAs/AlGaAs QWs. Therefore it is concuded that the decrease of the hole effective mass by the two-dimensional carrier confinement is prevented by the small spin–orbit splitting energy.

8.4.3 Subband in zincblende quantum wells

The valence subband structures in ZB GaN/Al$_{0.2}$Ga$_{0.8}$N QWs along the (a) [k00] and (b) [kk0] directions, with the well lengths 40 Å, are shown in Fig. 8.7, where the strain effect caused by the lattice mismatch between GaN and Al$_{0.2}$Ga$_{0.8}$N is neglected. In the subband structure, the energy bands were labeled as HH_i, LH_i, and SH_i, conventionally (see Fig. 8.3(c)). The energy dispersions in the k_x–k_y plane are isotropic in the very close vicinity of the Γ point but not isotropic in the longer wavelength region than $|\mathbf{k}| \sim 3 \times 10^6$ cm^{-1}. This is very different from WZ GaN/AlGaN QWs and conventional ZB GaAs/AlGaAs QWs. For any well length, since the LH_i bands are strongly coupled with the SH_i bands even at $\mathbf{k} = 0$, the energy dispersions show more

8.4 Subband structures of GaN / AlGaN quantum wells

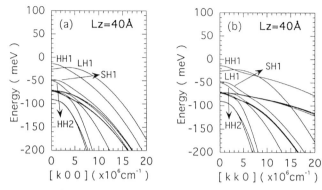

Fig. 8.7 Valence subband structures of the unstrained zincblende GaN/Al$_{0.2}$Ga$_{0.8}$N single quantum wells along (a) the [k00] and (b) the [kk0] directions, with the well lengths L_z being 40 Å.

complicated behavior. On the other hand, the HH$_i$ bands do not couple with the LH$_i$ and the SH$_i$ bands at $\boldsymbol{k}=0$ and little at $\boldsymbol{k}\neq 0$. Furthermore, in ZB GaN-based QWs, the heavy hole mass becomes light and the light hole mass becomes heavy due to symmetry lowering. This is the same as in conventional ZB QWs. However, the heavy hole mass in ZB GaN-based QWs is still quite heavy because both heavy and light hole masses in bulk ZB GaN are much heavier than those in conventional ZB III−V semiconductors like GaAs.

Here, as an important remark, we would like to mention the labeling of the ZB valence subbands under the conduction of strong LH$_i$–SH$_i$ coupling, or that of the weak spin−orbit coupling. Conventionally, the subbands are labeled according to their character at $\boldsymbol{k}=0$. It is well known that the LH$_i$ band is coupled to the SH$_i$ band more effectively than the HH$_i$ band near the Γ point. However, most authors labeled the subbands as 'LH$_i$' or 'SH$_i$' to describe the subbands. These labels cannot correctly express the subband structure, especially for such semiconductors as GaN and AlGaInP with the small spin−orbit splitting energy. If the coupling effects are considered, there are no pure LH$_i$ or SH$_i$ bands, even at the Γ point. It may be more appropriate to label the coupled subband states between LH$_i$ and SH$_i$ bands as LS$_i$ (light-hole and spin−orbit split-off coupled) and SL$_i$ (spin−orbit split-off and light-hole coupled), where the first latter denotes the main component of the wavefunction. Moreover, if there is little spin−orbit interaction, it must be more reasonable that the subbands are labeled as X_i, Y_i, Z_i. In this chapter, however, we follow the conventional notation to avoid confusion.

8.4.4 Strained quantum well structure

Experimentally, we can introduce a built-in strain to an epitaxial layer by growing it on a lattice-mismatched substrate. As long as the mismatched epitaxial layer is below the critical thickness, the produced strain is uniform and no dislocations are induced. As a result, the in-plane lattice constant of the epitaxial layer is fitted to that of the substrate, and the out-of-plane lattice constant is adjusted to a new lattice constant according to Hooke's law. Then, the subband structure is modified by introducing a built-in strain, and the strain has a dramatic influence on the electronic and optical properties of the system. By incorporating strain in QW lasers, some important advantages can be obtained. Thus, we explain the motivation for incorporating strain in the QWs.

There are some motivations for using a strain in conventional ZB QW lasers, as follow:

- Bandgap tunability: it is possible to adjust the bandgap (energy level) to reach a certain emission wavelength on a short scale, compared with alloying.
- Threshold reduction: the in-plane strain removes the degeneracy at the top of valence bands. It causes the decreased hole mass and the reduced DOS around the top of valence bands. As a result, the population inversion, namely the laser oscillation, can be reached at a lower injected carrier density.
- Polarization control: the HH state couples only to the TE mode. The LH state couples to the TM mode four times as strongly as it coupled to the TE mode. If the separations between the HH and the LH states and their occupations are changed, it is possible to control the optical polarization of the laser light.

These possibilities are available to us when conventional ZB QW lasers for specific applications are designed, and so are the III–V nitride QW lasers.

8.4.5 Subband in strained wurtzite quantum wells

The valence subband structures in WZ GaN/$Al_{0.2}Ga_{0.8}$N QWs with a compressive biaxial strain are shown in Fig. 8.8, where the well lengths are (a) 30 Å and (b) 50 Å. The compressive biaxial strain is induced only into the (0001) plane of

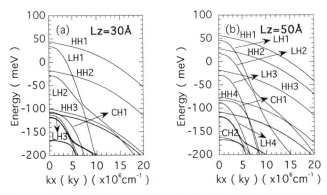

Fig. 8.8 Valence subband structures of the strained wurtzite GaN/$Al_{0.2}Ga_{0.8}$N single quantum wells with 0.5% compressive biaxial strain, with the well lengths L_z being (a) 30 Å and (b) 50 Å.

8.4 *Subband structures of GaN/AlGaN quantum wells* 329

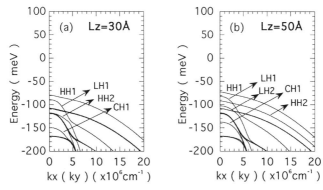

Fig. 8.9 Valence subband structures of the strained wurtzite GaN/Al$_{0.2}$Ga$_{0.8}$N single quantum wells with 0.5% tensile biaxial strain, with the well lengths L_z being (a) 30 Å and (b) 50 Å.

the GaN well layer due to the lattice mismatch with the Al$_{0.2}$Ga$_{0.8}$N barrier layers, and it is 0.5% in the (0001) plane. Furthermore, the valence subband structures in WZ GaN/Al$_{0.2}$Ga$_{0.8}$N QWs with a tensile biaxial strain are shown in Fig. 8.9, where the well lengths are (a) 30 Å and (b) 50 Å. The tensile biaxial strain is also introduced only into the GaN well layer, virtually by 0.5%. The main feature of hole subbands in the biaxial strained WZ QWs is the same as that in the unstrained WZ QWs. The difference between them can be qualitatively understod by the change of the well's depth and the strain effect on the bulk. Considering compressive (tensile) biaxial strains, we find that the QWs become deep (shallow) and the number of subbands tends to increase (decrease). For $L_z = 30$ Å, the HH$_3$ and the LH$_3$ bands also come into the QWs as the bound states. For $L_z = 50$ Å, on the other hand, the CH$_2$ band as well as the HH$_4$ and LH$_4$ bands come into the QWs. Then the LH$_1$–CH$_2$ coupling causes the non-parabolicity of the LH$_1$ band as well as of the LH$_2$ band. For both well lengths, furthermore, the energy difference between the HH$_1$ and HH$_2$ bands becomes large (small) due to the deeper (shallower) well's depth, and the energy difference between the LH$_i$ and the CH$_i$ bands also becomes large (small) due to the biaxial strain effect as well as in the bulk. This is the reason why the DOS around the VBM is smaller (larger) in the compressive (tensile) biaxial strained QWs that in the unstrained QWs. Namely, the compressive (tensile) biaxial strain plays a decreasing (increasing) role in the DOS around the VBM. However, the change of the DOS is quite small because the symmetry is the same as the bulk and there is no further removal of the degeneracy.

Figure 8.10 shows the valence subband structure in WZ GaN/Al$_{0.2}$Ga$_{0.8}$N QWs with a uniaxial strain. The uniaxial strain is introduced into both the GaN well layer and the Al$_{0.2}$Ga$_{0.8}$N barrier layers, with 1.0% tensile strain along the *x*-direction, where the GaN well layer is still lattice-matched to the Al$_{0.2}$Ga$_{0.8}$N barrier layers. This situation corresponds to epitaxial growth on the uniaxial strained hexagonal substrate, namely orthorhombic (C_{2v}) or monoclinic (C_{2h}) substrates. In the uniaxial strained QWs, each energy dispersion of the HH$_i$,

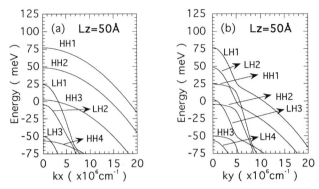

Fig. 8.10 Valence subband structure of the strained wurtzite GaN/Al$_{0.2}$Ga$_{0.8}$N single quantum wells, with 1.0% uniaxial tensile strain along the x-direction, along (a) the k_x- and (b) the k_y-directions. The well length L_z is 50 Å.

LH$_i$, and CH$_i$ bands is similar to the unstrained QWs. However, the energy splittings between the HH$_i$ and LH$_i$ bands are not isotropic in the k_x–k_y plane due to the symmetry lowering from C$_{6v}$ to C$_{2v}$. The tensile (compressive) strain along the x-direction (y-direction) makes not only the LH$_i$ bands along the k_y-direction but also the HH$_i$ bands along the k_x-direction move to the higher energy side. Therefore, the DOS at the VBM is more remarkably reduced in the uniaxial strained QWs than in the biaxial strained QWs, and the population inversion would be more easily realized. This feature is almost the same as that of bulk WZ GaN due to the weak spin–orbit coupling.

8.4.6 Subband in strained zincblende quantum wells

Figure 8.11 show the valence subband structures along the [k 00] direction in ZB GaN/Al$_{0.2}$Ga$_{0.8}$N QWs with (a) compressive and (b) tensile biaxial strains, with the well lengths being 40 Å. The compressive biaxial strain is induced into the

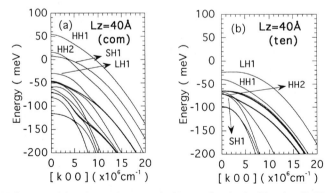

Fig. 8.11 Valence subband structures of the strained zincblende GaN/Al$_{0.2}$Ga$_{0.8}$N single quantum wells with (a) 0.5% compressive and (b) 0.2% tensile biaxial strains, with the well length L_z being 40 Å.

(001) plane of the GaN well layer by 0.5%, which corresponds to the lattice mismatch between the GaN well layer and the $Al_{0.2}Ga_{0.8}N$ barrier layers. The tensile biaxial strain is introduced by 0.2%, virtually the same as in the case of WZ QWs. Here, since 0.5% tensile biaxial strain causes almost zero barrier potential, we have no interest in it. In the biaxial strained ZB nitride QWs, the energy dispersions near the Γ point of the HH_i, LH_i and SH_i bands are similar to those in the unstrained QWs. Because the HH_i bands couple little with the LH_i and SH_i bands, even in the unstrained QWs, due to the weak spin–orbit interaction, there is no further removal of the degeneracy due to the same D_{2d} symmetry as that of the unstrained QWs. However, the biaxial strain makes the order among the energy levels changed. Under the condition with the compressive (tensile) biaxial strain, the HH_i and SH_i bands move to the higher (lower) energy side and the LH_i bands more to the lower (higher) energy side. In the compressive biaxial strained QWs, the SH_1 band is higher than the LH_i. Then the top of valence bands is still HH_1, arising from mostly $|X\rangle$ and $|Y\rangle$ states. On the other hand, in the tensile biaxial strained QWs, the LH_1 band, which is a mostly $|Z\rangle$ character, is higher than the HH_1 and becomes the top of the valence bands. In both cases, the biaxial strain cannot cause the decrease of the hole mass at the VBM. However, since the HH_1 and LH_1 bands are more separated from each other, the DOS around the VBM is considerably reduced. This is the reason why the biaxial strains are more effective on the reduction of the DOS at the VBM, as well as in conventional ZB QWs, than those in WZ QWs. Here, although the biaxial strain effect on ZB nitride QWs looks quite strange, compared with conventional ZB QWs, it seems to be essentially the same as that on conventional ZB QWs. The difference between them is caused by the weak spin–orbit coupling of nitrogen.

8.5 Optical gains of bulk GaN and GaN / AlGaN quantum wells

8.5.1 Optical gain and crystal symmetry

In this section, we discuss the optical gain on the assumption of the quasi-Fermi level. The optical gain is defined as the negative of the absorption coefficient. When the optical gain is balanced with the laser cavity loss, the lasing operation starts. Therefore the optical gain is one of the important factors deciding the threshold carrier density of semiconductor lasers. At first, we discuss the relation between optical gain and crystal symmetry. The difference between the WZ and the ZB structures is merely the relative positions of the third neighbors and beyond. Due to this difference, the crystal symmetry of the former is hexagonal and that of the latter cubic. Generally, optical gain is expressed as

$$g(\omega) = \frac{2\pi \bar{n}}{\hbar c} \sum_k \delta(\omega - E_k^e - E_k^h)$$
$$\times \left| \langle e, k | \frac{e}{m_0 c} \mathbf{A} \cdot \mathbf{p} | h, k \rangle \right|^2$$
$$\times \{ f_e(E_k^e) + f_h(E_k^h) - 1 \}, \qquad (8.5.1)$$

where the ks are quantum numbers in the bulk and c is the velocity of light. $\bar{n}$ is the refractive index, as the value of which the observed value for WZ GaN ($\bar{n} = 2.67$) (Ejder 1971) was used in both crystal structures. f_e and f_h are electron and hole distribution functions, respectively. The distribution functions are defined as

$$f_{e(h)}(E^{e(h)}) = \{1 + e^{(E^{e(h)} - E_F^{e(h)})/k_B T}\}^{-1}, \qquad (8.5.2)$$

where E_F^e and F_F^h are the quasi-Fermi levels in the conduction and valence bands, respectively. The calculations were performed under the condition of room temperature ($T = 298$ K). Here, if the intraband relaxation time τ_{in} is considered, the δ-function in the first factor is changed as follows:

$$\delta(\omega - E_k^e - E_k^e - E_k^h) \Rightarrow \frac{1}{\pi} \frac{\hbar/\tau_{in}}{(\omega - E_k^e - E_k^h)^2 + (\hbar/\tau_{in})^2}. \qquad (8.5.3)$$

In the calculations, we assumed that the intraband relaxation time τ_{in} is the same as that of conventional ZB semiconductors, i.e. $\tau_{in} = 0.1$ ps, though this is not sure for GaN.

The optical gain consists of three parts: the joint density of states (JDOS), the optical transition matrix elements, and the occupation factor, in that order. Since the electron masses are much smaller than the hole masses, the JDOS is governed by the electron masses. In both the WZ and the ZB structures, they are not so different because of the s-like states at the CBM. Thus, the JDOS seem to be not so sensitive to the crystal symmetry. The optical transition matrix elements are obtained by sandwiching an optical polarization between the conduction and valence band states. Since the splitting of the p-like states at the VBM is affected by the crystal-field, the optical transition matrix elements are strongly related to the crystal symmetry. As for the occupation factor, the positive sign indicates the occurrence of a population inversion. The quasi-Fermi level in f_h is dependent on the density of states at the VBM, where the degeneracy is governed by the symmetry. Thus, the occupation factor is also affected by the crystal symmetry.

In the ZB case, the bulk's crystal symmetry is cubic, and there are three equivalent symmetry axes. The p-like six-fold degenerated VBM is separated into a four-fold degenerated Γ_8 state and a two-fold degenerated Γ_7 state by the spin–orbit coupling. When the QW structure is grown along the [001] direction, the crystal symmetry is changed to the tetragonal. Then there are two equivalent symmetry axes, and the degeneracy of the Γ_8 state is removed. In general, the removal of the degeneracy causes the reduced DOS. Thus, it is expected that the DOS is reduced by using the QW structure. In the WZ case, the crystal symmetry of the QW structure grown along the [0001] direction is the same hexagonal as that of the bulk. Then there is one symmetry axis, and the VBM is separated into three two-fold degenerated states, one Γ_9 state and two Γ_7 states. Since it makes no difference in the symmetry whether the QW structure is adopted, the band mixing is almost unchanged. Thus, adopting the QW structure is not so useful for reducing the DOS. This is one of the salient features in the difference between the WZ and the ZB structures.

8.5.2 Optical gain of bulk GaN

In WZ GaN, the small spin–orbit coupling makes it possible to express the three hole bands approximately as follows:

$$\left|\Gamma_9^6(\text{HH}); \pm\tfrac{3}{2}\right\rangle = \frac{1}{\sqrt{2}}\left|X \pm iY, \pm\tfrac{1}{2}\right\rangle \quad (8.5.4)$$

$$\left|\Gamma_7^6(\text{LH}); \pm\tfrac{1}{2}\right\rangle \sim \frac{1}{\sqrt{2}}\left|X + iY, \mp\tfrac{1}{2}\right\rangle \quad (8.5.5)$$

$$\left|\Gamma_7^1(\text{CH}); \pm\tfrac{1}{2}\right\rangle \sim \left|Z, \pm\tfrac{1}{2}\right\rangle. \quad (8.5.6)$$

The basis $|\Gamma_i^j; m_J\rangle$ on the left-hand side indicates the eigenstates at the Γ point and the quantum number of the z-component of a total angular momentum operator J. The basis $|L, m_\sigma\rangle$ on the right-hand side indicates the orbital character and the quantum number of the z-component of a spin angular momentum operator σ_z. Along the k_z-direction, the hole masses of the upper two bands are very heavy, and that of the lowest CH band only is light, due to the strong coupling with the CBM through the $k_z p_z$ perturbation. Along the k_x-direction, on the other hand, the CBM is strongly coupled with the mixed state $\frac{1}{\sqrt{2}}(|\Gamma_9^6; \pm\tfrac{3}{2}\rangle + |\Gamma_7^6; \mp\tfrac{1}{2}\rangle) \sim |X, \pm\tfrac{1}{2}\rangle$, and the hole mass of the only mixed band is light. Here, the lowest CH band is mostly composed of the $|Z\rangle$ character, which yields the optical gain for the TM mode. This is the reason why a large carrier density is needed to obtain the TM mode optical gain and why the TE mode optical gain is dominant in bulk WZ GaN.

In ZB GaN, the three hole bands can be expressed by

$$\left|\Gamma_8^{15}(\text{HH}); \pm\tfrac{3}{2}\right\rangle = \frac{1}{\sqrt{2}}\left|X \pm iY, \pm\tfrac{1}{2}\right\rangle \quad (8.5.7)$$

$$\left|\Gamma_8^{15}(\text{LH}); \pm\tfrac{1}{2}\right\rangle = \frac{1}{\sqrt{6}}(\left|X \pm iY, \mp\tfrac{1}{2}\right\rangle \mp 2\left|Z, \pm\tfrac{1}{2}\right\rangle) \quad (8.5.8)$$

$$\left|\Gamma_7^{15}(\text{SH}); \pm\tfrac{1}{2}\right\rangle = \frac{1}{\sqrt{3}}(\left|X \pm iY, \mp\tfrac{1}{2}\right\rangle \pm \left|Z, \pm\tfrac{1}{2}\right\rangle). \quad (8.5.9)$$

Since the hole masses of the upper two bands are very heavy along any k direction, the DOS at the VBM is larger than that in WZ GaN, as along as $\Delta_{\text{cr}} > \Delta_{\text{so}}$. Here, the LH band as well as the SH band includes the $|Z\rangle$ character, whose coefficients are independent of the spin–orbit splitting energy. Thus, the TM mode optical gain will start with the TE mode one at the same carrier density. This is very different from WZ GaN. Figure 8.12(a) shows the maximum optical gain of bulk GaN. The transparent carrier density of the TE mode in WZ GaN is lower than that in ZB GaN due to the small DOS at the VBM. However, the transparent carrier density of the TM mode in WZ GaN is much higher than that in ZB GaN due to $\Delta_{\text{cr}} > \Delta_{\text{so}}$, and it is not observed in the range of Fig. 8.12.

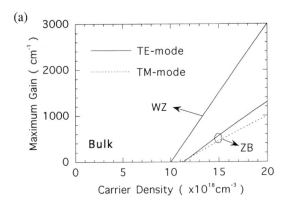

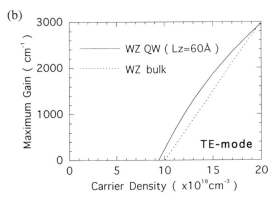

Fig. 8.12 (a) Optical gain of the wurtzite and the zincblende bulk GaN. (b) Optical gain of the wurtzite bulk GaN and the wurtzite GaN/Al$_{0.2}$Ga$_{0.8}$N single quantum wells with well lengths $L_z = 60$ Å.

8.5.3 Optical gain of wurtzite quantum wells

The optical gain of the QW structure is given by

$$g(\omega) = \frac{2\pi \bar{n}}{\hbar c} \sum_{n,m} \sum_{k_\perp} \delta\left(\omega - E^e_{n,k_\perp} - E^h_{m,k_\perp}\right)$$

$$\times \left|\langle e, n, k_\perp | \frac{e}{m_0 c} \mathbf{A} \cdot \mathbf{p} | h, m, k_\perp \rangle\right|^2$$

$$\times \left\{f_e\left(E^e_{n,k_\perp}\right) + f_h\left(E^h_{m,k_\perp}\right) - 1\right\}, \qquad (8.5.10)$$

where $m(n)$ represents the valence (conduction) subband label, and $k_\perp$ is the two-dimensional wavevector. Using the subband structures and the bulk's momentum matrix elements, we have calculated the optical gains with respect to the interband transition with the same wavevector k. The heterojunctions are perpendicular to the [0001] directions.

8.5 Optical gains of bulk GaN and GaN/AlGaN quantum wells

In the bulk WZ nitrides, the hybridization of the CH band with the HH and LH bands is negligible at the VBM, due to the small spin−orbit splitting energies. Thus, the eigenstates along the k_x (k_y) direction can be approximately expressed as

$$|HH\rangle \sim |Y(X)\rangle, |LH\rangle \sim |X(Y)\rangle, |CH\rangle \sim |Z(Z)\rangle.$$

In WZ QWs, the CH_i bands are more split off from the HH_i and LH_i bands. However, the band mixing is the same as that in the bulk GaN due to no change of the symmetry. Thus, the HH_1 mass is still heavy, and the DOS at the VBM is not so reduced. Then, the $|Z\rangle$ character, which yields the optical gain for the TM mode, is very low at the VBM. This is the reason why the optical gain for the TE mode is dominant in WZ GaN-based QWs. The following discussion is limited to the TE mode, and we take the optical polarization for the electric field as the y-direction in the calculations for TE mode.

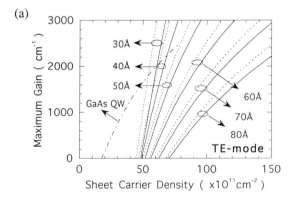

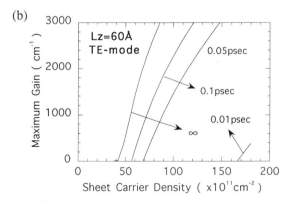

Fig. 8.13 (a) Optical gain of the wurtzite GaN/Al$_{0.2}$Ga$_{0.8}$N single quantum wells, with the well lengths $L_z = 30 \sim 80$ Å. The solid and dotted lines stand for the results without and with 0.5% compressive biaxial strain, respectively. The calculated result for the zincblende GaAs/Al$_{0.3}$Ga$_{0.7}$As single quantum wells with well length L_z being 80 Å is shown by the dash-dotted line. (b) Effect of intraband relaxation on optical gain of the wurtzite GaN/Al$_{0.2}$Ga$_{0.8}$N single quantum wells, with well lengths L_z being 60 Å.

336 *Electronic and optical properties of GaN-based quantum wells*

The maximum optical gain of bulk WZ GaN and WZ GaN/Al$_{0.2}$Ga$_{0.8}$N QWs with the well length $L_z = 60$ are shown in Fig. 8.12(b). As explained in the previous subsection, the threshold carrier density is not so reduced by adopting the QW structure, compared with the case of conventional ZB semiconductors. Figure 8.13(a) shows the well length dependence of the maximum optical gain of WZ GaN/Al$_{0.2}$Ga$_{0.8}$N QWs together with those of ZB GaAs/Al$_{0.3}$Ga$_{0.7}$As. In the unstrained QWs, as the well length becomes longer than 30 Å, a greater carrier density might be needed to obtain sufficient optical gain. This is quite obvious from the feature of the subband structure. As the well length becomes longer, the interval between HH$_1$ and HH$_2$ bands becomes shorter, and the DOS around the VBM becomes larger. This is the reason why it is more difficult to realize the population inversion for longer well length. However, even for $L_z = 30$ Å, the threshold carrier density would be higher than for ZB GaAs/AlGaAs QWs. According to a recent experimental report (Nakamura 1997), the threshold carrier density of InGaN MQW LDs is estimated to be $1 \sim 2 \times 10^{20}$ cm^{-3}. On the other hand, a typical value of our calculated transparent carrier density of the unstrained GaN/AlGaN SQWs is $1 \sim 2 \times 10^{19}$ cm^{-3}. Considering the loss of the injected carriers due to the irradiative recombination and the difference between MQWs and SQWs, our results seem to be reasonable. Further theoretical studies on structural dependence are necessary.

In order to discuss the effect of the intraband relaxation time τ_{in}, we have also calculated the optical gains for various values of τ_{in}. Figure 8.13(b) shows the calculated results of the maximum optical gain for TE mode with the well width $l_z = 60$ Å for $\tau_{in} = 0.01, 0.05, 0.1$ ps and ∞. As seen in Fig. 8.13(b), the value of the intraband relaxation time has a great influence on the threshold carrier density. The theoretical or experimental estimations of τ_{in} is a further problem for the III–V nitrides.

8.5.4 Optical gain of strained wurtzite quantum wells

Figure 8.14(a) shows the strain effect on the optical gain of WZ GaN/Al$_{0.2}$Ga$_{0.8}$N QWs. The well length dependence of the optical gain of WZ GaN/Al$_{0.2}$Ga$_{0.8}$N QWs with 0.5% compressive biaxial strain is shown in Fig. 8.13(a). In the compressive biaxial strained WZ QWs, the optical gain property is qualitatively improved for any well length, and the threshold carrier density would become a little lower than in the unstrained QWs. However, the reduction of the threshold carrier density is quantitatively not so effective. On the other hand, the tensile biaxial strain yields qualitatively negative effect on the threshold carrier density for any well length. Therefore no biaxial strain comes up to our expectations of reducing the threshold carrier density for laser oscillation in WZ GaN/AlGaN QWs.

Here, on the analogy of the bulk, if it were possible to introduce a uniaxial strain into the *c*-plane of WZ QWs it might cause a much larger reduction of not only the DOS at the VBM but also the threshold carrier density. According to our calculated subband structures, any uniaxial strain in the *c*-plane reduces

8.5 Optical gains of bulk GaN and GaN / AlGaN quantum wells

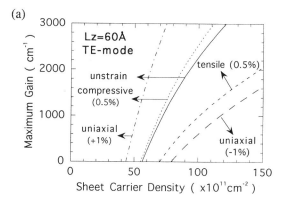

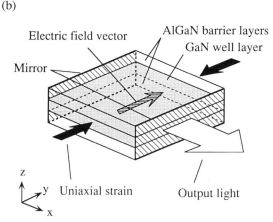

Fig. 8.14 (a) Strain effect on optical gain of the wurtzite GaN/Al$_{0.2}$Ga$_{0.8}$N quantum wells with well lengths L_z being 60 Å. The solid, dotted, short-dashed, dash-dotted, and long-dashed lines stand for the results without strain, with ±0.5% compressive biaxial strains, and with ±1.0% compressive uniaxial strains along the y-direction, respectively. (b) Schematic structure of the useful uniaxial strained wurtzite GaN/AlGaN single quantum wells.

the DOS at the VBM due to the anisotropic energy splittings in the k_x–k_y plane. Considering the optical polarization, the useful uniaxial strain directions are restricted to the following two cases. One is the compressive strain (ε_{uni}^{com}) parallel to the optical polarization for the electric field vector (**E**); that is, the y-direction ($\varepsilon_{uni}^{com} \parallel \mathbf{E}$). The other is the tensile one (ε_{uni}^{ten}) perpendicular to it; that, is the x-direction ($\varepsilon_{uni}^{ten} \perp \mathbf{E}$). Note that these two kinds of strain are equivalent in fact. Figure 8.14(b) shows the schematic structure of the useful uniaxial strained WZ GaN/AlGaN QWs, corresponding to the former case. On the other hand, the useless uniaxial strain directions are given by replacing the y-direction with the x-direction, namely, by changing the position of mirrors to the other lateral sides ($\varepsilon_{uni}^{com} \perp \mathbf{E}$, $\varepsilon_{uni}^{ten} \parallel \mathbf{E}$).

338 *Electronic and optical properties of GaN-based quantum wells*

The calculated optical gains for uniaxial strained WZ GaN/Al$_{0.2}$Ga$_{0.8}$N QWs are shown in Fig. 8.14(a), compared with those for biaxial strained QWs. If we induce the useful (useless) uniaxial strain, such as the compressive strain along the y-direction (x-direction), LH$_i$ bands along the k_y (k_x) direction and HH$_i$ bands along the k_x (k_y) direction move to the higher energy side. Then, the orbital component at the VBM becomes almost $|Y\rangle$ ($|X\rangle$) character (see Fig. 8.4). In other words, such useful (useless) uniaxial strains selectively isolate only the $|Y\rangle$ ($|X\rangle$) band at the VBM, which can (cannot) be coupled with the TE polarized light. This is the reason why the above useful uniaxial strains cause not only the reduced DOS but also the reduced threshold carrier density. Furthermore, the reduced DOS is caused by the exclusion of the little contributive bands, which are HH$_i$ bands along the k_y direction and LH$_i$ bands along the k_x direction, to the optical gain for the TE mode. Thus, the useful uniaxial strains give a larger differential optical gain. On the other hand, the useless uniaxial strains cause the reduced DOS but the increased threshold carrier density and smaller differential optical gain. Therefore, if it were possible to introduce a uniaxial strain in the c-plane of WZ QWs, the threshold carrier density would be more efficiently reduced and the differential optical gain would be increased, as long as the relation between the uniaxial strain's direction and the optical polarization is suitable.

8.5.5 Optical gain of zincblende quantum wells and strain effect

Figure 8.15 shows the maximum optical gain of ZB GaN/Al$_{0.2}$Ga$_{0.8}$N QWs with well width $L_z = 60$ Å. The heterojunctions are perpendicular to the [001] directions. In ZB QWs, since the symmetry is lowered from T$_d$ to D$_{2d}$, the degeneracy between the HH and LH bands at the Γ point is removed. Then the band mixing is modified such that the HH$_1$ mass becomes ligher than the LH$_1$ mass, and the DOS at the VBM is reduced. As a result, the transparent carrier density

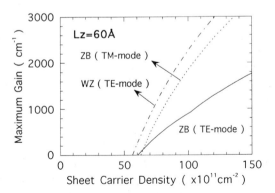

Fig. 8.15 Optical gain of the unstrained zincblende GaN/Al$_{0.2}$Ga$_{0.8}$N single quantum wells with well length L_z being 60 Å. The solid and dotted lines stand for the results for TE and TM modes, respectively. For comparison, the result for the TE mode of wurtzite GaN is shown by the dash-dotted line.

8.5 *Optical gains of bulk GaN and GaN / AlGaN quantum wells* 339

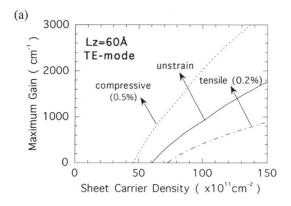

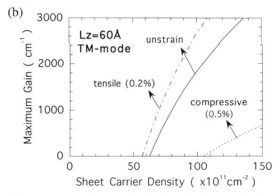

Fig. 8.16 Strain effect on optical gain of the zincblende $GaN/Al_{0.2}Ga_{0.8}N$ quantum wells with well lengths L_z being 60 Å. (a) and (b) correspond to the TE and TM modes, respectively. The solid, dotted, and dash-dotted lines stand for the results without strain, with 0.5% compressive baxial strain, and 0.2% tensile biaxial strain, respectively.

of the TE mode becomes low, compared with the bulk. However, since the small spin–orbit splitting enhances the mixing between the LH_1 and SH_1 bands, the transparent carrier density in ZB QWs is still higher than that in WZ QWs. Then enhanced DOS can be observed as the large TM mode optical gain after the occurrence of the population inversion.

Next, the compressive and tensile biaxial strain effects are discussed, because the biaxial strains are usually adopted to reduce the threshold carrier density in conventional ZB QWs. Figure 8.16 shows the maximum optical gain of biaxial strained ZB $GaN/Al_{0.2}Ga_{0.8}N$ QWs with well width $L_z = 60$ Å. The compressive biaxial strain strongly depresses the TM mode optical gain and enhances the TE mode optical gain. On the other hand, the effect of the tensile biaxial strain is the reverse of that of the compressive one. These results can be easily understood from the feature of bulk ZB GaN, where the compressive biaxial strain lifts the $|X\rangle$ and the $|Y\rangle$ states upper than the $|Z\rangle$ states, and the tensile one lifts the $|Z\rangle$ state upper than the $|X\rangle$ and the $|Y\rangle$ states (see Fig. 8.5). Thus, the compressive (tensile) biaxial strain is more effective on the reduction of the threshold carrier density for TE mode (TM mode) than the reduction in WZ QWs. This result in the ZB GaN/AlGaN QW system will be very useful in polarization control. Namely, one can obtain laser emission to have TE or TM polarized light with the help of appropriate strains.

8.6 Conclusions

Electronic and optical properties of GaN/AlGaN quantum wells (QWs) as well as of bulk GaN and AlN were investigated on the basis of the first-principles calculation and the effective mass theory. The information obtained provides a fundamental understanding of the III–V nitrides and leads to guidelines for material and device design with them.

The electronic band structures of bulk GaN and AlN were calculated for both the wurtzite (WZ) and zincblende (ZB) structures by the full-potential linearized augmented plane wave (FLAPW) method. The FLAPW band calculations for the deformed lattices were also performed, and the strain effects were investigated. In order to describe the electronic structure near the band edges, the effective mass Hamiltonian for the WZ structure was derived, taking the hexagonal C_{6v} symmetry into account. The physical parameters, such as effective masses, the Luttinger-like parameters, crystal–field and spin–orbit splitting energies, momentum matrix elements, and deformation potentials, of WZ GaN and AlN were determined by reproducing the FLAPW band structures with the effective mass Hamiltonian. Those of ZB structures were also determined. The parameters experimentally obtained so far for GaN are in agreement with our theoretical values.

The subband structures of GaN/AlGaN QWs were calculated for the first time on the basis of the effective mass approximation by using the physical parameters derived theoretically. In WZ GaN/AlGaN QWs, the subband dispersions are almost the same as the energy dispersions in bulk WZ GaN, and the heavy (light) hole mass is still heavy (light). This character originates from the hexagonal crystal symmetry and the weak spin–orbit coupling of the nitrogen atom, and it remains under biaxial stress. This is very different from conventional ZB GaAs/AlGaAs QWs, in which the heavy (light) hole mass becomes light (heavy) compared with bulk GaAs, due to symmetry lowering from T_d to D_{2d}. In ZB GaN/AlGaN QWs, the heavy hole mass certainly becomes light, but it is still much heavier than that of ZB GaAs/AlGaAs QWs. Furthermore, no biaxial strain reduces to much the density of states (DOS) around the valence band maximum (VBM) in both WZ and ZB GaN/AlGaN QWs. It was also found that the uniaxial strain in the c-plane of WZ QWs significantly reduced the DOS around the VBM, due to symmetry lowering from C_{6v} to C_{2v}.

As the first step towards obtaining a guideline for device design for the III–V nitride lasers, the optical gains with respect to the interband transition were evaluated for bulk GaN and GaN/AlGaN QWs. For both the WZ and the ZB structures, very heavy hole masses cause high threshold carrier density for laser oscillation, compared with conventional ZB III−V semiconductors. In WZ GaN/AlGaN QWs the compressive biaxial strain can reduce the threshold carrier density, but it is not so effective quantitatively. In ZB GaN/AlGaN QWs, the biaxial strain can reduce it more than in WZ QWs, but the threshold carrier density is almost the same or higher. It was suggested that the threshold carrier density would be considerably reduced if it were possible to introduce a uniaxial strain into the c-plane of WZ QWs.

Acknowledgements

The authors would like to thank Emeritus Professor A. Yanase of University of Osaka Prefecture for his helpful discussion and providing them with his FORTRAN source codes for the FLAPW band calculation.

References

Barker, A. S. and Ilegems, M. (1973). *Phys Rev. B* **7**, 743.
Bir, G. L. and Pikus, G. E. (1974). *Symmetry and Strain-Induced Effects in Semiconductors*. Wiley, New York.
Broido, D. A. and Sham, L. J. (1985). *Phys. Rev. B* **31**, 888.
Chichibu, S. Shikanai, A., Azuhata, T., Sota, T., Kuramata, A., Horino, K., and Nakamura, S. (1996). *Appl. Phys. Lett.* **68**, 3766.
Cunningham, R. D., Brander, R. W., Kuee, N. D., and Wickender, D. K. (1972). *J. Luminescence* **5**, 21.
Dingle, R. and Ilegems, M. (1971). *Solid State Commun.* **9**, 175.
Dingle, R., Sell, D. D., Stokowski, S. E., and Ilegems, M. (1971). *Phys. Rev. B* **4**, 1211.
Drechsler, M., Hofmann, D. M., Meyer, B. K., Detchprohm, T., Amano, H., and Akasaki I. (1995). *Jpn. J. Appl. Phys.* **34**, L1178.
Ejder, E. (1971). *Phys. Status Solidi A* **6**, 445.
Fiorentini, V., Methfessel, M., and Scheffler, M. (1993). *Phys. Rev. B* **47**, 13353.
Flores, G. R., Contreras, H. N., Martinez, A. L. Powell, R. C., and Greene, J. E. (1994). *Phys. Rev. B* **50**, 8433.
Gunnarson, O. and Lundqvist, B. I. (1976). *Phys. Rev. B* **13**, 4274.
Hedin, L. (1965), *Phys. Rev.* **139**, A796.
Hess, K., Bimberg, O., Lipari, N. O., Fischbach, J. K., and Altarelli, M. (1976). *Proc. 13th Int. Conf. on Phys. Semiconductors*, Rome, p. 142.
Hybertsen, M. S. and Louie, S. G. (1985). *Phys. Rev. Lett.* **55**, 1418.
Kamiyama, S., Ohnaka, K., Suzuki, M. and Uenoyama, T. (1995). *Jpn. J. Appl. Phys.* **34**, L821.
Kane, E. O. (1957). *J. Phys. Chem. Solids* **1**, 249.
Kosicki, B. B., Powell, R. J., and Burgiel, J. C. (1970). *Phys. Rev. Lett.* **24**, 1421.
Luttinger, J. M. and Kohn W. (1955). *Phys. Rev.* **97**, 869.
Maruska, H. P. and Tietjen, J. J. (1969). *Appl. Phys. Lett.* **15**, 327.
Meyer, B. K., Volm, D., Graber, A., Alt, H. C., Detchprohm, T., Amano, H., and Akasaki, I. (1995). *Solid State Commun.* **95**, 597.
Miwa, K. and Fukumoto, A. (1993). *Phys. Rev. B* **48**, 7897.

Mizuta, M., Fujieda, S., Matsumoto, Y., and Kawamura, T. (1986). *Jpn. J. Appl. Phys.* **25**, L945.
Monemar, B. (1974). *Phys. Rev. B* **10**, 676.
Nakamura, S. (1997). *Tech. Rep. IEICE.* OPE96-**144**, LQE96-**142** 31 (in Japanese).
Nakamura, S., Mukai, T., and Senoh, M. (1994). *Appl. Phys. Lett.* **54**, 1687.
Nakamura, S., Senoh, M., Iwasa, N., and Nagahama, S. (1995). *Jpn. J. Appl. Phys.* **34**, L797.
Nakamura, S., Senoh, M., Nagahama, S., Iwasa, N., Yamada, T., Matsushita, T., Kiyoko, H., and Sugimoto, Y. (1996). *Jpn. J. Appl. Phys.* **35**, L74.
Palummo, M., Reining, L., Godby, R. W., Bertoni, C. M., and Bornsen, N. (1994). *Europhys. Lett.* **26**, 607.
Pankove, J. I., Bloom, S., and Harbeke, G. (1975). *RCA Rev.* **36**, 163.
Rheinländer, B. and Neumann, H. (1974). *Phys. Stat, Sol. (b)* **64**, K123.
Rubio, A., Corkill, J. L., Cohen, M. L., Shirley, E. L., and Louie, S. G. (1993). *Phys. Rev. B* **48**, 11810.
Sheleg, A. U. and Savastenko, V. A. (1979). *Izv. Akad. Nauk SSSR, Neorg. Mat.* **15**, 1598.
Skolnick, M. S., Jain, A. K., Stradling R. A., Leotin, L., Ousset, J. C., and Ashkennazy, S. (1976). *J. Phys. C* **9**, 2809.
Stillman, G. E. Larsen, D. M., Wolfe, C. M., and Brandt, R. C. (1971). *Solid State Commun.* **9**, 2245.
Stillman, G. E., Wolfe, C. M., and Dimmock, J. O. (1970). *J. Phys, Chem. Solids* **31**, 1199.
Suzuki, M. and Uenoyama, T. (1995). *Jpn. J. Appl. Phys.* **34**, 3442.
Suzuki, M. and Uenoyama, T. (1996a). *Jpn. J. Appl. Phys.* **35**, 543.
Suzuki, M. and Uenoyama, T. (1996b). *Jpn. J. Appl. Phys.* **35**, 1420.
Suzuki, M. and Uenoyama, T. (1996c). *Jpn. J. Appl. Phys.* **35**, L953.
Suzuki, M. and Uenoyama, T. (1996d). *Appl. Phys. Lett.* **69**, 3378.
Suzuki, M. and Uenoyama, T. (1996e). *J. Appl. Phys.* **80**, 6868.
Suzuki, M. and Uenoyama, T. (1997). *Solid State Electron.* **41**, 271.
Suzuki, M., Uenoyama, T., and Yanase, A. (1995). *Phys. Rev. B* **52**, 8132.
Uenoyama, T. and Suzuki, M. (1995). *Appl. Phys. Lett.* **67**, 2527.
Wenchang, L., Kaiming, Z., and Xide, X. (1993). *J. Phys.: Condens. Matter* **5**, 875.
Williams, E. W. and Rehn, V. (1968). *Phys. Rev.* **172**, 798.
Wimmer, E., Krakauer, H., Weinert, M., and Freeman, A. J. (1981). *Phys. Rev. B* **24**, 864.
Yeh, C., Lu, Z. W., Froyen, S., and Zunger, A. (1992). *Phys. Rev. B* **46**, 10086.
Yim, W. M., Stofko, E. J., Zanzucchi, P. J., Pankove, J. I., Ettenberg, M., and Gilbert, S. L. (1973). *J. Appl. Phys.* **44**, 292.

9 Transistors and detectors based on GaN-related materials

Jean-Yves Duboz and M. Asif Khan

The recent explosion of research activity on GaN-related materials has given birth to many devices that are already highly performing. We describe here transistors and ultraviolet detectors and modulators.

9.1 Bipolar transistors

9.1.1 General principles

In order to show precisely the interest of using GaN and related compounds in bipolar transistors, we would like first to recall the main issues in bipolar transistors. A bipolar transistor is made of two back-to-back p–n junctions. The first one, called an emitter–base (E–B) junction, is forward-biased so that minority carrier are injected into the base. The second one, called a base–collector (B–C) junction is reverse-biased and minority carriers injected in the base can by diffusion or drift be collected in the collector (by definition, a proportion α of them actually are). The base current is due to recombination in the base, minority carrier injection into the emitter, and leakage current in the reverse-biased B–C junction. In a good transistor, almost all the carriers injected in the base are collected, so that $\alpha \simeq 1$. The power amplification arises from the B–C voltage being much larger than the E–B one. In the common emitter configuration, the ratio $\beta \simeq \alpha/(1-\alpha)$ between the collector and base currents can be important: it defines the current gain. The minority carrier mobility in the base has to be as large as possible in order to avoid recombinations in the base. As the electron mobility is larger than that of the holes, n–p–n transistors generally offer better performance than p–n–p transistors. In homojunction transistors, the hole injection in the emitter is reduced by reducing the base doping to a value below the emitter doping. As a consequence, the base access resistance becomes quite large for thin bases and the effective base width becomes sensitive to the reverse voltage applied to the C–B junction (Early effect). In heterojunction bipolar transistors (HBT), however, the hole injection is largely reduced by the valence band discontinuity as a larger bandgap material is used for the emitter. Then the base doping can be much higher, which allows the reduction of the base thickness without increasing the access resistance. For high-frequency operation, the transit time must be minimized. One important contribution is the base transit time. It is limited by diffusion, and high mobilities are required. It can be reduced by gradually changing the bandgap energy of the base material, thus creating an internal field in the base. This requires that a continuous tuning of the gap energy be possible using ternary alloys. A second important contribution arises from transport through the C–B junction. As the field is very high, the carrier velocity saturates and it is thus profitable that the material saturation velocity be large. Finally, as the power gain increases with the field in the C–B junction, the breakdown voltage must be as large as possible.

9.1.2 Why is GaN a good candidate for high-power HBTs?

First, the GaN bandgap energy is large (3.43 eV at 300 K, i.e. the largest among available materials after diamond). This has many positive implications: the breakdown field is large ($\geq 2 \times 10^6$ V/cm), the thermal generation of minority carriers is the junction is small, and the nonradiative recombination could in theory be reduced. On the contrary, let us note that the radiative recombination rate is larger than in smaller bandgap energy materials, as it varies as the third power of the gap energy. Second, the intervalley energy separation is large and the optical phonon energy is large (91 meV compared with 36 meV in GaAs). Hence the saturation velocity is large (predicted to be 2.5×10^7 cm/s). Moreover, for the same reasons, under a large electric field carriers may experience a large velocity overshoot, at least during part of their transit. Third, GaN can be combined with InGaN and AlGaN alloys to form heterostructures for the base and the E–B junction respectively. All these considerations were taken into account in a numerical evaluation of the potential of GaN versus other compounds for HBTs [Gao91]. The comparison showed that GaN (together with diamond) was the ideal candidate. Finally, the thermal conductivity is reasonably large (1.3 Wcm^{-1}K^{-1}), comparable to that of Si and much better than in other III–V compounds.

9.1.3 Remaining problems

The defect (dislocation...) density in GaN remains large and electron mobilities are smaller than in other III–V compounds. Moreover, GaN and related alloys are direct bandgap materials, which generally exhibit enhanced radiative recombination rates. Thus the radiative recombination in the base cannot be completely eliminated. The second problem is related to the acceptor level and the difficulty of obtaining p doping larger than 10^{18} cm^{-3}. We thus cannot fully profit from the heterostructure to highly dope the base. Finally, ohmic contacts must be improved. Back side contacts are not possible on sapphire substrates, which is in favor of GaN on SiC.

9.1.4 Demonstrated devices

To our knowledge, no HBT entirely made of III–V nitrides has yet been reported in the literature. However, a GaN/SiC HBT has been fabricated and characterized by Pankove *et al.* [Pankove94, Pankove96]. The bandgap energy of 6H–SiC is 2.9 eV at 300 K. Theory has predicted a GaN/SiC valence band

9.1 Bipolar transistors

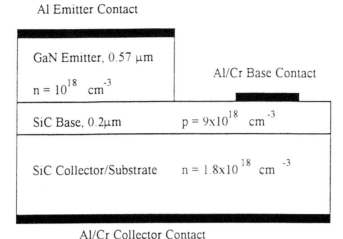

Fig. 9.1 Cross-section of the GaN/SiC heterojunction bipolar transistor (after [Pankove94].)

discontinuity between 0.2 and 0.25 eV [Schetzina96] while preliminary measurements gave 0.38 eV [Kong96]. In any case, the valence band discontinuity is well suited for the emitter–base junction. Moreover, SiC is indirect bandgap material and can be highly p-doped. Both factors made SiC a good candidate for the base. As a result, an n-GaN/p-SiC/n-SiC HBT was fabricated as shown in Fig. 9.1. The Al emitter contact served as a mask for self-aligning the emitter region. The GaN was etched down to the base by reactive ion etching in a CCl_2F_2 plasma. Actually, the SiC etching rate was four times larger than that for GaN, so that a substantial etching (one third) of the base could not be avoided. One fully appreciates the importance of the problem of the non-selectivity in the etching if one remembers that one key advantage of the GaInP/GaAs system over the AlGaAs/GaAs is the quasi-absolute selectivity in the emitter etching. Let us also note the large area ($0.25\,cm^2$) of the base–collector junction compared with the emitter contact area ($75\,\mu m \times 75\,\mu m$) due to the fact that devices were not isolated from each other. The current–voltage characteristics in common base mode measured at room temperature are shown in Fig. 9.2. For voltages less than 10 V, the current is almost independent of the collector to base voltage. This absence of the Early effect can be explained by the high base and collector dopings as already explained. Soft breakdown of the base–collector junction was observed at around 10 V. The breakdown was studied directly by measuring the base–collector current–voltage characteristics. The breakdown voltage, although not clearly defined, can be estimated to be of the order of 10 V, corresponding to a field of 10^6 V/cm. This value is smaller than the $2\text{--}3.7 \times 10^6$ V/cm which has been reported previously [Matus91]. The low breakdown voltage and the importance of leakage currents can be explained by the high dopings and the large area of the junction. These leakage currents prevented HBT measurement in the common emitter mode.

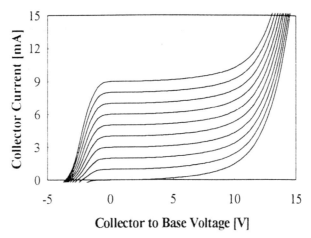

Fig. 9.2 Common base characteristics of a GaN/SiC HBT. The emitter current varies from 0 (bottom curve) to 9 mA (top curve). The emitter area is 75 μm × 75 μm. (After [Pankove94].)

It can be seen in Fig. 9.2 that the collector current is very close to that of the emitter, indicating a low level of recombination in the base and a large current gain dI_C/dI_B. The latter was measured at constant base–collector voltage and is presented in Fig. 9.3. The gain varies significantly with emitter current and collector–base voltage, reaching a maximum value above 10^5 at $V_{CB} = 2$ V and $I_E = 100$ mA. In the absence of composition gradient in the base, the electron motion in the base is dominated by the diffusion. The diffusion coefficient D_n can be deduced from the mobility (110 cm²/Vs) and the Einstein relation. The diffusion length L_n is then equal to $\sqrt{D_n \tau}$ where τ is the electron lifetime in the base, taken as 5 μs. Writing that the number of electrons reaching the end of the base (W_{base}) varies as $e^{-\frac{1}{2}W_{base}^2 L_n^{-2}}$ or $1 - \frac{1}{2}W_{base}^2 L_n^{-2}$, one calculates the base

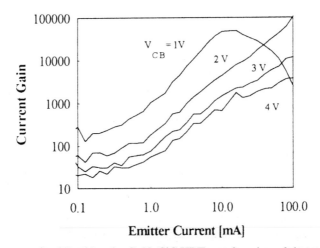

Fig. 9.3 Current gain, dI_C/dI_B of a GaN/SiC HBT as a function of the emitter current for different values of the collector to base voltage V_{CB}. (After [Pankove94].)

transport factor $\alpha = 0.999\,987$. In order to relate the electron base transport factor to the current gain, we need to take the hole injection from the emitter into account. The emitter efficiency is defined as the ratio γ of the electron current crossing the E–B junction to the total current in the same junction. An evaluation of the hole injection leads to $\gamma = 0.999\,999$. From $I_E = I_{nE} + I_{pE}$, $I_C = -\alpha I_{nE}$, and $I_B = -I_E - I_C$, one deduces that the current gain $\beta = \alpha\gamma/(1 - \alpha\gamma)$ is of the order of 8×10^4 in reasonable agreement with the experimental results. Note that, due to the random nature of the electron diffusion, the base transport varies as the square of the base thickness, while it varies linearly with the electron lifetime. This illustrates the advantage of reducing the base thickness, which requires a high base doping and a good etching selectivity.

The device was measured at higher temperatures [Pankove96]. At 260 °C for instance, the nature of the collector current in the common-base mode is essentially the same as in Fig. 9.2, except that in this case the soft breakdown occurs at a lower voltage, $V_{CB} = 5\,\text{V}$. The current gain at 260 °C is reduced compared with its value at room temperature. For $V_{CB} = 0.6\,\text{V}$ and $I_C = 6\,\text{mA}$, β was measured to be 1600. The gain measured as a function of the temperature was found to vary from 10^5 at 25 °C to 10^4 at 180 °C, 10^3 at 390 °C, and 10^2 at 500 °C. The Arrhenius plot of the gain showed that it varies as $e^{E_a/kT}$ with $E_a = 0.43\,\text{eV}$. This energy corresponds to the valence band discontinuity, which indicates that the hole emission at the E–B junction is the limiting phenomenon for high-temperature operation.

Power performances were also measured. At room temperature, a power density of $30\,\text{kW cm}^{-2}$ could be sustained by this HBT, while the maximum current density was as large as $1.8\,\text{kA cm}^{-2}$.

9.2 Field-effect transistors

9.2.1 General principles

Field-effect transistors (FETs) involve transport of one type of carrier only, which are most advantageously electrons, as the electron mobility is larger than that of holes. Electrons are flowing in a channel parallel to the device surface between two ohmic source and drain contacts. A third electrode (gate) placed in-between drain and source modulates the channel thickness and the electron density by a field effect. The source is grounded and the drain–source voltage is called V_{ds}, while the gate–source voltage is called V_{gs}. I_{ds} is the current between source and drain and I_g is the current entering the gate. The ability of the gate to modulate the source drain current is expressed by the transconductance $g_m = dI_{ds}/dV_{gs}$. An ideal transistor is characterized as follows: in order to give a large current in the ON-state, the density, mobility, and saturation velocity of electrons in the channel are large, and access resistances are small. To reduce the current in the OFF-state, the gate has to sustain large electric fields without leaking. For high frequency performance, the gate resistance and capacitance must be minimized. Decreasing the gate length will increase the resistance but lower the capacitance. It reduces the electron transit time in two ways. The first obvious effect is the reduction of transit length. Second, as the electric field is larger, the electron velocity is larger and actually reaches its saturation value. In very short channels, it is even larger than the steady state saturation velocity, and electrons experience a velocity overshoot that can be as large as a factor of 2. This overshoot in time relies on ballistic transport over short distances under a strong electric field. Ballistic transport ceases when the electron energy is large enough for inelastic collisions (such as optical phonons, intervalley scattering) to occur. This overshoot will thus be larger in materials with larger optical phonon energy, intervalley energies etc. For power applications, a large source to drain bias is needed which requires large breakdown fields. As shorter gates lead to higher cutoff frequencies and smaller breakdown voltages, a compromise for the gate length must be found. Finally, the thermal conductivity of the materials (including the substrate) must be as large as possible.

9.2.2 Different FETs and specific advantages of GaN-related materials

In MESFETs, the gate is a metal Schottky diode directly deposited on the channel. The channel is doped and conducting and its thickness is modulated by the depletion region of the Schottky diode. In JFETs, the gate is a p–n junction, which is in principle close to the previous case if one considers the metal as a p region where the doping is infinite. The barrier height is approximately equal to the bandgap in JFETs, while it is equal to the Schottky barrier height in MESFETs. This favours smaller gate leakage currents in JFETs compared with MESFETs. In both cases, using a wide bandgap material as GaN is interesting as the barrier height is also larger than the one obtained on other semiconductors. Thermal generation in the depletion region is also much smaller. The large saturation velocity and its large overshoot ($\hbar\omega_{LO} = 91\,\text{meV}$) are important advantages when short gates are used. Long gate devices are expected to perform poorly because of the low mobility. In order to increase the mobility, heterostructure FETs (HFETs) can be used. Here, the channel is made of a smaller bandgap energy material while the material (barrier) between the gate contact and the channel has a larger bandgap energy. The channel may be doped or not. In the latter case, the barrier is doped and electrons transfer to the channel. We thus have a modulation-doped transistor (MODFET). The coulombic interaction between donors and electrons being suppressed, the mobility is higher. Mobilities as high as $1700\,\text{cm}^2/\text{Vs}$ have been used measured in a two-dimensional electron gas [Shur96]. The advantage of using the GaN/AlGaN system lies in the available bandgap discontinuity. While the conduction band discontinuity in GaAs/AlGaAs or InGaAs/InAlAs does not allow for 2D electron densities larger than $10^{12}\,\text{cm}^{-2}$, the GaN/AlGaN conduction band discontinuity allows densities as large as $10^{13}\,\text{cm}^{-2}$ [Stengel96], the real space transfer of hot electrons from the channel to the barrier remaining negligible. Injection from the gate contact is also reduced by the larger Schottky barrier height. This injection can be further decreased by the use of an insulator for the barrier (MISFETs). In order to take full advantage of the large electron saturation velocity in GaN, FETs with short gates will be often used. In this case, one can make a simple and rather good estimation of some key parameters. The current I_{ds} per unit gate width is given by $I_{ds} = env_s$ and the charge below the gate is related to the gate capacitance and voltage by $en_s L_g = C(V_{gs} - V_{th})$, where V_{th} is the threshold gate voltage. The intrinsic transconductance per unit gate width is then equal to $g_{m0} = Cv_s/L_g$. As the capacitance is proportional to the gate length, g_{m0} does not depend on the gate length in the first-order approximation for gates short enough that the electron velocity in the channel is saturated. Actually, because this assumption is never strictly verified and because of the velocity overshoot, the transconductance generally increases with decreasing gate length. This trend should be particularly well verified in GaN-based devices, as velocity overshoot can be expected to be important, as already explained. The current gain, defined as the ratio between the drain–source and the gate currents is given by

$$h_{12} = \frac{g_{m0} V_{gs}}{j\omega C V_{gs}} = \frac{f_T}{jf}$$

where the current gain cutoff frequency is

$$f_T = \frac{g_{m0}}{2\pi C} = \frac{v_s}{2\pi L_g}$$

This frequency is thus simply the inverse of the electron transit time below the gate. The last parameter often used is the maximum oscillation frequency f_{max} describing the power gain. It is related to f_T and to passive elements of the device as access resistances and parasitic capacitances. Finally, the practical limitation for obtaining very large transconductances arises from the source access resistance. As part of the gate voltage will actually be applied across the source resistance, the extrinsic transconductance g_m is related to the intrinsic one g_{m0} by

$$g_m = \frac{g_{m0}}{1 + R_s g_{m0}}$$

As contact resistances tend to be higher on larger bandgap energy materials, it is clear that ohmic contacts will be one of the key issues for GaN based FETs.

Simulations have been performed in order to evaluate the potential of GaN-based material for FETs. Shin and Trew [Shin95] calculated the performance of GaN MESFETs. The model was first verified by comparing the calculated DC characteristics with experimental values [AsifKhan94a]. Then, the large signal RF performance was calculated for an optimized device and a gate length of 0.25 μm. The transconductance is around 93 mS/mm. Figure 9.4 shows the result of the simulation at room temperature with material parameters taken from the literature. At the operating frequency of 8 GHz the output

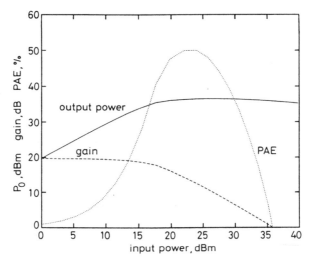

Fig. 9.4 Simulated large signal RF (8 GHz) performance of a GaN MESFET at 300 K. $V_{ds} = 40$ V and $V_{gs} = -3.5$ V. (After [Shin95].)

power reaches 4 W/mm with a Power Added Efficiency (PAE) of 50%. The PAE is the ratio of generated microwave (output−input) power to the DC power required to bias the device. At higher temperature (773 K), g_m is reduced to 25 mS/mm while the device still produces 1 W/mm with a PAE of 33%. These results are better than those obtained from a simulated 6H–SiC MESFET and comparable with those from a 4H–SiC MESFET.

As far as MODFETs are concerned, a more recent study was dedicated to the DC and small signal characteristics [Stengel 96]. The effects of the n-AlGaN layer thickness and doping, the insulating spacer AlGaN layer thickness, the Al content in these layers and the channel length on the electrical characteristics

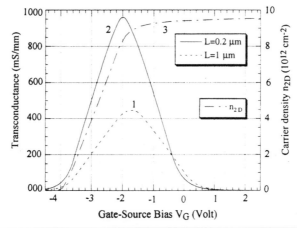

Fig. 9.5 Simulated variation of the 2DEG concentration and transconductance as a function of V_{gs} in an AlGaN/GaN MODFET. $T = 300$ K. (After [Stengel96].)

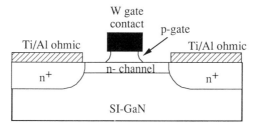

Fig. 9.6 Schematic diagram of an ion-implanted GaN JFET. (After [Zolper96].)

are presented. Figure 9.5 shows the electron sheet density and the transconductance for a 100 Å $Al_{0.3}Ga_{0.7}N$ layer doped to 5×10^{19} cm^{-3} and without any spacer. The 2D gas density almost reaches 10^{13} cm^{-2} while g_m increases from 450 mS/mm for a 1 μm long gate to almost 1000 mS/mm for a 0.2 μm gate. Note that such high g_m have up to now been measured in high-performance InAlAs/InGaAs MODFETs only.

9.2.3 Demonstrated field-effect transistors

All the GaN-based FETs have been grown on sapphire. Unless otherwise mentioned, the layers were grown by MOCVD.

9.2.3.1 Junction field-effect transistors

Up to now, there has been only one report of GaN JFET fabrication [Zolper96]. The device is sketched in Fig. 9.6. The n-channel, n$^+$ ohmic contacts, and the p-gate were realized by selective area ion implantation. The gate was etched by RIE. The p–n junction demonstrates a reasonable turn-on voltage around 2 V,

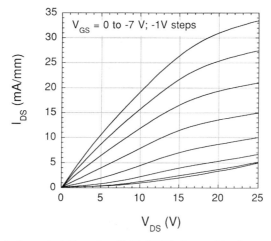

Fig. 9.7 Electrical characteristics of a GaN JFET. I_{ds} vs. V_{ds} for an area of 1.7 μm × 50 μm. V_{gs} varies from 0 (top curve) to −7 V (bottom curve). (After [Zolper96].)

but relatively large leakage currents in reverse bias (−10 V) of 1 mA/mm between gate and source. The JFET demonstrates good modulation characteristics, as can be seen in Fig. 9.7. Nearly complete pinch-off is obtained at $V_{gs} = -6$ V for low V_{ds}. The transconductance is modest (7 mS/mm), which can be explained by the large access resistances and by the electron mobility in the channel being degraded by the implantation. Improvements of the device may also imply a reduction of gate leakage currents that are possibly due to implantation defects and to a too-low p-doping level.

9.2.3.2 Metal semiconductor field-effect transistors

The first GaN MESFET was realized by Asif Khan et al. [AsifKhan93a]. Figure 9.8 shows the plan and the cross-sectional view of the device. The 0.6 μm thick GaN layer was unintentionally n-doped. Hall measurements gave an electron concentration of 10^{17} cm^{-3} and a mobility of 360 cm^2/Vs. Devices were electrically isolated using proton implantation. Ohmic contacts were done with 25 Å Ti/1500 Å Au annealed at 250 °C for 30 s. The contact resistance can be evaluated to be 15 Ω mm. Silver was used for the Schottky diode. Figure 9.9 shows I_{ds} as a function of V_{ds} for increasing values of V_{gs}. For $V_{gs} = 0$ V and $V_{ds} = 15$ V, I_{ds} reaches 15.5 mA, corresponding to a current density of 155 mA/mm. Complete pinch-off was observed for a gate reverse bias of −12 V. The extrinsic transconductance at $V_{gs} = -1$ V equals 23 mS/mm, corresponding to an intrinsic transconductance of around 35 mS/mm. The breakdown field was found to be much smaller than expected, probably due to insufficient isolation. The saturation velocity was estimated to be around 5×10^6 cm/s. Both findings prove that this device was not yet fully exploiting the GaN potential.

9.2.3.3 Heterostructure field-effect transistors

Shortly after, the same group demonstrated the first heterostructure transistor

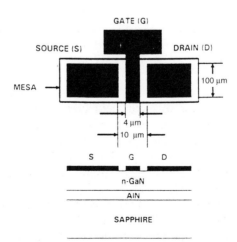

Fig. 9.8 Plan and cross-sectional view of a GaN MESFET. The gate length is 4 μm and source–drain spacing is 10 μm. (After [AsifKhan93a].)

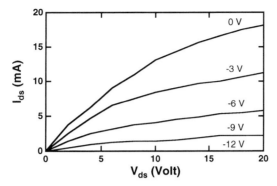

Fig. 9.9 Electrical characteristics of a GaN MESFET. The device is 100 μm wide. V_{gs} varies from 0 (top curve) to −12 V (bottom curve). (After [AsifKhan93a].)

(HFET) on GaN [AsifKhan93c]. The layer structure is the same as for the MESFET except for an additional n-doped $Al_{0.14}Ga_{0.86}N$ on top (1000 Å). The ohmic contact fabrication and the mesa isolation were also the same. The TiW gate was 4 μm long. Hall measurements at 300 K (77 K) confirmed the presence of the 2D electron gas with a density of $1.1 \times 10^{13} (7.6 \times 10^{12})$ cm^{-2} and a mobility of 563 (1517) cm^2/Vs. The transconductance was measured at $V_{gs} = +0.5$ V. At 300 K, g_m was found equal to 28 mS/mm. At 77 K, g_m was larger by a factor of 1.64, in agreement with the increase of the μn_s product at low temperature. A similar HFET was also fabricated with a thinner AlGaN layer (250 Å, n-doped to 4×10^{18} cm^{-3}) [AsifKhan94a]. The gate length was reduced for high-frequency performance while the distance between the source or the drain and the gate was reduced to 0.75 μm to reduce the access resistance. The latter was actually measured to be 28 Ω mm. The transconductance g_m was equal to 27 mS/mm. This relatively small value is in a large part due to the source resistance. One can calculate $g_{m0} = 110$ mS/mm. The high-frequency measurements gave $F_T = 11$ GHz and $f_{max} = 35$ GHz. From the saturation current at $V_{gs} = 0.75$ V, the threshold voltage V_T and the source resistance, one can deduce the electron density and velocity at the source. The velocity at the source was found to be around 1.35×10^6 cm/s. According to the relation between f_T and the saturation velocity, which is equal to or larger than the velocity at the source, one can calculate f_T as around 8.6 GHz, in agreement with the experimental result. It is thus clear that the main limitation now lies in the ohmic contacts. As far as power applications are concerned, another limiting phenomemon was observed in a similar structure [AsifKhan94b]. When biased at V_{ds} larger than 40 V, the I_{ds} current collapsed and the device conductance became smaller by a factor as high as 300. This state remained at 300 K for a few seconds before the normal state returned. This collapse removal could be activated by light illumination. This behavior is interpreted as real space transfer of hot electrons in the AlGaN barrier and their subsequent trapping. The detrapping time at room temperature was long enough that the collapse could be observed. Such a limitation on the V_{ds} swing should be eliminated by using a barrier with higher Al content or by reducing the trap density in the barrier.

For low-power electronics, enhancement and depletion mode FETs are desirable at room temperature [AsifKhan96b]. A thin (400 Å) unintentionally n-doped (7×10^{17} cm^{-3}) GaN channel was deposited on top of a thick insulating GaN layer. A thin (100 Å) Al$_{0.1}$Ga$_{0.9}$N layer was used as a barrier. Given the high Schottky barrier height on AlGaN (1.2 eV) and the thin barrier and channel, there is a complete pinchoff at 0 V and the FET is normally OFF. The threshold voltage V_T was found to be around +0.05 V. The measured transconductance on a 1 μm long gate HFET was rather small (23 mS/mm) due to the large source resistance. Nevertheless, this enhancement mode HFET was coupled to a depletion mode HFET $V_T = -0.4$ V) to give a direct coupled FET logic inverter. The next step, however, should be the inverter fabrication with both FETs on the same wafer. Selective implantation or barrier etching should make such a device obtainable.

9.2.3.4 MESFET at high frequency

Microwave performances of MESFETs have been also measured [Binari94]. A 0.25 μm thick GaN channel was grown on top of an isulating AlN barrier in order to confine electrons. The unintentional doping was on the order of 2.7×10^{17} cm^{-3} and the mobility was 400 cm^2/Vs. A 0.7–2 μm long Au gate was deposited between two Ti/Al ohmic contacts. Figure 9.10 shows the DC drain characteristics for a 1.4 μm long and 150 μm wide gate. The transconductance is 20 mS/mm only, but is little dependent on the gate bias, which is a positive point for large signal operation. This is attributed to the fact that most of the carriers are located near the GaN/AlN interface, as in a step-doped structure.

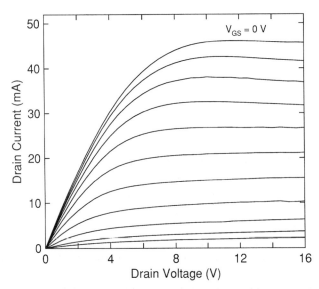

Fig. 9.10 Electrical characteristics of a GaN MESFET. The device is 150 μm wide. The gate length is 1.4 μm. V_{gs} varies from 0 (top curve) to −20 V (bottom curve). (After [Binari94].)

9.2 Field-effect transistors

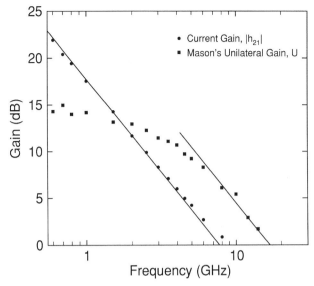

Fig. 9.11 High-frequency performance of a GaN MESFET. Short-circuit current gain $|h_{12}|$ and Mason Unilateral gain. The gate length is 0.7 μm. (After [Binari94].)

The S parameters of this MESFET were measured and gave $f_T = 8\,\text{GHz}$ and $f_{max} = 17\,\text{GHz}$, as can be seen in Fig. 9.11. Low-frequency dispersion was observed on this device, related to the presence of traps. The low-frequency dispersion was studied in an FET similar to the previous one. The main difference lies in the channel being thinner (0.1 μm). This led to a larger transconductance (45 mS/mm) and higher cutoff frequencies. It was observed that the g_m magnitude and phase depend on the frequency. In particular, the phase presented a peak as a function of frequency, indicating the response of a trap. The evolution of this trap response frequency with temperature could be related through an Arrhenius plot to a trap energy of 0.95 eV.

9.2.3.5 HFET at high temperature

As GaN-based devices are expected to work at elevated temperature, it became important to address this point. Asif Khan *et al.* [AsifKhan95a] studied the performances of an HFET from room temperature up to 300 °C. The device structure resembles the one fabricated by the same authors [AsifKhan94a] and described in Section 9.2.3.3. The AlGaN barrier is somewhat thicker (300 Å). Figure 9.12 shows the measured current–voltage characteristics at 25, 200, and 300 °C. In addition to the expected decrease in the transconductance with temperature, we also observe a sharp increase in the shunt conductance. An Arrhenius plot of this shunt conductance gives an activation energy of 0.5 eV, independent of gate bias. This parasitic conduction is believed to originate from deep traps with a very large concentration. The transconductance is maximum at $V_{gs} = -0.75\,\text{V}$ at any temperature. The temperature dependence of the

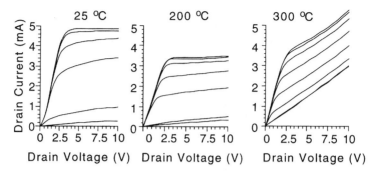

Fig. 9.12 Current–voltage characteristics of AlGaN/GaN HFETs at 25, 200, and 300 °C. Top curve is for $V_{gs} = 1$ V, step -0.5 V. (After [AsifKhan95a].)

transconductance is presented in Fig. 9.13. One observes that g_m decreases by a factor of 2 or so between 25 °C and 300 °C. The same plot for g_{m0} would show an even larger decrease from 55 to 26 mS/mm. From the relation between f_T and g_{m0}, one can expect that f_T decreases by the same amount with temperature. Figure 9.14 confirms this point. One notes the large value of f_{max} at 25 °C (70 GHz) and its fast roll-off with temperature. In order to improve the high-temperature performance, parasitic conductances must be suppressed. This requires an improvement of the material quality. As a matter of fact, a recent study on HFETs at high temperature has shown that a high and constant transconductance (in the range of 100 mS/mm) can be obtained from room temperature up to 200 °C while the cutoff frequency f_T remains constant at least up to 90 °C [AsifKhan96d]. In light of these results, one can be optimistic about the use of GaN electronic devices at moderately high temperature.

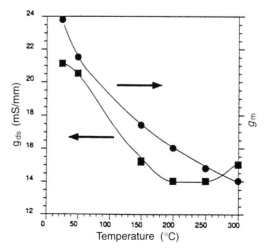

Fig. 9.13 Temperature dependencies of the maximum transconductance g_m and the drain conductance in the linear region g_{ds} of an AlGaN/GaN HFET. (After [AsifKhan95a].)

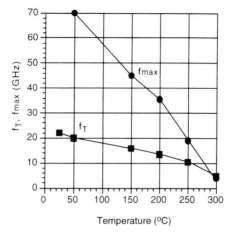

Fig. 9.14 Temperature dependencies of the cutoff frequencies f_T and f_{max} of an AlGaN/GaN HFET. $V_{ds} = 20$ V and $V_{gs} = -0.8$ V. (After [AsifKhan95a].)

9.2.3.6 Phototransistors

We have presented elsewhere in this chapter the optoelectronic properties of GaN-based materials. However, we would like to mention here some aspects related to the response under UV illumination of GaN FETs. Khan *et al.* [AsifKhan95c] used the HFET already described [AsifKhan94a] to compare the current–voltage characteristics in the dark and under UV illumination from a He:Cd laser ($\lambda = 0.325\,\mu$m). The I_{ds} current was found to increase with illumination as expected. The HFET responsivity as a function of gate voltage is shown in Fig. 9.15. As in the usual GaN photoconductors, the responsivity is large due to a high photoconductive gain. This gain originates from holes being rapidly captured while electrons, which have a much smaller capture cross-

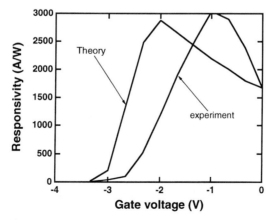

Fig. 9.15 Measured and calculated responsivity of a 0.2 μm gate AlGaN/GaN HFET photodetector illuminated with an He–Cd laser from the bottom sapphire–GaN interface. (After [AsifKhan95a].)

section than holes, remain free for a long time [Binet96b]. The responsivity variation with gate bias is similar to the $g_m(V_{gs})$ one. This comes as no surprise, as the UV illumination modulates the electron 2D gas density as the gate does. Under UV illumination, such HEFTs still present a complete pinchoff, but at a larger reverse gate bias than in the dark [AsifKhan95d]. This is in agreement with the illumination acting as an additional channel doping. As a result, the transconductance is improved by two phenomena. First, the 2D density and thus g_{m0} are increased. Second, the source resistance is lowered, which further enhances g_m.

Both effects were studied [Duboz96a] in a MISFET structure where a 300 Å thick SiO_2 layer was deposited on top of a 1 μm thick insulating GaN layer. The Au gate (1 to 40 μm long) was deposited on the oxide while Ti/Al ohmic contacts were deposited after a chemical etching of the oxide. In spite of the high layer resistivity, the contacts were almost ohmic but highly resistive. The source access resistance was as large as 1 MΩ for a 100 μm wide FET. The capacitance—voltage showed that an accumulation layer was formed at the oxide/GaN interface for positive gate voltages. However, due to the large access resistance, the measured extrinsic transconductance g_m was found to be extremely small, as can be seen in Fig. 9.16. For $V_{ds} = 12$ V, g_m equals 3 μS at $V_{gs} = +3.2$ V. The $g_m \times R_s$ product is larger than unity which proves that the source resistance is drain–source bias dependent and that the intrinsic transconductance may be reasonably large. A backside illumination allowed to enhance both the density and the access conductance as already explained. Reducing the access resistance without modifying too much the intrinsic transconductance was possible by a front side illumination. As the gate is thick and thus UV opaque

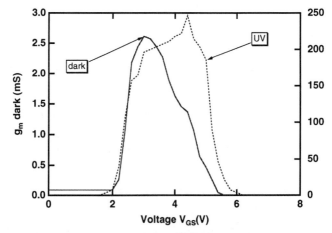

Fig. 9.16 Transconductance of a GaN/SiO_2 MISFET in the dark and under UV illumination as a function of gate bias. $T = 300$ K, $V_{ds} = 12$ V. The gate is 100 μm wide and 40 μm long. The GaN is highly resistive and access resistances are around 1 MΩ in the dark. The He–Cd laser is incident on the front side so that only the access region is illuminated. The extrinsic g_m increases as the access resistances decrease. (After [Duboz96a].)

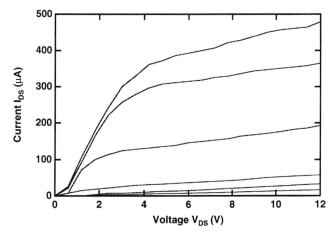

Fig. 9.17 Electrical characteristics of a GaN MISFET under UV illumination from the front side. As the channel is not illuminated, the MISFET is OFF at $V_{gs} = 0$ V (bottom curve) and ON at $V_{gs} = 5$ V (top curve), in accumulation. The I_{ds} saturation is due to the long channel pinchoff on the drain side of the gate. The performance is limited by the contact resistances below source and drain. (After [Duboz96a].)

and parasitic reflections are negligible, the channel below the gate is almost not illuminated. This is verified by the gate leakage current remaining almost as small as in the dark. Hence, n_s and g_{m0} are not modified. The source resistance is made of two terms. The contact resistance due to the metal/GaN interface below the metal is not changed by the illumination. On the contrary, the resistance due to the source–gate spacing decreases with illumination. As this term is very large in a highly resistive material, the effect of UV light may be important. This is exactly what Fig. 9.16 shows. The extrinsic transconductance g_m reaches 250 μS. Figure 9.17 presents the current–voltage characteristics. The slope at low V_{ds} depends very little on V_{gs}, indicating that the device is still limited by the access resistances (region below the ohmic contacts), of the order of 500 kΩ. By reducing this resistance with a doped GaN layer below the ohmic contacts and keeping an insulating channel below the gate and between contacts, one should be able to fabricate an FET that is in the OFF-state unless both a positive gate bias is applied and UV light is incident on the front side. This would allow for a double electrical and optical command of the FET.

9.2.3.7 *HFET versus MODFET for high performace*

As far as high performance is concerned, two different points of view exist on how to improve the transconductance and cutoff frequencies. According to Khan *et al.* [AsifKhan96c], the difference of mobility between 2D gases in doped and undoped GaN is negligible. As a consequence, the electron density can be increased by doping the channel. Furthermore, this will reduce the access resistances. Hence, a 0.1 μm thick n-GaN channel was grown on a 1 μm thick insulating GaN. On top of that, an $Al_{0.15}Ga_{0.85}N$ barrier was grown, with 30 Å

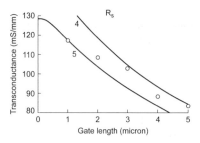

Fig. 9.18 Dependence of the DC transconductance on the gate length. The solid line is calculated from a modified unified HFET model. The calculation is performed for two different values of the series resistance, indicated in the figure in Ω mm. (After [Chen96].)

undoped and 300 Å n-doped. The resulting $n_s \mu$ product was as large as 1.2×10^{16}/Vs. The gate lengths were varied from 1 to 5 μm. Various samples were fabricated with different $n_s \mu$ products [Chen96]. The g_m dependence on gate length is shown in Fig. 9.18. One observes that g_m increases with decreasing gate length, as expected from our initial discussion of the saturation velocity. The second point is that g_m decreases with the $n_s \mu$ product. This and the contact resistance measurements indicate that g_m is limited by the source resistance. For the largest $n_s \mu$ product, R_s and g_m are measured to be 6.6 Ω mm and 120 mS/mm respectively. This corresponds to $g_{m0} = 576$ mS/mm at room temperature. As an additional illustration of the importance of ohmic contacts for HFETs, their very recent improvement ($R_s = 2.5 \,\Omega$ mm) led to enhanced performances: for instance, the transconductance g_m was increased to 142 mS/mm and the drain–source saturation current was as large as 1 A/mm [AsifKhan96d].

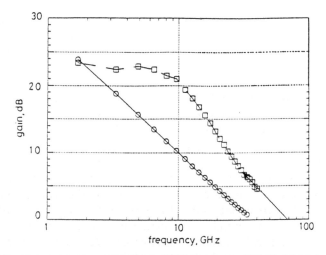

Fig. 9.19 Microwave measurements for a 0.25 μm gate AlGaN/GaN-doped channel HFET. $f_T = 36.1$ GHz and $f_{max} = 70.8$ GHz. (After [AsifKhan96a].

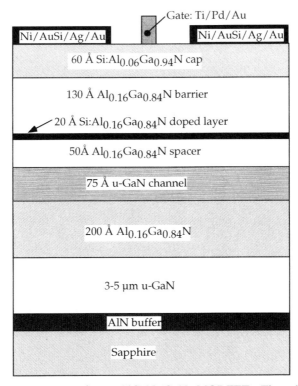

Fig. 9.20 Layer structure of an AlGaN/GaN MODFET. The electron density and mobility in the modulation doped channel are 7.3×10^{12} cm^{-2} and 680 cm^2/Vs respectively. (After [Burm96].

The microwave operation of these doped channel HFETs was studied [AsifKhan96a, AsifKhan96c, AsifKhan96d]. An f_T close to 20 GHz was found for 1 μm long gate devices [AsifKhan96c, AsifKhan96d]. On decreasing the gate length to 0.25 μm, cutoff frequencies around 36 GHz were reported while f_{max} is in the range of 70 to 80 GHz, as can be seen in Fig. 9.19.

On the other hand, it is argued that the doping modulation should lead to better performances. A modulation doped structure is shown in Fig. 9.20 [Burm96]. A very thin undoped GaN channel is grown between two Al$_{0.16}$Ga$_{0.84}$N barriers. The top barrier is modulation doped with Si. A 50 Å thick undoped AlGaN spacer was used in order to reduce coulombic interactions between electrons in the channel and ionized Si. The measured Hall mobility was 680 cm^2/Vs while the sheet charge density was 7.3×10^{12} cm^{-2}. Although this is not precisely studied, one must admit that even the unintentionally doped AlGaN layers must be doped in order to yield such a high sheet carrier density. The measured g_m was around 40 mS/mm, which can be explained by the rather large source resistance. The contact resistance was 9.2 Ω mm, mostly due to the specific contact resistance of 5×10^{-4} Ω cm^2. This arises from the contact being deposited on top of AlGaN. As the channel is not doped, etching away the

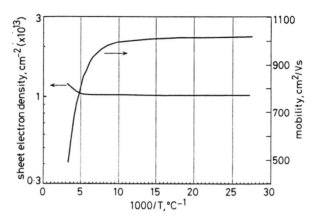

Fig. 9.21 Temperature dependencies of the electron mobility and sheet density in an AlGaN/GaN MODFET. Values were deduced from Hall effect measurements (After [Ozgur95].)

barrier would probably not help. This problem clearly limits the performance and has to be solved in order to take advantage of the MODFET structure. Nevertheless, high-frequency measurements gave $f_T = 21.4$ GHz and $f_{max} = 77$ GHz for a 0.25 μm gate. At a few GHz, a power of 0.5 W/mm was measured [Eastman96] and it is conjectured that a power 10 times larger should be obtained.

Contrary to this normally ON MODFET, normally OFF MODFET was also realized [Ozgur95]. Hall measurements were performed on the layer before the FET process. Figure 9.21 shows that the electron sheet density (1.2×10^{13} cm^{-2}) is one order of magnitude higher than those obtainable in GaAs/AlGaAs systems, exploiting here the very large conduction band discontinuity at the GaN/AlGaN interface. The density goes with a reasonably large mobility of 490 cm^2/Vs at 300 K. The characteristics of a 3 μm long gate MODFET are presented in Fig. 9.22. Due to the large Schottky barrier height, the channel is pinched off at $V_{gs} = 0$ V. A positive V_{gs} then gives rise to a large current (300 mA/mm) and a large transconductance of 120 mS/mm. This good performance for a long gate device was partly due to a low source resistance (2.5 Ω mm). The same group largely improved their MODFET performance the next year [Fran96]. This new sample was grown by reactive molecular beam epitaxy as opposed to previously described samples that were grown by MOCVD. In order to improve the crystallographic quality of the GaN channel, an AlN/GaN supperlattice was grown below the channel. The 1 μm thick channel was undoped. The AlGaN barrier had a high Al content in order to improve the carrier confinement and to reduce the gate leakage current. The first 50 Å were undoped while the last 100 Å were doped to 5×10^{18} cm^{-3}. The sheet carrier density was 3×10^{12} cm^{-2} and the mobility close to 700 cm^2/Vs. A threshold voltage of -1 V was measured so that best operation was achieved with positive gate biases. The saturation current was as large as 490 mA/mm and the transconductance equal to 186 mS/mm. These performances were obtained at a

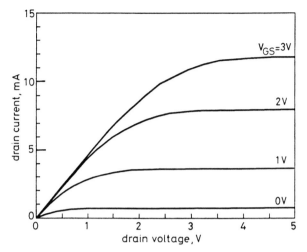

Fig. 9.22 Electrical characteristics of an AlGaN/GaN MODFET. Gate length $L_g = 3\,\mu$m, gate width $W_g = 40\,\mu$m and source–drain separation $L_{ds} = 5\,\mu$m. (After [Ozgur95].)

low $V_{ds} = 4$ V for a gate length of 2 μm. This is attributed to the high mobility that decreases the critical field, leading to a large electron velocity close to its saturation value below the gate, and the very low access resistances. The transconductance variations with the gate length are shown in Fig. 9.23. According to the authors, the drain–source voltage is kept constant and equal to 4 V whatever the gate length. It is then clear that the velocity in the long gate FETs is not saturated and g_m decreases as μ/L_g. This figure also presents the breakdown voltage as a function of gate length and drain–gate spacing (inset).

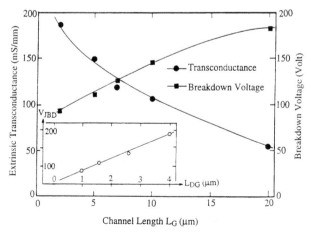

Fig. 9.23 Variations of extrinsic transconductance g_m and the junction breakdown voltage V_{JBV} as functions of the gate length L_g of various AlGaN/GaN MODFETs. g_m is measured at $V_{ds} = 4$ V. The variation of V_{JBV} with drain–gate distance L_{dg} is shown in the inset. (After [Fan96].)

As expected, the breakdown voltage decreases with decreasing distances. For the most performing MODFET ($L_g = 2\,\mu$m and $L_{dg} = 1\,\mu$m), it is still larger than 80 V, which translates to a field of 0.8 MV/cm. These performances were finally improved recently [Mohammad96b]. Using a 1.5 μm gate, the saturation current reached an impressive value of 626 mA/mm with a transconductance of 210 mS/mm. These are the largest published values for any GaN based FET. They show the high potential of this material and of MODFETs in particular as soon as contact resistances are lowered enough.

9.2.3.8 MODFET power generation

In terms of power generation, recent impressive studies have also been performed. Wu *et al.* [Wu96b] reported the fabrication and characterization of MODFETs. Thanks to the high modulation doping in the barrier (the background donor density in the 0.3 μm channel was around 4×10^{16} cm^{-3}), the electron sheet density reached 8×10^{12} cm^{-2} while the mobility was as high as 1500 cm^2/Vs. As a consequence, the device with a 1 μm long gate exhibits a transconductance of 140 mS/mm at room temperature, as can be deduced from the characteristics presented in Fig. 9.24. It could have been even larger with more conductive contacts. Since the metal–GaN alloy depth was only 15 nm, the 30 nm thick Al$_{0.15}$Ga$_{0.85}$N barrier layer was removed below the drain and source ohmic contacts, leaving a relatively resistive GaN layer under the ohmic metal. As a result, the contact resistance could not be decreased below 3 Ω mm. Recently, the same group [DenBaars96] succeeded in decreasing the contact resistance to 0.44 Ω mm by a regrowth process of highly doped GaN below the ohmic contacts. As a consequence, the transconductance was increased to 170 mS/mm. The gate-to-drain breakdown voltage was very large (340 V) in agreement with a breakdown field of 1.1×10^6 V/cm. The characteristics show the current saturation and decrease with increasing V_{ds}. The decrease is due to the channel heating and the subsequent degradation of the electron mobility

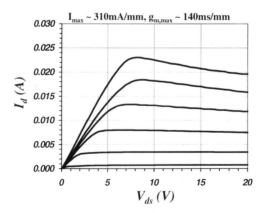

Fig. 9.24 Electrical characteristics of an AlGaN/GaN MODFET. Gate length $L_g = 1 - 1.5\,\mu$m, gate width $W_g = 75\,\mu$m and source–drain separation $L_{ds} = 3\,\mu$m. Only the low voltage part is shown. The breakdown voltages are much larger (170 V for $L_{dg} = 2\,\mu$m). (After [Wu96b].)

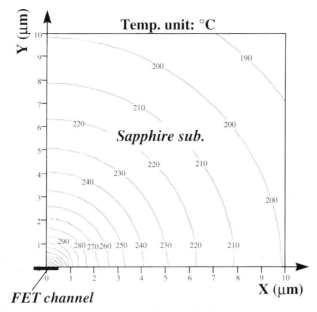

Fig. 9.25 Calulated temperature contours in the vicinity of an AlGaN/GaN MODFET. Gate length $L_g = 1\,\mu$m, gate width $W_g = 500\,\mu$m and source-drain separation $L_{ds} = 4\,\mu$m. The sapphire substrate is 325 μm thick and its back face is kept at 30 °C. The simulation was made at the maximum output power of 4 W/mm. (After [Wu96a].)

and saturation velocity. This channel heating is due to the poor thermal conductivity of the sapphire substrate [Wu96a]. A simulated temperature map of

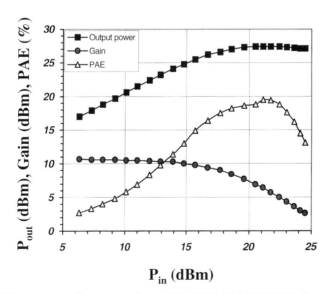

Fig. 9.26 RF power performance of an AlGaN/GaN MODFET. $L_g = 1\,\mu$m, $W_g = 500\,\mu$m and $L_{ds} = 4\,\mu$m. DC bias conditions: $V_{ds} = 26$ V, $V_{gs} = -0.3$ V, $I_d = 66$ mA. Frequency: 2 GHz. Maximum power output: 1.1 W/mm. (After [Wu96a].)

the MODFET in high-power operation is presented in Fig. 9.25. The channel temperature is found to reach 360 °C. Let us note that thermal problems could be at least partially solved by thinning the sapphire substrate (325 µm here) or by growing the GaN layers on SiC (this seems to be the choice of the Santa Barbara group [DenBaars96]). The power performances were evaluated. The bias point was set at $V_{ds} = 26$ V and $V_{gs} = -0.3$ V so that a total voltage swing of 38 V was possible. The output power, small signal gain, and the PAE measured at 2 GHz are shown in Fig. 9.26. A power of 1.1 W/mm is obtained with a PAE of 18.6%. Both Wu *et al.* [Wu96a] and Eastman [Eastman96] expect that a power around 5 W/mm should be obtained in GaN MODFETs at 2 GHz.

9.3 Ultraviolet photoconductors based on GaN

The history of the study of the ultraviolet radiation as well as its contemporary application fields (spectroscopy, astronomy...) are presented in a recent review by Razeghi and Rogalski [Razeghi96]. All the different types of UV detectors are described. We concentrate here on GaN-based devices only.

9.3.1 General principles

When an electric field is applied in a bulk conductive material, the induced current density is given by $J_{dark} = e(nv_n + pv_p)$, where $n(p)$ and $v_n(v_p)$ are the electron (hole) density and velocity respectively. If the material is a low doped semiconductor, then densities are low and so is the dark current. When illuminated by a flux ϕ of photons whose energy is larger than the gap energy, additional electrons δn and holes δp are photocreated. The time evolution of total carrier densities is driven by the generation and recombination. The generation comprises thermal and optical terms. The recombination includes the direct electron–hole recombination and the separate electron and hole recombination on defects (in the bulk or on surfaces). This important issue will be addressed in detail later on. As a phenomenological approach, one describes the excess carrier density relaxation with lifetimes τ_n and τ_p. Hence the steady state excess densities under illumination are given by $\delta n = \alpha\phi\tau_n$ and $\delta_p = \alpha\phi\tau_p$ where α is the absorption coefficient (in cm^{-1}). The photocurrent density is thus $J_{phot} = e(\delta n v_n + \delta p v_p)$. Depending on the applied field, carrier velocities are equal to the mobility-field product or to the saturation value. For a device with thickness z, width W and length L between contacts, the photocurrent equals $I_{phot} = J_{phot} zW$ while the received power is $h\nu\phi LW$. The responsivity is defined as the ratio of the photocurrent upon the received power:

$$R = \frac{e\alpha z(\tau_n v_n + \tau_p v_p)}{h\nu L}$$

In most cases, the photocurrent is mainly due to one type of carrier. As the electron velocities are larger than those of holes, we consider here electrons only. Then, the responsivity is given by

$$R = \frac{e\alpha z\tau_n v_n}{h\nu L}$$

9.3 Ultraviolet photoconductors based on GaN

The total absorption neglecting the reflection is given by $A = 1 - e^{-\alpha z} \simeq \alpha z$ for thin films. For thick films, we can integrate the absorption and the photocurrent along the z-axis. Recognizing the transit time $t = L/v_n$, one thus obtains

$$R = \frac{e}{h\nu} \times A \times \frac{\tau_n}{t}$$

The ratio τ_n/t defines the photoconductive gain g, which can be smaller or larger than unity. The noise (current fluctuation) in a photoconductor is usually dominated by the carrier density fluctuations arising from random generations and recombinations. The spectral density of current noise i_{noise} is equal to $\sqrt{4egI}$, where I is the total current under illumination, $I_{\text{dark}} + I_{\text{phot}}$. Finally, the detectivity that features the signal-to-noise ratio is defined as R/i_{noise}. It thus varies as $A\sqrt{g/I}$. Detector performance is improved if A is maximum (i.e. unity), g is large, and the dark current low. One thus uses a low-doped material with the largest bandgap energy (compatible with the detection spectral window), the thinnest possible material that gives a near unity absorption, a short distance between contacts, and the longest lifetimes compatible with the bandpass requirements. Indeed, the device's intrinsic response time is given by the carrier lifetime. This is true in a trap-free material. In the case of traps in thermal equilibrium with free carriers, the response time may be substantially longer, as trapped carriers need to be thermally excited to the conduction or the valence band before they recombine. Note that in this case the response time depends exponentially on temperature, with an activation energy equal to the trap depth.

In these regards, GaN with a strong absorption coefficient ($4 \times 10^4 \text{ cm}^{-1}$) for wavelengths smaller than 0.36 μm should be a good candidate for UV photoconductive detectors. The remaining issues are how low the residual doping can be obtained and how long the carrier lifetimes are. The third interesting point is the spectral selectivity and the ability to tailor the detection wavelength. These are the main points that we would like to address now.

9.3.2 First demonstration of a GaN photoconductor

Asif Khan *et al.* fabricated the first GaN-based UV detector in 1992 [AsifKhan92]. A 0.8 μm thick GaN layer was deposited by a switched atomic layer epitaxy procedure using MOCVD on sapphire, with a 0.1 μm thick AlN buffer layer. Thick gold electrodes were fabricated in an interdigitated geometry with a 10 μm spacing. This geometry offers a large area together with a short distance between contacts. It is equivalent to a detector that would be 10 μm long and 5×10^4 μm wide. The dark current was extremely small (2 nA for a 200 V bias). Taking a mobility of 100 cm^2/Vs, the device conductance in the dark corresponds to a carrier density in the range of 10^6 cm^{-3}. Even with a mobility as low as 1 cm^2/Vs, one finds a density of 10^8 cm^{-3}. The actual density is likely to be larger than that if one takes into account that contacts are rectifying. However, this result fully answers the question of how low the doping in GaN can be. The responsivity for a 5 V bias is shown in Fig. 9.27. The first observation is the clear cutoff for energies below the bandgap energy that makes

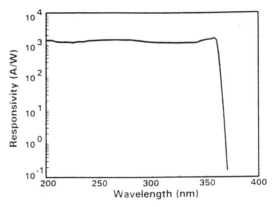

Fig. 9.27 Spectral responsivity of a GaN photoconductor biased at 5 V. The spacing between interdigitated electrodes is 10 μm. (After [AsifKhan92].)

this detector visible-blind. The second observation is the huge responsivity around 1000 A/W. The absorption being close to unity, this gives a gain on the order of 5000. The field is about 5000 V/cm and gives an electron velocity of 5×10^5 cm s^{-1}. Hence the transit time equals 2 ns. The photoconductive gain would then imply an electron lifetime in the range of 10 μs. The response time was directly measured and found to be around 1 ms. The discrepancy (factor 100) between the carrier lifetime deduced from the responsivity and the device response time might be smaller than the one calculated here if the actual field in the layer is smaller (resistive contacts) or the mobility lower. However, it seems that a factor 100 cannot be accounted for. One possible explanation would imply the presence of traps, as previously described. The temperature dependence of the response time is not reported, so we cannot check this assumption. Another possibility is that some carriers recombine very fast and actually do not participate in the current. Surfaces are known to provide such recombination centers and we should also keep in mind the large density of defects such as dislocations in these layers. These important issues related to recombinations are discussed in more detail in the next section.

9.3.3 Recombinations and time response of GaN photoconductors

A photoconductive detector was fabricated on n-GaN by MOCVD on sapphire [Kung95]. Although doping of 10^{17}–10^{18} cm^{-3} is not adapted for high performance, it gives information on the kinetics. The geometry was a van der Pauw-type cloverleaf. A bias of 6 V was applied on two adjacent contacts and the voltage signal was measured on the two other ones. Figure 9.28 shows the responsivity as a function of the frequency of chopping of the He:Cd laser. We first observe that the response decreases with the frequency, leading to a response time of 20 ms for the lowest excitation density. Second, we observe that the responsivity decreases with increasing incident power density. As an additional indication of the complex nature of recombinations in GaN, the response

9.3 Ultraviolet photoconductors based on GaN

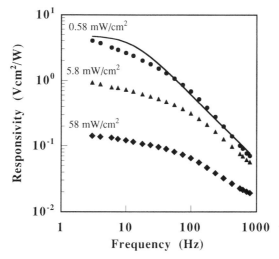

Fig. 9.28 GaN photoconductor responsivity as a function of chopping frequency for three different excitation power densities of the He–Cd laser light. (After [Kung95].)

to an optical pulse was found to be non-exponential. The nature of recombinations in GaN/sapphire was studied in more detail the following year [Binet96a]. A 2 μm thick semi-insulating GaN layer grown by MOCVD on sapphire with an AlN buffer was used. A 100 μm sided square contact separated from the surrounding contact by 15 μm was deposited. The contacts were found to be ohmic, although resistive, so that about one half of the bias is actually applied to the GaN layer. From the dark current and a typical value of 120 cm^2/Vs for the mobility, the electron density was estimated to be in the 10^{13} cm^{-3} range. The spectral response was close to the one described in Fig. 9.27. The responsivity at low frequency reached 4500 A/W at the band edge. Below the gap energy, it fell off very abruptly and then decreased nearly exponentially and was as small as 3×10^{-3} A/W at $\lambda = 0.5\,\mu$m. This responsivity was measured at low incident power (Xe lamp) and for an applied field of 3 kV/cm. At higher power (He:Cd) laser), the responsivity varies as shown in Fig. 9.29. We observe that it remains almost constant for power densities up to 40 W cm^{-2}. Above this point, it decreases as $P^{-0.5}$. The time reponse of the detectors was studied in both the low-power and high-power regimes. As can be seen in Fig. 9.30, the photocurrent decays exponentially with time at low power density with a characteristic time of about 4 ms. On the contrary, the high-power decay is first non-exponential but rather hyperbolic. It then becomes exponential with a characteristic time of 2.6 ms, which is shorter than in the low-power regime. Figures 9.29 and 9.30 clearly show that the recombinations at low power are monomolecular, due to recombinations on defects, while at high power they become bimolecular due to direct electron–hole recombination. We have already mentioned that the presence of traps in equilibrium with free carrier bands should give response times that vary with temperature. The measured response time was found to be independent of the temperature. Hence, the response time is not due to a slow

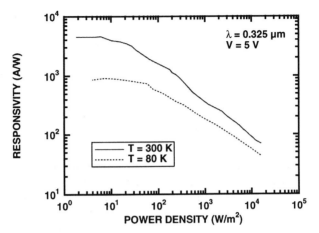

Fig. 9.29 Responsivity vs. optical power density of an He–Cd laser light for a GaN photoconductor at a 5 V bias. (After [Binet96a].)

detrapping but is mainly due to the carrier lifetime. A model was proposed in order to describe the observed slow recombinations. It does not exclude that part of the photoexcited carriers (those which are close to dislocations, surfaces...) may recombine very fast and do not participate in the current. This phenomenon is not taken into account in the model but can be accounted for by an additional collection factor smaller than unity in the photoresponsivity formula.

Let N_r be the total density of recombination centers and n_r the density of those occupied by a trapped electron. Let β_n and β_p be the electron and hole trapping coefficients and γ the electron–hole direct recombination coefficient.

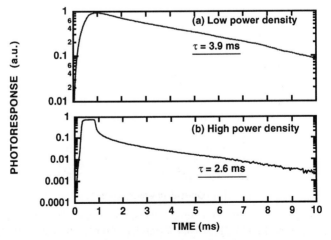

Fig. 9.30 Temporal photoresponse of a GaN photoconductor at a 5 V bias at low optical power density (a) and high optical density (b) (After [Binet96a].)

9.3 Ultraviolet photoconductors based on GaN

The time evolution of free electrons and holes and occupied traps is given by:

$$\frac{dn}{dt} = G_{opt} + G_{thn} - \gamma np - \beta_n(N_r - n_r)n$$

$$\frac{dp}{dt} = G_{opt} + G_{thp} - \gamma np - \beta_p n_r p$$

$$\frac{dn_r}{dt} = G_{thp} - G_{thn} + \beta_n(N_r - n_r)n - \beta_p n_r p$$

Thermal generation terms for electrons G_{thn} and holes G_{thp} are added here for the only purpose of balance between generation and recombination in the dark. They are not related to the thermal emission from the recombination centers. In the presence of light, they are small compared to the optical generation term G_{opt} and thermal emission can be neglected. The optical term is related to the incident flux and the absorption. In the high-power regime, the electron and hole densities are equal and given by the bimolecular recombination term. One thus verifies $G_{opt} = \gamma \times n^2$. A value of 1.3×10^{-8} cm^3 s^{-1} is deduced for γ. As deep traps are assumed to be responsible for the electrical compensation of this layer while GaN layers frequently exhibit residual electron densities above 10^{16} cm^{-3}, a conservative estimation for the trap density is around 10^{16} cm^{-3}. Then, a fit of the experimental results leads to $\beta_n = 2.2 \times 10^{-14}$ cm^3 s^{-1} and $\beta_p = 10^{-9}$ cm^3 s^{-1}. The physical interpretation of the large gain and long response times observed in GaN detectors is then the following: holes are captured very rapidly while electrons contribute efficiently to the current for long times before recombining. Although GaN is a wide direct

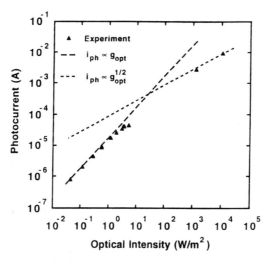

Fig. 9.31 Photocurrent vs. incident optical density of a GaN photoconductor at a 4 V bias. Dotted lines represent the predicted dependence for the bimolecular recombination model ($i_{ph} \propto \sqrt{g_{opt}}$) and for the monomolecular recombination model ($i_{ph} \propto g_{opt}$). (After (Stevens95]).

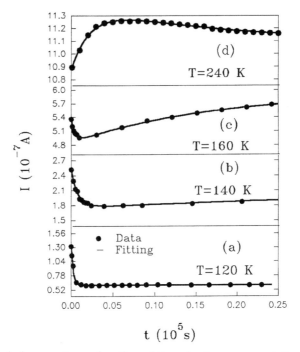

Fig. 9.32 The dark current as a function of time for four representative temperatures. The sample was cooled from room temperature to a given temperature T, and $t = 0$ is defined as the time when the temperature T was attained by the sample. (After [Johnson96].)

bandgap and should lead to a fast direct electron–hole recombination, the latter is limited as holes recombine nonradiatively. At high optical power, however, the nonradiative channel becomes saturated and radiative recombinations prevail. These features seem to be quite characteristic of GaN, even grown under very different conditions. For instance, Stevens *et al.* [Stevens95] had already obtained similar results in Mg-doped GaN grown by MBE on Si. They clearly evidenced the two recombination regimes, as can be seen in Fig. 9.31. At high power hyperbolic decays followed by long exponential decays were also measured. A fast hole capture and a slow electron capture was also given as an interpretation for these results in p-type GaN on Si, similarly to what was observed in compensated GaN. However, p-type GaN may also present additional phenomena as metastability. Johnson *et al.* [Johnson96] measured the time evolution of the current in p-type GaN grown on sapphire by MOCVD. As can be seen in Fig. 9.32, the dark current was varying on a very long time-scale before reaching a steady state value. These variations are temperature dependent and are explained in terms of a bistable impurity. The dark current after an optical pulse was also showing responses on a time-scale of hours. In terms of detection, these very slow phenomena lead to an increase of low-frequency noise. For video frequencies, this may not degrade the image quality too much if a calibration of the reference level can be made periodically.

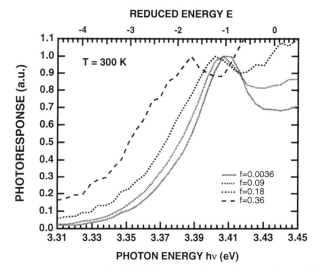

Fig. 9.33 Photocurrent spectra of a GaN photoconductor for varying applied fields. The field is defined in reduced units by $f = eFa/R$ where e, F, a, and R are the electron charge, the field, the exciton Bohr radius and the exciton Rydberg respectively. Thus, f compares the field potential on the exciton and the exciton binding energy. The currents are normalized to unity at the exciton peak. $T = 300$ K. (After [Binet96b].)

9.3.4 Excitonic effects in GaN photoconductors

Optical properties near the band edge in GaN are dominated at low temperature but also at higher temperature by excitons. It has been shown that the absorption even at room temperature exhibits a strong excitonic peak [Binet96b]. As its binding energy is 28 meV, i.e. close to kT at room temperature, the free exciton is almost completely ionized (this is not inconsistent with the excitonic peak still being observable in the absorption spectrum). Thus, a very small electrical field will separate electrons and holes and produce a photocurrent. As a consequence, the photoresponse at 300 K should exhibit an excitonic peak due to the absorption. As the field increases, it was shown that the excitonic absorption is quenched by the Franz–Keldysh field effect [Binet96b]. Hence, the excitonic peak in the response spectrum should also be quenched. This is exactly what can be seen in Fig. 9.33. On the other hand, at low temperature, the exciton is not thermally ionized. It of course still dominates the absorption spectrum, but the exciton is so stable that it cannot be ionized by a small electric field. Thus, the exciton is barely visible in the response at low temperature and voltages, as shown in Fig. 9.34. In order to enhance the excitonic peak in photoresponse, one has to increase the applied bias in order to field-ionize the exciton, as can be verified in Fig. 9.34.

Field effects on the excitonic absorption in GaN can also be used for electro-optic modulation. When the electric field in the GaN layer increases, the absorption peak shifts towards smaller energies. Thus, the absorption coefficient at a given energy increases or decreases, depending on whether the energy is

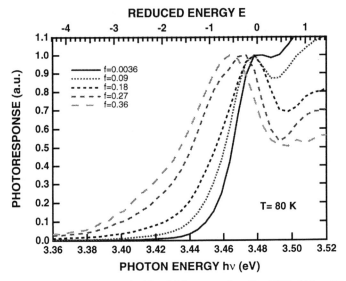

Fig. 9.34 Same measurements as in Fig. 9.33 except that $T = 80$ K. (After [Binet96b].)

smaller or larger than the exciton energy at zero applied field. Fig. 9.35 shows the transmission measured at three different energies as a function of the applied field. We observe that the transmission on the high-energy side of the exciton increases with the electric field while it decreases on the low-energy side of the exciton. This was the first demonstrated electro-optic modulator in the ultraviolet region based on GaN. The modulation depth reached 50%, but the ON state transmission is very low, around 5%. Improvements of the performance (increased modulation depth, reduced insertion losses) should be possible

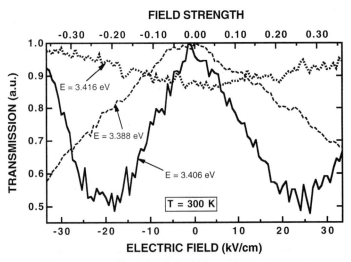

Fig. 9.35 Transmission of GaN as a function of the applied electric field for different photon energies. The GaN exciton peak at zero field is at an energy of 3.407 eV. For each energy, the maximum transmission is normalized to unity. $T = 300$ K. (After [Binet96b].)

in structures with amplified excitonic effects (quantum wells) or in different geometries (cavities etc.).

9.3.5 Spectral selectivity in GaN photoconductors

One important point for applications is the spectral selectivity, and more precisely the below-gap response. Figure 9.27 shows that the response below the gap drops by four orders of magnitude. Binet *et al.* [Binet96a] found a selectivity of six orders of magnitude between the UV and red ($\lambda = 0.5\,\mu$m) while Kung *et al.* [Kung95] found three orders of magnitude for the same spectral ranges. This response below the gap arises from defects and its magnitude is an indication of the density of defects. Its spectrum should also lead to the defect distribution in the gap. Qiu *et al.* [Qiu95] compared p-type and undoped samples. They measured spectral selectivities between $\lambda = 0.35\,\mu$m and $\lambda = 0.8\,\mu$m of 10^3 to 10^4. From the photoconductivity spectra, it was conjectured that the below-gap response in n-type GaN is due to the valence band tail, while the one in p-type GaN is due to the conduction band tail.

9.3.6 AlGaN photoconductors

A key advantage in GaN materials resides in the possibility of growing InGaN and AlGaN alloys which allows one to vary the bandgap energy from 1.9 to 6.2 eV. In terms of photodetection, it implies that a spectral range from 650 nm to 200 nm can be covered by GaN-related materials. On the high-energy side, precisely, there is considerable interest for reducing the cutoff wavelength below 300 nm. Indeed, stratospheric ozone limits the wavelengths of the solar radiation reaching the surface of the Earth to about 300 nm. Therefore, a detector that detects UV radiation below 300 nm only (called solar blind) will be able to observe UV emitting objects (hot reactors etc.) without being saturated by the solar radiation. Using a linear Vegard's law, which is reasonable for AlGaN,

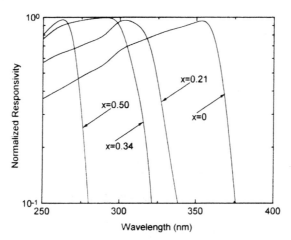

Fig. 9.36 The spectral response of the $Al_xGa_{1-x}N$ photoconductors. (After [Walker96].)

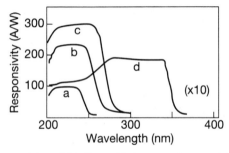

Fig. 9.37 Spectral responsivity of $Al_xGa_{1-x}N$ photoconductors. (a) $x = 0.61$, (b) $x = 0.55$, (c) $x = 0.46$, (d) $x = 0.05$. (After [Lim96].)

around 25% of Al should lead to a solar blind detector. Recently, Walker et al. [Walker96] and Lim et al. [Lim96] reported on AlGaN photoconductors grown by MOCVD on sapphire. Figure 9.36 shows the normalized responsivity spectra of AlGaN. The cutoff wavelength decreases from 365 nm for GaN to 260 nm for $Al_{0.5}Ga_{0.5}N$. The cutoff at the band edge remains sharp. The responsivities on the order of 0.1 A/W for an Al content of 0, 0.21, or 0.34 are quite low compared to the one usually measured in GaN detectors as described in Sections 9.3.2 and 9.3.3. This may be explained by the high n-doping of the layer (10^{18} cm^{-3}) which enhances the direct electron–hole recombination thus preventing the separate electron and hole capture that was responsible for the high photoconductive gain. Surprisingly, the responsivity for $Al_{0.5}Ga_{0.5}N$ was found to be as small as 3×10^{-6} A/W. Such a dramatic decrease of the responsivity for high Al contents is in disagreement with the results of Lim et al. [Lim96] as can be easily seen in Fig. 9.37. Contrary to the previous case, the layers used were highly resistive and the measured responsivities were of the same order of magnitude as in usual GaN detectors, between 100 and 300 A/W. Correlatively, response times were measured to be around 1 ms, confirming the high photoconductive gain of these AlGaN detectors.

9.3.7 Performances of GaN photoconductors

Measuring the noise figure of a device is of course important for application purposes. Very often, is also gives precious information on the physical phenomena occuring in the device. The noise figure of the GaN photoconductor presented in Fig. 9.33 was measured for frequencies between 1 Hz and 100 kHz. Figure 9.38 shows the current noise spectral density i_n of the photoconductor biased at 0 and 9 V. The device resistance is of the order of 0.46 MΩ, corresponding to a Johnson noise $\sqrt{4kT/R} = 2 \times 10^{-13}$ A/$\sqrt{Hz}$ at room temperature, in agreement with the measured white noise for 0 V. At a bias of 9 V, we observe that the noise is varying with frequency as $f^{-1/2}$. The noise power thus follows a $1/f$ variation. Such a behavior is usually attributed to surface defects. An extrapolation of the curve shows that the noise at 9 V becomes comparable with the noise at 0 V for frequencies equal to a few MHz. Beyond this

9.3 Ultraviolet photoconductors based on GaN

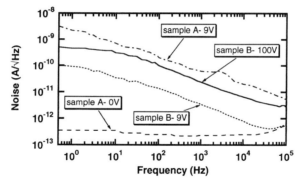

Fig. 9.38 Spectral density of current noise in two GaN photoconductors with and without applied bias. Sample B is more resistive than sample A by a factor of 40. $T = 300$ K. (After [Duboz96b].)

frequency, the noise is dominated by carrier velocity fluctuations and not by density fluctuations. One thus deduces an estimation of the shortest trapping and emitting times in the device in the μs range. The same behaviour was noted in other GaN detectors. The noise spectra of photoconductor fabricated on a more resistive layer has been added in Fig. 9.38. The device resistance was around 20 MΩ, leading to a Johnson noise equal to 3×10^{-14} A/$\sqrt{\text{Hz}}$. The noise measured at 9 and 100 V decreases with increasing frequency, as already noted. Extrapolating the 9 V curve, we again observe that the noise should be dominated by Johnson noise for frequencies above a few MHz, thus leading to the same estimation for trapping times as before. The 100 V noise is larger than the 9 V noise by a factor of 5 or so. The dark currents at 9 and 100 V were 0.47 μA and 2.8 μA respectively. Thus, the noise current increases linearly with the current I and not as $\sqrt{I}$. This again is characteristic of a $1/f$ behaviour.

The normalized detectivity of a detector with an area A can be deduced from noise and responsivity measurements by:

$$D^* = \frac{R \times \sqrt{A}}{i_n}$$

The responsivity at a bias of 9 V equals 8000 A/W for frequencies below 10 Hz and decreases at higher frequency. Above 100 Hz, it roughly decreases as $f^{-1/2}$. Thus, best detectivities are obtained for frequencies between 10 and 100 Hz. At 50 Hz, the detectivity is on the order of 2×10^{11} cm$\sqrt{\text{Hz}}$/W. Let us note that the responsivity, the current, and the noise current increase linearly with the bias so that similar detectivities would be achieved for a different bias. Finally, let us compare both detectors presented in Fig. 9.38. The noise in the more resistive device is 30 times smaller than in the other one. However, the responsivity was also found to be smaller by roughly the same amount, leading to similar detectivities. This is probably due to the compensation of these layers that reduces both the dark current and the response.

9.3.8 Photoconductor arrays

The fabrication of photoconductor arrays implies that two contacts are provided for every pixel on the front side. In addition, the interdigitated geometry makes the fabrication of very small pixels difficult as the contact lines have to be made very narrow. For a small number of pixels with large dimensions, this is, however, possible. Such an array was fabricated and tested by Huang *et al.* [Huang96]. The area is made of 16 pixels and the area of each of them is about $0.1\,\text{mm}^2$. The same responsivities and time responses as already described were observed for individual pixels. Although the authors do not provide any quantitative data, the array uniformity is reported as being good.

9.4 Ultraviolet photovoltaic detectors based on GaN

In a photovoltaic detector, there is an internal field due to an asymmetry of the structure. This field separates photoexcited holes and electrons, thus creating an internal photocurrent. If the device is connected to a small load impedance, the current is directly measured. On the contrary, in an open circuit geometry, a photovoltage appears between the two contacts of the device. This voltage equals the photocurrent times the internal impedance of the detector. One can also consider that the photovoltage arises from the charge separation and corresponds to an induced field screening the internal field. As contacts do not inject charges, the photoconductive gain is equal to unity. The structure asymmetry may be of different nature: p–n junction, composition gradient, metal–semiconductor interface etc.

9.4.1 Schottky diodes

9.4.1.1 *Carrier diffusion*

At the metal–semiconductor interface, the position of the metal Fermi level relative to the semiconductor bands is determined by the thermodynamics (difference of work functions) and also by surface states due to the material's discontinuity and to impurities. This position results in a potential barrier at the interface called Schottky barrier between the metal Fermi level and the conduction band in n-type material or the valence band in p-type material. This potential induces in the semiconductor an electrostatic field which is screened by the semiconductor free charges. Thus, its extension varies as $N^{-1/2}$ where N is the density of ionized impurities. As an example, a barrier of $1\,\text{eV}$ and a doping density of $10^{16}\,\text{cm}^{-3}$ lead to a depletion layer of $0.4\,\mu\text{m}$. As the absorption length in semiconductors is of the same order of magnitude, one can use the depletion region as the absorbing region of a photovoltaic detector. Moreover, this layer thickness can be increased by applying a reverse bias to the diode.

The first photovoltaic detector based on a Schottky barrier on GaN was reported by Khan *et al.* [AsifKhan93b]. A $2\,\mu\text{m}$ thick p-type GaN layer was

9.4 Ultraviolet photovoltaic detectors based on GaN

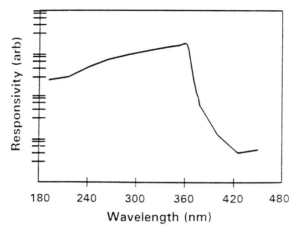

Fig. 9.39 Spectral responsivity of a Schottky barrier photovoltaic detector on p-type GaN. (After [AsifKhan93b].)

grown by MOCVD and Mg doped to $7 \times 10^{17}\,\text{cm}^{-3}$. A Ti/Au Schottky contact and Cr/Au ohmic contacts were deposited. Although a 1.5 V turn-on voltage was reported, the Schottky barrier height was not precisely measured. The detector was back side illuminated through the transparent sapphire substrate. The spectral response is shown in Fig. 9.39. The responsivity is nearly constant (slightly decreasing with decreasing wavelengths) for wavelengths from 200 to 365 nm. It falls sharply by two orders of magnitude for wavelengths longer than 365 nm. At 320 nm, the responsivity was measured to be 0.13 A/W. Given the photon energy and a unity photoconductive gain in this photovoltaic detector, this responsivity agrees with theoretical expectations. As the layer is thicker than the absorption length, charges are photoexcited outside the depletion region. Electrons thus have to diffuse towards the Schottky barrier before the internal field separates them from holes. The responsivity value seems to indicate that this diffusion is not limiting the electron collection.

The opposite case of a Schottky diode on n-type GaN has also been studied [Binet96c]. A 2.6 μm thick non-intentionally doped GaN layer was grown on sapphire by MOCVD. The residual donor density was evaluated to be $3 \times 10^{16}\,\text{cm}^{-3}$. The spectral response measured from back side illumination is shown in Fig. 9.40. It is compared to the absorption coefficient in GaN. Starting from low energies, the reponsivity first increases with energy as the absorption does. Surprisingly, it then falls above 3.35 eV although the absorption still increases and remains very weak at higher energies while the absorption is very large. The other remarkable point is the low responsivity value. The spectrum in Fig. 9.40 was recorded at non-zero frequency. At zero frequency, the responsivity was higher by a factor between 5 and 10. Nevertheless, this remains one order of magnitude below expectations. Both the spectral responsivity and its absolute value show that the carrier collection efficiency is low. As the responsivity is low when the absorption length is short (which means that charge are photocreated close to the GaN/sapphire interface), one deduces that the diffusion limits the

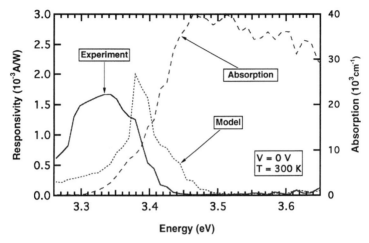

Fig. 9.40 Spectral responsivity of an n-type GaN Schottky barrier photovoltaic detector at zero bias. The experimental responsivity spectrum is compared with a calculated spectrum from a model based on diffusion. The experimental absorption coefficient spectrum in the GaN layer is added in the figure. (After [Binet96c].)

collection. In the case of Schottky diodes on n-type material, holes have to diffuse along the z-axis through the layer before reaching the metal, where they are collected. A simple model based on diffusion with an exponential decay for holes along the z-axis and using the experimental variations of α has been developed. A rough agreement between the experimental and the calculated responsivity spectra was obtained for a hole diffusion length of $0.1\,\mu$m. The results for Schottky diodes on p-type and n-type GaN show the asymmetry between electron and hole diffusion. A related asymmetry between electron and hole capture had already been reported [Binet96a].

The problem of carrier diffusion can be eliminated if the absorption occurs in the depletion region itself. This can be obtained by using very thin films in the back side illumination geometry. It can also be achieved by the front side illumination through a transparent Schottky contact. Such Schottky diodes were fabricated with a thin Pd film deposited on a n^-/n^+ GaN structure [Chen97]. The n^+ layer is used for the bottom ohmic contact. Figure 9.41 shows that the responsivity is in agreement with expectations: The responsivity is constant for energies larger than the GaN bandgap energy and equal to $0.17\,$A/W.

9.4.1.2 Internal photoemission

Photovoltaic detectors based on Schottky diodes exhibit a photoresponse at energies far below the bandgap energy. In order to clarify the origin of this below gap response, Fig. 9.42 shows the Fowler plot of the responsivity. As can be seen, it follows a linear dependence on the photon energy which indicates that the low energy response is due to internal photoemission of electrons from the metal to the GaN conduction band. The extrapolation at zero response gives a Schottky barrier height near $0.86\,$eV, not corrected for the Schottky barrier

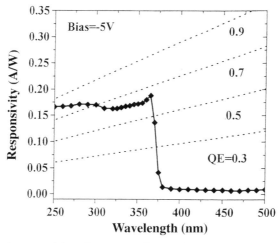

Fig. 9.41 Spectral responsivity of an n-type GaN Schottky barrier detector. The diode is front side illuminated through a semi-transparent Pd contact. (After [Chen97].)

lowering. The numerical estimation of the responsivity due to internal photoemission agrees well with the measured value. At a wavelength of 0.63 μm (He–Ne laser), the external quantum efficiency is calculated to be around 0.25%, corresponding to a responsivity of 10^{-3} A/W. Comparing with the maximum theoretical responsivity at the band edge (0.3 A/W), we can see that the internal photoemission severely limits the spectral selectivity to 300 between the UV and the red.

9.4.1.3 *Response time*
As photovoltaic detectors do not rely on charge injection, they are expected to

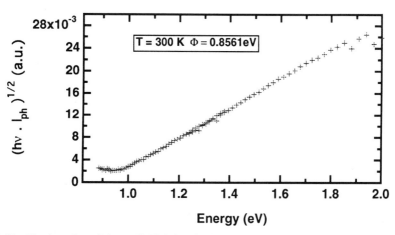

Fig. 9.42 Fowler plot of the n-GaN Schottky barrier photovoltaic detector of Fig. 9.40. The Fowler plot represents the square root of the detector responsivity multiplied by the photon energy vs. photon energy. The response below the gap is due to internal photoemission and the extrapolation of the linear part of the curve gives the Schottky barrier height. (After [Binet96c].)

exhibit short response times, as explained in Section 9.4.3. In the case of Schottky diodes on p-GaN, response times in the μs range were observed with a pulse nitrogen laser [AsifKhan93b]. However, we have already pointed out that the response time at lower optical power may be substantially longer. In the case of Schottky diodes on n-GaN, two different results have been reported. In a back side illuminated Schottky diode, long response times (ms) have been observed and were related to charge emission from traps in the depletion region [Binet96a]. It is important to recall here that this diode exhibited a very small band-edge responsivity due to limited carrier diffusion. On the contrary, semi-transparent front side illuminated Schottky diodes with an n^+ back contact were measured and found to have response times as short as 100 ns [Chen97]. These improvements of the response time may be related to the reduction of the access resistance by the use of a highly doped bottom layer, to the fact that the diffusion does not limit the response in this geometry or to improved material quality. More work is needed to clarify the importance of all these possible contributions.

9.4.1.4 Noise

Noise figures have been measured on the Schottky diode on n-type GaN [Duboz96b]. As the leakage current in this diode is quite large ($2 \times 10^{-4}\,\text{A cm}^{-2}$ at $-0.5\,\text{V}$), applying large reverse voltages actually transforms this photovoltaic detector into a photoconductor. The same noise figures ($1/f$ noise) as previously described are then measured. On the contrary, for small reverse biases (between 0 and $-0.2\,\text{V}$) the noise is almost independent of the frequency. Figure 9.43 presents the noise spectral density of the current measured at 13 kHz as a function of the applied bias. We observe a slight decrease of the noise for

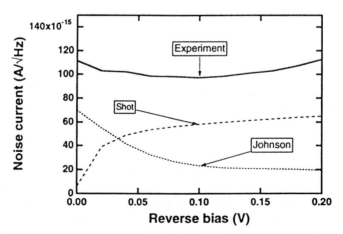

Fig. 9.43 Spectral density of noise measured at 13 kHz of the n-GaN Schottky barrier photovoltaic detector of Fig. 9.40 as a function of reverse bias. The calculated Johnson and shot noises are compared to the experimental noise. (After [Duboz96b].)

increasing bias from 0 to −0.1 V. Below −0.1 V, the noise increases with bias. As the current I was also measured, one can calculate the differential resistance R. The Johnson noise $\sqrt{4kT/R}$ and the shot noise $\sqrt{2eI}$ are plotted in the same figure. We thus see that the noise is dominated at low bias by the Johnson noise, characterizing a device without injection. As the bias increases, the shot noise takes over. The transition between both regimes actually describes the transition from a photovoltaic detector to a photoconductor-like device. The better the Schottky diode, the smaller the leakage current and the larger the transition bias. As the noise is minimum at this transition bias, this is a good polarization point in terms of detectivity. Increasing the reverse bias reduces the carrier sweeping time. It also increases the depletion layer thickness, which both reduces problems associated with the diffusion and decreases the RC time constant. This shows the interest of fabricating diodes with smallest possible leakage currents. As a matter of fact, Schottky diodes have been fabricated [Schmitz96, Mohammad96a, Duboz96c, Wang96] with leakage currents as small as a few $10^{-8}\,\text{A}\,\text{cm}^{-2}$, i.e. near four orders of magnitude less than in the diode studied here.

9.4.1.5 *Photovoltaic detector arrays*

Arrays of Schottky diodes on n-GaN have been realized [Lim97]. 8×8 mesa type detectors were fabricated on a 1 μm thick semi-insulating GaN layer grown on a 1 μm thick n$^+$ GaN layer. The Pd contact was thin enough (400 Å) to allow for front side illumination through the metal. The responsivity spectrum is close to the one shown in Fig. 9.41 measured on a similar device. The absolute responsivity is, however, somewhat less: about 0.046 A/W with a deviation of 1.91%. Two pixels only were found to be shorted. These results show that a good uniformity can already be obtained in a GaN detector array in spite of the still very large defect density.

9.4.2 p–n junctions

The asymmetry of the p–n junction is even larger than the one in a Schottky diode as the potential step is of the order of the bandgap energy if the p and n doping levels are high enough. The first photovoltaic detector based on a GaN p–n junction was realized by Chen *et al.* [Chen95]. Two structures were grown on sapphire by MOCVD. They consisted of a 1 μm thick n-GaN layer (p-GaN) followed by a 0.5 μm thick p-GaN layer (n-GaN). In this way, studies can be carried out with light incident through the sapphire from either the n or p side. Spectral responses of both detectors are shown in Fig. 9.44. In both detectors, the responsivity is maximum at the band edge (360 nm). At lower energies, there is no noticeable photoresponse. At higher energies, the responsivity tends to decrease, which is interpreted in terms of reduced charge collection due to the diffusion. Contrary to expectations, the diffusion of electrons in the p-type GaN seems to limit the responsivity more seriously than hole diffusion in the n-type GaN does. Also, diffusion lengths seem to be much larger than those found

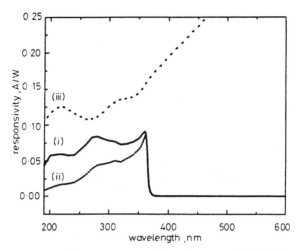

Fig. 9.44 Spectral responsivity of GaN p–n photodetectors. (i) Light from the n-side, (ii) light from the p-side, (iii) comparison with a UV-enhanced Si p–n detector. (After [Chen95].)

from measurements in Schottky diodes. The absolute responsivity is around 0.1 A/W, which is not so far from the theoretical value in an ideal detector with a unity photoconductive gain.

Different results were obtained by Zhang *et al.* [Zhang95]. Different p–n junctions were grown by MOCVD on sapphire with n-type GaN thicknesses ranging from 0.1 μm to 0.5 μm. The p-type GaN layer thickness was around 2 μm. The electron and hole densities are of the order 10^{18} cm^{-3} and 3×10^{16} cm^{-3} respectively. The light was incident through the substrate on the n-GaN so that hole diffusion plays a major role. In these regards, it was interesting to study the influence of the n-GaN thickness on the responsivity. The latter is shown for a 0.5 μm thick n-GaN layer in Fig. 9.45. The responsivity is sharply peaked at a wavelength of 362 nm corresponding to the band edge. At higher energies, it falls off quite strongly. It was noticed that the peak was narrower in detectors with a thicker n-GaN layer. As in Schottky diodes, the small responsivity at energies larger than the band edge is interpreted in terms of hole diffusion. A model taking into account the energy dependence of the absorption coefficient, surface recombinations, and the diffusion of electrons and holes could fit the experimental data with a few parameters. One of them was the hole diffusion length, which was found to be close to 0.1 μm. In addition, the negative peak in Fig 9.45 was interpreted in terms of internal photoemission. Hence these results seem to indicate that hole diffusion is a limiting phenomenon in p–n junctions as well as in Schottky diodes. Again, internal photoemission is likely to alter the spectral selectivity. However, improvements in ohmic contacts on p-type GaN should eliminate this problem in future devices.

Arrays of photovoltaic detectors have been fabricated [VanHove96]. A GaN p–i–n structure was grown by MBE on sapphire. A 1 μm thick highly (10^{18} cm^{-3})

9.4 Ultraviolet photovoltaic detectors based on GaN

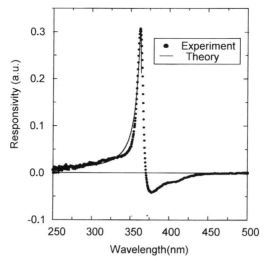

Fig. 9.45 Spectral responsivity of a GaN p–n photodetector. Light is incident from the n-side. The band-edge responsivity is fitted by a model based on the diffusion. (After [Zhang95b].)

doped n-GaN was grown first and served as a bottom ohmic contact layer, followed by a 0.5 μm thick undoped GaN layer and finally a 0.3 μm thick p-GaN layer doped to 5×10^{17} cm^{-3}. The light was incident from the top through the open ohmic contact. Responsivities in the 0.1 A/W range were measured at 360 nm, with a decrease by a factor of 10 at higher energies. The UV to visible rejection ratio was around 10^4. Attempts to measure the response from the back side were not successful, which was interpreted as a limited hole diffusion. The hole diffusion length was estimated to be 0.1 μm. The array had 10 pixels only and the uniformity was quite poor as the leakage current at -3 V was varying from 50 to 170 μA/mm^2. We also note that the current density is large for a p–i–n structure. The response times were of the order of 8 ms, indicating extrinsic recombinations. Finally, AlGaN p–i–n devices were also realized with Al contents of 5 and 10% [VanHove96]. Consequently, cutoff wavelengths were reduced to 343 and 313 nm respectively.

9.4.3 High-speed photodetector

Metal–semiconductor–metal (M–S–M) structures are symmetrical detectors. However, in some cases their behavior is close to the one observed in p–i–n junctions. In order to be in these cases, metal contacts have to be blocking and prevent any charge injection. This is achieved with perfect Schottky contacts. Under an applied field, electrons are collected at the forward-biased Schottky contact (anode) while holes are collected at the reverse-biased Schottky contact (cathode). Hence, the anode in the M–S–M structure (cathode) plays the role of the n-region (p-region respectively) in the p–i–n structure. In both cases, there is no charge injection. As we focus here on high-speed detectors, we will describe structures where a second approximation can be made. If the inter-electrode spacing is short enough, carriers do not have the time to recombine or to be trapped before being collected. In ideal defect-free materials this can be easily verified. In practical materials, this may imply a very short inter-electrode spacing and high fields. Nevertheless, let us consider that these conditions are verified. Then, the response time of an M−S−M photodetector is given by the carrier transit time. Obviously, the smaller the inter-electrode spacing, the shorter the transit time. For submicron devices, one may also take advantage from the velocity overshoot as already explained for transistors. Again, for transit time shorted than the carrier trapping time, carrier recombinations may be neglected. As high fields are applied, an accurate description of the transport requires that many phenomena be taken into account (phonon, inter-valley scattering) and Monte Carlo simulations are well suited to such a study [Joshi94a, Joshi94b]. Thanks to the large optical phonon energy (91 meV), ballistic transport can be achieved in a metal–semiconductor-metal structure with a 0.25 μm inter-electrode spacing. As the breakdown field is very large, the depletion region below the Schottky contacts can be made large which reduces the capacitance. It was shown that the actual limitation arises from the internal electrical polarization created by the spatial separation of electrons and holes [Joshi94a]. Consequently, the transit time was found to increase with the optical excitation intensity, from 8 ps for a photogeneration density of 10^{15} cm^{-3} to 20 ps for a density of 10^{16} cm^{-3}. One way to somehow reduce this polarization effect at high optical power was to increase the applied bias. A typical result is

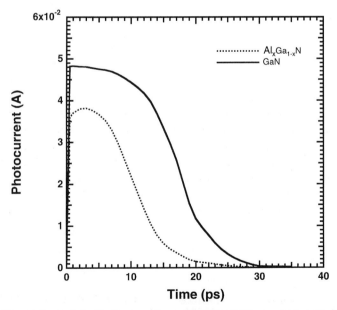

Fig. 9.46 Comparison of simulation results at 300 K for 0.25 μm graded Al$_x$Ga$_{1-x}$N and uniform GaN metal/semiconductor/metal structures. (After [Joshi94b].)

shown in Fig. 9.46. As the total response time is dominated by the hole transit time, it is necessary to break the symmetry of the structure in order to efficiently reduce the hole transit time and increase the electron one. This can be done by using a composition gradient [Josh94b]. It is assumed now that the material varied from GaN at the cathode to $Al_{0.15}Ga_{0.85}N$ at the anode. This corresponds to a 400 meV gradual increase of the gap energy. The incident beam is assumed to be monoenergetic with an excess energy of 0.1 eV above the GaN band gap. Hence carriers will be created in the first quarter of the device, on the cathode side. As holes are swept by the electric field towards the cathode, their transit time is considerably reduced. In addition, the internal field increases the total field so that the benefit of the gradient is twofold. On the other hand, it may be objected that mobilities are lower in a ternary alloy. However, transport in small devices under a high field is not controlled by mobility but is rather ballistic. As a result, the improvement due to the gradient can be appreciated in Fig. 9.46. Response times of the order of 10 ps for a photogeneration density of 10^{16} cm^{-3} are obtained. The technical difficulty of fabricating such a detector must unfortunately be emphasized, as the composition gradient is perpendicular to the surface.

9.5 Conclusion and perspectives

The nitride technology is recent and based on materials that are still very far from being perfect. There are difficulties related to the absence of adapted substrate, the intrinsically difficult growth, or the depth of p-type acceptors, etc. And yet, thanks to a very important research effort, very well performing devices have already been obtained, proving the not yet fully exploited potential of GaN based materials. Compared with other semiconductors, where defects so drastically degrade device properties, it remains not clearly understood why devices based on GaN are so little sensitive to defects. This interesting point will probably be addressed in the near future. Another very likely evolution of the subject is a larger use of heterostructures in devices, especially in photodetectors. Nitride heterostructures have already been largely used for electronics, in large part due to the analogy with the GaAs based devices. Bipolar transistors have not yet attracted great attention and we cannot state whether they will be able to compete with field effect transistors for high-power–high-temperature applications. For thermal reasons, the latter will likely move from sapphire to SiC substrates. Finally, there are undoubtely many new devices and integration schemes that will emerge in the future.

References

[AsifKhan92] M. Asif Khan, J. N. Kuznia, D. T. Olson, J. M. V. Hove, M. Blasingame, and L. F. Reitz. High-responsivity photoconductive ultraviolet sensors based on insulating single-crystal epilayers. *Appl. Phys. Lett.*, **60**:2917–9, 1992.

[AsifKhan93a] M. Asif Khan, J. N. Kuznia, A. Bhattarai, and D. T. Olson. Metal semiconductor field effect transistor based on single crystal GaN. *Appl. Phys. Lett.*, **62**:1786, 1993.

[AsifKhan93b] M. Asif Khan, J. N. Kuznia, D. T. Olson, M. Blasingame, and A. R. Bhattarai. Schottky barrier photodetector based on Mg-doped p-type GaN films. *Appl. Phys. Lett.*, **63**:2455–6, 1993.

[AsifKhan93c] M. AsifKhan, J. N. Kuznia, A. R. Bhattarai, and D. T. Olson. High electron mobility transistor based on GaN–$Al_xGa_{1-x}N$ heterojunction. *Appl. Phys. Lett.* **63**:1214, 1993.

[AsifKhan94a] M. Asif Khan, J. N. Kuznia, D. T. Olson, W. J. Schaff, J. W. Burm, and M. Shur. Microwave performance of a 0.25 μm gate AlGaN/GaN heterostructure field effect transistor. *Appl. Phys. Lett.* **65**:1121–3, 1994.

[AsifKhan94b] M. Asif Khan, M. S. Shur, Q. C. Chen, and J. N. Kuznia. Current/voltage characteristic collapse in AlGaN/GaN heterostructure insulated gate field effect transistors at high drain bias. *Electron. Lett.*, **30** 2175, 1994.

[AsifKhan95a] M. Asif Khan, M. S. Shur, J. N. Kuznia, Q. Chen, J. Burm, and W. Schaff. Temperature activated conductance in GaN/AlGaN heterostructure field effect transistors operating at temperatures up to 300 °C. *Appl. Phys. Lett.*, **66**:1083–5, 1995.

[AsifKhan95c] M. Asif Khan, M. S. Shur, Q. Chen, J. N. Kuznia, and C. J. Sun. Gated photodetector based on GaN/AlGaN heterostructure field effect transistor. *Electron. Lett.*, **31**:398–400, 1995.

[AsifKhan95d] M. Asif Khan, M. S. Shur, and Q. Chen. High transconductance AlGaN/GaN optoelectronic heterostructure field effect transistor. *Electron. Lett.*, **31**:2130–1, 1995.

[AsifKhan96a] M. Asif Khan, Q. Chen, M. A. Shur, B. T. Dermott, J. A. Higgins, J. Burn, W. Schaff, and L. F. Eastman. Short-channel GaN/AlGaN doped channel heterostructure field effect transistors with 36.1 GHz cutoff frequency. *Electron. Lett.*, **32**:357–8, 1996.

[AsifKhan96b] M. Asif Khan, Q. Chen, C. J. Sun, J. W. Yang, M. Blasingame, M. S. Shur, and H. Park. Enhancement and depletion mode GaN/AlGaN heterostructure field effect transistors. *Appl. Phys. Lett.*, **68**:514, 1996.

[AsifKhan96c] M. Asif Khan, Q. Chen, J. W. Yang, M. S. Shur, B. T. Dermott, and J. A. Higgins. Microwave operation of GaN/AlGaN doped channel heterostructure field effect transistors. *Electron. Dev. Lett.*, **17**:325, 1996.

[AsifKhan96d] M. Asif Khan, M. S. Shur, Q. Chen, J. W. Yang, R. Gaska, M. Blasingame, A. Ping, L. Adesida, J. Burm, and L. F. Eastman. Microwave performance of 0.25 micron doped channel GaN/AlGaN herstructure field effect transistor at elevated temperatures *and* High pinch-off voltage AlGaN–GaN heterostructure field effect transistor. *Invited talk IEDM*, San Franciso, Ca, USA, 1990.

[Binari94] S. C. Binari, L. B. Rowland, W. Kruppa, G. Kelner, K. Doverspike, and D. K. Gaskill. Microwave performance of GaN MESFETs. *Electron. Lett.* **30**:1248, 1994.

[Binet96a] F. Binet, J. Y. Duboz, E. Rosencher, F. Scholz, and V. Harle. Electric field effects on excitons in gallium nitride. *Phys. Rev.*, **54**:8116, 1996.

[Binet96b] F. Binet, J. Y. Duboz, E. Rosencher, F. Scholz, and V. Harle. Mechanisms of recombination in GaN photodetectors. *Appl. Phys. Lett.*, **69**:1202–4, 1996.

[Binet96c] F. Binet, J. Y. Duboz, N. Laurent, E. Rosencher, O. Briot, and R. L. Aulombard. Properties of a photovoltaic detector based on an n-type GaN Schottky barrier. *J. Appl. Phys.* **81**:6449, 1997.

[Burm96] J. Burm, W. J. Schaff, L. F. Eastman, H. Amano, and I. Akasaki. 75 Å GaN channel modulation doped field effect transistors. *Appl. Phys. Lett.*, **68**:2849–51, 1996.

[Chen95] Q. Chen, M. Asif Khan, C. J. Sun, and J. W. Yang. Visible–blind ultraviolet photodetectors based on GaN p–n junction. *Electron. Lett.*, **31**:1781–2, 1995.

[Chen96] Q. Chen, M. Asif Khan, J. W. Yang, C. J. Sun, M. S. Shur, and H. Park. High transconductance heterostructure field-effect transistors based on AlGaN/GaN. *Appl. Phys. Lett.*, **69**:794, 1996.

[Chen97] Q. Chen, J. W. Yang, A. Osinski, S. Gangopadhyay, B. Lim, M. Z. Anwar, M. Asif Khan, D. Kuksenkov, and H. Temkin. Schottky barrier detectors on GaN for visible–blind ultraviolet detection. *Appl. Phys. Lett.* **70**:2277, 1997.

[DenBaars96] S. DenBaars, U. K. Mishra, Y. F. Wu, D. Kapolnec, P. Kozodoy, M. Mack, J. S. Speck, and S. Keller. MOCVD growth of high quality GaN heterostructure materials and their application to high power electronic devices. *MRS Symposium on Nitrides*, vol. 449, eds. Ponce *et al*. MRS, Boston, 1996.

[Duboz96a] J. Y. Duboz and F. Hamann. Phototransistor based on GaN. Unpublished, 1996.

[Duboz96b] J. Y. Duboz and F. Binet. Photodetectors based on GaN. Unpublished, 1996.

[Duboz96c] J. Y. Duboz, F. Binet, N. Laurent, E. Rosencher, F. Scholz, V. Harle, O. Briot, B. Gil, and R. L. Aulombard. Influence of surface defects on the chatacteristics of GaN Schottky diodes. *MRS Symposium on Nitrides*, Vol. 449, eds. Ponce *et al*., MRS Boston, 1996.

[Eastman96] L. F. Eastman. Research on GaN MODFETs. In *First European GaN Workshop*, Rigi, Switerland, 1996.

[Fan96] Z. Fan, S. N. Mohammad, O. Aktas, A. E. Botchkarev, A. Salvador, and H. Morkoç. Suppression of leakage currents and their effects on the electrical performance of AlGaN/GaN modulation doped field-effect transistors. *Appl. Phys. Lett.*, **69**:1229–31, 1996.

[Gao91] G. Gao and H. Morkoç. Material-based comparison for power heterojunction bipolar transistors. *Trans. Electron. Dev.*, **38**:2410, 1991.

[Huang96] Z. C. Huang, J. C. Chen, D. B. Mott, and P. K. Shu. High performance GaN linear array. *Electron. Lett.*, **32**:1324, 1996.

[Johnson96] C. Johnson, J. Y. Lin, H. X. Jiang, M. Asif Khan, and C. J. Sun. Metastability and persistent photoconductivity in Mg-doped p-type GaN. *Appl. Phys. Lett.*, **68**:1808–10, 1996.

[Joshi94a] R. P. Joshi, A. N. Dharamsi, and J. McAdoo. Simulations for the high-speed response of GaN metal–semiconductor–metal photodetectors. *Appl. Phys. Lett.*, **64**:3611–3, 1994.

[Joshi94b] R. P. Joshi. Simulations for the transient response of graded $Al_xGa_{1-x}N$ submicron photodetectors. *J. Appl. Phys.*, **76**:4434–6, 1994.

[Kong96] H. S. Kong, M. Leonard, G. Bulman, G. Negley, and J. Edmond. AlGaN/GaN/AlGaN double-heterostructure blue LEDs on 6H–SiC substrates. In *Mat. Res. Soc. Symp. Proc.*, vol. 395, Pittsburgh, PA, USA, 1996.

[Kung95] P. Kung, X. Zhang, D. Walker, A. Saxler, J. Piotrowski, A. Rogalski, and M. Razeghi. Kinetics of photoconductivity in n-type GaN photodetector. *Appl. Phys. Lett.*, **67**:3792–4, 1995.

[Lim96] B. W. Lim, Q. C. Chen, J. Y. Yang, and M. Asif Khan. High responsivity intrinsic photoconductors based on $Al_xGa_{1-x}N$. *Appl. Phys. Lett.*, **68**:3761–2, 1996.

[Lim97] B. Lim, S. Gangopadhyay, J. W. Yang, A. Osinski, Q. Chen, M. Z. Anwar, and M. Asif Khan. An 8×8 GaN Schottky barrier photodiode array for visible–blind imaging. *IEEE Electron. Lett.* **33**:633, 1997.

[Matus91] J. A Powell L. G. Matus and C. S. Salupo. High voltage 6H–SiC p–n junction diodes. *Appl. Phys. Lett.*, **59**:1770, 1996.

[Mohammad96a] S. N. Mohammad, Z. Fan, A. E. Botchkarev, W. Kim, O. Aktas, A. Salvador, and H. Morkoç. Near-ideal platinum–GaN Schottky diodes. *Electron. Lett.*, **32**:598–9, 1996.

[Mohammad96b] S. N. Mohammad, Z. Fan, A. Salvador, O. Aktas, A. E. Botchkarev, W. Kim, and H. Morkoç. Photoluminescence characterization of the quantum well structure and influence of optical illumination on the electrical performance of AlGaN/GaN

modulation-doped field-effect transistors. *Appl. Phys. Lett.*, **69**:1420, 1996.
[Ozgur95] A. Özgür, W. Kim, Z. Fan, A. Botchkarev, A. Salvador, S. N. Mohammad, B. Sverdlov, and H. Morkoç. High transconductance-normally-off GaN MODFETs. *Electronics Lett.*, **31**:1389–90, 1995.
[Pankove94] J. Pankove, S. S. Chang, H. C. Lee, R. J. Molnar, T. D. Moustakas, and B. V. Zeghbroeck. High temperature GaN/SiC heterojunction bipolar transistor with high gain. In *International Electron Device Meeting Technical Digest*, ed. IEEE, p. 389–92, 1994.
[Pankove96] J. Pankove, S. S. Chang, M. Leksono, C. Walker, and B. V. Zeghbroeck. High power, high temperature heterobipolar transistor with GaN emitter. In *First European GaN Workshop*, Rigi, Switerland, 1996.
[Qiu95] C. H. Qiu, C. Hoggart, W. Melton, M. W. Leksono, and J. I. Pankove. Study of defect states in GaN films by photoconductivity measurement. *Appl. Phys. Lett.*, **66**:2712–4, 1995.
[Razeghi96] M. Razeghi and A. Rogalski. Semiconductor ultraviolet detectors *J. Appl. Phys.*, **79**:7433–73, 1996.
[Schetzina96] J. F. Schetzina. Growth and properties of III–V nitride films, quantum well structures and integrated heterostructure devices. In *Mat. Res. Soc. Symp. Proc*, vol. 395, Pittsburgh, PA, USA, 1996.
[Schmitz96] A. C. Schmitz, A. T. Ping, M. Asif Khan, Q. Chen, J. W. Yang, and I. Adesida. Schottky barrier properties of various metals on n-type GaN. *Semicond. Sci. Technol.*, **11**:1464–7, 1996.
[Shin95] M. W. Shin and R. J. Trew. GaN MESFETs for high power and high-temperature microwave applications. *Electron. Lett.*, **31**:498, 1995.
[Shur96] M. Shur, B. Gelmont, and M. A. Khan. Electron mobility in two-dimensional electron gas in AlGaN/GaN heterostructures and in bulk GaN. *J. Elec. Mat.*, **25**:777, 1996.
[Stengel96] F. Stengel, S. N. Mohammad, and H. Morkoç. Theoretical investigation of electrical characteristics of AlGaN/GaN modulation doped field-effect transistors. *J. Appl. Phys.*, **80**:3031, 1996.
[Stevens95] K. S. Stevens, M. Kinniburgh, and R. Beresford. Photoconductive ultraviolet sensor using Mg-doped GaN on Si(111). *Appl. Phys. Lett.*, **66**:3518–20, 1995.
[VanHove96] J. M. Van Hove, P. Chow, R. Hickman, A. Wowchak, J. Klassen, and C. Polley. Visible blind UV GaN photovoltaic detector arrays grown by RF atomic nitrogen plasma MBE. *MRS Symposium on Nitrides*, vol. 449, eds. Ponce *et al.*, MRS, Boston, 1996.
[Walker96] D. Walker, X. Zhang, P. Kung, A. Saxler, S. Javadpour, J. Xu, and M. Razeghi. AlGaN ultraviolet photoconductors grown on sapphire. *Appl. Phys. Lett.*, **68**:2100–1, 1996.
[Wang96] L. Wang, M. I. Nathan, T. H. Lim, M. A. Khan, and Q. Chen. High barrier height GaN Schottky diodes: Pt/GaN and Pd/GaN. *Appl. Phys. Lett.*, **68**:1267, 1996.
[Wu96a] Y.-F. Wu, B. P. Keller, S. Keller, D. Kapolnek, S. P. Denbaars, and U. K. Mishra. Measured microwave power performance of AlGaN/GaN MODFET. *IEEE Elec. Dev. Lett.*, **17**:455, 1996.
[Wu96b] Y.-F. Wu, B. P. Keller, S. Keller, D. Kapolnek, P. Kozodoy, S. P. Denbaars, and U. K. Mishra. Very high breakdown voltage and large transconductance realized on GaN heterojunction field effect transistors. *Appl. Phys. Lett.*, **69**:1438, 1996.
[Zhang95] X. Zhang, P. Kung, D. Walker, J. Piotrowski, A. Rogalski, A. Saxler, and M. Razeghi. Photovoltaic effects in GaN structures with p–n junctions. *Appl. Phys. Lett.*, **67**:2028–30, 1995.
[Zolper96] J. C. Zolper, R. J. Shul, A. G. Baca, R. G. Wilson, S. J. Pearton, and R. A. Stall. Ion-implanted GaN junction field effect transistor. *Appl. Phys. Lett.*, **68**:2273–5, 1996.

10 III–V nitride-based short-wavelength LEDs and LDs*

Shuji Nakamura

10.1 Introduction

Much research has been done on high-brightness blue light-emitting diodes (LEDs) and laser diodes (LDs) for use in full-color displays, full-color indicators, and light sources for lamps with the characteristics of high efficiency, high reliability, and high speed. For these purposes, II–VI materials such as ZnSe [1, 2], SiC [3], and III–V nitride semiconductors such as GaN [4] have been investigated intensively for a long time. However, it was impossible to obtain high-brightness blue LEDs with a brightness over 1 cd and reliable LDs. Much progress has been achieved recently on green LEDs and LDs using II–VI-based materials [5]. The short lifetimes prevent II–VI-based devices from commercialization at present. It is considered that the short lifetime of these II–VI based devices is caused by crystal defects at a density of $10^3/cm^2$, because one crystal defect would cause the propagation of other defects leading to failure of the devices. SiC is another wide bandgap material for blue LEDs. The brightness of SiC blue LEDs is only between 10 mcd and 20 mcd because of the indirect bandgap of this material. Despite this poor performance, 6H–SiC blue LEDs have been commercialized for a long time because there has been no competition for blue light-emitting devices [6].

On green devices, the external quantum efficiency of convential green GaP LEDs is only 0.1 % due to the indirect bandgap of this material, and the peak wavelength is 555 nm (yellowish green) [7]. As another material for green emission devices, AlInGaP has been used. The present performance of green AlInGaP LEDs is an emission wavelength of 570 nm (yellowish green) and maximum external quantum efficiency of 1 % [7, 8]. When the emission wavelength is reduced to the green region, the external quantum efficiency drops sharply because the band structure of AlInGaP becomes nearly indirect. Therefore high-brightness pure green LEDs, which have a high efficiency above 1% at the peak wavelength between 510–530 nm with a narrow full-width at half-maximum (FWHM), have not been commercialized yet.

Another type of short-wavelength-emitting device, the blue LD, is currently required for a number of applications, including full-color electroluminescent displays, laser printers, read–write laser sources for high-density information storage on magnetic and optical media, and sources for undersea optical

---

*An expanded version of this chapter has been published in: Shuji Nakamura and Gerhard Fasol, *The Blue Laser Diode* (1997), Springer-Verlag, Berlin. Reproduction of the figures and part of the text in this chapter by kind permission of Springer-Verlag.

communications. GaN and related materials, such as AlGaInN, are III–V nitride semiconductors with the wurtzite crystal structure and a direct energy band structure which is suitable for light-emitting devices. The bandgap energy of AlGaInN varies between 6.2 and 1.95 eV depending on its composition at room temperature. Therefore, these III–V nitride semiconductors are useful for light-emitting devices, especially in the short wavelength regions. Recent research on III–V nitrides has paved the way for the realization of high-quality crystals of GaN, AlGaN, and InGaN, and of p-type conduction in GaN and AlGaN [9–12]. The mechanism of acceptor compensation, which prevents us from obtaining low-resistivity p-type GaN and AlGaN, has been elucidated [13–16]. In Mg-doped p-type GaN, Mg acceptors are deactivated by atomic hydrogen, which is produced from NH_3 gas used as the N source during GaN growth. High-brightness blue LEDs have been fabricated on the basis of these results, and luminous intensities over 1 cd have been achieved [17, 18]. These LEDs are now commercially available. Also, high-brightness single quantum well structure (SQW) blue, green, and yellow InGaN LEDs with a luminous intensity above 10 cd have been achieved and commercialized [19–21]. By combining these high-power and high-brightness blue InGaN SQW LEDs, green InGaN SQW LEDs, and red AlInGaP LEDs, many kinds of applications, such as LED full-color displays and LED white lamps for use in place of light bulbs or fluorescent lamps, are now possible with characteristics of high reliability, high durability, and low energy consumption. Also, a white LED made by combining a blue InGaN SQW LED and yttrium aluminum garnet (YAG) phosphor, which is less expensive than a white LED composed of three primary color LEDs, has been developed [22].

On the blue LDs, recent developments have yielded an optically pumped stimulated emission from GaN films [23, 24], InGaN films [25, 26], AlGaN/InGaN double heterostructures [27], and GaN/AlGaN double heterostructures [28, 29]. However, stimulated emission had been observed only with optical pumping, not current injection. The first current-injection III–V nitride-based LDs were fabricated by the present authors using the InGaN multi-quantum-well (MQW) structure as an active layer [30]. The laser emission wavelength (417 nm) was the shortest one ever generated by a semiconductor LD. The mirror facet for the laser cavity was formed by etching of III–V nitride films due to the difficulty in cleaving the (0001) C-face sapphire substrate. The etched facet surface was relatively rough (approximately 500 Å). Also, sapphire substrate with ($11\bar{2}0$) orientation (A-face) was used to fabricate the InGaN MQW LDs because A-face sapphire could be cleaved along ($1\bar{1}02$) (R-face) [31]. The InGaN MQW LD structures were also fabricated on spinel ($MgAl_2O_4$) substrates, which had a small lattice mismatch (9.5%) in comparison with that (13%) between GaN and sapphire [32, 33]. Then characteristics of these InGaN MQW LDs were improved from RT pulsed operation [30–36] to low-temperature CW operation [37]. The recombination of localized excitons has been proposed as a spontaneous emission mechanism for these InGaN quantum-well-structure LEDs [38]. Also, the radiative recombination of the InGaN MQW LEDs and LDs was attributed to excitons (or carriers) localized at deep traps (100–250 meV) which originated from the In-rich region in the InGaN wells acting as quantum dots [39, 40]. Finally, RT continuous-wave (CW) operation of the InGaN MQW LDs with a lifetime of 27–35 hours has been achieved [41, 42]. Here, the present status of III–V nitride-based LEDs and LDs is described.

10.2 Blue InGaN / AlGaN double-heterostructure (DH) LEDs

Si and Zn co-doped blue InGaN/AlGaN DH LEDs with an output power of 1–3 mW were developed in 1994 for the first time [17, 18, 43–46]. The active layer was InGaN co-doped with Si and Zn to enhance the blue emission. The blue emission intensity was a maximum at an electron carrier concentration of 1×10^{19} cm^{-3}. The need for co-doping suggests that the high efficiency of this InGaN/AlGaN DH LED is the result of impurity-assisted (for example free-carrier-acceptor (FA)) recombination. A p-type GaN layer was used as the contact layer for the p-type electrode in order to improve the ohmic contact. After the growth, N_2 ambient thermal annealing at a temperature of 700 °C was performed to obtain highly p-type GaN and AlGaN layers. Fabrication of LED chips was accomplished as follows: the surface of the p-type GaN layer was partially etched until the n-type GaN layer was exposed. Next, an Ni/Au contact was evaporated on to the p-type GaN layer and a Ti/Al contact on to the n-type GaN layer. The wafer was cut into a rectangular shape (350 μm × 350 μm). These chips were set on the lead frame, and were then molded. The characteristics of LEDs were measured under DC-biased conditions at room temperature.

Figure 10.1 shows the electroluminescence (EL) spectra of the InGaN/ AlGaN DH blue LEDs at forward currents of 0.1 mA, 1 mA, and 20 mA. The carrier concentration of the InGaN active layer in this LED was 1×10^{19} cm^{-3}. Typical peak wavelengths and values of FWHM of the EL at 20 mA were 450 nm and 70 nm, respectively. The peak wavelength shifts to shorter wavelengths with increasing forward current. The peak wavelength is 460 nm at 0.1 mA, 449 nm at 1 mA, and 447 nm at 20 mA. At 20 mA, a narrower, higher-energy peak emerges at around 385 nm, as shown in Fig. 10.1. This peak, due to band-to-band recombination in the InGaN active layer, becomes resolved at injection levels, where the impurity-related recombination is saturated. The output power of the InGaN/AlGaN DH blue LEDs is 1.5 mW at 10 mA, 3 mW at 20 mA, and 4.8 mW at 40 mA. The external quantum efficiency is 5.4% at 20 mA. The typical on-axis luminuous intensity of InGaN/AlGaN LEDs with a 15° conical viewing angle is 2.5 cd at 20 mA. The forward voltage is 3.6 V at 20 mA. Using these high-brightness InGaN/AlGaN blue, GaP green, and GaAlAs red LEDs, full-color LED displays, especially for outdoor use, can be fabricated. Plate II shows the actual LED full-color display (10 m × 10 m) which was demonstrated in Tokyo, Japan, in 1994 for the first time. The color range of this full-color display is not so wide because the color of the green GaP LEDs is yellowish green (555 nm), not pure green. This means that this LED full-color display cannot express natural color, especially in the green region. For pure green LEDs emission wavelengths between 510 nm and 530 nm are required. Also, the luminous intensity of the green GaP LEDs (0.1 cd) is too low in comparison with that of

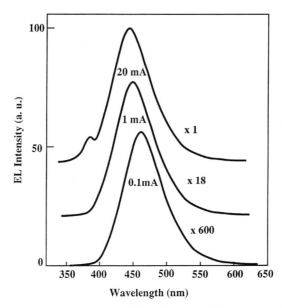

Fig. 10.1 EL spectra of the InGaN/AlGaN DH blue LEDs under different forward currents.

blue InGaN/AlGaN DH LEDs (2 cd) and red GaAlAs LEDs (2 cd) to fabricate high-brightness LED full-color displays. In order to obtain high-brightness pure blue and green LEDs, InGaN SQW LEDs have been developed as below.

10.3 Blue/green/yellow InGaN single quantum well (SQW) structure LEDs

High-brightness blue InGaN/AlGaN DH LEDs with a luminuous intensity of 2 cd have been fabricated and are now commercially available, as mentioned above [17, 18, 43–46]. In order to obtain blue and blue–green emission centers in these InGaN/AlGaN DH LEDs, Zn and Si co-doping into the InGaN active layer was performed. Although these InGaN/AlGaN DH LEDs produced high-power light output in the blue and blue–green regions with a broad emission spectrum (FWHM = 70 nm), green or yellow LEDs with peak wavelengths longer than 500 nm have not been fabricated [43]. The longest peak wavelength of the electroluminescence (EL) of InGaN/AlGaN DH LEDs achieved thus far is 500 nm (blue–green) because the crystal quality of the InGaN active layer of DH LEDs becomes poor when the indium mole fraction is increased in order to obtain a green band-edge emission.

In conventional green GaP LEDs the external quantum efficiency is only 0.1% due to the indirect transition bandgap material, and the peak wavelength is 555 nm (yellowish green) [7]. As another material for green emission devices, AlInGaP has been used. The present green AlInGaP LEDs have an emission

10.3 Blue/green/yellow InGaN single quantum well (SQW) structure LEDs

Green InGaN SQW LEDs

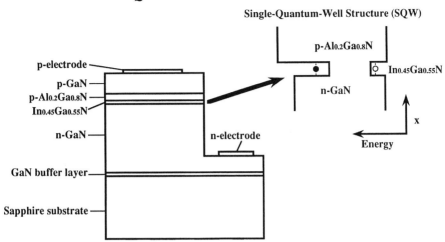

Fig. 10.2 The structure of a green InGaN SQW LED.

wavelength of 564 nm (yellowish green) and an external quantum efficiency of 0.6% [47]. When the emission wavelength is reduced to the green region, the external quantum efficiency drops sharply because the band structure of AlInGaP approached the indirect region. Therefore, high-brightness pure green LEDs, having a high efficiency of above 1% and a peak wavelength of between 510 and 530 nm with a narrow FWHM, have not been commercialized yet.

When the InGaN active layer becomes thin, it is expected that the elastic strain of the active layer is not relieved by the formation of misfit dislocations and that the crystal quality of the InGaN active layer improves. High-quality InGaN multi-quantum-well structures (MQW) with 30 Å well and 30 Å barrier layers were grown [48]. Here, the quantum well structure (QW) LEDs which have a thin InGaN active layer (about 30 Å) in order to obtain high-power emission in the region from blue to yellow with a narrow emission spectrum are described [19–21].

The green InGaN SQW LED device structures (Fig. 10.2) consist of a 300 Å GaN buffer layer grown at a low temperature (550°C), a 4 μm thick layer of n-type GaN:Si, a 30 Å thick active layer of undoped $In_{0.45}Ga_{0.55}M$, a 1000 Å thick layer of p-type $Al_{0.2}Ga_{0.8}N$:Mg, and a 0.5 μm thick layer of p-type GaN:Mg. The active region is an SQW structure consisting of a 30 Å thick $In_{0.45}Ga_{0.55}N$ well layer sandwiched by 4 μm thick n-type GaN and 1000 Å thick p-type $Al_{0.2}Ga_{0.8}N$ barrier layers.

Figure 10.3 shows the typical EL of the blue, green, and yellow SQW LEDs with different indium mole fractions of the InGaN well layer at a forward current of 20 mA [19–21]. These LEDs were grown only by changing indium mole fractions of the InGaN active layers. The peak wavelength and the FWHM of the typical blue SQW LEDs are 450 nm and 20 nm, respectively, those of the green SQW LEDs are 525 nm and 30 nm, and those of yellow are 600 nm and

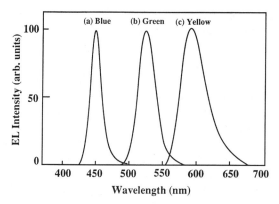

Fig. 10.3 Electroluminescence of (a) blue, (b) green, and (c) yellow InGaN SQW LEDs at a forward current of 20 mA.

50 nm, respectively. When the peak wavelength becomes longer, the FWHM of the EL spectra increases, probably due to inhomogeneities of the InGaN layer

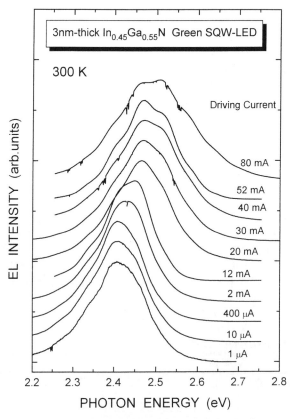

Fig. 10.4 Room-temperature EL of typical green InGaN SQW LEDs with different driving currents.

10.3 Blue / green / yellow InGaN single quantum well (SQW) structure LEDs

or the strain between well and barrier layers of the SQW, which is caused by mismatch of the lattice and thermal expansion coefficients between the well and barrier layers. At 20 mA, the output power and the external quantum efficiency of the blue SQW LEDs are 5 mW and 9.1%, respectively. Those of the green SQW LEDs are 3 mW and 6.3%, respectively. A typical on-axis luminous intensity of the green SQW LEDs with a 10° cone viewing angle is 10 cd at 20 mA. These values of output power, external quantum efficiency, and luminous intensity of blue/green SQW LEDs are the highest ever reported for blue/green LEDs. By combining these high-power and high-brightness blue InGaN SQW, green InGaN SQW, and red GaAlAs LEDs, many kinds of application, such as LED full-color displays and LED white lamps for use in place of light bulbs or fluorescent lamps, are now possible with characteristics of high reliability, high durability, and low energy consumption.

Figure 10.4 shows the typical EL of green SQW LEDs. The EL peak energy of the green SQW LEDs shows blue shifts by about 100 meV when increasing the driving current from 1 μA to 80 mA. A similar blue shift is found in both blue

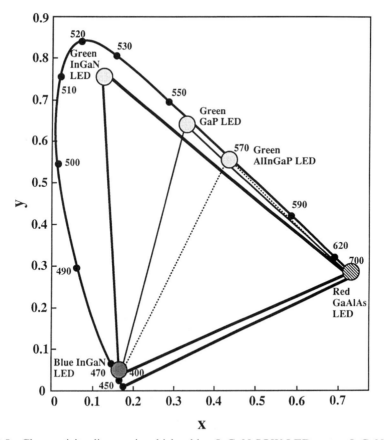

Fig. 10.5 Chromaticity diagram in which a blue InGaN SQW LED, green InGaN SQW LED, green GaP LED, green AlInGaP LED, and red GaAlAs LED are shown.

and yellow SQW LEDs. These blue shifts are due to a filling effect of localized states where carriers or excitons are recombined for emission with increasing currents. The localized states are formed by indium composition fluctuation in the InGaN well layer due to a phase separation of the InGaN during growth.

Figure 10.5 is a chromaticity diagram in which the blue and green InGaN SQW LEDs are shown. Commercially available green GaP LEDs, green AlInGaP LEDs, and red GaAlAs LEDs are also shown. The color range of light emitted by a full-color LED lamp in the chromaticity diagram is shown as the region inside each triangle, which is drawn by connecting the positions of three primary color LED lamps. Three color ranges (triangles) are shown for differences only in the green LED (green InGaN, green GaP, and green AlInGaP LEDs). In this figure, the color range of lamps composed of a blue InGaN SQW LED, a green InGaN SQW LED, and a red GaAlAs LED, is the widest. This means that the InGaN blue and green SQW LEDs show much better color and color purity in comparison with other blue and green LEDs. Using these blue and green SQW LEDs, a much more beautiful LED full-colour display was demonstrated in Tokyo, Japan, 1996, as shown in Plate III. In this display, blue and green LEDs are InGaN SQW LEDs, and only red LEDs are GaAlAs LEDs. Also, the green InGaN SQW LEDs have been used as traffic lights, as shown in Plate IV. Traffic lights may prove to be a great application for green LEDs. Total power consumption by traffic lights reaches the gigawatt range in Japan. The LED traffic lights, with an electrical power consumption of only 10% that of present incandescent bulb traffic lights, promise to save vast amounts of energy. With its extremely long lifetime (more then 10^6 hours), the repalcement of burned-out traffic light bulbs will be dramatically reduced. Using these green InGaN SQW LEDs, safe and energy-efficient roadway and railway signals can be achieved.

10.4 White LEDs

There are many applications for white light sources, for instance illumination and light sources, for backlighting of full-color liquid crystal displays, instead of conventional bulb lamps or fluorescent lamps. However, it has been impossible to obtain white LEDs until recently due to the lack of highly efficient blue and green LEDs. Now, highly efficient InGaN-based blue and green LEDs have become commercially available [19–21]. Using these LEDs, it is possible to fabricate white LEDs by mixing light from LEDs of the three primary colours, such as green and blue InGaN SQW and red GaAlAs LEDs. In this case, we need at least three LEDs, one of each primary color, and must adjust each supply current using a special circuit in order to control the intensity of each color. Therefore, the price of the white LED is much more expensive in comparison with that of a single-color LED. We can also fabricate white LEDs by exciting phosphors using blue LEDs because the excitation energy in the blue range is high enough to excite some phosphors. In this case, only one blue LED chip is required. This means that the price of a white LED can be almost the

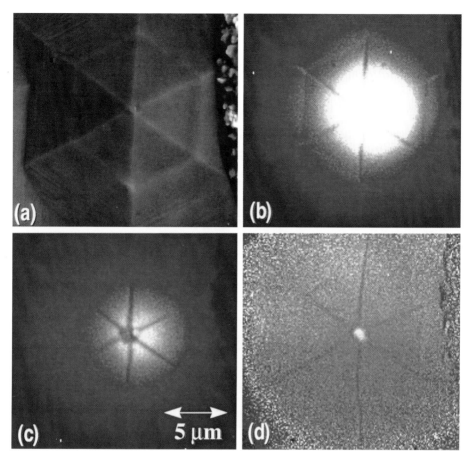

Plate I (Chapter 4) Hexagonal feature in homoepitaxial GaN/GaN grown on the [00$\bar{1}$] N-terminated side. (a) Secondary electron image taken at 30kV; (b) Raman scattering image at 735 cm^{-1}; (c) background reference image at 765 cm^{-1}; and (d) flat-field image obtained by dividing the images at 735 and 765 cm^{-1}.

Plate II (Chapter 10) The full-color LED display that was demonstrated in Tokyo, Japan, in 1994 for the first time. The blue InGaN/AlGaN LEDs, green GaP LEDs, and red GaAlAs LEDs are used as three primary color LEDs.

Plate III (Chapter 10) The full-color LED display that was demonstrated in Tokyo, Japan, in 1996. The blue InGaN SQW LEDs, green InGaN SQW LEDs, and red GaAlAs LEDs are used as three primary color LEDs.

Plate IV (Chapter 10) The LED traffic light that was demonstrated in Sweden in 1996, using green InGaN SQW, yellow AlInGaP, and re AlInGaP LEDs.

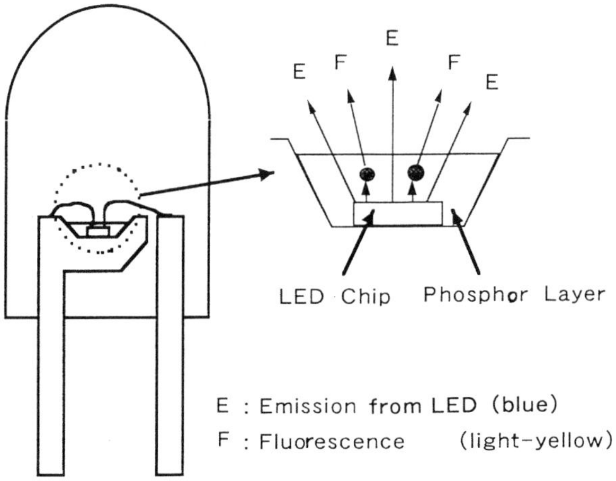

10.4 White LEDs

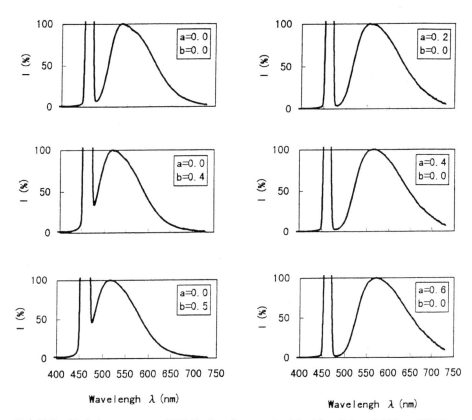

Fig. 10.6 Structure of white LEDs.

same as that of a single blue LED. Here, we describe a white LED which was fabricated using a blue LED and a phosphor [22].

Fig. 10.7 Emission spectra of YAG phosphors excited by blue light of 460 nm. YAG = $(Y_{1-a}GD_a)_3(Al_{1-b}Ga_b)_5O_{12}$:Ce.

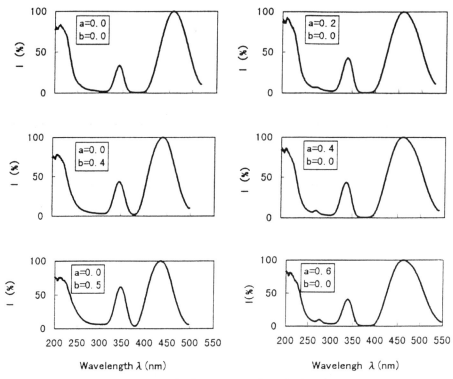

Fig. 10.8 Excitation spectra of YAG phosphors.
YAG = $(Y_{1-a}Gd_a)_3(Al_{1-b}Ga_b)_5O_{12}$: Ce.

Figure 10.6 shows the structure of the white LED. The structure is almost the same as that of a blue SQW LED, except for the phosphor layer on top of the blue LED chip [19–21]. When current is suppplied to the SQW blue LED chip, blue emission appears from the chip. The phosphor is then excited by this blue emission and emits yellow fluorescence. The mixture of the blue emission from the blue LED chip and the yellow emission from the phosphor results in a white emission.

Figure 10.7 shows the emission spectra of $Y_3Al_5O_{12}$ (yttrium aluminum garnet) phosphor (YAG phosphor) which was excited by blue emission with a peak wavelength of 460 nm. We can change the peak wavelength of the emission spectrum of phosphor between 510 nm and 570 nm by changing the composition of the YAG phosphor, as shown in Fig. 10.7. This means that we can control the color of the white LEDs.

Figure 10.8 shows the excitation spectra of YAG phosphors with various compositions. The excitation spectra show three peaks. The main peak is in the blue region between the wavelengths of 430 nm and 460 nm. Emission of these wavelengths is easily obtained by changing the indium composition of the InGaN active layer of the blue SQW LED [19–21]. Thus excitation of YAG phosphors by blue SQW LEDs of specific wavelengths can easily be achieved.

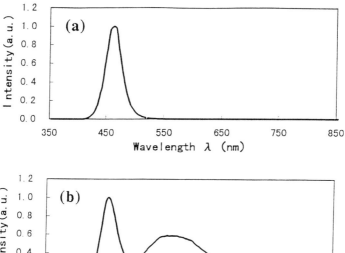

Fig. 10.9 Emission spectra of (a) a blue InGaN SQW LED and (b) a white LED.

Figure 10.9 shows the typical emission spectra of the blue SQW LED and the white LEDs. The blue LED emission is at around 465 nm. The white LED shows two peaks which correspond to the blue emission of the SQW LED and the yellow emission (555 nm) of the YAG phosphor.

Figure 10.10 shows the CIE color chromaticity diagram on which the positions of the white LED, the YAG phosphor, and the blue LED are shown. the color region of the white LED is obtained by connecting the positions of the YAG phosphor and the blue LED. The color position of the YAG phosphor can be changed by changing the composition of the phosphor, as shown in Fig. 10.7. Therefore the color range of the white LED can be controlled within the fan-shaped region. Inside the fan-shaped region, the color temperatures are also shown.

Figure 10.11 shows the relative luminuous intensity of the white LEDs as a function of the forward current. White LEDs typically show a luminuous intensity of 3 cd with a viewing angle of 30°, output power of 2 mW, luminous efficiency of 5 lm/W, average color rendering index of 85, and color temperature of 8000 K at a forward current of 20 mA.

Table 10.1 shows the characteristics of various visible LEDs. The luminous intensity of white, blue, bluish-green, and green LEDs, which were fabricated using InGaN-based materials, is higher than 3.6 lm/W, which is much greater than those of conventional GaP and InGaAlP green LEDs. The luminous efficiencies of conventional bulb lamps and fluorescent lamps are 1 lm/w and 20 lm/W, respectively. The luminous efficiency of 5 lm/W of white LEDs is

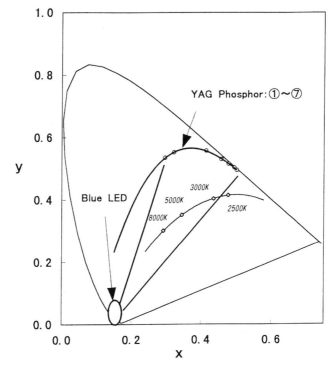

Fig. 10.10 CIE chromaticity diagram where the color range of blue LEDs, white LEDs, and YAG phosphors with various compositions are shown. The color of the white LEDs can be changed within the fan-shaped region by changing the composition of the YAG phosphor. YAG = $(Y_{1-a}Gd_a)_3(Al_{1-b}Ga_b)_5O_{12}$: Ce.

much higher than that of bulb lamps. Therefore, the white LEDs are superior to conventional bulb lamps in terms of efficiency and reliability, such as long lifetime (more than 10^6 hours). In the future, by improving the efficiency of blue LEDs, white LEDs lamps may be used for many kinds of light sources instead of conventional bulb or fluorescent lamps.

10.5 Emission mechanism of SQW LEDs

Static (DC) electroluminescence (EL), photovoltage (PV), and modulated-electroabsorption (EA) spectra were measured on the above mentioned SQW LEDs and MQW LDs [38]. Figure 10.12 summarizes room-temperature EL, PV, and EA spectra green, blue SQW LED, and MQW LD structure. The EL spectrum of the MQW LD structure was measured below the threshold current density. The lasing emission of this MQW LD appeared at 3.052 eV (406 nm) at the threshold current density of 11.3 kA/cm^2. In general, the low-field EA monitors exciton resonance absorption rather than band-to-band transition, even at RT in a widegap semiconductor such as GaN [49] under the condition that the modulation field is smaller than that needed to dissociate excitons. The

10.5 Emission mechanism of SQW LEDs

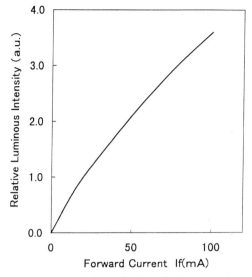

Fig. 10.11 Relative luminous intensity of white LEDs as a function of the forward current.

EA spectra were measured using the rectangular modulation bias of $-2\,\mathrm{V}$ to $+1.95\,\mathrm{V}$ to maintain the field strength of the upper level smaller than the dissociation field strength of excitons. Therefore the structures observed in the EA spectra are due to room-temperature free exciton resonances in the quantum wells.

The PV spectra were taken by irradiating a monochromated light, and the open-circuit voltage of the device was measured spectroscopically. The PV peaks at 3.21, 2.91, and 2.93 eV for MQW, blue, and green SQW structures correspond to exciton absorption in the quantum well, because the energies agree with those in the EA spectra. It is recognized that FWHM of the PV peak increases

Table 10.1 Characteristics of various visible LEDs. These measurements were performed at a forward current of 20 mA at room temperature.

Material	Color	Chromaticity Coordinates		Peak Wavelength (nm)	Half-width (nm)	External Quantum Efficiency (%)	Lumious Efficiency (lm/W)
		x	y				
InGaN + YAG	White	0.29	0.30	460/555	–	3.5[a]	5.0[a]
InGaN	Blue	0.13	0.08	465	30	5.6[a]	3.6[a]
InGaN	Bluish-green	0.08	0.40	495	35	5.0[a]	8.0[a]
InGaN	Green	0.17	0.70	520	40	4.0[a]	12[a]
GaP	Yellowish-green	0.37	0.63	555	30	0.1	0.6
GaP:N	Yellow green	0.45	0.55	565	30	0.4	2.4
InGaAlP	Greenish-yellow	0.46	0.54	570	12	1	6
InGaAlP	Orange	0.57	0.43	590	15	5	20
InGaAlP	Red	0.68	0.32	625	18	6	20
GaAlAs	Red	0.72	0.28	655	25	15	6.6

[a] These are commercially available InGaN-based LEDs

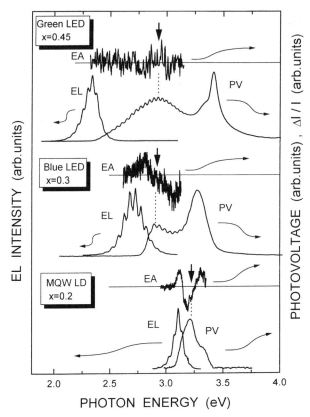

Fig. 10.12 EL, PV, and EA spectra for InGaN green (510 nm) and blue (450 nm) SQW LED structures and an MQW laser structure, whose lasing wavelength is 406 nm. The EL spectrum of the MQW structure was measured below the threshold current. The In composition in the $In_xGa_{(1-x)}N$ quantum well for green, blue, and MQW LEDs is 0.45, 0.3, and 0.2, respectively. The structure in the EA spectra correspond to the free-exciton resonances.

with increasing x. The PV peak energy decreases from 3.21 to 2.91 eV with increasing x from 0.2 to 0.45. However, the peak energy is almost unchanged for $x = 0.3$ and $x = 0.45$. This implies that InGaN does not form perfect alloys [50] due to a phase separation of InGaN during the growth, but forms compositional tailing, especially for larger x. Such a compositional tailing in the quantum well plane can produce two-dimensional potential minima.

The EL peak energy is smaller by 100, 215, and 570 meV than the free exciton resonance energy from MQW LDs, and blue and green SQW LEDs, respectively. All EL peaks are located at the low-energy tail of the free exciton resonance. Such low-energy tails of the exciton structure reflect a presence of certain potential minima in the quantum well plane. These EL emissions are originated from recombination of carriers or excitons at the deep localized states in the quantum well, which are almost equivalent to zero-dimensional quantum dot states.

10.6 Bluish-purple InGaN multi-quantum-well (MQW) structure LDs

Here we describe the characteristics of InGaN MQW LDs under RT CW operation. The InGaN MQW LD structures were grown by the two-flow metalorganic chemical vapor deposition (MOCVD) method. Details of two-flow MOCVD have been described elsewhere [51]. The growth was conducted at atmospheric pressure, and (0001) C-face sapphire was used as the substrate. The InGaN MQW LD device consisted of a 300 Å thick GaN buffer layer grown at a low temperature of 550 °C, a 3 μm thick layer of n-type GaN:Si, a 0.1 μm thick layer of n-type $In_{0.05}Ga_{0.95}N$:Si, a 0.5 μm thick layer of n-type $Al_{0.08}Ga_{0.92}N$:Si, a 0.1 μm thick layer of n-type GaN:Si, an $In_{0.15}Ga_{0.85}N/In_{0.02}Ga_{0.98}N$ MQW structure consisting of four 35 Å thick undoped $In_{0.15}Ga_{0.85}N$ well layers forming a gain medium separated by 70 Å thick undoped $In_{0.02}Ga_{0.98}N$ barrier layers, a 200 Å thick layer of p-type $Al_{0.2}Ga_{0.8}N$:Mg, a 0.1 μm thick layer of p-type GaN:Mg, a 0.5 μm thick layer of p-type $Al_{0.08}Ga_{0.92}M$:Mg, and a 0.5 μm thick layer of p-type GaN:Mg. The 0.1 μm thick n-type and p-type GaN layers were light-guiding layers. The 0.5 μm thick n-type and p-type $Al_{0.08}Ga_{0.92}N$ layers acted as cladding layers for confinement of the carriers and the light emitted from the active region of the InGaN MQW structure. The structure of the ridge-geometry InGaN MQW LD was almost the same as that described previously [34].

First, the surface of the p-type GaN layer was partially etched until the n-type GaN layer and p-type $Al_{0.08}Ga_{0.92}N$ cladding layer were exposed, in order to form a ridge-geometry LD [34]. A mirror facet was also formed by dry etching, as reported previously [30]. The area of the ridge-geometry LD was 4 μm × 550 μm. High-reflection facet coatings (30%) consisting of two pairs of quarter-wave TiO_2/SiO_2 dielectric multilayers were used to reduce the threshold

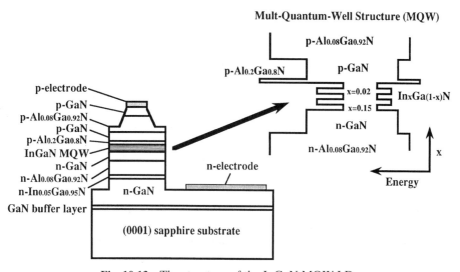

Fig. 10.13 The structure of the InGaN MQW LD.

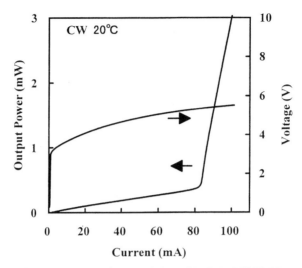

Fig. 10.14 Typical $L-I$ and $V-I$ characteristics of InGaN MQW LDs measured under CW operation at RT.

current. An Ni/Au contact was evaporated on to the p-type GaN layer, and a Ti/Al contact was evaporated on to the n-type GaN layer. The electrical characteristics of the LDs fabricated in this way were measured under a direct current (DC). The structure of the InGaN MQW LDs is shown in Fig. 10.13.

Figure 10.14 shows typical voltage-current ($V-I$) characteristics and the light output power per coated facet of the LD as a function of the forward DC current ($L-I$) at RT. No stimulated emission was observed up to a threshold current of 80 mA, which corresponded to a threshold current density of 3.6 kA/cm^2, as shown in Fig. 10.14. The operating voltage at the threshold current was 5.5 V. We were able to reduce the operating voltage significantly in comparison with values obtained previously (about 20–30 V) by adjusting the growth, ohmic contact, and doping profile conditions [30–35].

Figure 10.15 shows the results of a lifetime test of CW-operated LDs carried out at RT, in which the operating current is shown as a function of time under a constant output power of 1.5 mW per facet controlled using an autopower controller (APC). The operating current gradually increases due to the increase in the threshold current from the initial stage and sharply increases after 35 hours. This short lifetime is probably due to the large heat generation resulting from the high operating currents and voltages. Breakdown of the LDs occurred after a period of more than 35 hours due to the formation of a short circuit in the LDs.

The emission spectra of the LDs were measured under RT CW operation at an output power of 1 mW. An optical spectrum analyzer (ADVANTEST Q8347) which utilized the Fourier transform spectroscopy method by means of a Michelson interferometer was used to measure the spectra of the LDs with a resolution of 0.001 nm. At $J = 1.0 J_{th}$, where J is the current density and J_{th} is the threshold current density, longitudinal modes with many sharp peaks with a

10.6 Bluish-purple InGaN multi-quantum-well (MQW) structure LDs

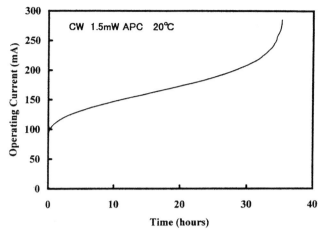

Fig. 10.15 Operating current as a function of time under a constant output power of 1.5 mW per facet controlled using an autopower controller. The LD was operated under DC at RT.

peak separation of 0.042 nm ($\Delta E = 0.3$ meV, where ΔE was the mode separation energy) were observed, as shown in Fig. 10.16(a). If these peaks arise from the longitudinal modes of the LD, then the mode separation $\Delta \lambda$ is given by

$$\Delta \lambda = \lambda_0^2/(2Ln_{\text{eff}}), \tag{10.1}$$

where n_{eff} is the effective refractive index and λ_0 is emission wavelength (405.83 nm). L is 0.055 cm. Thus, n_{eff} is calculated as 3.6, which is relatively

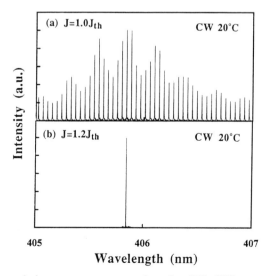

Fig. 10.16 Laser emission spectra measured under RT CW operation with current densities of (a) $J = 1.0 J_{\text{th}}$; (b) $J = 1.2 J_{\text{th}}$.

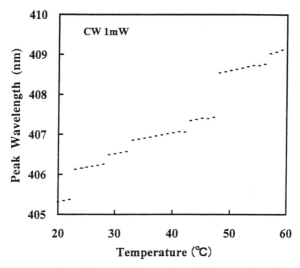

Fig. 10.17 Temperature dependence of the peak emission wavelengths of InGaN MQW LDs under CW operation with a constant output power of 1 mW.

large due to the wavelength and temperature dependence of the refractive indices of GaN and InGaN. Also, other periodic subband emissions are observed with a peak separation of 0.25–0.29 nm ($\Delta E = 1.8$–2.1 meV). The origin of these subband emissions has not yet been clarified. However, it is possible that these emissions result from transitions between quantum well or quantum dot subband energy levels, as mentioned previously [31, 34, 35]. Several peaks with a different peak separation (energy separation of 1–5 meV) from that of the longitudinal mode appeared under pulsed current operation, as described in our previous reports [31, 33–35]. These subband emissions with an energy separation of 1–5 meV are probably caused by mode hopping between adjacent quantum well or quantum dot subband energy levels due to the pulsed current operation, as described later. At $J = 1.2 J_{th}$, the main peak at 405.83 nm becomes dominant, as shown in Fig. 10.16(b).

The temperature dependence of the emission spectra of the LDs was measured between 20 °C and 60 °C under CW operation with a constant output of 1 mW, as shown in Fig. 10.17. Large mode hopping of the peak emission wavelength with an energy step of 1–7 meV is observed, which results from the temperature dependence of the gain profile. Mode hopping is probably a result of the change of the transitions between quantum dot subband energy levels. The change in the actual emission spectra with temperature between 47 °C and 48 °C is shown in Fig. 10.18. When the temperature is increased from 47 °C to 48 °C, the peak wavelength varies from 407.428 nm to 408.523 nm (with an energy difference of 7 meV) due to the change in the gain profile.

Figure 10.19 shows the emission spectra of InGaN MQW LDs with various operating currents under RT CW operation. The threshold current and voltage of this LD were 160 mA and 6.7 V, respectively, under RT CW operation. The threshold current density was 7.3 kA/cm². At a current of 156 mA, many

10.6 Bluish-purple InGaN multi-quantum-well (MQW) structure LDs

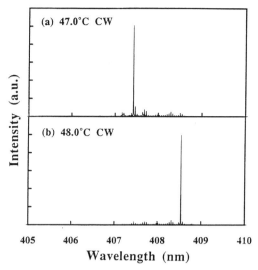

Fig. 10.18 Optical spectra of InGaN MQW LDs measured under CW operation at temperatures of (a) 47 °C and (b) 48 °C. The intensity scales for these two spectra are in arbitrary units, and each one is different.

longitudinal modes are observed with a mode separation of 0.042 nm, which is relatively small in comparison with the calculated value of 0.05 nm, probably due to the refractive index change from the value used (2.54) for the calculation. Periodic subband emissions are observed with a peak separation of about 0.025 nm ($\Delta E = 2$ meV). The origin of these subband emissions has not yet been clarified in detail. However, it is possible that these emissions result from transitions between quantum dot-like subband energy levels, as mentioned previously [31, 34, 35]. On increasing the forward current from 156 mA to 186 mA, the laser emission becomes a single mode and shows mode hopping of the peak wavelength toward higher energy, and the peak emission is at the center of each subband emission. Figure 10.20 shows the peak wavelength of the laser emission as a function of the operating current under RT CW operation. A gradual increase of the peak wavelength is observed, probably due to bandgap narrowing of the active layer caused by the temperature increase. At certain currents, a large mode hopping of the peak wavelength toward higher energy is observed with increasing operating current.

Under pulsed current operation, laser emission appears from all of these subband emissions because the operating current is modulated from zero to the peak with a pulse width of 0.5 μs and duty ratio of 20 %, as shown in Fig. 10.21. Figure 10.22 shows the peak wavelength of the maximum peak as a function of the operating pulsed current. The gradual increase of the peak wavelength with increasing operating current seen in Fig. 10.20 is not observed in Fig. 10.22 because the generation of heat is small under pulsed operation. The peak wavelength also shows the same mode-hopping toward higher energy with increasing operating current under pulsed operation as shown in Figs 10.19 and

410   *III–V nitride-based short-wavelength LEDs and LDs*

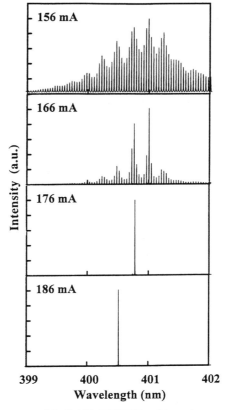

Fig. 10.19 Emission spectra of InGaN MQW LDs with various operating currents under RT CW operation.

10.20. Several peaks with different peak separations (energy separation of 1–5 meV) from that of the longitudinal mode appeared under pulsed current

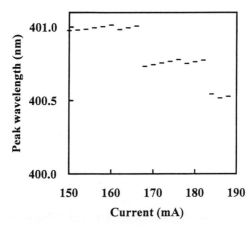

Fig. 10.20 Peak wavelength of the emission spectra of the InGaN MQW LDs as a function of the operating current under RT CW operation.

10.6 Bluish-purple InGaN multi-quantum-well (MQW) structure LDs 411

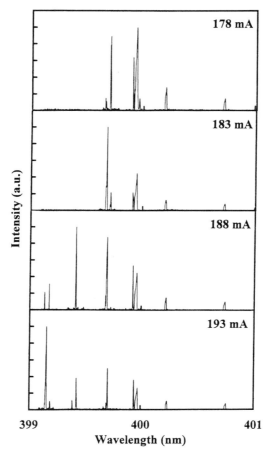

Fig. 10.21 Emission spectra of InGaN MQW LDs with various operating currents under RT pulsed operation. The pulse width and duty ratios of the pulsed currents were 0.5 μs and 20%.

operation, as described in our previous reports [31, 33–35]. The origin of these subband emissions has not yet been clarified exactly. These subband emissions with an energy separation of 1–5 meV are caused by mode-hopping, probably between adjacent quantum dot-like subband energy levels, as shown in Figs. 10.21 and 10.22.

Next, the delay time of the laser emission of the LDs as a function of the operating current was measured under pulsed current modulation using the method described in Ref. [35] in order to estimate the carrier lifetime (τ_s). The delay time t_d is given by

$$t_d = \tau_s \ln[I/(I - I_{th})], \tag{10.2}$$

where τ_s is the minority carrier lifetime, I is the pumping current, and I_{th} is the threshold current. Figure 10.23 shows the delay time t_d of the laser emission as a function of $\ln[I/(I - I_{th})]$. From this figure, τ_s was estimated to be 10 ns,

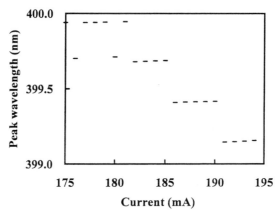

Fig. 10.22 Peak wavelength of the emission spectra of the InGaN MQW LDs as a function of the operating current under RT pulsed operation.

which was relatively large in comparison with the previous value of 3.2 ns [35]. The threshold carrier density (n_{th}) was estimated to be $2 \times 10^{20}/cm^3$ for a threshold current density of 3.6 kA/cm², a carrier lifetime of 10 ns, and an active layer thickness of 150 Å [35]. The thickness of the active layer was determined as 140 Å, assuming that the injected carriers were confined in the InGaN well layers in the active layer. Typical values are $\tau_s = 3$ ns, $J_{th} = 1$ kA/cm and $n_{th} = 2 \times 10^{18}/cm^3$ for AlGaAs lasers and $n_{th} = 1 \times 10^{18}/cm^3$ for InGaAsP lasers. In comparison with these values for convention lasers, n_{th} for our structure is relatively large (two orders of magnitude higher), probably due to the large density of states of carriers resulting from their effective masses [35].

Figure 10.24 shows typical near-field radiation patterns (NFP) for the InGaN MQW LDs in the planes parallel and perpendicular to the junction. The beam full width at half-power (FWHP) values for the parallel and perpendicular NFPs were 1.6 μm and 0.8 μm, respectively. The beam width, shown by $1/e^2$, was

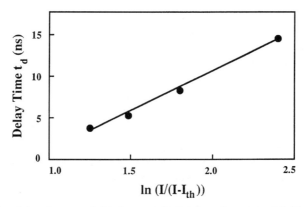

Fig. 10.23 The delay time t_d of the laser emission as a function of $\ln[I/(I - I_{th})]$. I is the pumping current and I_{th} is the threshold current.

10.6 Bluish-purple InGaN multi-quantum-well (MQW) structure LDs 413

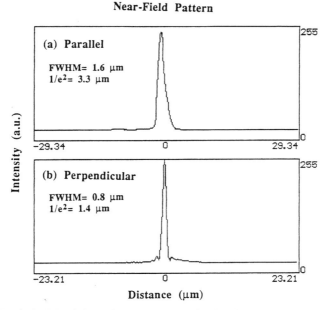

Fig. 10.24 Typical NFP of the InGaN MQW LDs in the planes parallel and perpendicular to the junction under RT CW operation with an output power of 2 mW.

3.3 μm, which was almost the same as the ridge width (4 μm). The transverse optical confinement resulting from the ridge geometry is relatively good using this ridge waveguide. The astigmatism of the laser diodes depends on the optical mode profile, which in turn is determined by the laser structure. A typical value of the astigmatism was 2 μm.

Typical far-field radiation patterns (FFP) of the InGaN MQW LDs in the planes parallel and perpendicular to the junction are shown in Fig. 10.25. The FWHP values for the parallel and perpendicular FFPs are 6.8° and 33.6°, respectively. From Figs 10.24 and 10.25 only fundamental transverse mode operation with an output power of 2 mW was observed in the laser emission. When the output power was changed, the NFP and FFP were almost the same as those in Fig 10.24 and 10.25.

The Stokes shift of energy differences between the exciting and emission energies of the InGaN MQW LDs was as large as 100–250 meV at RT, as shown in Fig. 10.12 [38–40]. This means that the energy depth of the localized state of the carriers is 100–250 meV in the InGaN MQW LDs. Both the spontaneous emission and the stimulated emission of the LDs originated from these deep localized energy states [38–40]. Using high-resolution cross-sectional transmission electron microscopy (TEM), a periodic indium composition fluctuation was observed, which was probably caused by InGaN phase separation during growth [39, 40]. Based on these results, the laser emission is considered to originate from the InGaN quantum dot-like states. The many periodic subband emissions probably result from the transitions between the subband energy levels of the

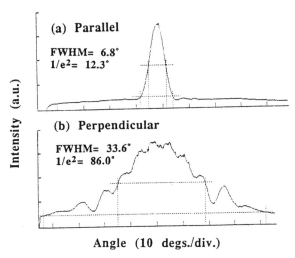

Fig. 10.25 Typical FFP of the InGaN MQW LDs in the planes parallel and perpendicular to the junction under RT CW operation with an output power of 2 mW.

InGaN quantum dots formed from In-rich regions in the InGaN well layers. We estimated the size of the InGaN dots to be approximately 35 Å using high-resolution cross-sectional TEM [39, 40]. These InGaN quantum dots were self-organized due to the phase separation during InGaN growth. Therefore, it is difficult to control the size of InGaN dots which are composed of adjacent In-rich and -poor regions. The energy separation of each subband emission in Figs. 10.16 and 10.19 is only about 2 meV, which is considered to be relatively small in comparison with the energy difference between the $n = 1$ and $n = 2$ subband energy transitions of the quantum dots. These periodic subband energy levels are probably caused by the transitions of the $n = 1$ subband energy levels of a number of quantum dots with different dot sizes.

10.7 Summary

Superbright blue and green InGaN SQW LEDs were fabricated. By combining high-power, high-brightness blue InGaN SQW LEDs, green InGaN SQW LEDs, and red GaAlAs LEDs, many kinds of application, such as LED full-color displays and LED white lamps for use in place of light bulbs or fluorescent lamps, are now possible with the characteristics of high reliability, high durability, and low energy consumption. Also a white LED made by combining a blue InGaN SQW LED and YAG phosphor, which is less expensive than a white LED composed of three primary color LEDs has been developed.

RT CW operation of InGaN MQW LDs was demonstrated with a lifetime of 35 hours. The laser emission was fundamental single mode emission with a peak wavelength of 405 nm. The carrier lifetime and the threshold carrier density were estimated to be 10 ns and $2 \times 10^{20}/cm^3$, respectively. The emission spectra of InGaN MQW LDs under CW operation at RT showed periodic subband emissions with an energy separation of 2 meV. The peak wavelength showed mode-hopping toward higher energy with increasing operating current. These periodic subband emissions probably result from the transitions between the subband energy levels of the InGaN quantum dot-like states formed from In-rich regions in the InGaN well layers due to a phase separation of the InGaN during the growth.

References

1. W. Xie, D. C. Grillo, R. L. Gunshor, M. Kobayashi, H. Leon, J. Ding, A. V. Nurmikko, G. C. Hua, and N. Otsuka, *Appl. Phys. Lett.* **60**, 1999 (1992).
2. D. E. Eason, Z. Yu, W. C. Hughes, W. H. Roland, C. Boney, J. W. Cook, Jr., J. F. Schetzina, G. Cantwell, and W. C. Harasch, *Appl. Phys. Lett.* **66**, 115 (1995).
3. J. Edmond, H. Kong, and V. Dmitrieve, *Inst. Phys. Conf. Ser.* **137**, 515 (1994).
4. J. I. Pankove, E. A. Miller, and J. E. Berkeyheiser, *RCA Review* **32**, 283 (1971).
5. H. Okuyama and A. Ishibashi, *Microelec. J.* **25**, 643 (1994).
6. K. Koga and T. Yamaguchi, *Progr. Crystal Growth Charact.* **23**, 127 (1991).
7. M. G. Craford, *Circuits & Devices*. 24 September (1992).
8. H. Sugawara, K. Itaya, and G. Hatakoshi, *Jpn. J. Appl. Phys.* **33**, 5784 (1994).
9. H. Morkoç, S. Strite, G. B. Gao, M. E. Lin, B. Sverdlov, and M. Burns, *J. Appl. Phys.* **76**, 1363 (1994).
10. H. Amano, M. Kito, K. Hiramatsu, and I. Akasaki, *Jpn. J. Appl. Phys.* **28**, L2112 (1989).
11. S. Nakamura and T. Mukai, *Jpn. J. Appl. Phys.* **31**, L1457 (1992).
12. M. A. Khan, J. N. Kuznia, D. T. Olson, M. Blasingame, and A. R. Bhattarai, *Appl. Phys. Lett.* **63**, 2455 (1993).
13. S. Nakamura, N. Iwasa, M. Senoh, and T. Mukai, *Jpn. J. Appl. Phys.* **31**, 1258 (1992).
14. M. Rubin, N. Newman, J. S. Chan, T. C. Fu, and J. T. Ross, *Appl. Phys. Lett.* **64**, 64 (1994).
15. M. S. Brandt, N. M. Johnson, R. J. Molnar, R. Singh and T. D. Moustakas, *Appl. Phys. Lett.* **64**, 2264 (1994).
16. J. M. Zavada, R. G. Wilson, C. R. Abernathy, and S. J. Pearton, *Appl. Phys. Lett.* **64**, 2724 (1994).
17. S. Nakamura, *Nikkei Electronics Asia* **6**, 65 (1994).
18. S. Nakamura, T. Mukai, and M. Senoh, *Appl. Phys. Lett.* **64**, 1687 (1994).
19. S. Nakamura, M. Senoh, N. Iwasa, and S. Nagahama, *Jpn. J. Appl. Phys.* **34**, L797 (1995).
20. S. Nakamura, M. Senoh, N. Iwasa, and S. Nagahama, *Appl. Phys. Lett.* **67**, 1868 (1995).
21. S. Nakamura, M. Senoh, N. Iwasa, S. Nagahama, T. Yamada, and T. Mukai, *Jpn. J. Appl. Phys.* **34**, L1332 (1995).
22. K. Bando, Y. Moguchi, K. Sakano, and Y. Shimizu, *Tech. Digest, Phosphor Res. Soc.* 264th Meeting, 29 November 1996 (in Japanese); S. Nakamura, *Proceedings of SPIE* **3002**, 26 (1997).
23. H. Amano, T. Asahi, and I. Akasaki, *Jpn. J. Appl. Phys.* **29**, L205 (1990).
24. A. S. Zubrilov, V. I. Nikolaev, D. V. Tsvetkov, V. A. Dmitriev, K. G. Irvine, J. A. Edmond, and C. H. Carter, *Appl. Phys. Lett.* **67**, 533 (1995).
25. M. A. Khan, S. Krishnankutty, R. A. Skogman, J. N. Kuznia, and D. T. Olson, *Appl. Phys. Lett.* **65**, 520 (1994).

26. S. T. Kim, H. Amano, and I. Akasaki, *Appl. Phys. Lett.* **67**, 267 (1995).
27. H. Amano, T. Tanaka, Y. Kunii, K. Kato, S. T. Kim, and I. Akasaki, *Appl. Phys. Lett.* **64**, 1377 (1994).
28. R. L. Aggarwal, P. A. Maki, R. J. Molnar, Z. L. Liau, and I. Melngailis, *J. Appl. Phys.* **79**, 2148 (1996).
29. T. J. Schmidt, X. H. Yang, W. Shan, J. J. Song, A. Salvador, W. Kim, O. Aktas, A. Botchkarev, and H. Morkoç, *Appl. Phys. Lett.* **68**, 1820 (1996).
30. S. Nakamura, M. Senoh, S. Nagahama, N. Iwasa, T. Yamada, T. Matsushita, H. Kiyoku, and Y. Sugimoto, *Jpn. J. Appl. Phys.* **35**, L74 (1996).
31. S. Nakamura, M. Senoh, S. Nagahama, N. Iwasa, T. Yamada, T. Matsushita, H. Kiyoku, and Y. Sugimoto, *Jpn. J. Appl. Phys.* **35**, L217 (1996).
32. S. Nakamura, M. Senoh, S. Nagahama, N. Iwasa, T. Yamada, T. Matsushita, H. Kiyoku, and Y. Sugimoto, *Appl. Phys. Lett.* **68**, 2105 (1996).
33. S. Nakamura, M. Senoh, S. Nagahama, N. Iwasa, T. Yamada, T. Matsushita, H. Kiyoku, and Y. Sugimoto, *Appl. Phys. Lett.* **68**, 3269 (1996).
34. S. Nakamura, M. Senoh, S. Nagahama, N. Iwasa, T. Yamada, T. Matsushita, Y. Sugimoto, and H. Kiyoku, *Appl. Phys. Lett.* **69**, 1477 (1996).
35. S. Nakamura, M. Senoh, S. Nagahama, N. Iwasa, T. Yamada, T. Matsushita, Y. Sugimoto, and H. Kiyoku, *Appl. Phys. Lett.* **69**, 1568 (1996).
36. K. Itaya, M. Onomura, J. Nishino, L. Sugiura, S. Saito, M. Suzuki, J. Rennie, S. Nunoue, M. Yamamoto, H. Fujimoto, Y. Kokubun, Y. Ohba, G. Hatakoshi, and gM. Ishikawa, *Jpn. J. Appl. Phys.* **35**, L1315 (1996).
37. S. Nakamura, M. Senoh, S. Nagahama, N. Iwasa, T. Yamada, T. Matsushita, Y. Sugimoto, and H. Kiyoku, *Appl. Phys. Lett.* **69**, 3034 (1996).
38. S. Chichibu, T. Azuhata, T. Sota, and S. Nakamura, *Appl. Phys. Lett.* **69**, 4188 (1996).
39. Y. Narukawa, Y. Kawakami, Sz. Fuzita, Sg. Fujita, and S. Nakamura, *Phys. Rev. B*, **55**, 1938R (1997).
40. Y. Narukawa, Y. Kawakami, M. Funato, Sz. Fujita, Sg. Fujita, and A. Nakamura, *Appl. Phys. Lett.* **70**, 981 (1996).
41. S. Nakamura, M. Senoh, S. Nagahama, N. Iwasa, T. Yamada, T. Matsushita, Y. Sugimoto, and H. Kiyoku, *Appl. Phys. Lett.* **70**, 1417 (1997).
42. S. Nakamura, *IEEE J. Quantum Electron.* **QE-3**, 435 (1997).
43. S. Nakamura, T. Mukai, and M. Senoh, *J. Appl. Phys.* **76**, 8189 (1994).
44. S. Nakamura, *Microelec. J.* **25**, 651 (1994).
45. S. Nakamura, *S. J. Cryst. Growth* **145**, 911 (1994).
46. S. Nakamura, *J. Vac. Sci. Technol. A* **13**, 705 (11995).
47. F. A. Kish, F. M. Steranka, D. C. DeFevere, D. A. Vanderwater, K. G. Park, T. D. Osentowski, M. J. Peanasky, J. G. Yu, R. M. Fletcher, D. A. Steigerwald, M. G. Craford, and V. M. Robbins, *Appl. Phys. Lett.* **64**, 2839 (1994).
48. S. Nakamura, T. Mukai, M. Senoh, S. Nagahama, and N. Iwasa, *J. Appl. Phys.* **74**, 3911 (1993).
49. S. Chichibu, T. Azuhata, T. Sota, and S. Nakamura, *J. Appl. Phys.* **79**, 2784 (1996).
50. K. Osamura, S. Naka, and Y. Murakami, *J. Appl. Phys.* **46**, 3432 (1975).
51. S. Nakamura, *Jpn. J. Appl. Phys.* **30**, 1620 (1991).

11 Cubic group III nitrides

Oliver Brandt

11.1 Introduction

The Group III nitrides[1] are, because of their large ionicity, among the most extreme representatives of compound semiconductors which, at equilibrium, condense in the wurtzite (α) structure only (Lawaetz 1972). It was, consequently, thought to be virtually impossible to force any of the Group III nitrides into the metastable zincblende (β) modification. This belief was substantially hardened by the lack of substrate materials with a reasonably close lattice match which otherwise could provide the required epitaxial constraint. In striking contrast to these considerations, researchers using molecular beam epitaxy (MBE) or metalorganic vapor phase epitaxy (MOVPE) have succeeded within the last decade in synthesizing the zincblende modifications of GaN, AlN, and InN on various, highly mismatched cubic substrates such as Si, MgO, 3C–SiC/Si, and GaAs.

Such an epitaxial stabilization of a metastable phase is, in general, based on a specific orientation relationship between the lattices of the deposit and the substrate which is characterized by a low lattice mismatch. However, the recent studies of the growth of, e.g., GaN on GaAs, demonstrate that the metastable cubic phase may form despite a lattice mismatch of 20%. This fascinating phenomenon constitutes one reason for the interest in cubic Group III nitrides, as it is one of the issues whose understanding has been identified to be crucial for the further development of heteroepitaxy (Peercy et al. 1990).

Besides their archetypal rôle in the field of heteroepitaxy, cubic Group III nitrides are of further interest since they represent essentially novel materials, the properties of which still remain to be explored. In fact, many of the most basic properties of these materials are known with little accuracy or not at all. This statement is vividly illustrated by the current state of knowledge for both β-AlN and β-InN: except for their lattice constants (4.38 and 4.98 Å, respectively), virtually nothing is known from experiments about these two compounds, not even their bandgaps. There has, however, been theoretical work (for an overview, see Edgar 1994). One of the most accurate calculations available (Wright and Nelson 1995) indicates that the direct gaps of all cubic III–N compounds are smaller than those of their hexagonal counterparts (namely, by

---

[1] In all that follows, we mean AlN, GaN, and InN when speaking about Group III nitrides. BN is excluded from our discussion since both its structural and electronic properties are more closely related to C than to either of the other III–N compounds.

0.06, 0.15, and 0.31 eV for AlN, GaN, and InN, respectively). Further valuable information for the experimentalist has become available from the recent work of Kim et al. (1996a) who calculated the elastic properties of the cubic Group III nitrides, including the elastic stiffness coefficients and the band-edge deformation potentials.

The lack of experimental knowledge is, first of all, related to the fact that cubic Group III nitrides are still much in their infancy compared with the 70 years of history of their hexagonal relatives: β-GaN was synthesized first by Mizuta et al. in 1986, β-AlN by Petrov et al. in 1992, and β-InN by Strite et al. in 1993. Second, the crystal quality of the cubic Group III nitrides is generally quite poor, particularly so for AlN and InN, and often does not permit meaningful measurements of the crystal's properties.

The situation is, however, slightly better in the case of β-GaN. This compound has proven to be easier to prepare than AlN or InN. Besides the obstacles common for both the hexagonal and cubic modifications in preparing single-crystal AlN and InN (namely, the high melting point and the high reactivity with O for AlN, and the low thermal stability of InN), there is an additional fundamental reason for the relative ease with which β-GaN may be obtained: GaN is the least ionic[2] of the Group III nitrides (i.e. it has the smallest deviation from the ideal c/a axial ratio) and is thus more sympathetic to the zincblende modification. Cubic GaN layers of decent quality were thus obtained by several groups (Paisley et al. 1989, Strite et al. 1991, Lei et al. 1992, Powell et al. 1993, Liu et al. 1993, Brandt et al. 1995a, Cheng et al. 1995, Schikora et al. 1996). Most investigations of β-GaN have been focused on the study of its optical properties. The values for the bandgap resulting from these studies scatter significantly, but appear to have recently converged towards a value of about 3.2 eV, i.e. about 200 meV smaller than that of α-GaN (see Section 11.6). It is this finding which makes β-GaN also an interesting material for applications: the visible (blue/green) spectral region may be reached with significantly smaller amounts of In alloyed to GaN then required in the case of the hexagonal phase. Less In, of course, means a reduced tendency of phase separation within the (In,Ga)N alloy (Ho and Stringfellow 1996) and lower strain between the active (In,Ga)N layer and the (Al,Ga)N barrier layers in a light-emitting diode.

In the context of device structures for optoelectronic applications, it is mandatory to investigate the fabrication of the required (In,Ga)N and (Al,Ga)N alloys and their heterostructures. Essential properties of these materials (e.g. their bandgaps, refractive indices, and effective masses as a function of composition) should be measured and catalogued. Furthermore, both n- and p-type doping of the cladding layers has to be investigated systematically. None of these vital demands has been met in the field of cubic Group III nitrides. Except for the work of Abernathy et al. (1995a), who reported the growth of cubic (In,Ga)N layers, and our own recent work (Yang et al. 1997, Müllhäuser et al. 1997), the synthesis of alloys made up by cubic Group III nitrides has not been attempted, let alone of their heterostructures. Only one group (Kim et al. 1994a) has

---

[2] According to the definition of Lawaetz (1972).

11.1 Introduction

reported the n-type doping of β-GaN by Si. Their amazingly good results, however, seem to have been influenced by parallel conductance by the 3C–SiC/Si substrate (Rode and Gaskill 1995). p-type doping of β-GaN by Mg has been reported by Lin et al. as early as 1993, and recently by Abernathy et al. (1995b) using C. A novel and perhaps promising approach to p-type doping of Group III nitrides has been reported recently by us (Brandt et al. 1996b), where the co-doping of β-GaN by Be and O was shown to result in high p-type conductivity (see Section 11.7).

The above statements are meant to illustrate the current situation in the field of cubic Group III nitrides. Cubic Group III nitrides are a novel class of materials whose fabrication and investigation is basic materials science at its best. Sure enough, these materials also hold promise for applications, but, equally surely, they are (as yet) far from being employed for device structures such as fabricated by Nakamura et al. (1994, 1996), comprising dozens of layers with well-controlled composition and doping. Will, then, cubic Group III nitrides ever reach maturity and be taken seriously for applications? One desires, of course, an answer at the very outset of studying these materials, and this answer must inevitably be related to the quality of the material.

One of the most commonly quoted figures of merit for the crystal quality of a material is the linewidth of the X-ray diffraction profile taken in the ω-mode across the lowest order symmetric reflection. The linewidth of this reflection is in fact almost two orders of magnitude larger for β-GaN (30') compared with that measured for α-GaN (1') of similar thickness.[3] Even when considering that symmetric reflections are insensitive to the most common defect type in hexagonaly films, namely, threading dislocations characterized by a Burgers vector along $\langle 11\bar{2}0 \rangle$ and running along [0001], asymmetric reflections are still as narrow as 3' for state-of-the-art hexagonal GaN and thus still an order of magnitude below the best values for cubic GaN.

One should, however, also bear in mind that the crystal quality of hexagonal GaN is not even remotely close to that of, say, GaAs. Threading dislocation densities in excess of 10^{10} cm^{-2} are reported for layers grown on sapphire substrates, and conventional knowledge suggests that such layers are far from being of device quality (Kressel 1981). Yet light-emitting diodes having an impressively high quantum efficiency are fabricated on the basis of this highly defective material. Phenomenologically, it is thus clear that structural defects are much less detrimental in the case of α-GaN than for other semiconductor materials, although the actual reason for this fortunate fact is as yet unknown. We may presume that the electronic impact of defects depends on their type and their atomic structure, and thus in turn on the symmetry of the host crystal as well as on the epitaxial relationship with the substrate. It is hence not clear at all whether the defects present in cubic Group III nitrides on a given substrate are of a similarly forgiving nature as those in their hexagonal counterparts.

This review will, given the above remarks, concentrate on β-GaN despite its

---

[3] Remarkably, similar values are measured for β-GaN layers on substrates with vastly differing lattice mismatches (for a compilation, see Brandt et al. 1995b and Yang et al. 1996b).

more general title. We will, nevertheless, attempt to find an answer to the question provoked implicitly in the last paragraph: are cubic Group III nitrides viable materials for applications? In this attempt we will focus on our own results rather than on those reported by other groups for one main reason. Most of our samples have been characterized by a fairly complete set of techniques, and we thus feel that we possess a more coherent understanding of their properties than for samples for which selected properties only have been reported in the literature.

The paper is organized as follows: We first briefly describe the growth equipment used at the Paul-Drude-Institut for fabricating the GaN layers whose properties are discussed in the following sections. Section 11.3 is devoted to the nucleation and growth of β-GaN on GaAs(001). Section 11.4 discusses the morphology of β-GaN/GaAs(001) layers thus obtained. In Section 11.5, results concerning the structural properties of these layers are presented. Section 11.6 presents the optical properties of β-GaN, and Section 11.7 discusses the electrical properties of β-GaN prepared in our laboratory. Except for the growth equipment, the experimental and theoretical methods are described at the relevant place, not in a separate section. Each section begins or is accompanied by a brief (and, certainly, incomplete) consideration of the work of other groups.

11.2 Growth equipment and *in situ* characterization

We do not want to enter the discussion about the ideal N source for use in an MBE system; however, a few comments are appropriate at this place. Cubic GaN has been fabricated using ECR, DC, and RF plasma sources, and, by analogy with results obtained for α-GaN (see e.g. Molnar and Moustakas 1994 and references therein), it is plausible to assume that plasma damage may affect the quality of the layers detrimentally. The different species of N produced by the various kinds of plasma discharge may also have a profound impact on the growth kinetics, depending on whether the impinging species is, say, an ionized molecule (N_2^+) or atomic N in an excited state (N^*). The inherent complexity of plasma discharges makes these and related effects a topic of study in its own right. Finally, the sputtering action of the N plasma may lead to the incorporation of impurities from the plasma chamber (most noticeably, Si and O if the chamber consists of SiO_2) and from the internal surfaces of the MBE system (Molnar and Moustakas 1994). Most researchers in this field would probably agree that plasma-activated processes are somehow contradictory of the simplicity of the MBE process. In fact, the use of NH_3 which dissociates at the growth front in a surface catalytic process has recently become fashionable after it was shown that high-quality layers could be obtained with this method (Yang *et al.* 1996e, Kim *et al.* 1996b).

The GaN layers discussed in this review are synthesized by MBE on semi-insulating GaAs substrates oriented along [001] and [113]A. The base pressures reached in our custom-made two-chamber MBE system are 10^{-8}

and 2×10^{-10} torr in the loadlock and the growth chamber, respectively. The system is equipped with elemental charges of Ga and As. Active N is generated via dissociating 6N purity N_2 gas by a high voltage ($\simeq 1.5$ keV) plasma glow-discharge in an SiO_2 plasma chamber. The plasma power is kept constant for all experiments at 30 W. The actual N species involved in growth are unknown.

The surface of the growing crystal is monitored *in situ* by reflection high-energy electron diffraction (RHEED), using an incident angle between 1° and 3° and an acceleration voltage of 15 kV. The RHEED intensity is detected by a CCD camera and analyzed by an image processing system.

All temperature values quoted in this paper are thermocouple readings following a three-point calibration, assuming (i) the (2×4)-c(4×4) reconstruction transition of GaAs to occur under an As_4 flux of 8×10^{-6} torr at 490°C, (ii) the desorption of the native oxide of GaAs to take place at 580°C, and (iii) the growth rate of GaAs as measured by RHEED intensity oscillations to drop at temperatures above 630°C.

The Ga flux used in this study, as determined by RHEED intensity oscillations during the GaAs buffer layer growth under an N_2 background pressure of 5×10^{-4} torr, ranges from $0.1–1.0 \times 10^{14}$ atoms/(cm^2 s), corresponding to nominal growth rates of 0.01–0.1 monolayers (ML)/s for GaN. The N flux is obtained via *ex situ* measurements of the thickness of films grown under Ga rich conditions (Yang *et al.* 1996a).

11.3 Nucleation and growth

11.3.1 Nucleation

GaN, regardless of being hexagonal or cubic, is most frequently grown on structurally and chemically dissimilar substrates. Wetting is thus difficult to achieve, and nucleation takes place generally by the formation of isolated nuclei which may exhibit a substantial orientational spread with respect to each other. Further deposition of GaN almost invariably leads to textured polycrystalline growth. The understanding and control of the nucleation stage is thus of utmost importance for the reproducible growth of high-quality GaN layers.

For enforcing wetting, a strategy has been adopted from early studies of heteroepitaxial growth which consists in the deposition of a quasi-amorphous 'wetting' layer of GaN or AlN at low temperature, possibly preceded by an initial annealing of the substrate in an N atmosphere ('nitridation'), followed by solid-phase epitaxial regrowth at elevated temperature. This *brute force* procedure seems to work well in the case of sapphire substrates, and has in turn been applied by many researchers to other substrates, such as GaAs, Si, and 3C–SiC, as well.

However, for the case of GaN on GaAs, this strategy does not suffice for more demanding purposes. Nitridation of GaAs does not take place homogeneously, but proceeds along {111} facets into the underlying GaAs layer (Kuwano *et al.* 1994, Okumura *et al.* 1996b). The faceted GaN/GaAs interface leads to the nucleation of the hexagonal phase and eventually to polycrystalline growth. Recrystallization of an amorphous GaN deposit at 680 °C is incomplete and leads to randomly oriented grains with predominantly wurtzite structure within a highly disordered polycrystalline matrix. Further deposition on this kind of template results in the well-known columnar growth of α-GaN. It is clear that, for achieving solid phase epitaxy, a higher temperature is needed than that we can reach in view of the limited thermal stability of GaAs.

In this section, we present an alternative approach for the nucleation of GaN on GaAs(001). Prior to nucleation, an atomically smooth surface is established by the deposition of a 100 nm thick GaAs buffer layer at 580–620 °C followed by a 2 min annealing under As_4 flux at the growth temperature. The nucleation of GaN is initiated on the (2×4) reconstructed GaAs surface by closing the As shutter and opening the Ga and N shutters. The RHEED pattern then invariably switches to an apparently N-induced (3×3) reconstruction which is commensurate to GaAs. The atomic configuration of this surface reconstruction is still an open question. We think, however, that in view of the huge lattice mismatch it is unlikely to originate from the formation of a coherently strained GaN monolayer as proposed by Hauenstein et al. (1995). Rather, it could correspond to the formation of an N-adsorption layer as detected by X-ray photoelectron spectroscopy employed in recent nitridation studies (Masuda et al. 1995). The substrate temperature is set to 580–620 °C at the beginning of GaN nucleation, while the N/Ga ratio is varied between 1 and 10 by changing the Ga flux and keeping the N flux constant at 0.1 ML/s.

For N-rich (Ga flux 0.1 ML/s) and stoichiometric (Ga flux 0.1 ML/s) conditions the RHEED pattern during nucleation exhibits polycrystalline (ring segments) and hexagonal (c-axis $\parallel$ [001]) reflections, respectively. The first of these conditions effectively corresponds, due to the low growth rate, to a nitridation of GaAs, while the second one is that used during growth and is aimed to maximize the Ga adatom mobility. The RHEED results show that neither the nitridation of GaAs nor a high adatom mobility of Ga leads to the formation of a cubic template for further growth. Figure 11.1 shows the RHEED pattern along the [$\bar{1}$10] azimuth observed after deposition of $\simeq$7 ML of GaN under optimum conditions, namely for a growth rate of 0.03 ML/s at a temperature of 620 °C. Transmission reflections at the positions expected for β-GaN are evident in this pattern. The same type of pattern is observed also for the [110] and [100] azimuths. The diffuse streaks connecting the {113} reflections with the (004) reflection are due to stacking faults.

An in-plane linescan across the {113} reflections of this RHEED pattern is shown in Fig. 11.2. The fit of this intensity profile by pseudo-Voigt functions yields a lattice constant (at growth temperature) of 4.55 ± 0.1 Å, which is identical to that obtained for thick GaN films except for the lower precision due to the large linewidth of these reflections. This linewidth in reciprocal space corresponds to a real-space dimension of $\simeq$1 nm. Empirically, we found that the width of the reflections in reciprocal space is a reliable measure of the success of nucleation: the broader these reflections are, the fewer polycrystalline features are present in the RHEED patterns. It thus seems that the optimum

11.3 Nucleation and growth 423

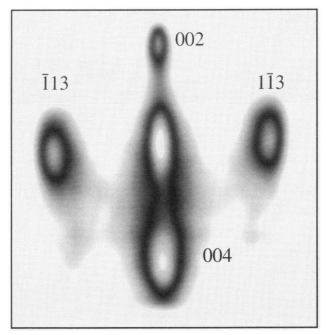

Fig. 11.1 RHEED pattern along the [$\bar{1}10$] azimuth after deposition of $\simeq 7$ ML GaN on GaAs(001). The pattern is Fourier filtered to suppress artifacts due to digitization and gray-scale encoded to enhance contrast.

conditions result in minimized nuclei size, and, consequently, to maximized nuclei density.

Figure 11.3 shows a cross-sectional high-resolution transmission electron

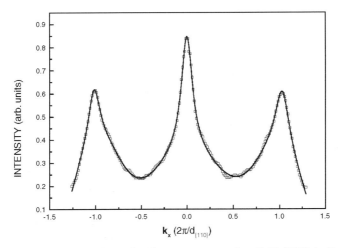

Fig. 11.2 In-plane RHEED intensity linescan across the GaN {113} bulk reflections (taken from raw data). The solid line shows a lineshape fit by pseudo-Voigt functions.

Fig. 11.3 ⟨110⟩-cross-sectional HREM micrograph of a ≃7 ML thick GaN nucleation layer on GaAs(001) in low magnification. This and the following images were acquired by a JEOL 4000 FX microscope operating at 400 kV. The ⟨110⟩-oriented cross-sections are prepared using conventional ion-milling techniques.

microscopy (HREM) micrograph of a nucleation layer deposited under these conditions. This micrograph shows that the 7 ML GaN deposit forms a quasi-connected, epitaxial layer even on a mesoscopic ($\leq \mu$m) scale. In fact, atomic force microscopy (AFM) reveals a quite smooth surface (rms roughness 1 nm on 10 μm^2), particularly on a sub-μm scale (rms roughness 2 Å on 0.01 μm^2). A magnified micrograph as shown in Fig. 11.4 further demonstrates that the mismatch between GaN and GaAs is relieved by pure edge dislocations formed

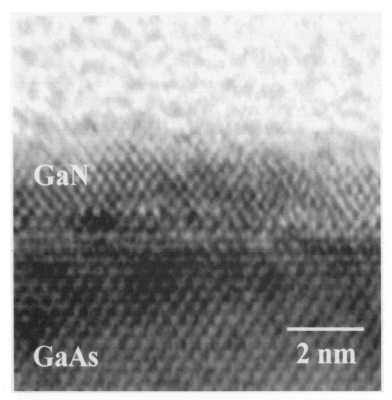

Fig. 11.4 Magnified portion of the HREM micrograph shown in Fig. 11.3. Note the regular edge-type dislocations at the interface.

every 5 interatomic distances. The phenomenon of a regular periodic spacing of these dislocations is observed along the $\langle 110 \rangle$ directions. The periodicity is broken just at those locations from which stacking faults originate. This highly regular dislocation array provides a 5:4 coincidence lattice matching between GaN and GaAs, and its formation seems to require both a slow growth rate and limited Ga adatom kinetics.

We have found that the pure edge-type misfit dislocations visible in Fig. 11.4 are instantaneously formed *during* nucleation of GaN islands via the incorporation of extra {110} lattice planes into the edge of the growing island without any climb or glide mechanism (Trampert *et al.* 1997). The classical theories of misfit dislocation formation (Matthews 1975) can thus not be applied to this case. It is, however, clear that the epitaxial interface is energetically more favorable than an interface involving the thermodynamically stable hexagonal lattice. Being thus confident that the strain energy represents the main portion of the total interface energy, we explain our results by a coincidence lattice or a 'magic mismatch' occurring between GaN and GaAs. True coincidence between the epilayer lattice (a_e) and substrate lattice (a_s) would occur when $a_e/a_s = m/n$, where m and n are integers. If $m = n \pm 1$, there is one extra lattice plane, i.e. a pure edge dislocation, in each unit cell of the coincidence lattice. In general, however, the epitaxial system is not expected to be at true coincidence, and the *coincidence-lattice mismatch* f_0 expresses this deviation from true coincidence as $f_0 = (ma_s - na_0)/ma_s$. This deviation introduces elastic strain at the interface in addition to the strain accommodated by the misfit dislocations. Therefore the energies of heteroboundaries are expected to be small if and only if f_0 does not deviate substantially from true coincidence. For the heterosystem investigated, we obtain $f_0 = -(0.02 \pm 0.2)\%$ by taking the most accurate values for the lattice constants available at growth temperature, namely, $a_{\text{GaN}} = 0.455 \pm 0.01$ nm and $a_{\text{GaAs}} = 0.568\,886$ nm. Thus, this system is close to true coincidence and an array of pure edge dislocations with a period of 5 GaN lattice planes will indeed account for the entire misfit. The occurrence of a 'magic mismatch' between GaN and GaAs provides an explanation of the phenomenon of epitaxial growth for a strain for which epitaxy of covalently bonded materials is usually no longer achieved.

The subsequent growth and the evolution of the defect microstructure detected in the HREM micrographs is determined by the initial nucleation of the islands. Stacking faults and microtwins are generated during the coalescence stage of the nuclei as observed in the HREM images of the nucleation layer (Fig. 11.4). In fact, since the spacing of the individual nuclei will not necessarily be in phase with respect to their dislocation array, the periodicity of the dislocation array will, in general, be broken upon their coalescence. These locations are centers of very high local strain (a 6/5 ratio between GaN and GaAs lattice planes, for example, corresponds to a residual misfit f_0 of -4%), and we believe that these local strain concentrations are responsible for the secondary defects in our layers, namely the stacking faults. Because a change in the stacking sequence may easily induce the energetically favored wurtzite structure, these defects may be quite stable during further layer growth. This problem of the generation of a high density of secondary defects is, within our understanding, an inevitable consequence of the statistical nature of nucleation and seems difficult to overcome, as its solution would amount to a spatial synchronization of the dislocation arrays formed independently in individual nuclei.

To conclude this section, we should add a few comments about the growth of β-GaN on other substrates. The choice of the 'best' substrate is a hotly discussed topic in the field of α-GaN.[4] The most frequently quoted criteria for the selection of a substrate are those directly related to growth, such as lattice and thermal mismatch, and those having implications for technology, such as electrical and thermal conductivity. The price and the ease with which a given substrate can be prepared prior to growth and processed afterwards are further issues which are considered to be important. Following these criteria, sapphire seems by all means the worst choice one could possibly make: it has a huge lattice and thermal mismatch to all Group III nitrides, it is an electrical insulator and a poor thermal conductor, and last but not least, it is hard to process and cannot be cleaved. Yet, both from the point of view of growth and technology, sapphire has been the most successful substrate for the growth of Group III nitrides.

A similar situation occurs when comparing the substrates used for the growth of cubic Group III nitrides. There are of course good reasons to use GaAs or Si, as they are both electrically conductive and have a well-established technology. However, their lattice mismatch to β-GaN is huge (20.0% and 16.8%, respectively). In this section, we have shown that due to a fortunate coincidence epitaxial growth of β-GaN on GaAs(001) is much more easy to achieve than one would have expected. The quality of such layers, regardless of being grown in our laboratory or by others, is comparable to or better than that of layers grown on substrates providing a superior lattice match, foremost among them 3C–SiC (−3.2%). In particular, room temperature luminescence (see section 11.6) has been observed exclusively, to the best of our knowledge, from β-GaN layers grown on GaAs(001). This fact shows that we may not always be as lucky as in the case of GaAs. Taking Si(001) as an example, it becomes clear that lattice and thermal mismatch are not the only criteria determining the quality of epitaxial growth. The reactivity of Si with N may lead to the formation of patches of amorphous Si_xN_y at the initial stage of nucleation, which obviously will disturb the growth of single-crystal material. However, to avoid the formation of antiphase domain boundaries, vicinal substrates are to be employed, but the vertical mismatch at the step edge may result in the generation of additional defects. It is just these conflicting requirements that make the choice of a substrate quite a complex business, and the *a priori* prediction of its suitability for growth almost impossible.

11.3.2 Growth kinetics

The growth behavior of crystals is, in general, determined by both bulk thermo-

---

[4] See, for example, the corresponding discussion forum in the *MRS Internet Journal of Nitride Semiconductor Research*.

dynamics and surface kinetics. However, for nonequilibrium conditions such as established during MBE or MOVPE, surface kinetics plays a major and often dominating rôle in growth. The reason for this is that the surface constitutes a two-dimensional phase in its own right which has properties much distinct from the underlying bulk phase (Moison *et al*. 1987, Copel *et al*. 1989, LeGoues *et al*. 1990). It is this surface phase which is responsible for the growth kinetics of the material (Tournié *et al*. 1995).

Studies of the surface kinetics of GaN are definitely required to obtain an improved understanding of its growth. MBE growth of β-GaN is commonly carried out at quite low temperatures ($0.3 \times T_M$, where T_M is the melting point) compared with the MOVPE case. Such low temperatures might be an actual prerequisite for stabilizing the cubic modification, but are also expected to impose limitations on the morphology and crystal quality one is able to obtain (Yang and Flynn 1989).

In fact, we have found that the surface kinetics has a very profound and, actually, unexpected influence on the growth of β-GaN. Single-phase β-GaN is obtained only if (i) nucleation of a cubic template succeeds as discussed in the previous paragraph, and (ii) the growth conditions during further growth (i.e. after the initial 5–10 ML) are adjusted such that the surface exhibits either a (2×2) or a $c(2 \times 2)$ reconstruction (Brandt *et al*. 1995a). The former reconstruction pattern was reported first by Strite *et al*. (1991) and was later also observed by Liu *et al*. (1993). We have shown that both these reconstructions are Ga-terminated and correspond to a slight N excess and near-stoichiometric conditions at the growth front, respectively (Brandt *et al*. 1995a, Yang *et al*. 1996c). An N excess during growth, such as used during the nucleation stage, leads to the disappearance of the reconstruction and triggers the co-nucleation of hexagonal domains oriented such that [0001] ∥ [111]B. The same phenomenon has been observed recently by Schikora *et al*. (1996) who used an RF plasma source for N_2 dissociation. This finding indicates that the nucleation of hexagonal domains in the presence of excess N is intrinsic to GaN and is not related to the species of N (ionic or neutral) impinging on the growth front. In particular, as RF sources are known to mainly produce neutral species of N, ion damage as the origin of the cubic-hexagonal phase-transformation may be ruled out.

Figure 11.5 displays the RHEED patterns obtained along two of the major azimuths after ≃100 nm of GaN growth under slight N excess (upper row) and near-stoichiometric conditions (bottom row). The patterns reveal a mixture of two-dimensional and three-dimensional growth: while the diffraction intensity is maximized at the position of the transmission reflections, the tendency of these reflections to elongate and to merge along the in-plane direction is evident. The (2×2) reconstruction visible in the upper row is characterized by comparatively intense half-order diffraction streaks along the $\langle 110 \rangle$ azimuths, and quite weak half-order diffraction streaks along [100]. In contrast, the $c(2 \times 2)$ reconstruction visible in the bottom row exhibits half-order diffraction streaks only along [100], which are about two times stronger than those for the (2×2) reconstruction. Since surface reconstructions are, generally, related to specific stoichiometry ranges in the surface phase diagram, this observation opens the way towards the *in situ* control of surface stoichiometry. In fact, as outlined below, clear surface

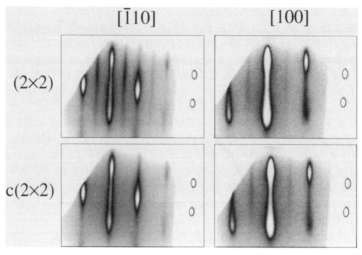

Fig. 11.5 RHEED patterns taken during growth along the [1̄10] and [100] azimuths for two different V/III ratios. The conditions employed for the patterns in the upper row correspond to a slight N excess, whereas those shown in the bottom row were recorded at near stoichiometry. The former patterns constitute a (2 × 2) and the latter ones a c(2 × 2) reconstruction. The patterns are gray-scale encoded for better visibility. The two spots on the right-hand side of each pattern stem from light-emitting diodes signaling the 'open' state for both the Ga and N shutters.

phase transitions are observed depending on the amount of Ga and/or N impinging on the surface (Brandt et al. 1996a). These experiments have led us to the conclusion that the (2 × 2) and c(2 × 2) reconstructions correspond to surfaces covered with 0.5 and 1.0 ML of Ga. From these experimentally determined surface terminations and reconstruction symmetries, simple surface structure models were constructed (Brandt et al. 1995a). The (2 × 2) reconstruction was thought to be formed by Ga dimer rows along the [110] direction which are separated by one missing dimer row. Filling of the missing dimer rows at the center positions of the four adjacent Ga dimers would then correspond at full monolayer coverage to the c(2 × 2) reconstruction.

An impinging flux of either Ga or N on the (2 × 2) surface leads to a transition to a c(2 × 2) or a (1 × 1) reconstruction, respectively. Measuring the intensity of the half-order reconstruction streak along a ⟨110⟩ azimuth allows us to record this phenomenon in real time. Figures 11.6 and 11.7 show selected examples of such time scans upon the pulsed supply of 1 ML N (Fig. 11.6) and 0.5 ML Ga (Fig. 11.7) for two different temperatures. The maximum intensity corresponds in each case to that of the (2 × 2) reconstruction, while the intensity drop is associated with the transition towards either the (1 × 1) or c(2 × 2) reconstructed surfaces (the former surface remains stationary for any amount of N supplied in excess of 1 ML). It is obvious from these experiments that the half-order streak first vanishes upon the impinging flux of both N and Ga, but recovers in a finite time once the supply has ceased. Note that the recovery time is substantially shorter at higher temperature. The simplest explanation for this

11.3 Nucleation and growth

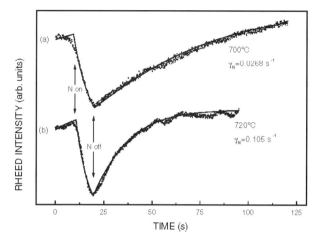

Fig. 11.6 RHEED intensity transient upon a 1.0 monolayer N dose at (a) 700 °C and (b) 720 °C. Solid squares represent experimental data, and solid lines show the best fit of eqn (11.4). The time interval of N supply is indicated in the figure, as well as the desorption rates deduced from the fits.

effect consists in the initial adsorption of either N or Ga, thus forming surface phases distinct in coverage and symmetry from the (2×2) surface phase, followed by the isothermal desorption of the species building up these phases.

Assuming that these reconstruction transitions are caused by adsorption, diffusion, and desorption of the corresponding surface species (Kreuzer 1988), we next develop a—largely phenomenological—model for the surface kinetics of GaN. The model is required to account for adsorption and desorption of both Ga and N as well as for the formation of excess Ga, which eventually will form droplets, and for GaN growth in the case of the simultaneous presence of Ga

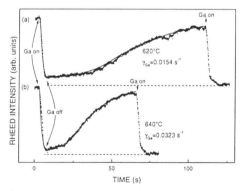

Fig. 11.7 RHEED intensity transient upon a 0.5 monolayer Ga dose at (a) 620 °C and (b) 640 °C. Solid squares represent experimental data, and solid lines show the best fit of eqn (11.4). The dashed lines show the zero level, which is defined by the half-order beam intensity equaling the background intensity. The time intervals of Ga supply are indicated in the figure, as well as the desorption rates deduced from the fits.

and N. Moreover, there are three surface reconstructions distinct in both coverage and symmetry that we have to deal with. Among them, the (2 × 2) and c(2 × 2) related phases are supposed to be built up by one and the same surface species, namely Ga dimers. The stability difference between these phases could phenomenologically be described by a coverage-dependent desorption rate for Ga. Alternatively, and mathematically equivalent, we denote these two distinct surface constitutions of GaN by θ_1 and θ_2, respectively, where $0 < \theta_i < 0.5$ ML is to be satisfied. Our model, in units of ML, is thus described by the equations[5]

$$\frac{d}{dt}[\theta_1(t)] = 2j_{Ga}(\tfrac{1}{2} - \theta_1) + \hat{D}_{Ga}\frac{(\tfrac{1}{2} - \theta_1)}{\theta_2}n$$
$$+ 2\gamma_N(\tfrac{1}{2} - \theta_1) - j_N(\theta_1 - \theta_2), \quad (11.1)$$

$$\frac{d}{dt}[\theta_2(t)] = 2j_{Ga}(\theta_1 - \theta_2) + \hat{D}_{Ga}\frac{(\theta_1 - \theta_2)}{\theta_2}n$$
$$- 2\gamma_{Ga}\theta_2 - 2j_N\theta_2, \quad (11.2)$$

$$\frac{d}{dt}[n(t)] = 2j_{Ga}\theta_2 - \hat{D}_{Ga}\frac{(\tfrac{1}{2} - \theta_2)}{\theta_2}n - k_n n^{2/3}, \quad (11.3)$$

where θ_1 and θ_2 denote the Ga dimer coverage related to (2 × 2) and c(2 × 2) domains, n the amount of excess (liquid) Ga, j_{Ga} and j_N the Ga and N fluxes, $\hat{D}_{Ga}$ the diffusion rate of excess Ga adatoms impinging on θ_2, γ_{Ga} and γ_N the rates of Ga and N desorption, and k_n the rate coefficient of excess Ga desorption. The first term of each equation accounts for adsorption of Ga, building up the θ_1 phase and subsequently the accompanying phases θ_2 and n. In much the same way, the second terms account for diffusion of Ga adatoms impinging on θ_2 domains having a size proportional to the coverage. In other words, the diffusion rate is defined as the ratio of diffusion coefficient D_{Ga} and domain size L^2. The third terms describe the desorption of N, Ga dimers (θ_2) and excess Ga (n). The latter phase has been assumed eventually to form hemispherical Ga droplets where desorption takes place from the droplets' surface, which is of order 2/3. The last term in eqns (11.1) and (11.2), finally, stand for the incorporation of Ga and N adatoms into the crystal, i.e. actual growth.[6] Note that the model described by eqns (11.1)–(11.3) is a linearized and

---

[5] With respect to a previous version (Brandt et al. 1995c), the model has been reformulated and simplified since experiments have shown some terms to be of minor importance. Furthermore, the terms describing adsorption of N have been modified as our recent experiments suggest that N adsorption is, to a certain degree, selective and takes place preferentially on θ_2.

[6] For the experiments presented in this work, eqns (11.1)–(11.3) actually simplify considerably. Since only one species, say N, is impinging on an initially (2 × 2) reconstructed surface, most terms of the equation system are actually zero and the temporal evolution of the surface is exclusively determined by j_N in the adsorption stage and by γ_N in the desorption stage. The equivalent statement applies for the case of Ga impinging on the (2 × 2) surface, except that in this case the formation of excess Ga (n) may come into play.

11.3 Nucleation and growth

highly simplified version of microscopic theories of the surface kinetics such as developed by Kreuzer (1988). This simplification is necessary to reduce the number of free parameters and thus to guarantee stable fits, but also means that, while our model is perhaps the most simple one accounting for the phenomena we wish to describe, there certainly are a variety of more refined models which could equally well describe the data.

Finally, we have to relate the surface coverage calculated by means of eqns (11.1)–(11.3) to the quantity experimentally observed, namely the RHEED intensity. Although the intensity in RHEED is, in general, to be calculated by dynamical diffraction theory, we here use the kinematical approximation in order to be able to perform the simulations in a reasonable amount of time.[7] Within the kinematical approximation (Lagally et al. 1988) we thus write for the intensity of the half-order reconstruction streak along a $\langle 110 \rangle$ azimuth

$$I_{\langle 110 \rangle}(t) = [\theta_1(t) - \theta_2(t)]^2. \tag{11.4}$$

Simulations based on eqns (11.1)–(11.4) are shown in Figs. 11.6 and 11.7 together with the corresponding experimental data, the latter of which were normalized to one, with the zero level defined by the simultaneously recorded background intensity. The simulations are fits to the data by a numerical least squares routine (conjugate gradient), and the fits were found to be stable regardless of the values of the initial parameters. Since these RHEED transients are actually a measure of the isothermal desorption rate as a function of coverage, the good agreement of the experimental and simulated transients shows that desorption of both N and Ga is a first-order process. A closer examination of Fig. 11.7 reveals that during the first (0.5 ML) supply of Ga the RHEED intensity does not actually drop to zero (dashed lines), i.e. the coverage of θ_2 is incomplete. Apparently, excess Ga is formed parallel to the θ_2 phase before the latter is completed. This finding indicates a comparatively small diffusion rate of Ga adatoms.

Figure 11.8 shows RHEED transitions at (a) 620 °C and (b) 680 °C upon the supply of 1.5 ML Ga on a logarithmic scale. The amount of excess Ga is larger and, thus, the influence of its diffusion is more pronounced than for the 0.5 ML dose shown above. The experimental data are compared with simulations where all parameters have been kept constant except for the diffusion rate $\hat{D}_{Ga}$. The sensitive dependence of the transient behavior on this parameter is evident.

Figure 11.9(a) shows the desorption fluxes derived from our simulations in an Arrhenius representation. The volatility of both Ga- and N-related surface phases is, in the investigated temperature range, of the same order of magnitude as that observed for the free evaporation of bulk GaN. The stability of all surface phases is many orders of magnitude higher than expected under equilib-

---

[7] Calculations based on a semi-kinematical model which includes dynamical diffraction effects by the inner potential of the crystal were also performed and were found to be in good agreement with the kinematical calculations presented throughout this paper. The adatom coverage θ was simulated by constructing supercells of up to ten times the unit mesh size, covering the range between 0.1–0.9 ML Ga coverage. The computation time required for calculating the RHEED intensity for one specified coverage varies between 10 and 100 min on a Sun Sparc 10/42 workstation, depending on the size of the supercell.

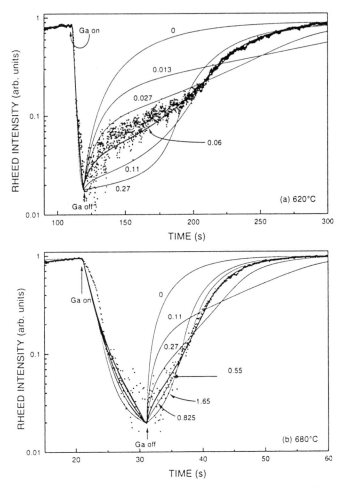

Fig. 11.8 Logarithmic display of RHEED intensity transients upon a 1.5 ML Ga dose at (a) 620 °C and (b) 680 °C. Solid squares represent experimental data, and the bold solid lines show the best fit of eqn (11.4) using diffusion rates of (a) 0.06/s and (b) 0.55/s. The other lines show simulations assuming various other values for the diffusion rate $\hat{D}_{Ga}$ as indicated for each curve. The time intervals of Ga supply are also indicated in the figures.

rium conditions, showing that in all cases desorption is greatly kinetically hindered. Interestingly, at temperatures below 760 °C, Ga is the more volatile species at the growth front rather than N. In Fig. 11.9(b), we present the temperature dependence of the diffusion coefficient D_{Ga}.[8] The diffusivity is

---

[8] The determination of $D_{Ga} = \hat{D}_{Ga} L^2$ requires knowledge of the domain size L, which is acquired here by scanning tunnelling microscopy investigations of our GaN films (see below). These studies show that the surface reconstruction is disordered on an atomic scale. As a worst-case approximation, the surface domain size L^2 has been assumed to be of the order of one unit cell. This assumption results in a pre-exponential term of 0.007 cm^2/s, which is in fact consistent with the expected 'universal' $D_0 \sim \nu/4N_0$, where ν is the frequency of thermal vibrations and N_0 is the surface site density.

11.3 Nucleation and growth 433

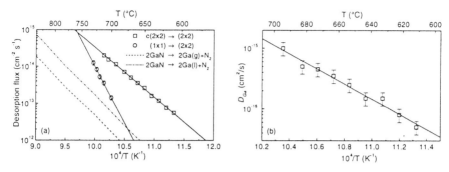

Fig. 11.9 Arrhenius representation of (a) the N and Ga desorption fluxes and (b) the diffusion coefficient D_{Ga}. The points are experimental data. For (a), open squares refer to Ga, while the open circles indicate values for N. The solid lines are least squares fits of single exponentials to our data. The dashed and dash-dotted lines in (a) represent the data of Munir and Searcy (1965) and Groh et al. (1974) for the free surface evaporation of GaN. For the Ga and N desorption flux, the fits results in Arrhenius parameters ln Z_0 and E_a (ln Z = ln $Z_0 - E_a/k_BT$) of 64.5 ± 0.6 and 2.69 ± 0.05 eV and 102 ± 2 and 6.1 ± 2 eV, respectively. For the diffusion coefficient D_{Ga}, we obtain $D_0 = 0.007$ cm^2/s and $E_d = 2.48$ eV, where $D_{Ga} = D_0 \exp(-E_d/k_BT)$.

characterized by an activation energy of 2.48 eV. Note that this activation energy may originate from the temperature dependence of the domain size L^2, which is determined by diffusion of Ga on the (2 × 2) reconstructed surface, rather than by D_{Ga}, which accounts for diffusion of excess Ga on the c(2 × 2) domains.

The results discussed above have several important implications.

(i) The surface phases of GaN represent a large kinetic barrier for evaporation, particularly the (2 × 2) reconstructed phase. The large difference in the pre-exponential terms and in the apparent energies of evaporation between Ga- and N-related surface phase means that these kinetic barriers are of both entropic and energetic character. The stability of the N-terminated surface is perhaps related to a steric effect: dimerization of N, which would greatly enhance the volatility of this surface, is prevented by the remoteness of chemisorbed N adatoms relative to their covalent radii.[9]

(ii) The diffusivity of Ga on (001) GaN is governed by a large energy of activation. This finding is consistent with arguments predicting a scaling of the surface diffusivity with the cohesive strength of the material (Yeng and Flynn 1989).

(iii) Under MBE growth conditions, the surface stoichiometry during growth is severely influenced by Ga desorption. This phenomenon intricates the growth of single-phase β-GaN films where the slightest deviation from the optimum surface stoichiometry induces the nucleation of the hexagonal phase (Brandt et al. 1995a). Furthermore, the diffusivity of Ga at MBE growth temperatures is low, which presumably inhibits the formation of truly smooth surfaces.

---

[9] Recent *ab initio* calculations (Neugebauer and Scheffler 1997) suggest that, actually, an N-terminated surface does not exist in thermal equilibrium. Within this understanding, the (1 × 1) surface is Ga-terminated, and the transition towards the (2 × 2) surface occurs by the desorption of N from the plane *beneath* the surface.

It must be pointed out that some details of the results presented above depend on our presumptions concerning the surface coverage with Ga. The surface coverage has been determined on the basis of the observed symmetries of the respective RHEED patterns and the approximate amount of Ga needed to induce a transition between them. It is, however, well known that structures different in real space may be identical in reciprocal space, i.e. RHEED patterns are by no means a unique representation of a certain structure. For determining the actual structure and coverage of surfaces, a real-space technique such as scanning tunneling microscopy (STM) is desirable. However, STM has proven to be difficult to conduct on GaN, and no high-resolution images have been obtained so far (Garni et al. 1996). The main reason for this situation seems to be contamination of the surface during transfer through air and the inability to remove these contaminants (namely O and C) in the STM chamber.

However, we have found that, at least for β-GaN, contamination is not such an important issue. Samples exposed to air for several days could be used for further growth after an annealing at 700 °C, which led to the appearance of the (2×2) reconstruction characteristic for the β-GaN surface in vacuum. The exceptional stability of this surface in the absence of any impinging flux allows annealing for several hours at 700 °C and above without measurable decomposition of the GaN layer. In a separate X-ray photoelectron spectroscopy experiment (Ding et al. 1996) the surfaces were indeed found to be free from contamination after extended annealing in vacuum at about 700 °C. The reconstruction intensity could furthermore be enhanced considerably by depositing and, subsequently, thermally desorbing several ML of Ga. Besides an ordering of the surface reconstruction induced by this process, we speculate that contaminants may dissolve within the liquid Ga phase and desorb concurrently.

Samples for STM investigations are thus prepared in the following way. A 10 nm thick layer is grown on an n^+-GaAs(001) substrate, using the nucleation conditions described in the previous paragraph and conditions corresponding to a stable (2×2) reconstruction during further growth. The sample is then slowly cooled to room temperature and quickly transferred in air into another MBE system, which is connected to an STM operating at UHV conditions. After the thermal desorption of the thin oxide grown on the surface during the brief exposure to air, the RHEED pattern displays weak half-order streaks corresponding to the (2×2) reconstructed surface. Deposition of a few ML of Ga on this surface and thermally desorbing the excess Ga at a temperature of 610 °C gives rise to an intense (2×2) pattern. The $c(2 \times 2)$ RHEED pattern is observed after depositing 0.5 ML Ga on the (2×2) surface at 550 °C, after which the sample is quenched to room temperature. The patterns thus obtained are identical to those observed in the growth chamber. The GaN sample is then transferred into the STM chamber under a vacuum below 10^{-19} torr. The STM is equipped with etched W tips and operates at tunneling currents of 100–150 pA at positive and negative tip voltages of 2.5–3.5 V.

Figure 11.10(a) shows an empty state STM image of the (2×2) reconstruction. The lateral spacing between each white protrusion agrees with the expected value of 6.36 Å. The structural model of the (2×2) reconstruction as

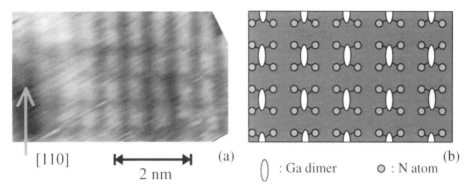

Fig. 11.10 (a) Empty state STM image of the β-GaN(001)-(2 × 2) surface reconstruction. (b) Ball-and-stick model of the (2 × 2) reconstruction.

proposed by us consists of Ga dimers that line up in rows along the [110] direction as shown in Fig. 11.10(b). The Ga coverage of this surface is 0.5 ML.

Figure 11.11(a) shows a filled state STM image of the Ga-terminated surface previously identified as a c(2 × 2) reconstruction. This surface exhibits domains with square symmetry oriented $37 \pm 2°$ with respect to each other and $19 \pm 1°$

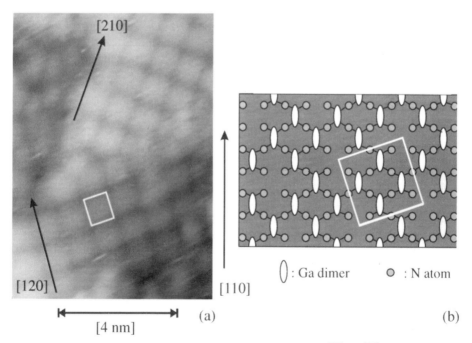

Fig. 11.11 (a) Filled state STM image of the β-GaN(001)-($\sqrt{10} \times \sqrt{10}$)$R$ 18.4° surface reconstruction. The unit mesh is marked by the tilted square. Reconstructed domains are oriented along the [120] and [210] directions. (b) Ball-and-stick model of the ($\sqrt{10} \times \sqrt{10}$)$R$ 18.4° reconstruction.

with respect to the [110] direction. A detailed line-scan analysis reveals a lateral periodicity within the domains of 10.02 ± 0.03 Å. The orientation along the [120] and [210] directions as well as the periodicity distinguishes this surface structure as a $(\sqrt{10} \times \sqrt{10})R\,18.4°$ reconstruction following the notation of Wood (1964).[10] The two different domains running along [120] and [210] can be constructed by a reflection of each domain at the (110) mirror plane.

Having identified the two-dimensional Bravais lattice of the apparent $c(2 \times 2)$ surface to be actually a $(\sqrt{10} \times \sqrt{10})R\,18.4°$ structure, knowledge of the basis of this reconstruction is now required for a complete structural determination. In Fig. 11.11(b) we show a ball-and-stick model of the $(\sqrt{10} \times \sqrt{10})R\,18.4°$ surface reconstruction, constructed as a vacancy structure of the originally proposed $c(2 \times 2)$ reconstruction by removing 1 out of 5 Ga dimers. These vacancies form the Bravais lattice of the $(\sqrt{10} \times \sqrt{10})R\,18.4°$ structure. It is important to note that both the originally proposed $c(2 \times 2)$ and the $(\sqrt{10} \times \sqrt{10})R\,18.4°$ surface reconstruction observed in STM cannot be distinguished by RHEED. Simulations of the corresponding RHEED patterns within the kinematical approximation in fact reveal essentially identical intensity distributions along the major azimuths for these reconstructions.

The distinct motif of the $c(2 \times 2)$ reconstruction, namely a Ga dimer surrounded by 4 Ga dimers in the $\langle 100 \rangle$ directions, is absent in the $(\sqrt{10} \times \sqrt{10})R\,18.4°$ surface reconstruction. In fact, this proposed structure is distinguished by having the highest possible Ga dimer coverage (0.8 ML) without exhibiting this structural element. The complete absence of the $c(2 \times 2)$ motif in the STM images led us to conclude that this structural element is indeed *forbidden*, for if it were allowed, inevitable lateral fluctuations in the Ga coverage should result in a domain structure which also contains $c(2 \times 2)$ reconstructed areas.

These STM results suggest that the results deduced from the RHEED transients shown above are quantitatively incorrect. In seems, in fact, impossible to construct a surface having full (1 ML) Ga coverage while simultaneously exhibiting a $(\sqrt{10} \times \sqrt{10}R\,18.4°$ symmetry. Hence, the model presented above also has to be modified taking into account the actual Ga coverage of the $(\sqrt{10} \times \sqrt{10})R\,18.4°$ reconstruction. However, this correction will only influence the absolute values for the desorption flux of Ga and its diffusion coefficient, and not the activation energies. As long as the actual coverage of these surfaces has not been confirmed by more detailed experiments, there seems no necessity for such modifications.

11.4 Morphology

Only a little information is available in the literature concerning the morphology

---

[10] The translational vector of this reconstruction with respect to the underlying cubic bulk having the lattice constant a is constructed by going three lattice constants along and one lattice constant perpendicular to the [110] direction. The resulting directions are rotated 18.4° towards [110] and the period amounts to $\sqrt{3^2 + 1^2}\,a = 10.06$ Å, in excellent agreement with the experimental results.

11.4 Morphology

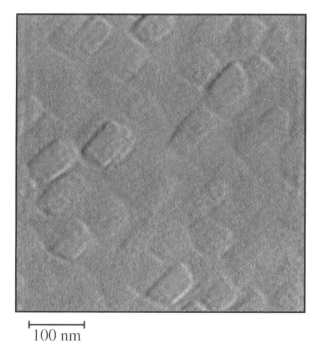

100 nm

Fig. 11.12 AFM micrograph of a 100 nm thick β-GaN layer on GaAs(001). The micrograph is presented as a surface relief, which shows the characteristic rectangular domains surrounded by coalescence boundaries particularly clearly. The coalescence boundaries are oriented along $\langle 110 \rangle$.

of β-GaN (Paisley *et al.* 1989, Lei *et al.* 1992). It is, however, clear from both the published RHEED patterns and scanning electron microscopy (SEM) micrographs that the surface of β-GaN is far from being atomically smooth.

Figure 11.12 shows an AFM micrograph which illustrates the characteristic surface morphology of β-GaN. This particular GaN layer was grown under the c(2 × 2) surface reconstruction (corresponding to an N/Ga ratio of one) to a thickness of 100 nm. Its surface, while being quite smooth ('mirror-like') on a macroscopic scale, is microscopically rather rough and consists of a densely-packed array of highly regular rectangular domains. Although the rms roughness measured from this and similar surfaces over a lateral scale of 1 μm is impressively low (0.5 nm), peak-to-valley values of up to 4 nm are observed between adjacent domains. It has to be pointed out that these domains do not correspond to a grain or columnar structure of the layer. Rather, these features are the residue of coalescence boundaries created in the nucleation stage. Both the lateral size and the height between these coalescence domains scale approximately linearly with the thickness of the layer. A 1 μm thick GaN layer, for example, generally exhibits coalescence domains with a lateral size of 500 nm and a height difference of 40 nm between adjacent domains. This scale-invariant behavior of the surface with time is one of the most serious issues which arises in the context of applications. Figure 11.13 displays an STM micrograph taken from the $(\sqrt{10} \times \sqrt{10})R\,18.4°$ reconstructed surface of the 10 nm thick GaN film

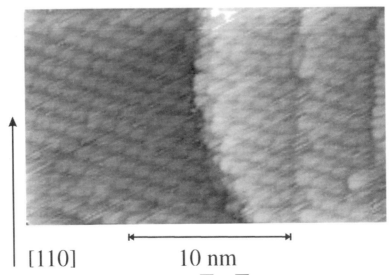

Fig. 11.13 Filled state STM image of the ($\sqrt{10} \times \sqrt{10} R\, 18.4°$ reconstructed β-GaN(001) surface showing the presence of mono- and bilayer steps. Note that the step edges run along [110].

already discussed above. This micrograph evidences the presence of surface roughness on a scale inaccessible by AFM. Both mono- and bilayer steps are present on this surface, and their distance is only 10–50 Å. The reconstruction domains visible in Fig. 11.11 are of similar extension. These findings are consistent with the low surface diffusivity of Ga deduced from our analysis of RHEED transients (*cf.* Fig. 11.9).

11.5 Structural properties

11.5.1 X-ray diffraction

As we have mentioned in the introduction, X-ray diffraction (XRD) is commonly employed for measuring the quality of GaN. Frequently, however, it is apparent that the diffraction conditions are specified incorrectly, i.e. scans along or perpendicular to the growth direction are confused with each other. Furthermore, one often finds the statement that the layer is purely cubic, although the data presented give little or no information regarding phase mixture.

In our laboratory, X-ray measurements are performed using a double-crystal diffractometer equipped with a $Cu_{K\alpha 1}$ anode and a Ge(004) monochromator. We speak of an $\omega-2\theta$ scan when the detector, whose angular position is denoted by θ and which has a 50 μm entrance slit, is rotated with twice the angular velocity of the sample, the angular position of which is denoted by ω. If the detector's angular position is fixed to twice the Bragg angle of the reflection

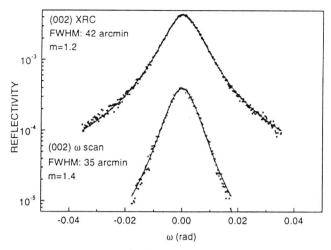

Fig. 11.14 X-ray profiles across the (002) reflection obtained from a 1 μm thick β-GaN layer on GaAs(001) in the XRC (upper curve) and ω (lower curve) modes. The linewidths are indicated in the figure, as are the values of m obtained from the fits of the reflectivity R by a Pearson function (thick solid lines) $R(\omega) \propto [1 + 4\omega(2^{1/m} - 1)/\sigma^2]^{-m}$, where σ is the width of the profile.

under consideration, we distinguish measurements taken either with a wide open detector (X-ray rocking curve (XRC)) or with the 50 μm slit (ω scan).[11] We point out that these commonly employed X-ray scans, regardless of being taken in the ω–2θ (i.e. along [00l]) or the ω (i.e. along [hh0]) modes, are insufficient for detecting hexagonal grains other than those oriented such that [0001] ∥ [001]. As already stressed by Lei et al. (1993), these simple scans may prove the existence of the cubic phase, but by no means the absence of hexagonal components. Reciprocal space-mapping (Brandt et al. 1995a) or, better, a complete texture analysis, is necessary for obtaining reliable information about the presence of the hexagonal phase.

Figure 11.14 shows the X-ray profiles across the (002) reflection obtained from a 1 μm thick GaN layer on GaAs(001) in the XRC (upper curve) and ω (lower curve) modes. Within the detection limit of X-ray diffraction and Raman spectroscopy (Giehler et al. 1995), we have found this layer to be purely cubic. The reflection linewidth in XRC includes contributions from mosaicity, lateral and vertical domain size, and inhomogeneous strain. The two latter contributions are absent for measurements taken in the ω mode and are those which determine the linewidth in conventional ω–2θ scans. In fact, the linewidth found for these latter measurements is typically 5–10′ for our layers, i.e. close to the difference in linewidth apparent from the two scans in Fig. 11.14. This finding is common for binary semiconductors and is usually attributed to a mosaic structure of the crystal.

---

[11] Note that this latter distinction is not commonplace and is introduced here only for the sake of clarity.

Current models of X-ray diffraction from mosaic crystals invariably predict Gaussian reflection profiles resulting from the statistical orientation spread of individual mosaic blocks (Vook 1975). Gaussian profiles are also obtained when assuming that the broadening is caused by the lateral domain size (or, more precisely, by the limited X-ray coherence length within the layer), even when considering a monotonic increase of this domain size with the thickness of the layer. More advanced models, which consider the strain fields of dislocations and their interactions, predict Gaussian profiles as well (Kaganer et al. 1997). However, the data shown in Fig. 11.14 cannot be described by Gaussians but rather by Lorentzians. A perfect fit is obtained using a Pearson function (thick solid lines) which mediates between a Lorentzian ($m = 1$) and a Gaussian ($m \to \infty$) by means of the parameter m. The values of m returned by the fit are close to one, meaning that the reflection profiles displayed in Fig. 11.14 are in fact close to Lorentzians. Note that, while such a lineshape is frequently observed (and expected for pure domain size broadening) even for polycrystalline samples in ω–2θ scans, it remains mysterious for the scans performed here. Using the linewidth of these reflection profiles as an absolute means to judge the quality of the layer seems, consequently, not well justified, as we do not understand the mechanisms responsible for both the peculiar shape of the profile and its width.

11.5.2 Transmission electron microscopy

The defect structure of β-GaN has been the subject of several excellent studies to which we have little to add. Regardless of whether the substrate has been chosen to be MgO (Powell et al. 1993), 3C–SiC (Paisley et al. 1989, Okumura 1996b), Si (Basu et al. 1994), or GaAs (Chandrasekhar et al. 1995, Okumura et al. 1996b), the most prevalent structural defects observed in β-GaN are stacking faults and micro-twins. The abundance of these defects is easily understood when considering the low stacking fault energy for GaN (which is equivalent to the small difference in energy between the zincblende and wurtzite modifications) and the fact that both intrinsic and extrinsic stacking faults as well as $\langle 111 \rangle$ rotational twin boundaries may be viewed as a planar inclusion of the wurtzite modification (Trampert et al. 1997). Since stacking faults which terminate within the material are bound by partial dislocations, they may also play a rôle in relieving the strain within the layer. However, stacking faults are certainly not the major source of strain relief (in fact, their density would be much too low) as sometimes stated in the literature. As we have already seen in our discussion of the nucleation stage, strain relief is achieved by pure edge-type dislocations running along the interface.

Figure 11.15 represents a typical selected area diffraction (SAD) pattern taken from the $\langle 110 \rangle$-cross-sectional sample of a GaN/GaAs(001) structure displaying the superposition of diffraction spots from the GaN and GaAs lattices. The pattern shows that the GaN layer is cubic with the same orientation

11.5 Structural properties

Fig. 11.15 SAD pattern of a $\langle 110 \rangle$-cross-sectional β-GaN/GaAs(100) heterostructure.

as the GaAs substrate. The sharpness and symmetrical intensity of the GaN diffraction spots—there is no elongation towards the GaAs spots—demonstrate the perfect alignment of the layer with regard to the substrate and the homogeneous strain distribution within the film. The streaks along the $\langle 111 \rangle$ directions originate from stacking faults. The distances between the diffraction spots can be measured to estimate the lattice parameter of the β-GaN epilayer by defining the diffraction spots of GaAs as a calibration standard. Setting $a_{\text{GaAs}} = 0.565\,325$ nm we obtain for the GaN lattice constant $a_{\text{GaN}} = 0.452 \pm 0.001$ nm, which is in good agreement with the values reported in the literature for β-GaN. These lattice constants correspond to a lattice mismatch of $(a_{\text{GaAs}} - a_{\text{GaN}})/a_{\text{GaAs}} = 0.200 \pm 0.002$ with regard to the GaAs reference lattice.

In bright-field micrographs (not shown here: see Brandt *et al.* 1997), it is seen that the GaAs buffer layer is free of strain and defects even close to the heterointerface, which appears to be remarkably sharp. The GaN layer contains a high density of planar defects propagating along the {111} planes. These planar defects primarily originate at the GaN/GaAs interface and have their highest density near the interface. The density is reduced by more than an order of magnitude within the first 40 nm of the layer. The remaining planar defects are frequently arranged in bundles extending through the entire epilayer. The absence of columnar or polycrystalline domains, however, is noteworthy, as is the exact orientation of the planar defects throughout the GaN layer. Furthermore, only very rarely are 60° threading dislocations detectable either in cross-section or plan-view micrographs on a μm length scale.

Details of the interface between GaN and GaAs and the nature of the planar

442 *Cubic group III nitrides*

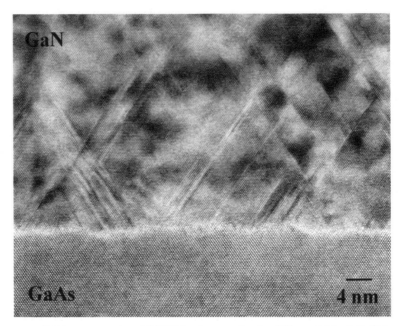

Fig. 11.16 ⟨011⟩-cross-sectional HREM micrograph of a 250 nm thick β-GaN layer on GaAs(001). The scale is given in the micrograph.

defects become evident from Fig. 11.16. The interface is flat on an atomic scale with an overall vertical roughness of only 4–5 monolayers. A more careful inspection of the interface structure in Fig. 11.16 reveals that the boundary in fact contains atomically flat parts. Within these regions no planar defects originate from the interface, and, simultaneously, regularly arranged pure edge dislocations appear at the interface. In fact, the cores of these dislocations are observed at every fifth GaN {111} lattice plane. In between the dislocation cores, the continuity of the {111} lattice planes across the interface is clearly visible. The planar defects penetrating into the GaN layer are seen to be predominantly stacking faults and microtwins. Local distortions in the vicinity of terminated stacking faults are visible in this micrograph, but the layer clearly possesses a single-crystal nature in that it retains its epitaxial relationship to the substrate throughout its thickness (Trampert *et al.* 1997).

11.6 Optical properties

Most information on the optical properties of β-GaN stems from either PL or CL investigations at low temperature and from transmission measurements at room temperature. Rarely (Lacklison *et al.* 1995) have these techniques been conducted in a comparative way. Equally rarely have the samples under investigation been sufficiently characterized by structural techniques for excluding phase mixture and residual strain induced by lattice or thermal mismatch. It is

Table 11.1 Low temperature (5 K) cathodoluminescence transition energies in eV observed for β-GaN single crystals. FX: free exciton; (D^0, X): donor bound exciton; (D^0, h): neutral donor to valence band; (A^0, e): conduction band to neutral acceptor; (D^0, A^0): donor–acceptor pair; XL: unknown; LO_{XL}: LO phonon replica of XL.

FX	(D^0, X)	(D^0, h)	(A^0, e)	(D^0, A^0)	XL	LO_{XL}
3.272	3.263	3.234	3.208	3.165	3.075	2.990

thus not surprising that the values reported for the bandgap and the related exciton and defect transitions of β-GaN scatter over a fairly wide range.

Much information collected in our laboratory on the emission properties of β-GaN at low temperatures has been obtained from comparatively large ($\simeq 1\,\mu$m) GaN single crystals which were grown inadvertently under Ga-rich growth conditions (Yang et al. 1996d). These crystals are found to be virtually free of extended defects and exhibit well-resolved low-temperature emission lines when excited selectively by spatially resolved cathodoluminescence. We have studied both the stationary and transient emission properties of these crystals (Klann et al. 1995, Menniger et al. 1996a, Menniger et al. 1996b, Müllhäuser et al. 1996). The transition energies found in the stationary experiments are compiled in Table 11.1. The transient properties were found to be similar to those observed for ternary semiconductors: the free exciton (FX) dominates the PL spectrum for a short time ($\simeq 30$ ps) after arrival of the initial excitation pulse, after which it relaxes to potential minima, presumably provided by inhomogeneous strain within the layer (Klann et al. 1995).

The low-temperature emission spectra of homogeneous layers grown in our laboratory are inevitably dominated by either the (A^0, e) or the (D^0, A^0) transitions. SIMS analyses of these layers show the dominant impurities to be O, Si, and C, each of which is present in the layer at a concentration of at least $10^{17}\,cm^{-3}$. It is noteworthy that most of the transition energies reported in the literature (see, e.g. Okumura et al. 1996a, Schikora et al. 1996) seem to closely correspond to those compiled in Table 11.1. At present, however, a chemical identification of the defects and impurities involved in these various transitions is not at hand. Since most of the β-GaN fabricated so far has been grown with fairly low rates (in our study, e.g. at 50 nm/h), such an identification may prove to be actually less important than it seems now. Material grown with higher rates will definitely be less affected by incorporation of extrinsic impurities, and may thus exhibit emission spectra dominated by intrinsic transitions. In a certain sense, the optical properties of β-GaN at room temperature are thus more fundamental than those at low temperature, and they are certainly more relevant for applications. It is for this reason that we concentrate in this section on the room temperature properties of β-GaN. Note that room temperature PL emission from β-GaN is not to be taken for granted. Ramírez-Flores et al. (1994), using layers grown on MgO which exhibit XRD linewidths among the lowest ever reported for β-GaN (28'), obtained PL emission only below 150 K.

The sample studied in this section is a 1 μm thick β-GaN layer on GaAs(001)

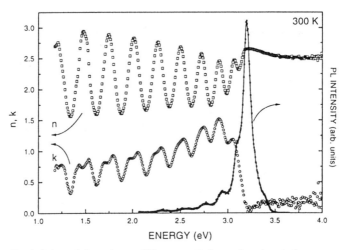

Fig. 11.17 Real (n) and imaginary (k) part of the refractive index as measured by spectroscopic ellipsometry and PL spectrum taken from the 1 μm thick β-GaN layer on GaAs(001) (cf. Fig. 11.14). The PL has been excited using the 325 nm line of an He–Cd laser.

which has been well characterized including AFM, XRD (cf. Fig. 11.14), Raman spectroscopy, and Hall measurements. Its surface is featureless under an optical microscope and exhibits an rms roughness of 8 nm in 3×3 μm^2 AFM scans. It has been found to be predominantly cubic (no hexagonal reflections have been observed in XRD reciprocal space-maps, and no E_2 has been detected by Raman spectroscopy). The background electron concentration as determined by Hall measurements is 8×10^{17} cm^{-3} and the electron mobility 75 Vs/cm^2.

Figure 11.17 presents some of the fundamental optical properties of cubic GaN at room temperature. The real (n) and imaginary (k) parts of the refractive index are measured by *ex situ* spectroscopic ellipsometry. The low-energy part of these spectra displays interference fringes which demonstrate the optically homogeneous thickness of the layer and which allow us to determine this thickness to be 0.97 ± 0.03 μm. The interference fringes die out when significant absorption sets in, i.e. at the layer's bandgap. Simulations of these spectra employing a transfer-matrix technique demonstrate that significant absorption occurs at 3.21 ± 0.01 eV with an exponential (Urbach) low-energy tail of 70 meV width. The simulations are in equally good agreement with the experiments if the primary absorbing agent is chosen to be a band-to-band transition (with a strength corresponding to an absorption coefficient of $\simeq 7 \times 10^4$ cm^{-1} just above the band edge), or if an exciton (having an oscillator strength of 0.15 nm^{-3}) is considered in addition. The absorption broadening is far too large to allow the detection of excitonic effects in these spectra. However, two quantities of value may be gained from these experiments: just above the band edge, the refractive index of β-GaN is 2.5, and its absorption coefficient (calculated from the measured value of k) is $\simeq 7 \times 10^4$ cm^{-1}.

The PL spectrum peaks, according to a Lorentzian fit, at 3.205 ± 0.002 eV. Its width is 110 meV, and it is superimposed by a tail to lower energies which is modulated by Fabry–Perot interferences. The existence of optical interferences evidences once again the macroscopic homogeneity of the layer and its transparency below the bandgap. It is interesting to note that the width of the highest energy peak in the PL spectrum taken at 5 K (which, in our understanding, corresponds to a superposition of free and bound exciton transitions) from this sample is 45 meV. The large thermal broadening of this near-band edge transition is characteristic for our samples (the β-GaN single crystals, for example, exhibit linewidths for the highest energy peak of 8 and 90 meV at 5 and 300 K, respectively). In fact, a thermal broadening from 3 to 100 meV between 5 and 300 K has also been observed for some α-GaN layers (Fischer et al. 1996). This behavior, however, is not a fundamental property of GaN, as much narrower lines (down to 30 meV, i.e. substantially below the theoretical linewidth of band-to-band transitions) are routinely obtained at 300 K for α-GaN. In contrast, the narrowest PL linewidth at room temperature reported so far for β-GaN is 70 meV (Nakadaira and Tanaka 1997). In any case, the sample investigated here exhibits PL emission at essentially the same energy as it happens to absorb, and we can thus be sure that this PL emission is intrinsic in nature. Note that the omnipresent yellow band around 2.2 eV is too low in intensity to be visible in these spectra. This finding may partly result from the moderately high excitation density of 100 W/cm^2 used here, which was chosen to be similar to that employed in light-emitting diodes. However, also at an excitation density of 1 W/cm^2 (at 300 K), the intensity of the yellow band is about two orders of magnitude below that of the exciton line.

Figure 11.18 shows PL spectra at three different temperatures taken from the 1 μm thick GaN layer on GaAs(001). At the high excitation density used here, the PL efficiency drops by only a factor of five between 20 °C and 220 °C, while

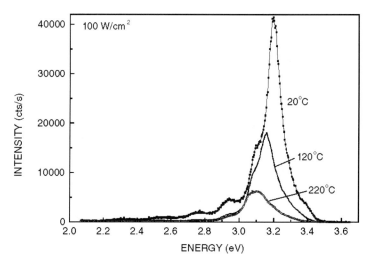

Fig. 11.18 PL spectra at three different temperatures taken from the 1 μm thick β-GaN layer on GaAs(001).

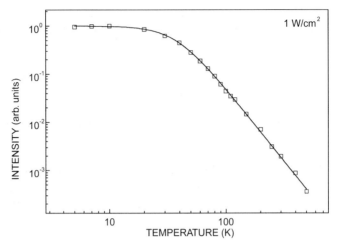

Fig. 11.19 Double logarithmic plot of the total PL intensity I (spectrally integrated between 2.9 and 3.3 eV) measured for the 1 μm thick β-GaN layer on GaAs(001) between 4 and 500 K. The solid line shows a fit of the data by the equation $I(T) = (1 + \alpha T^\beta)^{-1}$. The values for α and β returned from the least squares routine are 2.5×10^{-5} and 3, respectively.

the linewidth increases from 110 meV (13 nm) at 20 °C to 180 meV (22 nm) at 220 °C. Despite the large width of the XRD profile and the correspondingly high density of structural defects, as well as the additional handicaps of a free surface and an absorbing substrate, the absolute PL efficiency of this layer is quite comparable to that of considerably thicker α-GaN layers on Al_2O_3(0001) measured side-by-side.

The participation of nonradiative channels in recombination becomes more apparent when using a lower excitation density. Figure 11.19 shows a double logarithmic representation of the total PL intensity I measured for the same GaN layer between 4 and 500 K with an excitation density of 1 W/cm². The PL intensity drops by three orders of magnitude between 5 and 500 K, which, while being a common observation for bulk-like semiconductor layers, provides evidence for the importance of nonradiative recombination channels in this β-GaN layer. These data are *not* well represented by the standard formula $I(T) = [P_1 \exp(-E_{A1}/kT) + P_2 \exp(-E_{A2}/kT)]^{-1}$, where E_{A1} and E_{A2} are the thermal activation energies and P_1 and P_2 are the weights of the respective nonradiative dissociation channels. Having extended the temperature range to 500 K, the visually satisfactory fit obtained by this formula in previous work of us (Brandt *et al.* 1997) turned out upon a more careful examination to be an artifact of the specific (Arrhenius) representation of the data. On plotting the data in the way done in Fig. 11.19, it becomes apparent that the PL intensity actually drops according to a simple power law (namely, with T^3). At present, we cannot offer an explanation for this finding. We stress, however, that we have found the thermal quenching of the PL intensity from both β- and α-GaN to be more complex than that predicted by the simple thermal activation model.

The strong impact of excitation density on these processes has been studied previously in more detail (Müllhäuser et al. 1996). It has become clear that the nonradiative recombination channels responsible for the quenching of PL intensity are readily saturated at modest excitation density. Thanks to this fact, the threshold of stimulated emission should be only marginally affected by the presence of these channels (Brandt et al. 1996c). For investigating nonradiative recombination in β-GaN in some more detail and in the attempt to eventually obtain optical gain from β-GaN, we have studied the recombination dynamics of the β-GaN layer whose CW emission has been discussed above upon pulsed excitation up to fairly high excitation densities.

The emission spectra were recorded using 150 fs pulses from a frequency-doubled Ti:sapphire laser at a photon energy of 3.36 eV with a repetition rate of 76 MHz. The excitation power was reduced by neutral-density filters and controlled by a powermeter. For recording the surface emission, the sample was excited in backscattering geometry with the laser spot focused to a diameter of $\simeq 100\,\mu$m. For recording the edge emission, the excitation spot was focused to an $l \times 50\,\mu\mathrm{m}^2$ stripe on the top surface adjacent to a cleaved edge, where l denotes the excitation length. To measure the optical gain we used the method pioneered by Shaklee et al. (1973), in which l is varied with an adjustable slit. The emitted light was collected by a lens, dispersed by a 22 cm monochromator (600 lines/nm grating) and focused onto the photocathode of a streak tube allowing for time resolution down to 2 ps. The streak images were recorded by a cooled CCD array. The spectral resolution of the setup was about 3 meV. All measurements were performed at room temperature.

Figure 11.20 shows surface emission transients for different excitation densi-

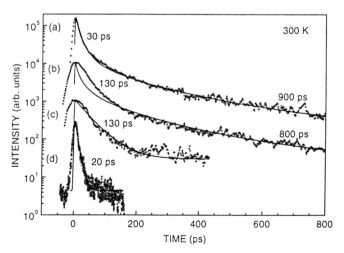

Fig. 11.20 Transients of the surface emission (symbols) in backscattering geometry of the 1 μm thick β-GaN layer on GaAs(001) at 300 K. The excitation fluences are 12, 3, 0.1, and 0.01 μJ cm^{-2} for transients (a), (b), (c), and (d), respectively. The traces are vertically shifted for clarity. The solid lines are fits of a recombination model. The major decay times are indicated.

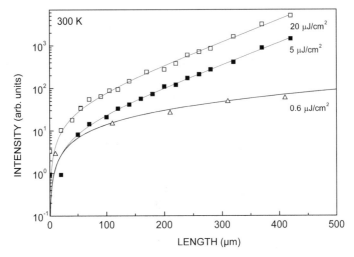

Fig. 11.21 Total edge emission vs excitation stripe length for different fluences of the 1 μm thick β-GaN layer on GaAs(001) at 300 K. The lines are fits of model calculations to deduce the respective gain values.

ties (Klann et al. 1997). The decay is mono-exponential and rapid ($\simeq 20$ ps) at low fluences. The decay time increases to values of $\simeq 1$ ns with increasing fluence. At the highest fluence used, the initial decay accelerates and reaches a value as short as 30 ps. The solid lines in Fig. 11.20 originate from simulations based on a recently published model (Brandt et al. 1996c), which considers carrier recombination by band-to-band and defect-mediated transitions in parallel. Within this model, the rapid decay at low fluences is understood as being dominated by defect-mediated recombination. The decay decelerates once these defects are saturated at sufficiently high fluences, and thus recombination approaches the purely radiative regime. However, the rapid initial decay at high fluences cannot be understood within this model. It can be described only when an additional bimolecular term, which is characterized by a carrier-density-dependent radiative recombination coefficient, is added to the equations. Although this phenomenological term is formally similar to that describing Auger recombination, we find the total (spectrally and temporally integrated) surface emission intensity to increase slightly superlinearly (exponent 1.2) with fluence. It thus seems more likely that the rapid initial decay originates from the onset of laterally amplified spontaneous emission (Kawase et al. 1993).

Figure 11.21 shows the total edge emission intensity (symbols) as a function of the excitation stripe length for fluences of 0.6, 5, and 20 μJ cm^{-2} at room temperature. For the two higher fluences, an exponential dependence of the edge emission intensity on excitation length is observed, revealing the presence of optical gain and thus of the onset of amplified spontaneous emission. In order to determine the gain coefficient g from the length dependence of the emission intensity I, we used the well-known formula

$$I = \frac{I_0}{g}(e^{gl} - 1), \tag{11.5}$$

where I_0 denotes the spontaneous emission rate per unit volume (Shaklee et al. 1973). The calculated gain curves corresponding to the best fit to the data points are shown as solid lines in Fig. 11.21. The gain coefficient increases from $5\,\text{cm}^{-1}$ at $0.6\,\mu\text{J}\,\text{cm}^{-2}$ to values between 110 and 120 for 5 and $20\,\mu\text{J}\,\text{cm}^{-2}$. The latter values are comparable to the gain coefficients measured for hexagonal $Al_{0.1}Ga_{0.9}N/GaN$ heterostructures (Kim et al. 1994b). Note that the carrier density created by an ultrashort optical pulse is proportional to the *fluence* of the pulse and not to its *intensity*. While the excitation intensities used here are significantly higher than in previous works, the fluences are not.

11.7 Electrical properties

Very few data have been published on the electrical propeties of cubic Group III nitrides. Most of the unintentionally doped β-GaN layers were reported to be highly n-type (10^{17}–10^{20}), showing that the cubic and hexagonal modifications share the problem of a high background electron concentration. As to the origin of the high background doping, most authors adopt the traditional point of view which holds the N vacancy as being responsible. In fact, there are experimental data which seem to confirm this suspicion, such as the dramatic increase in resistivity with N/Ga ratio as reported by Fanciulli et al. (1993) and by Powell et al. (1993). However, this increase in conductivity might also result from ion damage and increasing compensation, as concluded by Molnar and Moustakas (1994). In fact, recent theoretical work suggests that the N vacancy may be excluded as the source for the n-type conductivity of GaN, as its formation energy is much too high to be created in appreciable concentrations (Neugebauer and Van de Walle 1994).

The evaluation of the electrical properties of all samples discussed below was done by Hall measurements employing the van der Pauw technique, using either In or Zn contacts, or simply (for samples exhibiting metallic conductivity) by touching the GaN layer with In-coated copper tips. Internal short circuits caused by spatial inhomogeneities of the layer are inhibited by a pseudo-cloverleaf geometry created by scribing or reactive-ion etching. To exclude any influence of the semi-insulating GaAs substrate and the (optional) GaAs buffer layer, free-standing films were measured in addition after having removed all GaAs with an $H_2SO_4:H_2O_2:H_2O$ etch solution.[12] This method was applied also when measuring the current–voltage characteristics of p–n junctions formed by p-type GaN layers on undoped (hence n-type) GaN buffer layers. The ohmic contacts used for the latter measurements were made of the In/Ga liquid alloy. Secondary ion mass spectrometry (SIMS) was performed using a CAMECA IMS 4F system and either Cs^+ or O_2^+ primary ions with energies of 10 kV and 15 kV,

---

[12] The importance of this step cannot be overemphasized. Parallel conductance by both the (optional) p-type GaAs buffer layer and the semi-insulating GaAs substrate may lead to grossly overestimated values for the conductivity of the GaN layer. Moreover, diffusion of an intentional dopant from the GaN layer into the GaAs buffer layer could create a highly conductive channel dominating the sample's electrical properties.

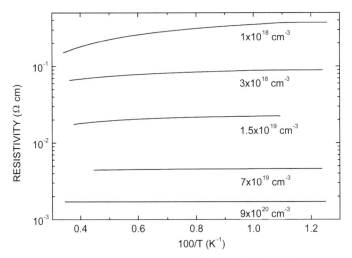

Fig. 11.22 Arrhenius representation of the resistivity of β-GaN layers grown consecutively after having opened the system. The number labelling each curve gives the electron concentration measured at room temperature.

respectively. The concentration standards were obtained from ion-implanted samples.

Undoped GaN layers grown in our system exhibit electron concentrations in the high 10^{17} to mid 10^{18} cm^{-3} range and mobilities around 50 cm^2/Vs, regardless of whether growth is carried out under Ga-rich or N-rich conditions. This independence of background electron concentration on surface stoichiometry during growth confirms the doubt about the commonly adopted interpretation as to which native defects, particularly, N vanancies, are responsible for the high background electron concentration in undoped GaN. In fact, when growing GaN layers immediately after having opened the system, electron concentrations up to 10^{21} cm^{-3} are measured. Figure 11.22 shows the resistivity of a series of layers grown subsequently after having opened the system. Successively grown samples show a monotonic decrease of electron concentration, with the standard background being reached after about ten growth runs. This decrease is correlated with a decrease of the H$_2$O partial pressure (measured by quadrupole mass spectrometry) in the MBE system, which drops from 10^{-9} torr immediately after having opened the system to 10^{-12} torr in standard conditions. This finding suggests that O, supplied from residual H$_2$O vapor in the growth chamber, is the actual reason for the background electron concentration in our samples. A rough estimate, assuming a Ga beam equivalent pressure of 10^{-8} torr at the growth rate used here, and unity sticking and dissociation coefficients for impinging H$_2$O molecules, yields atomic O concentrations between 0.01% at standard conditions and, potentially, 10% with the highest H$_2$O partial pressure.

The electron concentration measured for the sample grown immediately after the system has been exposed to air is 9×10^{20} cm^{-3}, and an O concentration of the same order (2%) is detected by SIMS. The same agreement is found for samples with lower electron concentration. We thus conclude that O is incorporated substitutionally on N sites and has an electrical activity close to unity. The strikingly high incorporation of O on substitutional (N) sites which is required for resulting in such high electron concentrations may be understood intuitively by considering the atomic properties of O with respect to N. The ionic radius of O is quite close to the values for the covalent and ionic radii of N, meaning that O will fit well on N sites without generating undue strain into the host lattice. Furthermore, the electronegativities of O and N are almost identical. From a chemical point of view, O is thus the ideal substitutional donor in GaN, and more detailed theoretical considerations arrived at the same conclusion (Jenkins and Dow 1989).

The method we have used here to achieve incorporation of O is obviously impractical. Furthermore, H_2O as the precursor for O doping is not desirable in MBE. However, O_2 introduced into the system by a precision leak valve and dissociated by a plasma discharge is pumped effectively and is thus a suitable precursor for use in an MBE system.

The achievement of p-type doping requires, first of all, an overcompensation of the high background electron concentration present in β-GaN. Lin et al. (1993) have succeeded in obtaining hole concentrations up to 8×10^{18} cm^{-3} by doping with Mg. No information, however, was given concerning either the background electron concentration or the absolute concentration of Mg required for overcompensation.

Since we are convinced that at least for our samples the vacuum rather than the growth conditions determine the background electron concentration, we made great efforts to obtain a good vacuum in our (two-chamber) system by extensive baking. Cubic GaN grown afterwards indeed showed electron concentrations in the mid 10^{17} cm^{-3} range despite the low growth rate (50 nm/h) employed. We have also found that the electron concentration is lower by a factor of two or three when growth takes place on the (113)A rather than on the (001) plane, which is why we concentrated initially on this orientation for the p-type doping experiments. We have chosen Be as the acceptor species instead of the commonly used Mg in view of the fact that Be is, in all other III–V semiconductors, a more shallow acceptor than Mg. Furthermore, Be has other advantages over Mg, such as less surface segregation, and is generally more easy to handle in MBE systems than Mg.

Despite the precautions described above, the results we have obtained when doping with Be under these 'optimum' conditions (regime I, see Table 11.2) were quite disappointing. p-type conduction was not observed for any of these layers, but rather n-type conductivity (even higher than the background of undoped samples) or high resistivity. At Be concentrations above 10^{20} cm^{-3}, finally, a transition to polycrystalline dendritic growth took place. In fact, Bernadini et al. (1997) have theoretically shown that this phenomenon is actually to be expected, since the solubility of Be in GaN is fairly low and exceeding it triggers the formation of Be_3N_2 precipitates.

Meanwhile, our system was not so well-conditioned after all, as a leak had opened due to the (overly) excessive baking (regime II). Not aware of this

Table 11.2 Compilation of the results obtained for different growth regimes distinguished by the nominal Be concentration and the vacuum conditions (p_{H_2O} is the partial pressure in the system as measured by quadrupole mass spectrometry). The values for the hole concentration p and the hole mobility μ are obtained at room temperature.

Regime	[Be] (cm^{-3})	p_{H_2O} (torr)	Type	p (cm^{-3})	μ (cm^2/Vs)
I	$\simeq 10^{20}$	$\ll 3 \times 10^{-11}$	–[a]	–	–
II	$< 10^{20}$	$\gg 3 \times 10^{-11}$	n	–	–
III	$\simeq 10^{20}$	$> 3 \times 10^{-10}$	SI	–	–
IV	$\simeq 10^{20}$	$< 2 \times 10^{-10}$	p	$\simeq 5 \times 10^{18}$	$\simeq 70$
V	$> 5 \times 10^{20}$	$> 3 \times 10^{-10}$	p	$\simeq 1 \times 10^{18}$	$\simeq 150$

[a] Layers not connected

failure, we continued our doping studies just to find, to our great delight, that the layers produced turned suddenly p-type and appeared from both their structural and morphological properties virtually unaffected by Be concentrations in excess of 10^{20} cm^{-3} (regime IV). XRD profiles exhibited a clear (113) reflection from cubic GaN, with a width comparable to undoped GaN layers. AFM micrographs revealed values for the rms surface roughness of about 5 nm, which again is comparable to undoped layers, meaning that the surface morphology did not degrade even at this doping level.

A summary of our experiences is shown in Table 11.2, classifying them according to the vacuum conditions, the Be concentration as measured by SIMS, and the hole concentration achieved (if any). The (at first glance) most remarkable finding is that layers doped with Be well above its solubility limit are largely unaffected in terms of their crystaline and morphological properties. SIMS measurements of these layers showed that in addition to Be, O is present in a comparable amount. Indeed, the calculations of Bernadini *et al.* (1997) predict that the formation of Be_3N_2 precipitates may be inhibited by the simultaneous incorporation of O, as the formation of Be–O ion pairs should be energetically favorable in this case.

Next, we should point out two more remarkable findings, namely (i) the high p-type conductivity of typically 50/Ω cm reached in regime IV, and (ii) the high value for the hole mobility obtained for these samples (the high conductivity of these layers directly results from the high value of the mobility). Strikingly, the mobility values obtained are about one to two orders of magnitudes higher than those reported in the literature for a significantly lower degree of compensation (Wang *et al.* 1993, Tanaka *et al.* 1994, Lin *et al.* 1993). Calculations of the influence of ionized and neutral impurity scattering on the mobility (Seeger 1981) in fact show that mobilities between 1 and 10 cm^2/Vs are to be expected at these impurity concentrations.

However, when assuming that the residual O donors and Be acceptors are spatially correlated in the form of ion pairs which occupy nearest-neighbors sites (Rimbey and Mahan 1985) the experimental mobility values are reproduced as the two individual, long-range Coulomb scatterers are replaced in this case by a

11.7 Electrical properties

single, short-range dipole scatterer having a minor impact on mobility. We are thus led to the conclusion that the high mobilities (i.e. conductivities) measured in our β-GaN layers are a result of the formation of ion pairs by ionized Be acceptors and O donors. In other words, it is the *co-doping* of GaN with acceptors and donors which allows high mobility and thus high conductivity to be obtained.[13]

Such a phenomenon is not actually surprising when considering the high reactivity of Be and O. Indeed, just as predicted by Bernadini *et al.* (1997), the incorporation of Be might cause an enhanced and spatially correlated incorporation of O, such that perfectly compensated layers are obtained when O is available in a virtually unlimited amount (see regime III in Table 11.2). However, once a surplus of Be is offered, p-type conductivity takes over (regimes IV and V). The important point here is that the number of isolated Coulomb scatterers equals just the surplus of Be, not the total concentration of Be and O.

These considerations led us to grow samples under quite extreme conditions, namely, for an intentionally produced (by allowing atmospheric leakage into the growth chamber) high H_2O partial pressure, and an extremely high Be flux (regime V). Figure 11.23 shows a SIMS profile of one of these samples for which Hall measurements yielded p-type conductivity with a hole concentration of 10^{18} cm^{-3} and hole mobilities of 150 cm^2/Vs. The SIMS profile shows that both Be and O are incorporated in almost equal amounts, namely about 5 ×

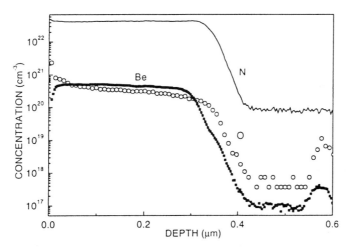

Fig. 11.23 SIMS depth profile of a β-GaN grown within regime V. Note the negligible diffusion of Be, which closely follows the reference signal of N, rigidly shifted by the thickness of the buffer layer (30 nm).

---

[13] Note that, while we have used Be as the acceptor species and O as the reactive donor, this concept of reactive co-doping is more general in nature and may well be extended to other reactive impurity combinations (such as Mg and O) and also to other host materials, such as, particularly, the hexagonal modification of GaN.

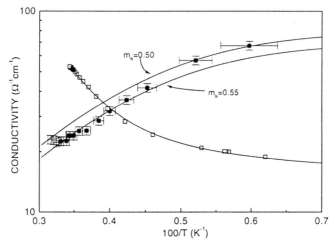

Fig. 11.24 Arrhenius representation of the conductivity of p-type β-GaN layers. The open and solid squares show the experimental data of layers grown under conditions characteristic for regimes IV and V, respectively. The solid lines show fits using a standard thermal activation model for the former and simulations including polar optical, deformation potential, acoustic phonon, piezo-electric, and ionized impurity scattering (with a temperature-independent concentration of 10^{18} cm^{-3}) fot the latter. The numbers labeling these latter curves indicate the different hole masses assumed for the calculations. The error bars shown for this data set apply equally well to the other one.

10^{20} cm^{-3}. Despite this high degree of compensation, the hole mobility measured for these samples is more than twice as high as the values obtained for samples grown under 'low' H_2O partial pressure (regime IV). This finding would remain a paradox if donor–acceptor pair formation is neglected.

Note that in contrast to a recent report (Cheng et al. 1996) we do not observe significant diffusion of Be into the GaN buffer layer despite the high Be concentration used here. The vertical current–voltage characteristics of these samples exhibit a clear rectifying behavior, as expected from a p–n junction formed by the Be doped layer on the n-type GaN buffer. This inhibited diffusion of Be might be related to the formation of Be–O pairs, which are expected to be immobile at growth temperature. Additionally, the built-in potential of the p–n junction opposes the diffusion of positively charged species, such as the higher mobile Be interstitials.

Figure 11.24 shows the Arrhenius representation of the conductivity of layers grown under conditions corresponding to regimes IV and V as defined above. Whereas the former exhibits an activated conductivity typical for impurity band conduction (Fritzsche 1955) the latter shows a behavior expected for comparatively pure semiconductors. The solid lines in Fig. 11.24 fitting the latter data show the conductivity theoretically predicted for GaN having a temperature-independent ionized impurity concentration of 1×10^{18} cm^{-3}. This absence of activated conductivity means that the conductivity of this sample is metallic in nature. The transition between these different conductivity mechanisms is expected to occur at dopant concentrations p sufficiently high that the wavefunctions of individual acceptors overlap and thus form a band within the forbidden gap which merges with the valence band. Mathematically, this phenomenon is expected at acceptor concentrations satisfying the condition $a_B p^{1/3} \sim 0.25$, where a_B is the Bohr radius of the (hydrogenic) acceptor, which can be estimated to be about 0.5 nm for Be. Correspondingly, the transition observed experimentally is expected to occur at a Be concentration of 1.25×10^{20} cm^{-3}, in agreement with the Be concentration in these layers.

11.8 Conclusion

The reader who has followed us through this review certainly will agree that the progress made within the last decade in the field of cubic Group III nitrides has been substantial. All of the binary III–N compounds have at least been demonstrated to be capable of being produced in their metastable zincblende modification. Cubic GaN is even fabricated in the form of (while highly defective) homogeneous epitaxial layers which permit detailed studies of their physical properties. As a consequence, we have obtained a clearer understanding of the growth of this material, enabling us to further improve the layers' quality, which is now high enough to allow, for example, optical gain to be achieved at room temperature. This progress is indeed highly satisfactory, and seems to demonstrate that β-GaN is approaching the state-of-the-art of its hexagonal relative. Yet, we should not be overly enthusiastic about this progress as long as so many unaddressed issues remain in the context of applications. Some of these issues are quite obvious and have been mentioned in the introduction, such as the almost complete lack of work on doping and alloys. Some others, however, are more subtle, and are related to the fact that we often do not know if an apparently characteristic property is inherent to β-GaN (on a given substrate) or merely results from the specific conditions employed for fabricating the material. We have just begun in our laboratory to conduct a comparative study of material grown under different conditions (using, most noticeably, different vacua, N species, and growth rates) and might so be able eventually to find an answer to these questions. This comparative study also includes the synthesis of cubic (In,Ga)N. The results we have obtained in this study were not presented in this review, as we regard them as preliminary, but they are noteworthy in this context. Using the MBE system with which all the layers discussed in this review were grown, we have obtained an In content of up to 11%, above which both the structural and morphological integrity of the layer were corrupted (Yang *et al.* 1997). None of these layers exhibited PL emission at room temperature. In contrast, we have obtained blue emission at room temperature from a cubic $In_{0.17}Ga_{0.83}N$ layer grown with an RF plasma source (Müllhäuser *et al.* 1997).

These last statements are meant to illustrate the trivial (but often disregarded) fact that the potential of cubic Group III nitrides may be judged correctly only when examining the properties of layers grown by different techniques. There are, for example, only a very few studies of the MOVPE growth of β-GaN, and yet the material with the narrowest reported PL linewidth of 70 meV (at room temperature) has been grown by MOVPE. An answer to the question provoked in the introduction—are cubic Group III nitrides viable materials for applications?—at this stage would thus be preliminary at best. At least, we have not yet found any fundamentally unsurmountable barriers for reaching 'device-quality' material out of cubic Group III nitrides. On the contrary: our recent results on cubic (In,Ga)N bring cubic Group III nitrides much closer to light-emitting devices than one would have dared think a few years ago.

Acknowledgments

The author of this review is much indebted to his co-workers at the PDI and elsewhere for their enthusiasm in our ongoing investigation of cubic Group III nitrides. In particular, I want to express my gratitude to J. Behrend, M. Giehler, H. T. Grahn, U. Jahn, B. Jenichen, R. Klann, H. Kostial, J. Menniger, J. R. Müllhäuser, M. Ramsteiner, A. Trampert, Y. Suzuki, M. Wassermeier, A. Yamada, B. Yang, and H. Yang for their invaluable contributions to this work. Special thanks are due to H.-P. Schönherr and P. Schützendübe for expert technical assistance and to L. Däweritz, K. H. Ploog, and C.-G. Schulz for continuous encouragement and support. I gratefully acknowledge financial support of part of this work by the Bundesministerium für Bildung und Forschung (BMBF) of the Federal Republic of Germany.

References

Abernathy, C. R. and MacKenzie, J. D. (1995a). *Appl. Phys. Lett.* **66**, 1632.
Abernathy, C. R., MacKenzie, J. D., Pearton, S. J., and Hobson, W. S. (1995b). *Appl. Phys. Lett.* **66**, 1969.
Basu, S. N., Lei, T., and Moustakas, T. D. (1994). *J. Mater. Res.* **9**, 2370.
Bernadini, F., Fiorentini, V., and Bosin, A. (1997). *Appl. Phys. Lett.* **70**, 2990.
Brandt, O., Yang, H., Jenichen, B., Suzuki, Y., Däweritz, L., and Ploog, K. H. (1995a). *Phys. Rev. B* **52**, R2253.
Brandt, O., Yang, H., Ramsteiner, M., Menniger, J., and Ploog, K. (1995b). *Int. Phys. Conf. Ser.* **145**, 33.
Brandt, O., Yang, H., and Ploog, K. (1995c). *Proceedings of the 1st International Conference on Semiconductor Heteroepitaxy*, Montpellier (France) 1995, B. Gil and R.-L. Aulombard, World Scientific, Singapore, p. 250.
Brandt, O., Yang, H. and Ploog, K. H. (1996a). *Phys. Rev. B* **54**, 4432.
Brandt, O., Yang, H., Kostial, H., and Ploog, K. H. (1996b). *Appl. Phys. Lett.* **69**, 2707.
Brandt, O., Yang, H., and Ploog, K. (1996c). *Phys. Rev. B* **54**, R5215.
Brandt, O., Yang, H., Müllhäuser, J., Trampert, A., and Ploog, K. (1997). *Mat. Sci. Eng. B* **43**, 215.

Chandrasekhar, D., Smith, D. J., Strite, S., Lin, M. E., and Morkoç, H. (1995). *J. Cryst. Growth* **152**, 135.
Cheng, T. S., Jenkins, L. C., Hooper, S. E., Foxon, C. T., Orgon, J. W., and Lacklison, D. E. (1995). *Appl. Phys. Lett.* **66**, 1509.
Cheng, T. S., Foxon, C. T., Jenkins, L. C., Hooper, S. E., Lacklison, D. E., Orton, J. W., Ya Ber, B., Merkulov, A. V., and Novikov, S. V. (1996). *Semiconf. Sci. Technol.* **11**, 538.
Copel, M., Reuter, M. C., Kaxiras, E., and Tromp, R. M. (1989). *Phys. Rev. Lett.* **63**, 632.
Ding, S. A., Neuhold, G., Weaver, J. H., Häberle, P., Horn, K., Brandt, O., Yang, H., and Ploog, K. H. (1996). *J. Vac. Sci. Technol. A* **14**, 819.
Edgar, J. H. (ed.) (1994). *Properties of Group III Nitrides*, Electronic Materials Information Service (EMIS) Datareviews Series, Institution of Electrical Engineers, London.
Fanciulli, M., Lei, T., and Moustakas, T. D. (1993). *Phys. Rev. B* **48**, 15144.
Fischer, S., Wetzel, C., Hansen, W. L., Bourett-Courchesne, E. D., Meyer, B. K., and Haller, E. E. (1996). *Appl. Phys. Lett.* **69**, 2716.
Fritzsche, H. (1955). *Phys. Rev.* **99**, 406.
Garni, B., Ma, J., Perkins, N., Liu, J., Kuech, T. F., and Lagally, M. G. (1996). *Appl. Phys. Lett.* **68**, 1380.
Giehler, M., Ramsteiner, M., Brandt, O., Yang, H., and Ploog, K. H. (1995). *Appl. Phys. Lett.* **67**, 733.
Groh, R., Gerey, G., Bartha, L., and Pankove, J. I. (1974). *Phys. Stat. Sol.* (a) **26**, 353.
Hauenstein, R. J., Collins, D. A., Cai, X. P., O'Steen, M. L., and McGill, T. C. (1995). *Appl. Phys. Lett.* **66**, 2861.
Ho, I. and Stringfellow, G. B. (1996). *Appl. Phys. Lett.* **69**, 2701.
Jenkins, D. W. and Dow, J. D. (1989). *Phys. Rev. B* **39**, 3317.
Kaganer, V. M., Köhler, R., Schmidbauer, M., Opitz, R., and Jenichen, B. (1997). *Phys. Rev. B* **55**, 1793.
Kawase, M., Gamire, E., Lee, H. C., and Dapkus, P. D. (1993). *IEEE J. Quantum Electron.* **QE-29**, 2306.
Kim, J. G., Frenkel, A. C., Liu, H., and Park, R. M. (1994a). *Appl. Phys. Lett.* **65**, 91.
Kim, S. T., Amano, H., Akasaki, I., and Koide, N. (1994b). *Appl. Phys. Lett.* **64**, 1535.
Kim, K., Lambrecht, W. R. L., and Segall, B. (1996a). *Phys. Rev. B* **53**, 16310.
Kim, W., Salvador, A., Bothkarev, A. E., Aktas, O., Mohammed, S. N., and Morkoç, H. (1996b). *Appl. Phys. Lett.* **69**, 559.
Klann, R., Brandt, O., Yang, H., Grahn, H. T., Ploog, K. H., and Trampert, A. (1995). *Phys. Rev. B* **52**, R11615.
Klann, R., Yang, H., Brandt, O., Grahn, H. T., and Ploog, K. H. (1997). *Appl. Phys. Lett.* **70**, 1076.
Kressel, H. (1981) in *Semiconductors and Semimetals*, vol. 16, eds. Willardson, R. K. and Beer, A. C., Academic Press, New York, p. 1.
Kreuzer, H. J. (1988) in *Chemistry and Physics of Solid Surfaces VII*, eds. Vanselow, R. and Howe, R. F., Springer, Berlin, Springer Ser. Surf. Sci. vol. 7, p. 259.
Kuwano, N., Nagatomo, Y., Kobayashi, K., Oki, K., Miyoshi, S., Yaguchi, H., Onabe, K., and Shiraki, Y. (1994). *Jpn. J. Appl. Phys.* **33**, 18.
Lacklison, D. E., Orton, J. W., Harrison, I., Cheng, T. S., Jenkins, L. C., Foxon, C. T., and Hooper, S. E. (1995). *J. Appl. Phys.* **78**, 1838.
Lagally, M. G., Savage, D. E., and Tringides, M. C. (1988) in *Reflection High-Energy Electron Diffraction and Reflection Electron Imaging of Surfaces*, eds. Larsen, P. K. and Dobson, P. J., Plenum Press, New York, NATO ASI Series B, vol. 188, p. 139.
Lawaetz, P. (1972). *Phys. Rev. B* **5**, 4039.
LeGoues, F. K., Kesan, V. P., Iyer, S. S., Tersoff, J., and Tromp, R. (1990). *Phys. Rev. Lett.* **64**, 2038.

Lei, T., Moustakas, T. D., Graham, R. J., He, Y., and Berkowitz, S. J. (1992). *J. Appl. Phys.* **71**, 4933.
Lei, T., Ludwig, Jr., K. F., and Mostakas, T. D. (1993). *J. Appl. Phys.* **74**, 4430.
Lin, M. E., Xue, G., Zhou, G. L., Greene, J. E., and Morkoç, H. (1993). *Appl. Phys. Lett.* **63**, 932.
Liu, H., Vrenkel, A. C., Kim, J. G., and Park, R. M. (1993). *J. Appl. Phys.* **74**, 6124.
Masuda, A., Yonezawa, Y., Morimoto, A., and Shimizu, T. (1995). *Jpn. J. Appl. Phys.* **34**, 1075.
Matthews, J. W., van der Merwe, J. H., and Ball, C. A. B. (1975) in *Epitaxial Growth*, Part B, ed. Matthews, J. W., Academic Press, New York.
Menniger, J., Jahn, U., Brandt, O., Yang, H., and Ploog, K. H. (1996a). *Phys. Rev. B* **53**, 1881.
Menniger, J., Jahn, U., Brandt, O., Yang, H., and Ploog, K. H. (1996b). *Appl. Phys. Lett.* **69**, 836.
Mizuta, M., Fujieda, S., Matsumoto, Y., and Kawamura, T. (1986). *Jpn. J. Appl. Phys.* **25**, L945.
Moison, J. M., Guille, C., and Bensoussan, M. (1987). *Phys. Rev. Lett.* **58**, 2555.
Molnar, R. J. and Moustakas, T. D. (1994). *J. Appl. Phys.* **76**, 4587.
Müllhäuser, J., Brandt, O., Yang, H., and Ploog, K. (1996) in *Proceedings of the 1st International Symposium on GaN and Related Materials*, Boston (USA), 1995, Mat. Res. Soc. Ser. vol. 395, eds. Dupuis, R. D., Ponce, F. A., Nakamura, S., and Edmond, J. A., Materials Research Society, Pittsburgh, p. 27.
Müllhäuser, J. R., Jenichen, B., Wassermeier, M., Brandt, O., and Ploog, K. H. (1997). *App. Phys. Lett.* **71**, 909.
Munir, Z. A. and Searcy, A. W. (1965). *J. Chem. Phys.* **42**, 4223.
Nakadaira, A. and Tanaka, H. (1997). Topical Workshop on III–V Nitrides, Nagoya (Japan) 1995, *Sol.-Stat. Electron.*, in press.
Nakamura, S., Mukai, T., and Senoh, M. (1994). *Appl. Phys. Lett.* **64**, 1687.
Nakamura, S., Senoh, M., Nagahana, S., Iwasa, N., Yamada, T., Matsushita, T., Sugimoto, Y., and Kiyoku, H. (1996). *Appl. Phys. Lett.* **69**, 1477.
Neugebauer, J. and Van de Walle, C. G. (1994). *Phys. Rev. B* **50**, 8067.
Neugebauer, J. and Scheffler, M. (1997). Personal communication.
Okumura, H., Ohta, K., Ando, K., Rühle, W. W., Nagamoto, T., and Yoshida, S. (1996a). *Inst. Phys. Conf. Ser.* **142**, 939.
Okumura, H., Ohta, K., Nagamoto, T., and Yoshida, S. (1996b). *J. Cryst. Growth* **164**, 149.
Paisley, M. J., Sitar, Z., Posthill, J. B., and Davis, R. F. (1989). *J. Vac. Sci. Technol. A* **7**, 701.
Peercy, P. S. *et al.* (1990). *J. Mater. Res.* **5**, 852; and references therein.
Petrov, I., Mojab, E., Powell, R., Greene, J., Hultman, L., and Sundgren, J. (1992). *Appl. Phys. Lett.* **60**, 2491.
Petroff, P. M., Logan, R. A., and Savage, A. (1980). *Phys. Rev. Lett.* **44**, 287.
Powell, R. C., Lee, N.-E., Kim, Y.-W., and Greene, J. E. (1993). *J. Appl. Phys.* **73**, 189.
Ramírez-Flores, G., Navarro-Contreras, H., Lastras-Martínez, A., Powell, R. C., and Greene, J. (1994). *Phys. Rev. B* **50**, 8433.
Rimbey, P. R. and Mahan, G. D. (1985). *J. Appl. Phys.* **57**, 2812.
Rode, D. L. and Gaskill, D. K. (1995). *Appl. Phys. Lett.* **66**, 1972.
Schikora, D., Hankeln, M., As, D. J., Lischka, K., Litz, T., Wang, A., Buhrow, T., and Henneberger, F. (1996). *Phys. Rev. B* **54**, R8381.
Seeger, K. (1981). *Semiconductor Physics*, Springer Series in Solid State Sciences vol. 40, eds. Cardona, M., Fulde, P., von Klitzing, K., and Queisser, H.-J., Springer, Berlin.
Shaklee, K. L., Nahory, R. E., and Leheny, R. F. (1973). *J. Lumin.* **7**, 284.

Strite, S., Ruan, J., Li, Z., Salvador, A., Chen, H., Smith, D. J., Choyke, W. J., and Morkoç, H. (1991). *J. Vac. Sci. Technol. B* **9**, 1924.
Strite, S., Chandrasekhar, D., Smith, D. J., Sariel, J., Chen, H., Teraguchi, N., and Morkoç, H. (1993). *J. Cryst. Growth* **127**, 294.
Tanaka, T., Watanabe, A., Amano, H., Kobayashi, Y., Akasaki, I., Yamazaki, S., and Koike, M. (1994). *Appl. Phys. Lett.* **65**, 593.
Tournié, E., Trampert, A., and Ploog, K. H. (1995). *Europhys. Lett.* **24**, 663.
Trampert, A., Brandt, O., Yang, H., and Ploog, K. H. (1997). *Appl. Phys. Lett.* **70**, 583.
Vook, R. W. (1975) in *Epitaxial Growth*, ed. Matthews, J. W., Academic Press, New York, p. 339.
Wang, C. and Davis, R. (1993). *Appl. Phys. Lett.* **63**, 990.
Wood, E. A. (1964). *J. Appl. Phys.* **35**, 1306.
Wright, A. F. and Nelson, J. S. (1995). *Phys. Rev. B* **51**, 7866.
Yang, B., Brandt, O., Jenichen, B., Müllhäuser, J., and Ploog, K. (1997). *Appl. Phys. Lett. J. Appl. Phys.* **82**, 1918.
Yang, H., Brandt, O., and Ploog, K. (1996a). *J. Electron. Mat.* **25**, 787.
Yang, H., Brandt, O., Trampert, A., and Ploog, K. H. (1996b). *Appl. Surf. Sci.* **104/105**, 461.
Yang, H., Brandt, O., and Ploog, K. (1996c). *Appl. Phys. Lett.* **68**, 244.
Yang, H., Brandt, O., and Ploog, K. (1996d). *Phys. Stat. Sol. (b)* **194**, 109.
Yang, M. H. and Flynn, C. P. (1989). *Phys. Rev. Lett.* **62**, 2476.
Yang, Z., Li, L. K., and Wang, W. I. (1996e). *J. Vac. Sci. Technol. B* **14**, 2354.

Index

absorption coefficient 444
absorption spectra 195–7
acceptor binding energy 52–3, 251–3
acceptor-bound excitons 245–6, 253–5
acceptor compensation 392
activation energies 433
  donors and donor-bound excitons 247–8
activation of Mg acceptors 57–9, 104–7, 210–11
adducts 74, 75–6
adsorption 429–31
AlGaN 107–11
  AlGaN/GaN heterostructures 114–15
  blue InGaN/AlGaN DH LEDs 393–4
  GaN/AlN solid solution 107
  high-speed photodetector 385–7
  modulation spectroscopy 168–73, 178–9
  MOVPE growth 107–11
  photoconductors 375–6
AlInGaP LEDs 391, 394–5
aluminum nitride 12, 20, 60–1, 125
  buffer layer thickness 88
  bulk single crystals 123–4
  cubic structure 418
  failure of doping 52–3
  GaN/AlN solid solution 107
  growth 43–4
  interface with sapphire 129–31
  interface with SiC 132–4
  properties 21
  RAS 178–9
  transition metals see transition metals
ammonia 420
  flow rate and p-type doping by Mg 57–8
  nitrogen precursor 71–4
  oxygen from and contamination of MOCVD process 96–8
  parasitic reactions with Group III precursors 75–6
annealingy see heat treatment
arrays
  photoconductors 378
  photovoltaic detectors 383, 384–5
arsenic 1–2
ASTEX CECR source 26
autodoping 53–4, 208–9

band-to-band transitions 163–5
bandgap 3–4, 8, 12
  effects of biaxial strain on excitons 173–7
  GaN 15, 344
  and interatomic distance 127–8
  offsets between GaN and AlN 299–301
  temperature dependence of bandgap energies 168–73
  tunability 328
band-structure parameters 193–5
beating 234–5
beryllium 59, 211–12
  doping of cubic Group III nitrides 451–5
biaxial strains
  effects on excitons 173–7
  electronic band structures 312–13
  wurtzite quantum wells 328–9, 336, 337
  zincblende quantum wells 330–31, 339–40
bimolecular recombination 369–72
binding energy
  acceptor 52–3, 251–3
  donor 52–3, 251–3
    residual donor 249–51
  donor-bound excitons 247–8
  exciton 185, 195–201
  impurity 185–6
bipolar transistors 343–7
  demonstrated devices 344–7
  GaN's suitability 344
    general principles 343–4
Bloch functions 164, 323
blue LDs 391–2, 392–3, 405–14
blue LEDs 391, 392
  InGaN/AlGaN DH LEDs 393–4
  InGaN SQW LEDs 394–8
  white LEDs 398–401
boron nitride 12
Bose–Einstein equation 170, 171–2
bound excitons 185–6
  acceptor-bound 245–6, 253–5
  donor-bound 244–6, 246–7, 247–8, 253–5
  linewidth 248–9
  magneto-optical investigations 253–5
  PLE bound exciton resonance 202–4
  undoped GaN epitaxial films 244–6
Brillouin zones 308–9
Brooklyn College photoreflectance system 159–60
buffer layers 79, 125
  MBE growth 32–43
  MOCVD on sapphire substrates 82–99
    buffer layer thickness 88–9
    deposition 83

462 *Index*

heat treatment 84, 90–2
  role of buffer layer 84–5
Burgers vectors 139–41, 144–6

cadmium 211, 253
calcium 104, 212
carbon 59, 96, 97
  contamination in MOCVD growth 98–9
carrier diffusion 378–80, 381
carrier transit time 386–7
cathodoluminescence (CL)
  cubic group III nitrates 442–3
  microstructure 152–4
CCD arrays 7
charge-transfer energies 285, 292, 299–301
chemical bond length 127–8
chemical properties 20–3
chemical vapor deposition (CVD) 10
chromaticity diagrams 397, 398, 401, 402
chromium 282, 285, 293–6, 298, 299, 300
cobalt 288
co-doping 453–4
coherent-transient degenerate FWM measurement 232, 233–5
coincidence-lattice mismatch 425
commercialization 4, 4–7, 16
complex dielectric function 163–5, 166, 167–8
composition control 107–9, 112–13
concentration, doping 209–10
conductivity 3, 454–5
contactless electroreflectance (CER) 161–2
  temperature dependence of energies and linewidths 168–73
contacts, ohmic 14
contamination
  effects in MOCVD process 96–9
  growth of cubic group III nitrides 434
continuous-wave (CW) operation of LDs 224, 225, 408–9, 410
conventional diffraction contrast 139, 140
convergent beam electron diffraction (CBED) 146–7
  large-angle (LACBED) 140–1, 144–6
copper bromide 2
crystal-field splitting 193–5, 310–11, 318–19
crystal quality 79, 82, 126, 219
  cubic group III nitrides 419
  effect of strain and lattice mismatch 47–8
  MBE growth 36–7, 45–7
  ternary alloys 109–11, 113–14
crystal structure 8, 15, 126–8
  common structures 20–2
  see also cubic structure; wurtzite structure

crystal symmetry: optical gain and 331–2
cubic structure 8, 15, 20–2, 126–8, 417–56
  cubic symmetric substrate 43
  electrical properties 449–55
  electronic band structures 308–12
    strain effect 313
  growth equipment and *in situ* characterization 420–21
  growth kinetics 426–36
  growth of mixed cubic and wurtzite structures 41
  morphology 437–8
  nucleation 421–6
  optical gain 332, 333–4
  optical properties 442–9
  quantum wells *see* zincblende quantum wells
  structural properties 438–42
  temperature dependence of energies and linewidths in GaN 168–73, 174
current gain
  bipolar transistor 346–7
  FETs 349
current injection laser diodes 222–6
current-voltage characteristics
  bipolar transistor 345–6
  FETs 359, 362, 363, 364
  InGaN MQW LEDs 406
cutoff frequencies 349, 353
  HFET vs MODFET for high performance 359–64
  and temperature in HFET 356–7

decay time 204
deep levels 17
defect spectroscopy 242–301
  deep structural defects in GaN 255–60
    yellow luminescence 257–60
  shallow impurities in GaN 243–55
    activation energies and binding energies 247–8
    bound-exciton linewidth 248–9
    donor-based excitons 246–7
    excitonic and impurity-related recombination 243–6
    localization and donor/acceptor binding energies 251–3
    magneto-optical investigations of bound excitons 253–5
    residual donor binding energies 249–51
  transition metals *see* transition metals
defects 123–54
  crystalline structure 126–8
  cubic group III nitrides 419, 424–6, 440–42
  deep structural defects 255–60
  dislocations in GaN 139–41

evolution of thin film morphology 125–6
growth and microstructure 123–5
inversion domains 146–8
lasing 219
lattice vibrations 149–52
MBE growth 45–52
　dislocations 49–51
　improvement of epitaxial crystal
　　structure 45–7
　native defects 51–2
　stacking fault defect 49
　strain and lattice mismatch 47–8
MOCVD thin films 135–8
nanopipes 143–6
polarity determination 141–3
recombinations related to 205–12, 448
role in widegap semiconductors 16–18
spatial variation of luminescence 152–4
substrate/epilayer interface 128–35
deformation potentials 173–7, 193–5, 196, 322
degenerate four-wave mixing experiments
　228–35
depletion mode FETs 354
desorption 429–34
desorption mass spectrometry 30, 31
diamond 10
dielectric function 163–5, 166, 167–8
diethylaluminum azide 75
diethylgallium azide 75
diffusion, carrier 378–80, 381
diffusion (growth kinetics) 429–34
diffusion limited growth 78–9
dimethylhydrazine 74
direct bandgap materials 8, 344
dislocations 17–18
　characterization in GaN 139–41
　cubic group III nitrides 424–5
　　dislocation array 425–6
　　pure edge dislocations 424–5, 442
　MBE growth 49–51
　MOCVD thin films 136–8
dissociation pressure 22–3
dissociation rate (ammonia) 72
donor-to-acceptor pair (DAP)
　transitions 206, 244
donor binding energy 52–3, 251–3
　residual 249–51
donor-bound excitons 244–6, 246–7,
　247–8, 253–5
dopant precursors 76–7
dopant-related luminescence 152
doping 8–9, 13–14, 125–6, 344
　AlGaN 111
　concentration 209–10
　during MBE growth 51–9, 60–1
　MOCVD growth of GaN 103–7
　radiative recombinations in doped GaN
　　209–12
　Raman imaging and dopant distribution
　　150–2
　II–VI compounds 11–12

double heterostructure (DH) 215, 216
　blue InGaN/AlGaN DH LEDs 393–4

edge emission intensity 448–9
effective lifetime 186–9
effective masses
　electron 319–20
　hole 320–21
elastic moduli 141
elastic stiffness constants 193–4 324
electrical properties
　cubic group III nitrides 449–55
　impact of buffer layer on GaN
　　epilayer 82
electric field-induced second harmonic
　(EFISH) 227, 228
electroabsorption spectra (EA) 402–3, 404
electroluminescence (EL) spectra
　InGaN/AlGaN DH LEDs 393, 394
　SQW LEDs 395–8, 402, 404
　successive generations of LEDs 14, 15
electromodulation (EM) 158, 159
　lineshape considerations 163–8
electron concentration 95, 449–51
electron cyclotron resonance (ECR)
　MBE 25–7
electron diffusion 378–9
electron effective masses 319–20
electronic band structures 308–13
　first-principles band calculations 308–9
　　results 309–12
　strain effect 312–13
electronic materials 7–9
electron spin resonance (ESR) 262–3
electro-optic energy 163
electro-optic functions 167
electro-optic modulation 373–5
emission intensity 219–20, 446–9
emission spectra 406–11, 412
enhancement mode FETs 354
envelope functions 323–4
epilayer/substrate interface 128–35
epitaxy 8, 11, 13, 124–5
　excitonic and impurity-related recombination
　　in undoped GaN films 243–6
　homoepitaxy 42, 99–100, 134–5
　MBE see molecular beam epitaxy
　MOCVD see MOCVD
　structure of substrate/epilayer interface
　　128–35
excitation density 446–9
excitation stripe length 219–20, 448–9
excited states 177–8
exciton localization energy 185–6,
　243, 247–8, 251–3
exciton-phonon interactions 173, 201–5,
　249

exciton resonance 174-7, 195-7, 231-5
excitons 8
  bound see bound excitons
  deep structural defects 256-7
  effects in GaN photoconductors 373-5
  free 184-5, 202-4, 244-6
  modulation spectroscopy 166
    biaxial strain 173-7
    excited states 177-8
    temperature dependence of energies and linewidths 168-73, 174
  optical transitions 183-205
    exciton binding energy 195-201
    exciton-phonon interactions 201-5
    general aspects of excitons 183-6
    recombinations 186-90
    strain effects 191-4
  recombination in undoped GaN epitaxial films 243-6

Fabry-Perot interference fringes 218-19
far-field radiation patterns (FFP) 413, 414
femtosecond four-wave mixing 231-5
field-effect transistors (FETs) 5, 347-66
  demonstrated FETs 351-66
  general principles 347-8
  types and advantages of GaN-related materials 348-51
V/III molar ratio 89-90, 94-6, 97
Fourier transform infrared (FTIR) absorption 250-51
four-wave mixing (FWM) 228-35
  femtosecond 231-5
  picosecond 228-31
Fowler plot 380-81
Frank equation 144-6
Franz-Keldysh oscillations (FKOs) 159, 166-8, 180
free excitons 184-5, 202-4, 244-6
frequency
  cutoff frequencies see cutoff frequencies
  maximum oscillation frequency 349, 360-61
  MESFET at high frequency 354-5
full-potential linearized augmented plane wave (FLAPW) method 308-9

GaAlN see AlGaN
GaInN see InGaN
gallium antimonide 2
gallium arsenide 1-2, 5, 6, 8, 17
  comparison with GaN 15
  substrates 40-1, 101-2, 426

gallium nitride 2
  comparison with GaAs 15
  crystal structures 20-1, 126-7
  cubic structure see cubic structure
  defect spectroscopy
    deep structural defects 255-60
    shallow impurities 243-55
    transition metals see transition metals
  defects
    characterization of dislocations 139-41
    role in device operation 16-18
  donor and acceptor binding energies 52
  doping 103-7
  electronic band structures of bulk GaN 308-13
  evolution of thin film morphology 125-6
  GaN/AlGaN heterostructures 114-15
  GaN/InGaN heterostructures 115-16
  growth 12-13
    bulk growth 123-4
    MBE 23-43
    MOCVD see MOCVD
  homoepitaxy 42, 99-100, 134-5
  interface with sapphire 129-32
  modulation spectroscopy
    CER 168-73
    excited states 177-8
    PR 173-7
    RAS 178-9
  optical gain of bulk GaN 333-4
  optical properties see optical properties
  phase diagrams 22-3
  photoconductors see photoconductors
  photovoltaic detectors see photovoltaic detectors
  quantum wells see quantum wells
  suitability for transistors 344, 348-51
  surface stabilization effect 73-4
gallium phosphide 40-1
gallium vacancies 206-9
gas phase composition 107-9, 112-13
gas source MBE (GSMBE) 30
gate length 347-8, 349, 359-64
Gaussian reflection profiles 440
germane 76, 103-4
germanium 1-2, 7, 103-4, 212
green LEDs 391, 394-8
group III precursors 70-1
  parasitic reactions with ammonia 75-6
growth 10, 10-11, 11-12, 12-13
  cubic group III nitrides
    equipment 420-21
    kinetics 426-36
  direction and quality of epitaxy 141-3, 144
  MBE see molecular beam epitaxy
  mechanisms 77-9
  and microstructure 123-5
  MOCVD see MOCVD
  modeling MBE-like growth 31-2
  rate and ion damage 26-7

temperature
   and buffer layer 87–8
   and 'main' GaN layer 92–4
GW approximation 309

Hamiltonian
   8×8 $\mathbf{k} \cdot \mathbf{p}$ Hamiltonian 314–15
   spin-Hamiltonian 262–3, 265
   2×2 and 6×6 $\mathbf{k} \cdot \mathbf{p}$ and strain Hamiltonian 315–17
harmonic generations 226–8
Haynes' rule 185–6 243, 252–3
heat treatment
   buffer layer 84, 90–2
   doping with Mg 57, 105–6, 210–11
   growth of cubic structure GaN 434
   vanadium doping and 0.82 eV luminescence 290–92
heterojunction bipolar transistors (HBTs) 343–7
heterostructure FETs (HFETs) 348, 352–4
   at high temperature 355–7
   vs MOSFETs for high performance 359–64
   response under UV 357–8
heterostructures 114–16
   double (DH) 215, 216, 393–4
   separate-confinement (SCH) 215–17, 224, 225
hexagonal structure *see* wurtzite structure
high-resolution transmission electron microscopy (HREM) 423–5
high-speed photodetector 385–7
hole diffusion 379–80
hole effective masses 320–21
hole mobility 452–3, 453–4
homoepitaxy 42, 99–100, 134–5
hydrazine 74–5
hydrazonic acid 75
hydride vapor phase epitaxy (HVPE) 124
hydrogen
   carrier gas flow rate 80–1
   doping 56
   passivation 210–11
hydrostatic deformation potentials 193–5, 196

impurities
   impurity argument for autodoping 54
   Raman imaging and distribution of 150–2
   recombinations associated with 205–12, 243–6
impurity binding energy 185–6
incorporation ratio 30, 31
indirect semiconductors 7–8

indium
   incorporation efficiency 45
   phase separation 222, 226
indium nitride 12, 52
   challenges in growth 44
   crystal structures 20–2
   cubic structure 418, 455
indium phosphide 2, 5
InGaN 111–14, 455
   cubic structure 455
   InGaN/AlGaN DH LEDs 393–4
   InGaN/GaN heterostructures 115–16
   MOCVD growth 112–14
   MQW structure LDs 222, 222–6, 392–3, 405–14
   SQW structure LEDs 394–8
initial nucleation density 79, 80–1, 82
interatomic separation 127–8
internal photoemission 380–81
internal reference rule 300–301
intraband relaxation time 332, 335, 336
inversion domains 146–8
invisibility criterion 140
ion damage 26
iron 259, 292
   band offsets between GaN and AlN 299, 300–301
   1.3 eV luminescence in GaN 260, 261–78

Jahn–Teller coupling 270, 283, 284, 289
Johnson noise 376–7, 382, 383
joint density of states (JDOS) 332
junction breakdown voltage 363–4
junction field-effect transistors (JFETs) 348, 351–2

kinetically limited growth 77–9
$\mathbf{k} \cdot \mathbf{p}$ theory 313–22
   8×8 $\mathbf{k} \cdot \mathbf{p}$ Hamiltonian 314–15
   numerical derivation of physical parameters 318–22
   quasi-cubic approximation 317–18
   2×2 and 6×6 $\mathbf{k} \cdot \mathbf{p}$ and the strain Hamiltonian 315–17

large-angle convergent beam electron diffraction (LACBED) 140–1, 144–6
laser diodes (LDs) 6, 12, 391–3
   current injection 222–6
   InGaN MQW LDs 405–14, 414–15

lasing 212–26
　optical pumping 213–22
　pumping threshold 215–19
lateral growth control 80, 81–2
lattice constants/parameters 20–2
　cubic GaN 441
　and GaN layer thickness 36
　strain effects on excitonic transitions 191–2
　substrates 33, 99
lattice mismatch 34, 127–8
　effect on crystal structure 47–8
　'magic mismatch' 425
lattice vibrations 149–52
LEEBI (low-energy electron beam irradiation) 104–5
lifetime test 406, 407
light-emitting diodes (LEDs) 6, 125–6, 391–404, 414
　characteristics 401–2, 403
　efficiency and dislocation density 138
　EL spectra of successive generations 14, 15
　InGaN/AlGaN DH LEDs 393–4
　SQW LEDs 392, 394–8, 402–4
　white LEDs 398–402
lithium gallate 102
local density functional approximation (LDA) 308, 309
local vibrational mode (LVM) 276–7
localization energy 185–6, 243, 247–8, 251–3
longitudinal optical (LO) phonons 149–52
　replicas 244–5
Lorentzians 440
luminescence
　deep center yellow 257–60
　deeply bound excitons 256–7
　spatial variation 152–4
　transition metals *see* transition metals
　see also cathodoluminescence; electroluminescence spectra; photoluminescence spectra
Luttinger-like parameters 319

'magic mismatch' 425
magnesium 13–14, 252–3, 451
　activation of Mg acceptors 57–9, 104–7, 210–11
　doping during MBE 56–9, 60
　doping during MOCVD 77, 104–7
　optical properties of doped GaN 209–11
magnesium aluminate 41, 102–3
magnesium oxide 41–2
melting points 22–3
metal semiconductor field-effect transistors (MESFETs) 348, 349–50, 352
　at high frequency 354–5

metal-semiconductor-metal (M-S-M) photodetectors 385–7
microstructure *see* defects
MISFETs 348–9, 358–9
MOCVD (metalorganic chemical vapor deposition) 124–5, 210–11
　alternative substrates 99–103
　direct growth of GaN on sapphire substrate 77–82
　　growth conditions' influence 79–82
　　growth mechanisms 77–9
　doping 103–7
　growth of GaInN 111–14
　growth of GaN on sapphire substrate using a buffer layer 82–99
　　MOCVD process 83–4
　　contamination effects 96–9
　　microstructure of thin films 135–8
　　precursors 70–7
　see also MOVPE
mode hopping 225, 408, 409
modeling growth 31–2
mode separation 407
modulation-doped FETs (MODFETs) 348, 350–51
　vs HFETs for high performance 359–64
　power generation 364–6
modulation spectroscopy 158–80
　lineshape considerations 158, 163–8
　results
　　biaxial strain 173–7
　　excited states 177–8
　　RAS 178–9
　　temperature dependence of energies and linewidths 168–73, 174
　techniques 159–62
molecular beam epitaxy (MBE) 8, 11, 19–61, 124–5, 210–11
　AlN 43–4
　defects 45–51
　doping during growth 51–9, 60–1
　growth system 23–31
　InN 44
　modeling growth 31–2
　substrates and buffer layers 32–43
　ternary and quaternary alloys 44–5
MOMBE 30–1
momentum matrix elements 321–2
monomethylhydrazine 74
monomolecular recombination 369–72
MOVPE (metalorganic vapor phase epitaxy) 11, 13
　AlGaN 107–11
　heterostructures 114–16
　see also MOCVD
multiphoton spectroscopy 226–8
multiple quantum well (MQW) LDs 222–6, 405–14, 414–15

nanopipes 143–6, 148
native defects 51–2, 206–9
near-band-edge luminescence 152–4, 186, 187
near-field radiation patterns (NFP) 412–13
nickel 288
nitridation 36, 85–7
nitrogen 22–3, 32, 420
  precursors 71–5
nitrogen vacancies 51–2, 54, 206–9, 449
noise
  photoconductors 376–7
  Schottky diode 382–3
nonlinear optical properties 226–35
  femtosecond FWM 231–5
  harmonic generations and multiphoton spectroscopy 226–8
  picosecond FWM 228–31
non-radiative recombinations 7, 17, 186–90, 371–2, 446–7
n-type doping 8–9, 76–7, 103–4, 212
  cubic group III nitrides 418–19, 449–51
  Schottky diode 379–80
nucleation
  control of initial nucleation density 79, 80–1, 82
  cubic group III nitrides 421–6, 427

occupation factor 332
ohmic contacts 14
optical cavity: difficulties in forming 218–19
optical gain 219–20, 221, 331–40, 448–9
  bulk GaN 333–4
  and crystal symmetry 331–2
  wurtzite quantum wells 334–8
  zincblende quantum wells 338–40
optically detected magnetic resonance (ODMR) 263–5
optical phonon replicas 244–5, 278, 279
optical properties 182–236
  cubic group III nitrides 442–9
  effect of microstructure 152–4
  nonlinearities of GaN 226–35
  optical transitions associated with excitons see excitons
    radiative recombinations involving impurities and defects 205–12
    stimulated emission and lasing 212–26
      current injection LDs 222–6
      optical pumping 213–22
optical transition matrix elements 332
optoelectronic materials 7–9
Oxford MPD 21 plasma source 27
oxide substrates 101, 102–3
oxygen 209, 252
  contamination in MOCVD 96–8
  cubic group III nitrides 450–51, 452–4
  doping 54–6

parasitic reactions 75–6
Peclet number 81
phase diagrams 22–3
phase separation 111–12, 222, 226, 413–14
phonon-exciton interactions 173, 201–5, 249
phonon-plasmon coupling 149–52
phonon replicas 244–5, 278, 279
phosphor 399–401, 402
photoconductors 366–78
  AlGaN 375–6
  arrays 378
  excitonic effects 373–5
  first demonstration of GaN 367–8
  general principles 366–7
  performances 376–7
  recombinations and time response 368–72
  spectral selectivity 375
photodetectors 6–7
  see also photoconductors; photo-voltaic detectors
photoluminescence excitation (PLE) 197–8
  exciton–phonon interactions 202–5
  and time-resolved PL 267–74
photoluminescence (PL) spectra
  buffer layer and 37, 38, 89–90
  cubic group III nitrides 443–6
  effect of nitridation of substrate 85–6
  excitonic recombinations in GaN 186–90
  GaN films and grid potentials 28–9
  influence of V/III molar ratio 95–6
  influence of growth temperature 93
  MG doping and annealing 105–6
  microstructure and 152–3
  radiative recombinations 205–12
  time-resolved PL 186–9, 267–74
photoreflectance (PR) 159, 159–61
  effects of biaxial strain 173–7
  excited states 177–8
  exciton binding energy 198–201
phototransistors 357–9
photovoltage spectra 403–4
photovoltaic detectors 378–87
  high-speed photodetector 385–7
  p–n junctions 383–4
  Schottky diodes 378–83
picosecond four-wave mixing 228–31
piezoreflectance 168
plasma-based MBE 25–9, 420–21
plasmon-phonon coupling 149–52
p–n junctions 383–5
point contact transistor 5
polarity determination 141–3, 144

polarization control 328
power added efficiency (PAE) 350
power generation: MODFET 364–6
precursors: MOCVD 70–7
 dopant precursors 76–7
 group III precursors 70–1
  parasitic reactions with ammonia 75–6
 nitrogen precursors 71–5
pressure
 energy of yellow luminescence and 207–8
 excitonic recombinations and 189–90
p-type doping 8–9, 16, 77
 cubic group III nitrides 418–19, 451–5
 by Mg during MBE 56–9
 during MOCVD 104–7
 radiative recombinations 209–12
 Schottky diodes 378–9
pumping power density 215–17
pyrolysis 71

quantum beats 234–5
quantum dots 221–2, 413–14
quantum size effect 325–6
quantum wells (QWs) 307–41
 electronic band structures of bulk GaN and AlN 308–13
 k · p theory with strains for wurtzite 313–22
 MQW LDs 222–6, 405–14, 414–15
 optical gains 331–40
  strained QWs 336–8, 339–40
  wurtzite QWs 334–8
  zincblende QWs 338–40
 SQW LDs 224–5
 SQW LEDs 392, 394–8, 402–4
 subband structures 322–31
  strained QWs 327–31
  wurtzite QWs 325–6, 328–30
  zincblende QWs 326–7, 330–31
quasi-cubic approximation 317–18, 319, 322
quasi-particle correction 309
quaternary alloys 44–5, 391, 394–5

radiative recombinations 7–8, 186–90, 371–2
 cubic group III nitrides 447–9
 impurities and defect states 205–12
  deep center yellow luminescence 257–60
  doped GaN 209–12
  nominally undoped GaN 205–9
radio-frequency (RF) MBE 25, 27–8
Raman spectra 149–52
rare earth 59

reactive-ion MBE (RMBE) 29–30
reactor efficiency 75–6
recombinations 7–8
 cubic group III nitrides 446–9
 excitonic in GaN 186–90
 non-radiative 7, 17, 186–90, 371–2
 radiative see radiative recombinations
 and time response of photoconductors 368–72
reconstructions, surface 427–30, 434–6
reflection anisotropy spectroscopy (RAS) 162, 168, 178–9
reflection high-energy electron diffraction (RHEED) 421, 422–3, 427–9, 431, 432, 434, 436
reflectivity: differential changes 163–5, 166, 167–8
refractive index 230–31, 444
residual donor binding energies 249–51
resistivity 450
resonances, exciton 174–7, 195–7, 231–5
response time
 high-speed photodetector 386–7
 photoconductors 367, 368–72
 Schottky diodes 381–2
responsivity
 photoconductors 366–7, 368–9, 370, 375–6, 377
 p–n junctions 383–4, 385
 Schottky diodes 378–80, 381
rock salt structure 20

sapphire 124, 426
 GaN interface with 129–32
 lattice structure 130
 MBE growth on 33–8
 MOCVD growth of GaN
  direct growth 77–82
  using a buffer layer 82–99
scaling-factor approximation (SFA) 276–7
scanning tunneling microscopy (STM) 434–6, 438
scattering efficiency 230–31
Schottky diodes 378–83
 arrays 383
 carrier diffusion 378–80
 internal photoemission 380–81
 noise 382–3
 response time 381–2
Schrödinger equation: time-dependent 164
secondary ion mass spectrometry (SIMS) 449–51, 453
second-harmonic generation spectroscopy 227–8

Index 469

selected area diffraction (SAD) pattern 440–41
semiconductor families 1–2, 9–12
   IV–IV compound 10–11
   group IV elements 9–10
   II–VI compounds 11–12
semiconductors: definition 2–3
separate-confinement-heterostructure (SCH) 215–17, 224, 225
Seraphin coefficients 163
shear deformation potential constants 173–7
shot noise 382, 383
silane 76, 103–4
silicon 1, 5, 6–7, 9, 9–10, 426
   binding energy 251–2
   doping 54, 76, 103–4, 212
   substrate in MBE 39–40
      cubic SiC-coated 43
silicon carbide 2, 10–11, 391
   AlN interface with 132–4
   substrates 24, 38–9, 100–1
      cubic silicon carbide-coated silicon 43
single quantum well (SQW) LDs 224–5
single quantum well (SQW) LEDs 392, 394–8
   emission mechanism 402–4
solid phase composition 107–9, 112–13
spectrally resolved (SR) FWM 233–4
spectral selectivity 375, 376
spectroscopic ellipsometry 444
spinel 41, 102–3
spin-Hamiltonian 262–3, 265
spin-orbit splitting 193–5, 310–11, 318–19
stacking faults 49, 50, 440, 441, 442
stimulated emission 212–26
   current injection LDs 222–6
   optical pumping 213–22
strain 37–8, 43
   biaxial see biaxial strains
   effects on crystal structure 47–8
   effects on excitonic transmissions 191–5
   electronic band structures 312–13
   quantum wells
      optical gain 336–40
      subband structure 327–31
      wurtzite QWs 328–30, 336–8
      zincblende QWs 330–31, 339–40
   uniaxial 312, 329–30, 336–8
strain energy 141, 142
strain Hamiltonian 315–17
structural defects see defects
structural film quality see crystal quality
structural properties: cubic group III nitrides 438–42
subband emissions (LDs) 408, 409–11, 413–14
subband structures 322–31
   method of subband calculations 322–4
   strained QWs 327–31

wurtzite QWs 325–6, 328–30
zincblende QWs 326–7, 330–31
substrate/epilayer interface 128–35
substrates 124, 426
   alternative substrates for MOCVD 99–103
   MBE growth 24, 32–43
   MOCVD growth on sapphire 77–82
      pre-treatment of substrate 83, 85–7
      using a buffer layer 82–99
surface decomposition 72–3, 106
surface emission transients 447–8
surface emitting laser action 220–21
surface kinetics 426–36
surface stabilization effect 73–4
symmetry, crystal 331–2

Tanabe–Sugano diagram 284
temperature 9
   dependence of emission spectra of LDs 408, 409
   dependence of energies and linewidths 168–73, 174
   HFET at high temperature 355–7
   MOCVD growth 92–4
      buffer layer 87–8
      buffer layer annealing temperature 91–2
   PL decay time 188–9
   PL spectra for cubic group III nitrides 445–6
ternary alloys 2
   growth 44–5, 107–14
   high-speed photodetectors 385–7
tetraethyl silane 76–7
thermal expansion coefficients 33, 99, 127–8, 191
thermal expansion mismatch 191–2
thermoreflectance 168
thickness: buffer layer 88–9
thin film morphology: evolution of 125–6
third-harmonic generation spectroscopy 226–7
threshold carrier density: optical gain and 331–40
threshold reduction 328
time-dependent Schrödinger equation 164
time-integrated FWM 232, 233–5
time-resolved photoluminescence 186–9, 267–74
titanium 285, 287–8, 293, 294–6, 299, 300
traffic lights 398
transconductance 347, 349, 353
   and gate length 359–64
   phototransistors 358–9
   and temperature in HFET 355–6

transistors 4, 5
  bipolar 343–7
  field-effect 5, 347–66
transition metals 98, 260–301
  in AlN 292–301
    1.297 eV luminescence 292–3
    1.20 eV luminescence 293–6
    1.043 eV luminescence 296
    0.943 eV luminescence 296–8
    0.797 eV luminescence 298–9
    band offsets between GaN and AlN 299–301
  in GaN 260–92
    1.3 eV luminescence 261–78
    1.19 eV luminescence 278–85
    1.047 eV luminescence 285–8
    0.93 eV luminescence 288–90
    0.82 eV luminescence 290–92
transit time, carrier 386–7
transmission electron microscopy 440–42
traps 367, 369–71
triethyl molecules 70–1
trimethyl molecules 70–1
two-color stimulation experiments 258–9

ultraviolet radiation
  FETs 357–9
  photoconductors see photoconductors
  photovoltaic detectors see photovoltaic detectors
uniaxial deformation potentials 193–5, 196
uniaxial strains 312, 329–30, 336–8
unit step function 167

vanadium 285, 287–92 passim, 297, 298–9, 299, 300
van der Waals epitaxy 43
vapor pressure 70–1
variable-stripe-length (VSL) excitation 219–20
Varshni equation 169–70, 171–2
velocity overshoot 348
'Volmer–Weber' island growth mode 79

Wannier–Mott approximation 184
wavelength, peak 408, 409, 410, 412

Wavemat MPDR ECR source 25–6
'wetting' problem 79, 421
white LEDs 392, 398–402, 403
widegap semiconductors 9
  role of defects 16–18
wurtzite quantum wells
  optical gain 334–8
  strained QWs 328–30, 336–8
  subband structures 325–6, 328–30
wurtzite structure 8, 20–2, 126–8
  electronic band structures 308–12
  strain effect 312
  growth of mixed wurtzite and cubic structures 41
  $\mathbf{k} \cdot \mathbf{p}$ theory with strains 313–22
  modulation spectroscopy 168–78
    biaxial strain 173–7
    excited states 177–8
    temperature dependence of energies and linewidths 168–73
  optical gain 332, 333–4

X-ray diffraction profiles 126
  bound-exciton linewidth 248–9
  quality of MBE growth 29–30, 36–7
  structural properties of cubic group III nitrides 419, 438–40
  temperature dependence of linewidths 168–73
  ternary alloys 109–10

yellow LEDs 394–8
yellow luminescence band 152–4
  deep center 257–60
  radiative recombinations 206–9
yttrium aluminum garnet (YAG) phosphors 399, 400–401

Zeeman experiments 253–5, 265–7, 274–5, 280–85, 297
zinc 59, 211, 252–3
zincblende quantum wells
  optical gain 338–40
  strained 330–31, 339–40
  subband structures 326–7, 330–31
zincblende structure see cubic structure
zinc oxide 39, 101, 301
zinc selenide 2, 11–12, 16, 17